Seaweed Cultivation

Seaweed Cultivation

Edited by:

Bruno Augusto Amato Borges

Delve
PUBLISHING

www.delvepublishing.com

Seaweed Cultivation

Bruno Augusto Amato Borges

Delve Publishing
2010 Winston Park Drive,
2nd Floor
Oakville, ON L6H 5R7
Canada
www.delvepublishing.com
Tel: 001-289-291-7705
001-905-616-2116
Fax: 001-289-291-7601
Email: orders@arclereducation.com

ISBN: 978-1-77361-039-9 (Hardcover)

DECLARATION

Some content or chapters in this book are open access copyright free published research work, which is published under Creative Commons License and are indicated with the citation. We are thankful to the publishers and authors of the content and chapters as without them this book wouldn't have been possible.

ABOUT THE EDITOR

Bruno Augusto Amato Borges

Bruno Augusto is Aquaculture Engineer from Federal University of Santa Catarina. His focus is on writing projects and development of new technologies in aquaculture. He is currently project service provider, consultant and develops research in the field of Biofloc Technology (BFT) for freshwater fish farming.

TABLE OF CONTENTS

SECTION III
BIOLOGY AND MANAGEMENT TECHNIQUES

SECTION IV
COMPOSITION, CONSTITUENTS AND USES

SECTION V

SEAWEED CULTIVATION AND ENVIRONMENT

LIST OF CONTRIBUTORS

Bruno Augusto Amato Borges
Aquaculture Engineer, Florianopolis, Brazil

Mark L. Wells
School of Marine Sciences, University of Maine, Orono, ME 04469, USA

Philippe Potin
Integrative Biology of Marine Models, Station Biologique Roscoff, CNRS-Université Pierre et Marie Curie, Place Georges Teissier, 29680 Roscoff, France

James S. Craigie
National Research Council of Canada, 1411 Oxford Street, Halifax, NS B3H 3Z1, Canada

John A. Raven
Division of Plant Sciences, University of Dundee (James Hutton Inst), Invergowrie, Dundee DD2 5DA, Scotland, UK
Plant Functional Biology and Climate Change Cluster, University of Technology Sydney, Ultimo, NSW 2007, Australia

Sabeeha S. Merchant
Department of Chemistry & Biochemistry, University of California-Los Angeles, 607 Charles E. Young Dr., East, Los Angeles, CA 90095-1569, USA

Katherine E. Helliwell
Department of Plant Sciences, University of Cambridge, Downing St., Cambridge CB2 3EA, UK
Present address: Marine Biological Association of the UK, Citadel Hill, Plymouth PL1 2PB, UK

Alison G. Smith
Department of Plant Sciences, University of Cambridge, Downing St., Cambridge CB2 3EA, UK

Mary Ellen Camire
School of Food and Agriculture, University of Maine, Orono, ME 04469, USA

Susan H. Brawley
School of Marine Sciences, University of Maine, Orono, ME 04469, USA

Pierrick Stévant
Møreforsking Ålesund AS, Ålesund, Norway
Department of Biotechnology, Norwegian University of Science and Technology (NTNU), Trondheim, Norway

Céline Rebours
Møreforsking Ålesund AS, Ålesund, Norway
Norwegian Institute for Bioeconomy Research (NIBIO), Ås, Norway

Annelise Chapman
Møreforsking Ålesund AS, Ålesund, Norway

Céline Rebours
Bioforsk, Norwegian Institute for Agricultural and Environmental Research, Ås, Norway
Arctic Agriculture and Land Use Division, Bioforsk, Bodø, Norway

Eliane Marinho-Soriano
Department of Oceanography and Limnology, Federal University of Rio Grande do Norte, Natal, Brazil

José A. Zertuche-González
Instituto de Investigaciones Oceanológicas, Universidad Autónoma de Baja California, Ensenada, Mexico

Leila Hayashi
Aquaculture Department, Federal University of Santa Catarina, Florianópolis, Brazil

Julio A. Vásquez
Facultad de Ciencias del Mar, Centro de Estudios Avanzados en Zonas Aridas, Universidad Católica del Norte, Coquimbo, Chile

Paul Kradolfer
PSW SA, Lima 16, Peru

Gonzalo Soriano
Soriano SA, Gaiman, Argentina

Raul Ugarte
Acadian Seaplants Limited, Dartmouth, Canada

Maria Helena Abreu
ALGAPlus Lda, Ílhavo, Portugal

Ingrid Bay-Larsen
Nordland Research Institute, Bodø, Norway

Grete Hovelsrud
Nordland Research Institute, Bodø, Norway

Rolf Rødven
Bioforsk, Norwegian Institute for Agricultural and Environmental Research, Ås, Norway

Daniel Robledo
Department of Marine Resources, CINVESTAV-IPN, Mérida, Mexico

Adibi M. Nor
School of Marine Science and Technology, Newcastle University, Ridley Building, Claremont Road, Newcastle upon Tyne, England NE1 7RU, UK

Tim S. Gray
School of Geography, Politics and Sociology, Newcastle University, Newcastle upon Tyne, England NE1 7RU, UK

Gary S. Caldwell
School of Marine Science and Technology, Newcastle University, Ridley Building, Claremont Road, Newcastle upon Tyne, England NE1 7RU, UK

Selina M. Stead
School of Marine Science and Technology, Newcastle University, Ridley Building, Claremont Road, Newcastle upon Tyne, England NE1 7RU, UK

Siti Fadilah
Research and Development Institute for Seaweed Aquaculture, Boalemo, Gorontalo 96265, Indonesia

Alimuddin
Department of Aquaculture, Faculty of Fisheries and Marine Sciences (FFMS), IPB, Bogor 16680, Indonesia

Petrus Rani Pong-Masak
Research and Development Institute for Seaweed Aquaculture, Boalemo, Gorontalo 96265, Indonesia

Joko Santoso
Department of Fish Processing Technology, FFMS, IPB, Bogor 16680, Indonesia

Andi Parenrengi
Research and Development Institute for Brackishwater Aquaculture, Maros, Sulawesi Selatan 90512, Indonesia

Christina Carl
MACRO – the Centre for Macroalgal Resources and Biotechnology, and School of Marine and Tropical Biology, James Cook University, Townsville, Queensland, Australia

Rocky de Nys
MACRO – the Centre for Macroalgal Resources and Biotechnology, and School of Marine and Tropical Biology, James Cook University, Townsville, Queensland, Australia

Rebecca J. Lawton
MACRO – the Centre for Macroalgal Resources and Biotechnology, and School of Marine and Tropical Biology, James Cook University, Townsville, Queensland, Australia

Nicholas A. Paul
MACRO – the Centre for Macroalgal Resources and Biotechnology, and School of Marine and Tropical Biology, James Cook University, Townsville, Queensland, Australia

Bénédicte Charrier
Centre National de la Recherche Scientifique, Sorbonne Université, UPMC Univ Paris 06, UMR 8227, Integrative Biology of Marine Models, Station Biologique de Roscoff, Roscoff, France

Elodie Rolland
Centre National de la Recherche Scientifique, Sorbonne Université, UPMC Univ Paris 06, UMR 8227, Integrative Biology of Marine Models, Station Biologique de Roscoff, Roscoff, France

Vishal Gupta
Seaweed Biology and Cultivation Group, Division of Marine Biotechnology and Ecology, CSIR-Central Salt and Marine Chemicals Research Institute, Bhavnagar, India

C. R. K. Reddy
Seaweed Biology and Cultivation Group, Division of Marine Biotechnology and Ecology, CSIR-Central Salt and Marine Chemicals Research Institute, Bhavnagar, India

Louisi Souza de Oliveira
Departamento de Biologia Marinha, Instituto de Biologia, Universidade Federal do Rio de Janeiro (UFRJ) Av. Carlos Chagas Filho, 373-CCS - IB - BLOCO A (ANEXO) A3-202, Rio de Janeiro 21941-599, Brazil.

Gustavo Bueno Gregoracci
Departamento de Biologia Marinha, Instituto de Biologia, Universidade Federal do Rio de Janeiro (UFRJ) Av. Carlos Chagas Filho, 373-CCS - IB - BLOCO A (ANEXO) A3-202, Rio de Janeiro 21941-599, Brazil.

Genivaldo Gueiros Zacarias Silva
Laboratório de Bioinformática e Biologia Evolutiva, Universidade Federal de Pernambuco. Av. Prof. Moraes Rego 1235, Cidade Universitária, Recife 50670-901PE, Brazil.

Leonardo Tavares Salgado
Instituto de Pesquisa Jardim Botânico do Rio de Janeiro, Rua Pacheco Leão, 915. Jardim Botânico, Rio de Janeiro 22460-030RJ, Brazil.

Gilberto Amado Filho
Instituto de Pesquisa Jardim Botânico do Rio de Janeiro, Rua Pacheco Leão, 915. Jardim Botânico, Rio de Janeiro 22460-030RJ, Brazil.

Marcio Alves-Ferreira
Departamento de Genética. Instituto de Biologia. Av. Prof. Rodolpho Paulo Rocco, s/n, CCS, Sala A2-93, Universidade Federal do Rio de Janeiro (UFRJ), Rio de Janeiro 21941-599RJ, Brazil.

Renato Crespo Pereira
Departamento de Biologia Marinha, Universidade Federal Fluminense (UFF). Morro do Valonguinho, s/n. Centro, Niteroi 24001-970RJ, Brazil.

Fabiano L Thompson
Departamento de Biologia Marinha, Instituto de Biologia, Universidade Federal do Rio de Janeiro (UFRJ) Av. Carlos Chagas Filho, 373-CCS - IB - BLOCO A (ANEXO) A3-202, Rio de Janeiro 21941-599, Brazil.

Muhammad Ikbal Illijas
Department of Aquaculture, Pangkep State Polytechnic of Agriculture, Indonesia

Arifuddin
Department of Aquaculture, Pangkep State Polytechnic of Agriculture, Indonesia

Luqman Saleh
Department of Aquaculture, Pangkep State Polytechnic of Agriculture, Indonesia

Yutaka Itabashi
Graduate School of Fisheries Sciences, Hokkaido University, Japan

Zhang Yi
School of Environmental and Chemical Engineering, Dalian Jiaotong University, Dalian 116028, China
School of Life Science and Biotechnology, Dalian University of Technology, Dalian 116024, China

Han Jinyuan
School of Environmental and Chemical Engineering, Dalian Jiaotong University, Dalian 116028, China

Mu Jun
School of Environmental and Chemical Engineering, Dalian Jiaotong University, Dalian 116028, China

Feng Yan
School of Environmental and Chemical Engineering, Dalian Jiaotong University, Dalian 116028, China

Gu Xiaojie
School of Environmental and Chemical Engineering, Dalian Jiaotong University, Dalian 116028, China

And Ji Yuanxi
School of Environmental and Chemical Engineering, Dalian Jiaotong University, Dalian 116028, China

Snehal Subhash Shitole
Central Institute of Fisheries Education, Mumbai, India

Amjad Khansaheb Balange
Central Institute of Fisheries Education, Mumbai, India

Shardul Sham Gangan
Khar Land Research Station, Panvel, India

Seema Patel
Department of Biotechnology, Lovely Professional University, Jalandhar, India

Andreas Grassauer
Marinomed Biotechnologie GmbH, Veterinaerplatz 1/HA, A-1210 Vienna, Austria
Laboratory of Tropical Veterinary Medicine, Veterinary University Vienna, Veterinaerplatz 1, A-1210 Vienna, Austria

Regina Weinmuellner
Marinomed Biotechnologie GmbH, Veterinaerplatz 1/HA, A-1210 Vienna, Austria

Christiane Meier
Marinomed Biotechnologie GmbH, Veterinaerplatz 1/HA, A-1210 Vienna, Austria
Laboratory of Tropical Veterinary Medicine, Veterinary University Vienna, Veterinaerplatz 1, A-1210 Vienna, Austria

Alexander Pretsch
Marinomed Biotechnologie GmbH, Veterinaerplatz 1/HA, A-1210 Vienna, Austria
Laboratory of Tropical Veterinary Medicine, Veterinary University Vienna, Veterinaerplatz 1, A-1210 Vienna, Austria

Eva Prieschl-Grassauer
Marinomed Biotechnologie GmbH, Veterinaerplatz 1/HA, A-1210 Vienna, Austria
Laboratory of Tropical Veterinary Medicine, Veterinary University Vienna, Veterinaerplatz 1, A-1210 Vienna, Austria

Hermann Unger
Laboratory of Tropical Veterinary Medicine, Veterinary University Vienna, Veterinaerplatz 1, A-1210 Vienna, Austria

Saowapa Rattaya
Department of Food Technology, Faculty of Agro-IndustryPrince of Songkla UniversityHat YaiThailand

Soottawat Benjakul
Department of Food Technology, Faculty of Agro-IndustryPrince of Songkla UniversityHat YaiThailand

Thummanoon Prodpran
Department of Material Product Technology, Faculty of Agro-IndustryPrince of Songkla UniversityHat YaiThailand

Adam D Hughes
Scottish Association for Marine Science, Scottish Marine Institute, Department of Ecology, Oban, Argyll PA37 1QA, Scotland.

Maeve S Kelly
Scottish Association for Marine Science, Scottish Marine Institute, Department of Ecology, Oban, Argyll PA37 1QA, Scotland.

Kenneth D Black
Scottish Association for Marine Science, Scottish Marine Institute, Department of Ecology, Oban, Argyll PA37 1QA, Scotland.
Scottish Association for Marine Science, Scottish Marine Institute, Marine BioEnergy Scotland, Oban, Argyll PA37 1QA, Scotland.

Michele S Stanley
Scottish Association for Marine Science, Scottish Marine Institute, Department of Ecology, Oban, Argyll PA37 1QA, Scotland.
Scottish Association for Marine Science, Scottish Marine Institute, Marine BioEnergy Scotland, Oban, Argyll PA37 1QA, Scotland.

Shi-Qi Ji
Key Laboratory of Biofuels, Shandong Provincial Key Laboratory of Synthetic Biology, Qingdao Institute of Bioenergy and Bioprocess Technology, Chinese Academy of Sciences, Qingdao, People's Republic of China

Bing Wang
Key Laboratory of Biofuels, Shandong Provincial Key Laboratory of Synthetic Biology, Qingdao Institute of Bioenergy and Bioprocess Technology, Chinese Academy of Sciences, Qingdao, People's Republic of China

Ming Lu
Key Laboratory of Biofuels, Shandong Provincial Key Laboratory of Synthetic Biology, Qingdao Institute of Bioenergy and Bioprocess Technology, Chinese Academy of Sciences, Qingdao, People's Republic of China

Fu-Li Li
Key Laboratory of Biofuels, Shandong Provincial Key Laboratory of Synthetic Biology, Qingdao Institute of Bioenergy and Bioprocess Technology, Chinese Academy of Sciences, Qingdao, People's Republic of China

Emily T. Kostas
The University of Nottingham, Loughborough, UK

Daniel A. White
Plymouth Marine Laboratory, Prospect Pl, Plymouth, UK

Chenyu Du
The University of Nottingham, Loughborough, UK

David J. Cook
The University of Nottingham, Loughborough, UK

Paul Bikker
Livestock Research, Wageningen University and Research Centre, De Elst 1, 6708 WD Wageningen, The Netherlands

Marinus M. van Krimpen
Livestock Research, Wageningen University and Research Centre, De Elst 1, 6708 WD Wageningen, The Netherlands

Piet van Wikselaar
Livestock Research, Wageningen University and Research Centre, De Elst 1, 6708 WD Wageningen, The Netherlands

Bwee Houweling-Tan
Food and Biobased Research, Wageningen University and Research Centre, Bornse Weilanden 9, 6708 WG Wageningen, The Netherlands

Nazareno Scaccia
Food and Biobased Research, Wageningen University and Research Centre, Bornse Weilanden 9, 6708 WG Wageningen, The Netherlands

Jaap W. van Hal
Biomass & Energy Efficiency, Energy research Centre of the Netherlands (ECN), Westerduinweg 3, 1755 LE Petten, The Netherlands

Wouter J. J. Huijgen
Biomass & Energy Efficiency, Energy research Centre of the Netherlands (ECN), Westerduinweg 3, 1755 LE Petten, The Netherlands

John W. Cone
Animal Nutrition Group, Wageningen University, De Elst 1, 6708 WD Wageningen, The Netherlands

Ana M. López-Contreras
Food and Biobased Research, Wageningen University and Research Centre, Bornse Weilanden 9, 6708 WG Wageningen, The Netherlands

E. James Hehre
Sea Around Us Project /Institute for oceans and Fisheries, University of British Columbia, Vancouver, BC, Canada

Jessica J. Meeuwig
School of Animal Biology and Centre for Marine Futures, University of Western Australia, Crawley, WA, Australia

Izabela Michalak
Department of Advanced Material Technologies, Faculty of Chemistry, Wrocław University of Science and Technology, Wrocław, Poland

Radosław Wilk
Department of Advanced Material Technologies, Faculty of Chemistry, Wrocław University of Science and Technology, Wrocław, Poland

Katarzyna Chojnacka
Department of Advanced Material Technologies, Faculty of Chemistry, Wrocław University of Science and Technology, Wrocław, Poland

PREFACE

The seaweeds are a large group of marine macroalgae that can be found in several coastal regions around the world. Though there are about 9200 species of seaweeds, only 221 species are economically important (Mohammed, 2013). There are three different groups of seaweed—red, brown, and green their characteristic colors and properties. Seaweed cultivation is a relatively old and extremely important activity for the human being, because of the various properties and utilities that it owns. There are records of use of this feature in the 4th century in Japan and in the 6th century in China. Both countries along with the Republic of Korea are the biggest consumers of seaweed in the world, contributed with 80% of global production (Mohammed, 2013).

Worldwide, the most extensive crops are performed in the open sea or estuaries, with little technology involved and in order to help in the income of the fishermen. However, in this system the biggest concern and effort involved is in management, mainly of the harvest, but also daily by the need of monitoring the situation and existing dynamics in these regions. Seaweeds have various uses and great commercial importance on the world market. Among them, we highlight the production of hydrocolloids (agar, carrageenans, and alginate), fertilizers, nutraceuticals, human food, water treatment, bioremediation, biofuels, medicinal uses, for spa and beauty, among others. The worldwide seaweed industry provides a wide variety of products for direct or indirect human uses that have an estimated total value of US$10 billion per year (Bixler & Porse 2011).

Since the emergence of the seaweed cultivation processes are in constant improvement, mainly related to knowledge, management and improvements within the activity. Naturally, the demand for seaweed products has generated the need for increased knowledge to resolve main problems in the development of seaweed aquaculture, as for example the selection of suitable environments; engineering of the seashore for farming;

protection of crops from pests and physical or biotic impacts; and proper transplantation and harvesting procedures (Buschmann et al., 1995).

Seaweeds are a highly renewable food resource. They can be grown and harvested all year round, in any marine environment. Production is low impact - seaweeds need no chemicals, fertilizers or pesticides, in addition to require no deforestation. Seaweed cultivation can still be added in different production systems as bioremediation and in treatment of effluents from industry, in order to contribute to the maintenance of the quality of the environment. In addition to bring benefits to the environment, the seaweed cultivation has contributed to increase income of the coastal communities that depend on fishing.

Although the cultivation of seaweed being a relatively simple activity with respect to processes, many improvements can be made from the point of view of management and production, so as to make easiest and most effective methods. Therefore, it is essential to maintain the focus and work on research that seeks the development and evolution of the area. In addition to being an important economic resource, is fundamental basis for the conservation of marine and coastal ecosystems worldwide.

SECTION I

HISTORY, DEVELOPMENT AND FUTURE PERSPECTIVES OF SEAWEED CULTIVATION

CHAPTER

1

HISTORY, DEVELOPMENT AND FUTURE PERSPECTIVES OF SEAWEED CULTIVATION

Bruno Augusto Amato Borges

Aquaculture Engineer, Florianopolis, Brazil

HISTORY AND CURRENT SITUATION

The seaweeds are a large group of marine macroalgae that can be found in several coastal regions around the world. Though there are about 9200 species of seaweeds, only 221 species are economically important (Mohammed, 2013). Like plants, they utilize photosynthesis to convert inorganic nutrients into organic biomass, an important niche that provides food, shelter, and habitat for other organisms. There are three different groups of seaweed—red, brown, and green. They are grouped according to their photosynthetic pigments, which give them their characteristic colors and properties.

Seaweed cultivation is a relatively old and extremely important activity for the human being, because of the various properties and utilities that it owns. There are records of use of this feature in the 4th century in Japan and in the 6th century in China. Both countries along with the Republic of Korea are the biggest consumers of seaweed in the world, contributed with 80% of global production (Mohammed, 2013). In Asia, seaweeds have been consumed as a vegetable since the beginning of time. Seaweed aquaculture has grown faster than any other marine production sector over the last 20 years, and has had an annual global growth rate of 7.7% since 1970 (Buchholz et al. 2012). More than 96% of global seaweed production (23.8 million t) is produced in aquaculture (FAO, 2016). On average; the Japanese eat 1.4 kg per person per year (Burtin, 2003). According to FAO, 2016 the above quantities do not include production of aquatic plants, mostly seaweeds, that reached 28.5 million tonnes in 2014, of which aquaculture produced 27.3 million tonnes (96 percent). Carrageenan seaweeds (including *Kappaphycus alvarezii* and *Eucheuma* spp.) are the main cultured seaweeds (11 million tonnes), followed by japanese kelp (7.7 million tonnes). While the capture of marine algae remained the same in the year 2005 until today, the world production from cultivation was 13 million tons in 2005 to the 27 million tonnes in the year 2014 (FAO, 2016). Overwhelmingly dominated by seaweeds in terms of volume, aquatic plant farming is practiced in about 50 countries. It expanded at 8 percent per year in the past decade, up from 6.2 percent in the previous decade, with output more than doubling in this period (FAO, 2016a). The main seaweed-producing countries currently are China, Indonesia, Philippines and Japan.

In addition to various uses and economic importance, the seaweeds are essential for the maintenance of life on the planet. They are the main responsible for the formation and maintenance of atmospheric oxygen, are made of fossil fuels, carbon remover, have high bioremediation potential and are main primary producers. Regarding the economic importance, the seaweeds are a source of food and polysaccharides.

Seaweeds contain large amounts of polysaccharides, notably cell wall structural polysaccharides that are extracted by the hydrocolloid industry: alginate from brown seaweeds, carrageenans and agar from red seaweeds (Burtin, 2003). These compounds are sulfated polysaccharide, which has preserving, stabilizer and thickener properties, plus features such as biocompatibility, biodegradability, high water retention capacity and mechanical strength of its gels (Prajapati et al., 2014). Such compounds are widely used in a number of food products such as cheese, ice cream, chocolates, jellies and gelatins, hams, sausages, burgers, beers, among others. They also can be used in the production of toothpaste.

In recent years, much attention has been focused on polysaccharides isolated from natural sources. Several algal species belonging to phaeophyta, rhodophyta and chlorophyta divisions have been recognized as crucial sources of sulfated polysaccharides. The biological features of the sulfated polysaccharides reported till now are antioxidant, antitumor, immunomodulatory, inflammation, anticoagulant, antiviral, antiprotozoan, antibacterial, antilipemic (Patel, 2012).

In General, cultivation does not require a high degree of management and technologies to work. The main deciding factor for success is the choice of the area to fit the desired species, as well as bureaucratic and legislation aspects present in most countries.

Worldwide, the most extensive crops are performed in the open sea or estuaries, with little technology involved and in order to help in the income of the fishermen. However, in this system the biggest concern and effort involved is in management, mainly of the harvest, but also daily by the need of monitoring the situation and existing dynamics in these regions. Seaweed farming is a viable alternative source of income for small-scale fisherman.

Currently, there is a growing quest for more sustainable crops that uses the available resources in a responsible manner with minimal waste. Various researches has been made in the pursuit of knowledge and improvements regarding the integrated more than one specie on cultivation, named Integrated Multi-Trophic Aquaculture (IMTA), system that reduces the environmental impact and waste of nutrients from feed (Samocha et al., 2015; Matos et al., 2005; Hadley et al., 2014; Abreu et al., 2013; Pereira & Abreu, 2012).

According to Buschmann et al. (2001), one of the common problems detected in farms is an abrupt drop in productivity, which is always preceded by 2–3 years of high yields. It is believed that this situation is the result of thallus aging, and seems to be influenced by the harvesting method. Typically, the method of collection is based on the removal of the stalks, being left behind the oldest for later production. Another very common problem in cultivation of seaweed is the loss of genetic variability and diversity among natural stocks and produced by aquaculture, which constantly reproduce among themselves in various aquatic environments.

In addition, there is a great deal of concern related to production and introduction of exotic species in some countries, which along with the lack of control, escape to the natural environments and competition with native species, causing local ecological imbalance.

I believe this is the path to be followed. The search for better use of resources and avoiding waste is essential, through good management practices and a responsible cultivation can be enlarged and diversified productions bringing countless benefits for producers, consumers and the entire production chain, always in search of wellbeing and human comfort.

MAIN SPECIES AND USES

In General, different species of marine algae are cultivated around the world. Each species has a commercial use that justifies your production (food, cosmetics, personal hygiene, agriculture), where it is inserted. Is a very important resource for many cultures and has great importance in the world market. That way, the search for improvements in the sector is fundamental to, in addition to meet the demand, developing new technologies that allow the use for different purposes.

The main genres of seaweed grown currently are *Kappaphycus, Eucheuma, Laminaria, Gracilaria, Porphyra* and *Undaria*.

The green algae (Division Chlorophyta) are present in the most diverse environments, and 90% of the species occur in freshwater and marine forms compose the benthos of tropical and subtropical waters. Are filamentous or parenchymal algae, their pigments are chlorophyll a and b and carotenoids and your booking is starch. Is extremely present in Asian countries and is consumed dry, ground, or in the form of pates.

The red algae (Division Rhodophyta) are mostly benthic marine algae and inhabit from equatorial to polar regions, with depths of up to 250 m in waters with high level of transparency. Has economic importance in human consumption in sushi (*Porphyra*-Nori) and salads (*Gracilaria, Palmaria, Chondrus*). Marine red alga *Kappaphycus striatum* (= *Ecucheuma striatum* = *Kappaphycus alvarezii*) is one of the best sources of kappa carrageenan as compared to other seaweeds, in addition is economically important tropical red seaweed and highly demanded for its cell wall polysaccharide, being the most important source of ê-carrageenan in the world (Bixler, 1996). According to Pickering (2006), *Kappaphycus* farming has been strongly promoted in the Pacific region because it requires a low level of technology and investment, can be operated at the family level, has relatively little environmental impact, does not require refrigeration or high-tech postharvest processing within the country, and is normally compatible with traditional fishing and other subsistence uses of the inshore environment. The genus *Gruciluriu* is widely distributed around the world. It occupies a variety of habitats, both in tropical and temperate latitudes, forming monospecific stands or multi-specific vegetational assemblages (Santelices & Doty, 1989). Several *Gracilaria* species form the basis of an expanding multi-million dollar phycocolloid industry (Johnson et al., 2014). Cultivation of the agarophyte, Gracilaria chilensis, has been a commercial activity since mid 80s along the Chilean coast (Buschmann et al., 2001).

Brown algae (Division Phaeophyta) are almost exclusively marine environments, and most inhabit cold waters. Covers the main species grown for direct human consumption: *Laminaria japonica* (kombu) and *Undaria-pinafitada* (wakame). It used for alginate extraction, mostly comes from Atlantic Ascophyllum *nodosum* and *Laminaria hyperborea*. Alginates, derivatives of alginic acids, are used commercially for toothpastes, soaps, ice cream, tinned meats, fabric printing, and a host of other applications. It forms a stable viscous gel in water, and its primary function in the above applications is as a binder, stabilizer, emulsifier, or moulding agent. *Saccharina japonica*, formerly *Laminaria*, and other species of the genus are grown on ropes in China, Korea and Japan for food and alginate production. *Undaria pinnatifida* is also cultivated in Japan, Korea and China for production of Wakame, valuable food kelp (Guiry, 2000-2017).

Seaweeds have various uses and great commercial importance on the world market. Among them, we highlight the production of hydrocolloids (agar, carrageenans, and alginate), fertilizers, nutraceuticals, human food, water treatment, bioremediation, biofuels, medicinal uses, for spa and beauty, among others. The worldwide seaweed industry provides a wide variety of products for direct or indirect human uses that have an estimated total value of US$10 billion per year (Bixler & Porse 2011).

Currently, it can be said that the industry is extremely dependent on the cultivation of seaweed, since most of the market is stocked by cultivation, not exploitation. Therefore, must be sought whenever evolution and development in the methods of cultivation and extraction of the desired components. In addition, it is important to continue the development of research to seek out new potential species, other components that could be used and other utilities that this feature can provide. All this in order to use your full, with minimal losses.

Due to the growth of the industry and demand for the use of fuels, has been quite discussed the search for alternative sources for your achievement. According to some studies, the seaweed has potential for production of biofuels and the processes to extract and obtaining the final product are already known. In addition, rising concerns over environmental pollution and global warming have encouraged the movement to alternate fuels; the world ethanol market is projected to reach 86 billion liters in 2013 year (Kraan, 2013).

The large scale production of marine macroalgae, mainly for human consumption, has given rise to their consideration as a non-lignocellulosic feedstock for the production of renewable fuels (Baghel et al., 2015). According to Ghadiryanfar et al. (2016) high levels of structural polysaccharides and low lignin contents make seaweed attractive feedstocks for production of liquid biofuels via fermentation and biogas production via anaerobic digestion.

Unlike most energy sources that make up the current world energy matrix, the marine algae are a renewable source and do not rely on external inputs for cultivation, just the choice of the area and proper management to obtain the final product.

One of the major problems with respect to use of seaweed to obtain biofuels is your low income in the production of methane gas, and that a project with that goal becomes feasible, must be implemented large-scale crops so that they can compete with the other existing fuel in the global market. In addition, the seaweeds currently are source for other products which may be more important on the world stage than the production of biofuel. According to Hannon et al. (2010) the algal biofuels production chain is outlined and shows that the major challenges include strain isolation, production management, harvesting, coproduct development, fuel extraction, refining and residual biomass utilization.

Therefore, the search for improvements in fuel procurement via seaweed should be discussed and evaluated, mainly related to your viability and market share. It is clear that renewable sources must be sought in the various industrial sectors, but before applying our must be sure that will work correctly.

Therefore, it can be observed that the cultivation of seaweed is very important to supply world demand of this feature, which has various uses in different sectors of industry. In addition, it is essential to continue the search for knowledge and new technologies in the sector, which are able to improve or optimize existing processes within the production chain, and to seek alternatives to the problems currently in question.

MANAGEMENT AND TECHNIQUES

Since the emergence of the seaweed cultivation processes are in constant improvement, mainly related to knowledge, management and improvements within the activity. However, when it comes to extensive crops, we can affirm the management methodology employed is practically the same for a long time, a relatively rudimentary management that ran for hundreds of years. Development of seaweed culture techniques, like that of agriculture, has gradual (Jackson, 1980). Also according to Jackson (1980), primitive seaweed harvesting is done by gathering drift algae throw on the shore or by collecting from natural stands. Techniques vary of hand picking of exposed algae at low tide to mechanical harvesting of giant kelp.

Naturally, the demand for seaweed products has generated the need for increased knowledge to resolve main problems in the development of seaweed aquaculture, as for example the selection of suitable environments; engineering of the seashore for farming; protection of crops from pests and physical or biotic impacts; and proper transplantation and harvesting procedures (Buschmann et al., 1995). According to Mazarrasa et al. (2014), we expect that the growth and distribution of seaweed related patents will vary among countries depending on their investment in research, their access to technologies and their tradition of seaweed aquaculture.

In addition to research and development investment, I believe it is essential that producers and researchers to change a little the mindset in relation to production, more responsibility and seriousness with everything that involves the culture, such as the environment, local ecology and the people involved. A few years here, has been discussed and increasingly used the concept of best management practices (BMP) in aquaculture. BMP consists in the realization of the various production processes in a responsible manner taking into account all aspects. Best management practices in feeding, in the treatment of effluents, in the choice of the area, in the management of harvest, in the settlement, are some examples of processes that can be improved and standardized in search of a responsible and sustainable aquaculture.

According to Chopin et al., (2001), in its search for best management practices, the aquaculture industry should develop innovative and responsible practices that optimize its efficiency and create diversification, while ensuring the remediation of the consequences of its activities to keep in the health of coastal waters.

Currently, there are three main cultivation techniques which are widely used around the world. Among them: off-bottom, floating line and long-line methods.

In the off-bottom method, nylon ropes with seaweed are tied between two wooden pegs. The pegs are from mangroves or land-based plants. The seaweed branches, usually about 100 g, are tied to the lines and allowed to grow for six weeks before they are harvested. When harvesting, farmers remove the lines and the seaweed and then tie-in new seaweed branches (Msuya, 2011).

Floating line technique where the seaweeds are planted in water of 2-5 m, depending on the tidal range, and the cast method where the seaweeds are bound to rocks using rubber bands and allowed to attach and grow (Msuya 2006).

The long-line method is similar to the raft or floating method in several ways. In fact, seaweed is hung to a rope which is suspended by floaters. In this method, however, the main rope is thicker; about 10 to 15 millimeters in diameter. This rope is tied to plastic floaters or bamboo set at 4 to 5 meters interval and anchored at each end. The pieces of raffia holding the seaweed can be tied to the main rope in two ways (Forscarini & Prakash, 1990).

In General, all methods of production has its peculiarities and are adapted primarily due to region and the place in which they are inserted. The main factors that determine the choice of a method of cultivation are the location, the availability of workforce, the level of technology employed, water quality, oceanographic factors and the species will be cultivated. It is important that all these factors are considered at the time of preparation of the project.

SEAWEED AND ENVIRONMENT

Like any business, there will always be challenges and methodologies that can improve or modify processes. Currently, it is remarkable the lack of work and projects that take environmental issues seriously reconciled production. The main focus is productivity; the environment is in the background.

The culture of seaweed is also significantly influenced by environmental factors in several countries. Turbidity, nutrient levels, phytoplankton blooms, temperature, salinity fluctuations are all significant factors in the successful development of culture areas (Phillips, 1990).

Seaweeds are a highly renewable food resource. They can be grown and harvested all year round, in any marine environment. Production is low impact - seaweeds need no chemicals, fertilizers or pesticides, in addition to require no deforestation.

On the contrary, the introduction of seaweed and structures can considerably enhance the productivity of invertebrates and fish much the same way as artificial reefs, due to increased availability of shelter and food organisms. Studies in Japan have shown that *Laminaria japonica* farms act as shelters increasing numbers of fish, sea urchins, sea cucumbers and abalones in giant kelp culture areas (Suo et al., 1986). Seaweeds are one of major components of primary biomass production in coastal maritime ecosystems and play an essential ecological role as habitat and substrata for invertebrates, fish, mammals, and birds (Graham et al., 2007). The attractiveness of seaweed farms can also cause problems for seaweed farmers, by attracting invertebrates and finfish which may predate on the seaweed (North, 1987).

In addition to promoting the development and restoration of local fauna, marine algae can act integrated to cultivation of marine fish and shrimp by taking advantage of the nutrients from the food waste and feces. By integrating fed aquaculture (finfish, shrimp) with inorganic and organic extractive aquaculture (seaweed), the wastes of one resource user become a resource (fertilizer or food) for the others. Such a balanced ecosystem approach provides nutrient bioremediation capability, mutual benefits to the co-cultured organisms, economic diversification by producing other value-added marine crops, and increased profitability per cultivation unit for the aquaculture industry (Chopin et al., 2001).

The term sustainability is gaining enough space over the last decade and has been inserted into some new ventures in a variety of ways; claims for the use of ecological engineering tools towards the development of a more sustainable aquaculture have increased (Chopin et al., 2008). In seaweed cultivation, one should opt for the cycling and reuse of effluent nutrients of cultivation through Integrated Multi-Trophic Aquaculture (IMTA) , the lowest possible use of water (through different systems of cultivation), deployment in that area do not undermine the local ecosystem, among other decisions aimed to avoid losses, waste land unnecessary spending. By cultivating seaweed in the effluent tanks the nutrient pollution from aquaculture systems can be alleviated. In addition, the seaweed can be harvested to produce another value-added product offering fish enterprises another economic incentive to improve environmental management (Kraan et. al., 2007).

Seaweed cultivation can still be added in different production systems as bioremediation and in treatment of effluents from industry, in order to contribute to the maintenance of the quality of the environment. According to Neori et al. (2004), as in certain traditional polyculture schemes, plants can drastically reduce feed use and environmental impact of industrialized mariculture and at the same time add to its income. These nutrient-assimilating photoautotrophic plants use solar energy to turn nutrient-rich effluents into profitable resources. According to Luning & Pang (2003), mass culture of commercially valuable seaweed species is likely to play an increasingly important role as a nutrient-removal system to alleviate eutrophication problems due to fed aquaculture.

In addition to bring benefits to the environment, the seaweed cultivation has contributed to increase income of the coastal communities that depend on fishing. For the past four decades, seaweed culture has been globally introduced to many maritime countries for experimental and commercial cultivation as a sustainable alternate livelihood for coastal villagers (Bindu & Levine, 2011). Currently, several communities around the world are organized into cooperatives to produce seaweed in partnership; since it is an activity that requires daily management, its important the participation of all. In addition to improving the family income, increases social relationships of families involved in the community.

Although the cultivation of seaweed being a relatively simple activity with respect to processes, many improvements can be made from the point of view of management and production, so as to make easiest and most effective methods. Therefore, it is essential to maintain the focus and work on research that seeks the development and evolution of the area. In addition to being an important economic resource, is fundamental basis for the conservation of marine and coastal ecosystems worldwide.

REFERENCES

1. Abreu, M.H. et al. (2013). New developments of IMTA in Europe: Fucusing on seaweed. 21st International Seaweed Symposium. Seaweed Science for Sustainable Prosperity. April, 2013. Bali, Indonesia.
2. Baghel, R.S., Trivedi, N., Gupta, V., Neori, A., Reddy, C.R.K., Lali, A., Jha, B. (2015). Biorefining of marine macroalgal biomass for production of biofuel and commodity chemicals. Green Chemistry, 17: 2436-2443. DOI: 10.1039/C4GC02532F.
3. Bindu, M.S., Levine, I.A. (2011). The commercial red seaweed Kappaphycus alvarezii – an overview on farming and environment. Journal of Applied Phycology, v. 23, Issue 4, pp 789-796. DOI: 10.1007/s10811-010-9570-2.
4. Bixler, H.J., Porse, H. (2011). A decade of change in the seaweed hydrocolloids industry. Journal of Applied Phycology, 23:321–335. DOI: 10.1007/s10811-010-9529-3.
5. Bixler, H.J. (1996). Recent developments in manufacturing and marketing carrageenan. Hydrobiol., 326/327: 35-37.
6. Buchholz, C.M., Krause G., Buck B.H. (2012). Seaweed and man. *Seaweed biology: new insights into ecophysiology, ecology and utilization*. Berlin, Germany: Springer, 471–493.

7. Burtin, P. (2003) Nutritional value of seaweeds. EJEAFChe, 2 (4), 498-503.
8. Buschmann AH,Westermeier R, Retamales CA. (1995). Cultivation of Gracilaria on the sea-bottom in southern Chile: a review. Journal of Applied Phycology; 7:291–301.
9. Buschmann, A.H., Correa, J.,Westermeier, R., Hernández-González, M., Norambuena, R., (2001). Cultivation of red algae in Chile: a review. Aquaculture 194, 203–220.
10. Chopin, T. et al. (2001). Integrating seaweeds into marine aquaculture systems: a key toward sustainability. Journal of Phycology, v. 37, Issue 6, 975-986. DOI: 10.1046/j.1529-8817.2001.01137.x.
11. Chopin, T., Robinson, S.M.C., Troell, M., Neori, A., Buschmann, A.H., Fang, J., (2008). Multitrophic integration for sustainable marine aquaculture. Ecological Engineering. Vol. 3. Encyclopedia of Ecology, vol. 5. Elsevier, Oxford, pp. 2463–2475.
12. FAO. 2016. FAO yearbook. Fishery and Aquaculture Statistics. 2014. Rome, Italy.
13. FAO (2016a). The State of World Fisheries and Aquaculture 2016. Contributing to food security and nutrition for all. Rome. 200 pp.
14. Foscarini, R., Prakash, J. (1990). Handbook on Eucheuma seaweed cultivation in Fiji. Fisheries and aquaculture department. FAO Corporate document repository. Suva, Fiji.
15. Ghadiryanfar, M., Rosentrater, K.A., Keyhani, A., Omid, M. (2016). A review of macroalgae production, with potencial applications in biofuels and bioenergy. Renewable and suatainable energy reviews, v. 54, pp 473-481.
16. Graham, M.H., Vásquez, J.A., Buschmann, A.H (2007). Global ecology of the giant kelp *Macrocystis*: from ecotypes to ecosystems. Oceanography and Marine Biology: An Annual Review, 45:39–88.
17. Guiry, M.D. (2000-2017). The seaweed site: information on marine algae. Brown Algae. Site access: 11 march, 2017.
18. Hadley, S., Wild-Allen, K., Johnson, C., Macleod, C. (2015). Modeling macroalgae growth and nutrient dynamics for integrated multi-trophic aquaculture. Journal of Applied Phycology, 27: 901-916.

19. Hannon, M., Gimpel, J., Tran, M., Rasala, B., Mayfield, S. (2010). Biofuel from algae: challenges and potential. Biofuels, 1 (5): 763-784.

20. Jackson, G.A. (1980). Biochemical and photosynthetic aspects of energy production. Marine biomass production through Seaweed Aquaculture – Chapter 2. Department of Biology, Indiana. ISBN 0-12-618980-3.

21. Johnson, R.B., Kim, J.K., Armbruster, L.C., Yarish, C., (2014). Nitrogen allocation of *Gracilaria tikvahiae* grown in urbanized estuaries of Long Island Sound and New York City, USA: a preliminary evaluation of ocean farmed *Gracilaria* for alternative fish feeds. Algae 29, 227–235.

22. Kraan, S. (2013). Mass-cultivation of carbohydrate rich macroalgae, a possible solution for sustainable biofuel production. Mitigation and adaptation strategies for global change, v. 18, Issue 1, pp 27-46.

23. Kraan, S., Salvesen, I., Wilkes, R. (2007). Seaweed: a natural means of reducing environmental impact. Irish Seaweed Centre, National University of Ireland.

24. Luning, K., Pang, S. (2003). Mass cultivation of seaweeds: current aspects and approaches.

25. Matos, J., Costa, S., Rodrigues, A., Pereira, R., Sousa Pinto, I. (2005). Experimental integrated aquaculture of fish and red seaweeds in Northern Portugal. Aquaculture, 252, 31-42.

26. Mazarrasa, I., Olsen, Y.S., Mayol, E., Marbà, N., Duarte, C.M. (2014). Global unbalance in seaweed production, research effort and biotechnology markets. Biotechnology Advances, v. 32, 1028-1036.

27. Mohammed, G. (2013). Seaweed farming. Calicut Research Centre of CMFRI. Kerala, India, pp 257-262.

28. Msuya, F.E. (2011). The impact of seaweed farming on the socioeconomic status of coastal communities of Zanzibar, Tanzania. World aquaculture, 42:3, pp 45-48.

29. Msuya, F.E. (2006). The Seaweed Cluster Initiative in Zanzibar, Tanzania. Proceedings of the 3rd Regional Conference on Innovation Systems and Innovative Clusters in Africa.. Pages 246-260.

30. Neori, A., Chopin, T., Troell, M., Buschmann, A.H., Kraemer, G.P., Halling, C., Shpiegel, M., Yarish, C. (2004). Integrated aquaculture: rationale, evolution and state of the emphasizing seaweed biofiltration in modern aquaculture. Aquaculture, v. 231, Issues 1-4, pp 361-391.

31. North, W.J. 1987. Oceanic farming of Macrocystis, the problems and non-problems. Seaweed Cultivation for Renewable Resources (ed. K.T. Bird and P.H. Benson, pp 39–68. Elsevier, Amsterdam.

32. Patel, S. (2012). Therapeutic importance of sulfated polysaccharides from seaweeds: updating the recent findings. 3 Biotech, v. 2, Issue 3, pp 171-185.

33. Pereira, R., Abreu, H. (2012). Seaweed aquaculture in on-shore IMTA systems: some limitation and future perspectives. Inter-regional network to promote sustainable development in the marine algal industry. November, 2012. San Sebastian, Spain.

34. Phillips (1990). Environmental aspects of seaweed culture. Technical resource papers regional workshop on the culture and utilization of seaweeds. FAO – Corporate document repository. Volume 2. Institute of Aquaculture, Scotland.

35. Pickering, T. (2006). Advances in seaweed aquaculture among the Pacific Island contries. Journalof Applied Phycology. 18: 227-234.

36. Prajapati, V. D.; Maheriya, P. M.; Jani, G. K.; Solanki, H. K. (2014). Carrageenan: A natural seaweed polysaccharide and its applications. Carbohydrate polymers, v. 105, p. 97- 112.

37. Samocha, T.M., Fricker, J., Ali, A.M., Shpigel, M., Neori, A. (2015). Growth and nutrient uptake of the macroalga *Gracilaria tikvahiae* cultured with the shrimp *Litopenaeus vannamei* in an Integrated Multi-Trophic Aquaculture (IMTA) system. Aquaculture, 446, 263-271.

38. Santelice, B., Doty, M.S. (1989). A review of *Gracilaria* farming. Aquaculture, 78: 95-133.

39. Suo, R. Liu, G., Shi, Z., Wang, F., Zhang, S. and Liu, T. 1986. Cultural test with submerged floating rope on giant kelp in South Huangcheng Island. Marine Fisheries Research, 7, 1–7.

SECTION II

GLOBAL IMPORTANCE AND CURRENT SITUATION

CHAPTER

2

ALGAE AS NUTRITIONAL AND FUNCTIONAL FOOD SOURCES: REVISITING OUR UNDERSTANDING

Mark L. Wells[1], Philippe Potin[2], James S. Craigie[3], John A. Raven[4,5], Sabeeha S. Merchant[6], Katherine E. Helliwell[7,8], Alison G. Smith[7], Mary Ellen Camire[9], and Susan H. Brawley[1]

[1]School of Marine Sciences, University of Maine, Orono, ME 04469, USA

[2]Integrative Biology of Marine Models, Station Biologique Roscoff, CNRS-Université Pierre et Marie Curie, Place Georges Teissier, 29680 Roscoff, France

[3]National Research Council of Canada, 1411 Oxford Street, Halifax, NS B3H 3Z1, Canada

[4]Division of Plant Sciences, University of Dundee (James Hutton Inst), Invergowrie, Dundee DD2 5DA, Scotland, UK

[5]Plant Functional Biology and Climate Change Cluster, University of Technology Sydney, Ultimo, NSW 2007, Australia

[6]Department of Chemistry & Biochemistry, University of California-Los Angeles, 607 Charles E. Young Dr., East, Los Angeles, CA 90095-1569, USA

[7]Department of Plant Sciences, University of Cambridge, Downing St., Cambridge CB2 3EA, UK

[8]Present address: Marine Biological Association of the UK, Citadel Hill, Plymouth PL1 2PB, UK

[9]School of Food and Agriculture, University of Maine, Orono, ME 04469, USA

ABSTRACT

Global demand for macroalgal and microalgal foods is growing, and algae are increasingly being consumed for functional benefits beyond the traditional considerations of nutrition and health. There is substantial evidence for the health benefits of algal-derived food products, but there remain considerable challenges in quantifying these benefits, as well as possible adverse effects. First, there is a limited understanding of nutritional composition across algal species, geographical regions, and seasons, all of which can substantially affect their dietary value. The second issue is quantifying which fractions of algal foods are bioavailable to humans, and which factors influence how food constituents are released, ranging from food preparation through genetic differentiation in the gut microbiome. Third is understanding how algal nutritional and functional constituents interact in human metabolism. Superimposed considerations are the effects of harvesting, storage, and food processing techniques that can dramatically influence the potential nutritive value of algal-derived foods. We highlight this rapidly advancing area of algal science with a particular focus on the key research required to assess better the health benefits of an alga or algal product. There are rich opportunities for phycologists in this emerging field, requiring exciting new experimental and collaborative approaches.

Keywords: Algal foods, Antioxidants, Arsenosugars, Experimental design, Microalgal supplements, Nutritional minerals, Omega-3-fatty acids, Polysaccharides, Sea vegetables, Vitamins

INTRODUCTION

Algae have been part of the human diet for thousands of years, based on archaeological evidence from 14,000 yBP in Chile (Dillehay et al. 2008) and early written accounts (e.g., in China, 300 A.D.; in Ireland, 600 A.D.; Newton 1951; Tseng 1981; Aaronson 1986; Turner 2003; Gantar and Svircev 2008; Craigie 2010). In North America, the Tsimshian First Nations' people named the month of May for the time of year when they harvested the important food crop of *Pyropia* (Fig. 1). More contemporaneously, the global harvest of seaweeds in 2013 was estimated at US $6.7 billion, and over 95 % was produced in mariculture, with

China and Indonesia being the top producers (FAO 2015). In addition to macroalgae, some microalgae are cultivated for foods and food additives (Switzer 1980; Jassby 1988; Fournier et al. 2005; Gantar and Svircev 2008; Chacón-Lee and González-Mariño 2010; FAO 2016). The FAO (2014) estimated that 38 % of the 23.8 million t of seaweeds in the 2012 global harvest was eaten by humans in forms recognizable to them as seaweeds (e.g., kelps, nori/laver), not counting additional consumption of hydrocolloids (e.g., agars, alginates, carrageenans) used as thickening agents in foods and beverages. Human consumption of algal foods varies by nation, with Japanese diets representing a recent (2010–2014) annual per capita consumption ranging from 9.6 (2014) to 11.0 (2010) g macroalgae day^{-1} (MHLW 2014). Overall, the trend towards increasing nutritional demand for algal products on a global basis stems from a greater focus on health and wider use of food additives.

Figure 1. a *Pyropia* spp. being dried in squares in the intertidal zone by First Nations' people at Pearce Island, British Columbia (2009). Harvesters would traditionally lay the seaweed out to dry on warm rocks while waiting for those fishing to return with the canoes (photo credit, Amy Deveau). b Checking the seaweed squares after transfer to cedar racks for final drying (photo credit, Victoria Wyllie-Echeverria),

In addition to their nutritional value, algae increasingly are being marketed as "functional foods" or "nutraceuticals"; these terms have no legal status in many nations but describe foods that contain bioactive compounds, or phytochemicals, that may benefit health beyond the role of basic nutrition (e.g., anti-inflammatories, disease prevention; Bagchi 2006; Hafting et al. 2012). The path from algal research to the launching of new food products or dietary supplements is strongly affected by industrial, regulatory, and nutritional considerations (e.g., see Borowitzka 2013a; Finley et al. 2014). The widespread interest in algal foods and/or their functional food potential is evident in numerous recent reviews (Warrand 2006; MacArtain et al. 2007; Kulshreshtha et al. 2008; Bocanegra et al. 2009; Mendes et al. 2009; Cottin et al. 2011; Harnedy and FitzGerald 2011; Holdt and Kraan 2011; Lordan et al. 2011; Pangestuti and Kim 2011; Stengel et al. 2011; Cornish et al. 2015; Hafting et al. 2015) and books (Rhatigan 2009; Mouritsen 2013; Tiwari and Troy 2015; Fleurence and Levine 2016). Many studies report the potential nutritional or bioactive content of different algae but many fewer studies quantify the bioavailability of nutrients and phytochemicals from algal foods. Our purpose is to review and assess what is known about different food components (i.e., proteins, polysaccharides, lipids, vitamins, minerals, and antioxidants, potential toxicants) in the context of improving knowledge about the efficacy of algal foods. There are rich opportunities for phycologists to collaborate with other scientists and clinicians in this emerging field from algal "prospecting" to defining nutritional value, bioaccessibility, and subsequent bioactivity, to the design and construction of mid-large cultivation systems for production of commercial-scale product.

DIGESTION AND BIOAVAILABILITY

In this article we use the term bioavailability, as defined by Carbonell-Capella et al. (2014) "as a combination of bioactivity and bioaccessibility," where bioaccessibility refers to the release from the food matrix, transformations during digestion, and transport across the digestive epithelium, while bioactivity encompasses uptake into tissues, metabolism, and physiological effects. Because of the difficulties, both practical and ethical in terms of measuring bioactivity, the fraction of a given compound or its metabolite that reaches the systemic circulation (Holst and Williamson 2008) can be considered bioaccessible, but not necessarily bioactive. Most published evaluations of bioactivity of algal foods are based on short-term in vitro tests using algal extracts that frequently are of ill-defined composition and purity, so a clear understanding of their food value is highly constrained. Particularly lacking is information on the behavior of algal food components in the gut. For example, can the purported active metabolites identified in in vitro studies be transferred from the gut lumen into the body? Likewise, are observed in vivo biological effects the consequence of biological uptake or instead indirect outcomes stemming from improved functionality or composition of the intestinal microbiome? It is important then to consider the process of digestion and transformation in the human system.

Digestion refers to the physical and biochemical degradation of foods and the nutrients therein in preparation for absorption into the body. Digestion begins in the mouth with chewing, which reduces particle size and mixes food with saliva (Lovegrove et al. 2015). The predominant salivary enzyme is alpha (α)-amylase, which is specific for α(1→4) glucose linkages, and human salivary amylase is more active than that from other primates (Boehlke et al. 2015). Hardy et al. (2015) hypothesized that cooking to increase digestibility and sensory quality of starch-rich foods helped drive human evolution by providing more glucose to growing brains. Studies of the effect of human saliva on algae and specifically algal starch are lacking, however. The relative importance of salivary versus pancreatic amylase in starch digestion also is not clear (Lovegrove et al. 2015). Pepsin and the pepsinogens begin protein digestion in the stomach, aided by hydrochloric acid that denatures proteins and releases nutrients from the food matrix. Lipases produced in the mouth and stomach begin the process of digesting triacylglycerols. The stomach also releases intrinsic factor that is essential for vitamin B_{12} absorption in the small intestine. Gastric peristalsis further reduces food particle size, preparing macronutrients for additional chemical breakdown and absorption in the small intestine. The pancreas discharges a mixture of trypsin, chymotrypsin, carboxypeptidases, α-amylase, lipase, and other enzymes that respectively digest proteins and peptides, starches, triacylglycerols, and other compounds in the small intestine (Gropper and Smith 2013). The mixture of proteases, amylase, and lipase are collectively known as pancreatin; porcine pancreatin is often used to model human digestion in in vitro systems. The small intestine itself releases a variety of enzymes acting on peptides, amino acids, monoacyglycerols, disaccharides, and α(1→4) and α(1→6) linkages in oligosaccharides, dextrins, and polysaccharides such as starch. Micronutrients such as vitamins and minerals also are absorbed in the small intestine once they are solubilized from the food matrix. Fucoxanthin, a key algal carotenoid, may be better absorbed if other lipids are present (Peng et al. 2011).

Humans lack the ability to digest β(1→4) linkages in glucan polysaccharides, as in cellulose and hemicelluloses such as xyloglucan, and this indigestible material is referred to as dietary fiber. The undigested materials continue on to the large intestine (colon) where microbial co-metabolism ferments substrates such as non-starch polysaccharides, resistant starch, and oligosaccharides to short-chain fatty acids, and proteins into a wider variety of compounds. These bacterial-dependent enzymatic processes are not considered "digestion," although the fermentation products can provide nutritional or functional benefits either by being absorbed and transported via the bloodstream or by shaping more healthful gut microbiomes and chemical conditions in the colon (MacFarlane and MacFarlane 2012). Indigestible, fermentable carbohydrates and sugar alcohols are referred to as FODMAP (fermentable, oligo-, di- mono-saccharides and polyols) (Gropper and Smith 2013). Algal proteins and carbohydrates that escape complete digestion in the small intestine may benefit humans by stimulating immune response indirectly via promotion of microbial responses (Cian et al. 2015). Dietary modulation of the colonic flora and the impact of bacterial fermentation products on human health are rapidly evolving areas of research (Duffy et al. 2015) and are likely to be especially important considerations in assessing the health benefits of algal-derived foods.

Not all human gut microbiomes have equal competencies, as algal polysaccharide fermentation differs among humans from different regions. The arsenal of polysaccharide-degrading enzymes exhibited in the common gut bacterium (*Bacteroides plebeius*) of Japanese people, but not Americans, appears to result from horizontal gene transfer (HGT) from *Zobellia galactanivorans* (Bacteroidetes), a marine bacterium inhabiting the surfaces of algae such as nori (Hehemann et al. 2010). HGT also may explain the presence of a gene cluster in Japanese gut *Bacteroides* that enables fermentation of alginates in brown algal cell walls (Thomas et al. 2012). Similarly, a small cohort of Spaniards possesses gut microbiomes with apparently HGT-provided porphyranases and agarases (Hehemann et al. 2012). Such striking differences emphasize the complex interactions among food customs, dietary history, and gut microbiomes that complicate study of the nutritional and functional benefits of algal foods (Paulsen and Barsett 2005; Costello et al. 2012; Gordon 2012; Nicholson et al. 2012).

The importance of assessing the biological availability of nutritional and functional food components cannot be underestimated. Bioavailability has critical relevance to both the proportional digestion and uptake of nutrients and functional food components, but also the degree of fermentation and nature of the host-microbial co-metabolism in the colon. While there exists a vast literature on the food content of microalgal and macroalgal foods and supplements, extrapolating these findings to assess their quantitative contribution to human health is more tenuous. The analytically determined concentration of constituents in food can differ, sometimes substantially, from that actually crossing from the digestive tract into the blood (i.e., the bioaccessible fraction). Moreover, current analytical approaches give even less insight to the complexity of interacting effects that regulate the bacterial flora of the colon, and hence the nature of fermentation products. Confounding issues stem from the food itself (e.g., the presence and nature of intact cell walls, soluble fiber characteristics, and the presence of other substances that may inhibit or facilitate the uptake of metabolites), the harvest season (e.g., altered metabolite and biomass composition, environmental variability of essential precursors, and anthropogenic factors), and the food preparation methods (Sensoy 2014). Analytical methods such as simulated gastrointestinal digestion (Moreda-Pineiro et al. 2011; Maehre et al. 2014), xenobiotic animal models, and molecular biological and genetic techniques can provide a sound basis for improved assessment of bioavailability; however, their use is not yet widespread in the study of foods of algal origin. As a consequence, and despite highly accurate and precise analytical determinations of food content, current knowledge of the nutritional or functional food value of algal products remains largely qualitative. The development of appropriate model systems and use of rigorous experimental design thus is essential in order to verify the bioavailability of nutritional and functional components of algae used in all foods.

PROTEINS

Protein content differs widely across groups of algae (Online Resource 1). The filamentous cyanobacterium *Arthrospira platensis* ("spirulina") and various commercial species of the unicellular green alga *Chlorella* (Fig. 2) contain up to 70 % dry wt protein; these microalgae also have an amino acid profile that compares well with egg, notably containing all of the essential amino acids (EAA) that humans cannot synthesize and must obtain from foods (Online Resource 2). Historically, "spirulina" was wild-harvested as a protein-rich whole food in many cultures outside Europe and North America (Gantar and Svircev 2008). Today, domesticated "spirulina" and *Chlorella* from several large producers have "GRAS" designations [Generally Recognized As Safe (FDA 2016)]. Large-scale production of both "spirulina" and *Chlorella* occurs throughout the world, and these well-domesticated crops are added to many foods to increase their protein and other nutritional contents (e.g., salad dressings, beverages, baked goods), and/or sold as protein supplements (e.g., Lubitz 1963; Ciferri 1983; Jassby 1988; Belay 1997; Gantar and Svircev 2008; Szabo et al. 2013; Safi et al. 2014).

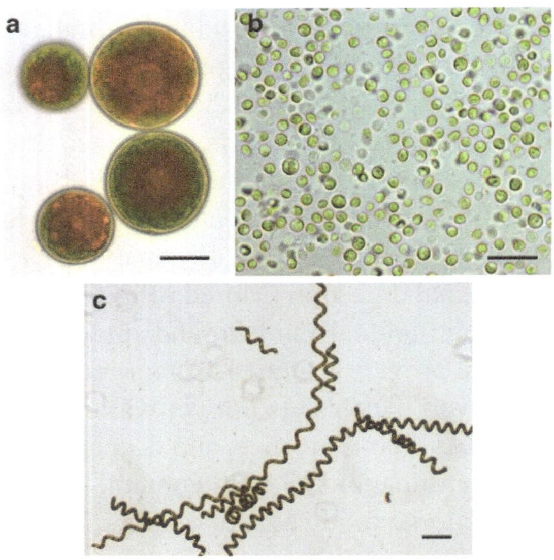

Figure 2. a *Haematococcus pluvialis* cells showing droplets of red astaxanthin within the cells; b *Chlorella vulgaris*; c*Arthrospira maxima* SAG 21-99 (also known as spirulina). *Scale bar* = 15 μm. (photo credits, Maria Zori).

Among the marine macroalgae, red and green algae [e.g., *Porphyra* spp. ("laver"), *Pyropia* spp. ("nori"), *Palmaria palmata* ("dulse"), *Ulva* spp. ("sea lettuce")] often contain high levels of protein (as % dry wt) in contrast to lower levels in most brown algae (Online Resource 1; Dawczynski et al. 2007; Holdt and Kraan 2011; Pereira 2011; Taboada et al. 2013; Angell et al. 2016). During periods of nutrient limitation such as during the summer stratification of coastal waters, however, macroalgal protein content decreases, and the relative proportions of amino acids change (Online Resource 2; Galland-Irmouli et al. 1999; Johnson et al. 2014; Schiener et al. 2015). Historic harvesting times and current harvesting regimes usually occur at times when protein contents are favorable (e.g., Butler 1936; Black 1950; Turner 2003), but there is remarkably poor documentation of seasonal changes in protein content and amino acid profiles. Strong conclusions about nutritional content also depend upon good biological sampling (=simultaneously collected replicates) combined with appropriate laboratory analyses (=analytical replicates of each biological sample). Galland-Irmouli et al. (1999) analyzed one dulse blade/month (except August) from the Brittany coast with three technical replicates and found ~15 % of dry mass as protein in a June blade while a November blade contained ~23 % protein. A seasonal study (October 2010–October 2011, 3–8 months sampled/species) of protein content of four kelps (*Laminaria digitata*, *Laminaria hyperborea*, *Saccharina latissima*, *Alaria esculenta*) based on three technical replicates/species recently demonstrated an inverse relationship between protein content—higher in winter—and polysaccharide content—higher in summer, as well as clearly showing the higher protein content of *Alaria esculenta*(Online Resource 1) compared to the other kelps (Schiener et al. 2015). There is a pressing need for better replication of protein and amino acid analyses, as for all nutritional components in macroalgal studies, as well as better definition of the natural intertidal or commercial sites from which analyzed samples were obtained (N.B. Shuuluka et al. 2013 as example). Some new sea vegetable products will benefit from complementary food constituents (Woolf et al. 2011; see vProtein software).

Protein concentration in algae is often estimated using a total nitrogen-to-protein (NTP) conversion factor (6.25) based on the assumption that most N in the sample occurs as protein. This conversion factor, however, often over-estimates the protein content because of the presence of variable amounts of non-protein-N in the sample (Lourenço et al. 2002; Safi et al. 2013; Angell et al. 2016). For example, the conversion factor calculated for crude biomass for *Chlorella vulgaris* (walled) was 6.35, whereas it was 5.96 based on direct protein extracts (Safi et al. 2013). Similar studies of 19 tropical marine algae yielded even lower average factors of 4.59 (red algae), 5.13 (green algae), and 5.38 (brown algae) (Lourenço et al. 2002), perhaps related to seasonally lower N inputs to tropical surface waters. Zhou et al. (2012 [see her Tables 3–5]) reported similar findings. These conversion factors certainly will vary with season based upon varying amino acid composition, emphasizing the need for protein and amino acid studies to determine the seasonal optima for harvest among algal foods. Angell et al. (2016) argued for a new universal conversion factor, after finding a median nitrogen-to-protein value of 5 in a literature-based meta-analysis of 103 macroalgae; however, the range of values in their analysis was high (see their Fig. 4). The algae have polyphyletic origins and this, too, is reflected in the absence of a universal N to protein conversion factor.

In most analyses of amino acid composition in marine algae, glutamic acid, and aspartic acid represent the highest proportions of amino acids (e.g., Fleurence 1999b; Lourenço et al. 2002; Online Resources 1, 2; Holdt and Kraan 2011). These amino acids occur as protein constituents and as free amino acids or their salts. For humans, glutamate is the major component of the savory, the fifth basic taste called umami from its characterization in kelp (Ninomiya 2002; Mouritsen 2013). Glutamic acid content may decrease after several successive harvests of *Pyropia yezoensis* (nori; Niwa et al. 2008). Other amino acids (alanine and glycine) also contribute to distinctive flavors of some marine algae (e.g., see Holdt and Kraan 2011).

The non-proteinaceous amino acid taurine is especially abundant in marine red algae (e.g., ~1–1.3 g taurine per 100 g DW of nori, Niwa et al. 2008). Although taurine is not an EAA for adults, it is a component of bile acids that complex and lower cholesterol in the bloodstream (Medeiros and Wildman 2015).

In general, protein in most algae is digested less completely than reference proteins such as casein (a milk protein) in in vitro model systems containing digestive enzymes such as pepsin, pronase, and pancreatin, with evidence that this is due especially to inhibitory soluble fibers (e.g., Fujiwara-Arasaki et al. 1984; Fleurence 1999a; Urbano and Goni 2002; Marrion et al. 2003, 2005; Wong and Cheung 2003; De Marco et al. 2014). Inclusion of pre-analytical steps such as freezing, milling, digestion of crude sample with polysaccharide-digesting enzymes, and/or osmotic rupture of cells to free intracellular compounds is an active area of research (e.g., Harnedy and FitzGerald 2013; Safi et al. 2014; Ursu et al. 2014; and references therein). Importantly, a recent study (Maehre et al. 2016) with excellent biological and technical replication shows the beneficial effect of cooking on amino acid availability from dried dulse (Online Resource 3); however, cooking did not significantly increase the total amino acids measured from *Alaria* (Online Resource 3). Furthermore, Maehre et al. (2016) demonstrated that the apparent amino acid bioaccessibility from both raw and 30 min-boiled dulse was higher than from an equivalent dry weight of wheat, rice, or corn flour in a simulated in vitro gastrointestinal digestion model with analysis at each sequential digestive step (amylase/saliva buffer; pepsin/gastric buffer; pancreatin/duodenal buffer) (Fig. 3). Future research on microalgal and macroalgal protein bioavailability might incorporate measures such as the protein digestibility-corrected amino acid score (PDCAAS), which involves urinary and fecal determinations of N absorption in rats, as well as the FAO recommended replacement of PDCAAS by the digestible indispensable amino acid score (DIAAS) (Medeiros and Wildman 2015; Rutherfurd et al. 2015).

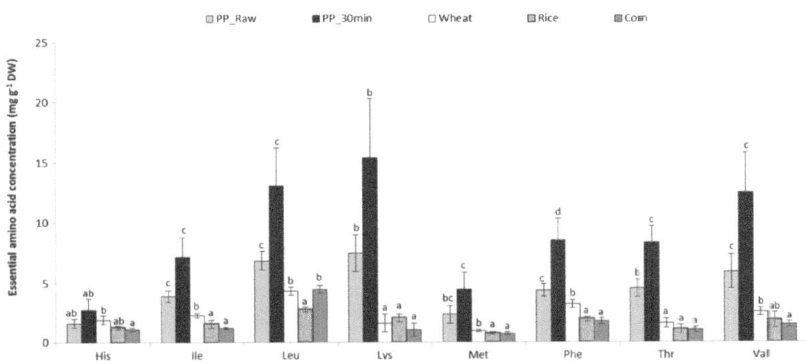

Figure. 3. A comparison of essential fatty acids liberated from 1 g dry weight of *Palmaria palmata*(raw and boiled for 30 min) wheat, rice, and corn flours in simulated gastrointestinal digestion. The mean values ± 1 SD ($n = 5$) are shown

in mg g^{-1}. Significant differences between species ($p>0.05$) are indicated by *different letters*. (Used with permission from Maehre et al. (2016)).

LIPIDS

Lipids are essential for all living organisms as components of membranes, energy storage compounds, and as cell signaling molecules (Eyster 2007). Although humans and other mammals synthesize lipids, some essential lipids must be obtained from dietary oils or fats. Phospho- and glycolipids, important for membrane function, contain a polar head group with two fatty acid chains, while the triacylglyceroles (TAGs), important energy stores in the cell, are non-polar (neutral) lipids containing three fatty acid chains (Fig. 4). Lipid membranes contain sterols such as fucosterol and β-sitosterol (Fahy et al. 2005) that also have reported health benefits (Arul et al. 2012). Embedded in algal lipid fractions are the nutritionally valuable carotenoid pigments that will be discussed in the "phytochemicals" section (below). TAGs have attracted great attention in recent years as a source for biodiesel, with some microalgae accumulating up to 40–60 % of their dry weight as TAGs (Georgianna and Mayfield 2012). However, marine macrophytes typically do not exceed 2–4.5 % dry wt as lipids, mainly as phospholipids and glycolipids (Holdt and Kraan 2011). Of these, the long-chain polyunsaturated fatty acids (PUFAs) and carotenoids are most noteworthy as functional foods (Holdt and Kraan 2011).

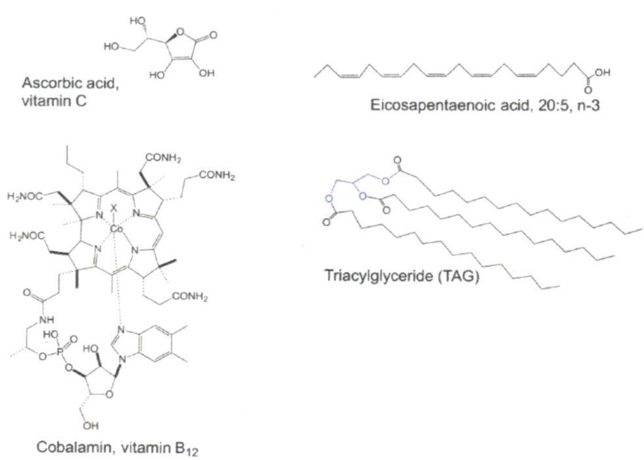

Figure. 4. Structures of some key vitamins and lipids mentioned in review.

Long-chain PUFAs

There are two general families of PUFAs: the linoleic acids (n-6 or omega 6 fatty acids) and the α-linolenic acids (n-3 or omega 3 fatty acids). Long-chain PUFAs comprise a substantial portion of marine algal lipids, with planktonic algae being the source of most omega fatty acids in fish. The most important of these PUFAs are the essential fatty acids (EFAs) eicosapentaenoic acid (EPA; 20:5 n-3) and docosahexaenoic acid (DHA; 22:6 n-3) along with their precursors α-linolenic acid (ALA; 18:3 n-3) and docosapentaenoic acid (22:5 n-3) (Cottin et al. 2011). The first product of ALA in the synthesis pathway to C20–22 PUFAs is stearidonic acid (SA, 18:4n-3), and this fatty acid can represent a significant portion of PUFAs in some edible macroalgae (sea vegetables) (Guil-Guerrero 2007). EPA is the predominant PUFA in many sea vegetables (Fig. 4), along with arachidonic acid (20:4 n-6), particularly in red algae (Norziah and Ching 2000; Wen et al. 2000; Ortiz et al. 2009) where EPA comprises up to 50 % of the total fatty acid content (e.g., *Palmaria palmata*, van Ginneken et al. 2011). Humans and other animals cannot convert ALA to EPA and DHA at required levels, so dietary sources of these EFAs are critically important for the health of humans (Cottin et al. 2011) and many animals (Li et al. 2009).

Numerous epidemiological and controlled interventional trials (N.B., the excellent reviews of Conquer and Holub 1996; Holub 2009; Cottin et al. 2011) support the health benefits to humans of DHA and EPA long-chain omega-3 fatty acids from fish oils and algal sources (mainly extracts). In contrast to most other algal food constituents, the bioaccessibility of DHA and EPA in algal-derived oils and extracts is well quantified for humans, ranging from ~50 to 100 % depending on the matrix (Haug et al. 2011; Schuchardt et al. 2011). While clinical research to date strongly supports a nutritional need for oils that are enriched in DHA and EPA, there is more understanding about the bioactivity of DHA than of EPA (Conquer and Holub 1996; Holub 2009; Cottin et al. 2011). There is a considerable literature (Cottin et al. 2011) on the cardioprotective effects of DHA-containing TAG from *Crypthecodinium cohnii* (a dinoflagellate, Mendes et al. 2009) and *Schizochytrium* sp. (a thraustochytrid stramenopile, Li et al. 2009; Barclay et al. 2010), and as a consequence, infant formula, infant foods, and certain other food categories (dairy, bakery, eggs, and non-alcoholic beverages) and marketed nutritional products now are supplemented with algal-derived DHA. There is evidence that enhanced DHA intake may improve infant cognitive performance and enhance visual acuity (Jensen et al. 2005, 2010; Imhoff-Kunsch et al. 2011), although more recent data raises question about this linkage (Delgado-Noguera et al. 2015). There also is some understanding about the bioaccessibility of DHA in different algal products. Algal oil capsules based on a patented commercial source (Martek) and cooked salmon are reported to represent nutritionally equivalent sources of DHA (see in Cottin et al. 2011). A similar human trial showed that DHA from two different strains of *Schizochytrium* sp. (DHASCO-T and DHASCO-S) supplied in capsules generated equivalent dose-dependent DHA levels in plasma phospholipids and erythrocytes (Arterburn et al. 2008). Fortified snack bars also delivered equivalent amounts of DHA on a DHA dose basis (Arterburn et al. 2007). A systematic review of plant omega-3 fatty sources by Lane et al. (2014) concluded that further research on algal sources was warranted based on promising preliminary work.

Nonetheless, the relative health benefits of commercial algal supplements that tend to be DHA-rich versus natural fish oils that contain both DHA and EPA are uncertain. Cottin et al. (2011) found that "Recent evidence from randomized controlled trials has produced equivocal results. Heterogeneity of the studies in terms of dosage, duration, population target, sample size, as well as the relative amounts of EPA and DHA used in supplements could account for the variability of the results." Even so, important trends stand out. While both EPA and DHA reduce TAG levels in humans (Wang et al. 2006; Bernstein et al. 2012), DHA appears responsible for the blood pressure and heart rate-lowering effect of fish oils (Valera et al. 2014). DHA also seems to be beneficial for endothelial and platelet function, although a direct role for EPA in regulating TAGs has not been ruled out. Algal DHA extracts can produce other cardiovascular protective effects in humans by altering plasma lipoproteins at reasonably small doses (2 g algal DHA day^{-1} over 4.5 months: Neff et al. 2011). The health benefits of algal DHA supplements for subgroups such as vegetarians, who otherwise may have low essential fatty acid intakes, remains a high research priority (Geppert et al. 2005; Cottin et al. 2011).

Fish oils also have demonstrated anti-inflammatory and insulin-sensitizing properties in vitro and in animal studies (Nauroth et al. 2010;

Cottin et al. 2011); however, human trials often yield conflicting findings. Neither EPA nor DHA alone showed any effects on inflammation in double-blind trials with cystic fibrosis patients (Van Blervliet et al. 2008) or insulin sensitivity in human subjects, despite indications for potency in vitro (critically reviewed in Cottin et al. 2011). Without better quantification of the biological uptake of EPA or DHA, the reason for this discrepancy remains unknown.

Microalgae are the primary sources of DHA and EPA for zooplankton, fish, and other multicellular organisms, and these essential fatty acids (EFAs) become increasingly concentrated up the food web (e.g., Legezynska et al. 2014). Therefore, fish oils are rich in both DHA and EPA because they represent the trophic integration of DHA-rich flagellates and EPA-rich diatoms in the food web (Viso and Marty 1993). There is emerging evidence that ocean acidification, the result of changing coastal processes and increased atmospheric CO_2, can negatively change the supply of these essential fatty acids to higher trophic levels (Rossoll et al. 2012). This and other factors affecting EFA production in algal assemblages will be an important area of future research (Chrismadha and Borowitzka 1994; Pasquet et al. 2014).

Concern over the sustainable supply of fish oils and the commercial dominance of algal-based DHA-only supplements has led to a large industry effort towards developing alternatives to fish oil-derived EPA (Zeller 2005). One example is Lovaza™, a prescription pharmaceutical containing purified DHA and EPA synthesized from fish oils that reportedly have anti-hyperlipidemic properties (Weintraub 2014), although there are some negative indicators for this product (Spindler et al. 2014). A new promising biotechnological source of EPA has been proposed by Řezanka et al. (2010) from the Eustigmatophyceae *Trachydiscus minutus*; however, its commercial production is not developed yet. Other biotechnological production of EPA is provided by the diatoms *Phaeodactylum tricornutum* grown in tubular photobioreactors (Chrismadha and Borowitzka 1994) or *Odontella aurita* co-cultivated in raceway ponds with the red macroalga *Chondrus crispus* in France by the Innovalg company (Braud 2006). Commercial production of DHA and EPA is one of the main targets of producers and has benefited from the development of microalgal cultivation via fermentation technology (Branger et al. 2003; Barclay et al. 2013).

Several recent studies analyzed the constituent fatty acids of large numbers of red, brown, and green macroalgae from polar (Graeve et al. 2002, 20 species), temperate (Schmid et al. 2014, 16 species; McCauley et al. 2015, 10 species), and tropical (Kumari et al. 2010, 27 species; Kumar et al. 2011, 22 species) habitats, and, despite some species variability, red (Rhodophyta) and brown (Phaeophyceae) macroalgae had a high proportion of total FAs in EPA and arachidonic acid across latitudes, whereas the green (Chlorophyta) algae had low EPA (as % of total FA) but some DHA, and, were enriched in C18 LC PUFA. Phytoplankton contain more PUFA, as expected, when grown at low temperature (e.g., DHA in *Crypthecodinium*, Jiang and Chen 2000), and higher temperatures good for maximal biomass production can be lowered for as little as 12 h to induce maximal EPA content in the diatom *Phaeodactylum* (Jiang and Gao 2004).

Whether omega-3 FA content can be manipulated by the timing of wild harvest or grow-out of sea vegetable crops in winter to increase EFA of whole foods needs much more work. Marine macrophytes generally contain low total lipid contents, so their comparative value as a food energy source likely is small (Holdt and Kraan 2011; Maehre et al. 2014), and at realistic daily consumption levels (e.g., 8 g dry wt., Blouin et al. 2006), even red algae such as *Porphyra umbilicalis* (laver) and *Palmaria palmata* (dulse) (Fig. 5) that have a high proportion of their total fatty acids as EPA (Mishra et al. 1993; Graeve et al. 2002; Blouin et al. 2006; Schmid et al. 2014) will not meet dietary recommendations for daily EFA alone (Blouin et al. 2006), although higher levels of macroalgae can support EFA needs for animal aquaculture feeds (Mulvaney et al. 2015; Wilke et al. 2015). Thus, relevant growth conditions should be manipulated for promising macroalgae in the laboratory to see if further increases are possible; unialgal cultures will be important because recent PUFA studies of green algae in a "green tide" in the Gulf of Finland showed that the high EPA content was largely due to epiphytic diatoms (Gubelit et al. 2015).

Figure 5. Sea vegetables used in European cuisine include marinated kelp (*Alaria esculenta*) in a cannelloni bean salad (a), laver/nori (*Porphyra umbilicalis/Pyropia yezoensis*) in chocolate molasses meringues (b), the traditional Welsh laver-bread cakes, with dulse (*Palmaria palmata*) crisps (c), and dulse-cheese scones (d). These additions add texture, protein, vitamins and minerals, and flavor. (Used with permission of Prannie Rhatigan from *The Irish Seaweed Kitchen*).

The bulk of research on omega-3 long chain PUFAs in microalgae and sea vegetables has been empirical, testing differences among species under different growth conditions. It will be important to understand their biosynthetic pathways and metabolic controls, and the increasing availability of microalgal genomes should provide excellent opportunities in this goal. One recent example is five genes functionally characterized in the haptophyte *Emiliania huxleyi* that are predicted to underpin omega-3 LC-PUFAs synthesis (Sayanova et al. 2011). Additional putative genes for functionally redundant pathways for the synthesis of omega-3 EPA and DHA were also annotated in the *E. huxleyi* pan genome sequence (Read et al. 2013).

Further investigation to produce biomass or extracts of sea vegetables containing EPA at a range of doses compatible with functional foods would enable research to examine the protective effects of consuming this source of long chain n-3 PUFA. Such trials could provide clear evidence for the clinical therapeutic potential of consuming EPA rich macroalgae in combination with supplementation of microalgal DHA.

Sterols

Algae vary in their total sterol content and in the variety of sterols present (Holdt and Kraan 2011). Older analytical techniques may have misidentified algal sterols as cholesterol since their structures are similar (Pereira et al. 2016). Fucosterol occurs in many algae, especially red and brown macroalgae (Pereira et al. 2016), and this compound may have value in treating complications of diabetes and hypertension, as well as other major health concerns (Abdul et al. 2016). However, like other studies of algal foods, the linkages are implied but little is known about the actual in vivo effects of fucosterol when algae are consumed by humans. Nonetheless, as for long-chain PUFAs, understanding the seasonal, environmental, and species-specific factors that alter the abundance and composition of algal sterols, such as in the recent studies in Antarctic seaweeds (Pereira et al. 2016), will be fundamental to understanding their potential effects in human diets.

POLYSACCHARIDES

Polysaccharides are used for energy storage and as structural elements in marine algae and terrestrial plants. Humans possess enzymes that degrade algal starches to mono-and di-saccharides for transport across the gut lumen, but generally cannot digest the more complex polysaccharides, as was first recognized more than a century ago (Saiki 1906). These resistant polysaccharides, known as dietary fiber, may be fermented in the large intestine to varying degrees depending on the enzymatic competence of the microbiome (Terada et al. 1995; Michel and MacFarlane 1996; Hehemann et al. 2010; Cian et al. 2015). Algal cell walls differ from those of terrestrial plants as they contain uncommon polyuronides and polysaccharides that may be methylated, acetylated, pyruvylated, or sulfated (Paulsen and Barsett 2005; Pal et al. 2014; Rioux and Turgeon 2015; Stiger-Pouvreau et al. 2016). It is fair to say that algal polysaccharides are the most widely, and often unknowingly, consumed food of algal origin. Small amounts are incorporated into beverages, meat and dairy products, and fillers (Cofrades et al. 2008; Gupta and

Abu-Ghannam 2011a, b; Griffin 2015) at levels generally deemed to be beneficial and safe by regulatory agencies (extensively reviewed in Mabeau and Fleurence 1993; MacArtain et al. 2007; Watson 2008; Holdt and Kraan 2011; Barlow et al. 2015; Fleurence and Levine 2016).

Edible macroalgae contain unusually high amounts of dietary fiber, ranging from 23.5 % (*Codium reediae*) to 64.0 % of dry weight in *Gracilaria* spp., values that frequently exceed those for wheat bran (Ruperez and Saura-Calixto 2001; McDermid et al. 2005; Benjama and Masniyom 2012). The nomenclature of food-derived fiber is confusing because there is no consensus on its definition among scientists and regulatory agencies. *Dietary fiber*, considered a nutrient in the USA under the Nutrition and Education Act of 1990 (Thomas.loc.gov/ H.R. 3562.ENR), comprises "nondigestible carbohydrates and lignin that are intrinsic in intact plants." Some fraction of this human-inert matter is considered by some as *Functional fiber*; that fraction of isolated, non-digested carbohydrates having apparent beneficial physiological effects beyond nutrition in humans (Institute-of-Medicine 2005; Medeiros and Wildman 2015). In this case, *Total fiber* is the sum of *dietary* and *functional* fiber (Institute-of-Medicine 2005; Medeiros and Wildman 2015). In contrast, the European Food Safety Authority, following the CODEX Alimentarius Commission definition of *dietary* fiber (Jones 2014), acknowledges that benefits beyond nutrition can occur but does not formally distinguish *functional* from *dietary*fiber because no analytical methods exist for this differentiation (EFSA 2010). Regardless of these semantics, non- or partially fermented fiber that generates physiological benefits, through either physical or chemical pathways, meets the definition of *dietary* fiber (Jones 2014).

"Soluble fiber" comprises 52–56 % of total fiber in commonly used green and red macroalgae and 67–85 % in brown macroalgae (Lahaye 1991). Much of it can be fermented to short-chain fatty acids (SCFAs) such as acetate, propionate, and butyrate (see Table 1 in Michel and MacFarlane 1996; Cantarel et al. 2012) which both nourish the epithelia of the large intestine and offer other benefits to the host (Terada et al. 1995; Michel and MacFarlane 1996). For example, acetate and propionate are transported in the blood to many organs where they are oxidized for energy or utilized in signaling to help regulate aspects of energy homeostasis and immune function (reviewed by Nicholson et al. 2012).

The fermentation process and SCFA products also nourish and modify the microbial consortia in the large intestine, thereby exerting prebiotic effects and influencing digestive outcomes (e.g., Fernando et al. 2008; O'Sullivan et al. 2010; Cian et al. 2015). Investigating the coupling of algal (and other) polysaccharides to the health of intestinal microbiomes and their animal and human hosts is an active and needed area of research (Bäckhed et al. 2005; Hehemann et al. 2010; Cantarel et al. 2012). These beneficial responses may include reduced risk of diabetes, hypertension, and cardiac heart disease (Institute-of-Medicine 2005). However, the complexity of interactions among functional and dietary fiber and the intestinal microbiome challenges efforts to demonstrate the functional food and biomedical benefits of algal polysaccharides (de Jesus Raposo et al. 2015; Dhargalkar 2015).

The evidence for bioactivity of algal polysaccharides derives largely from in vitro experiments using isolated oligomers/polymers, with fewer data on testing any whole alga in animal or human trials. Compositional analysis of *Chlorella* and similar microalgae began more than 60 years ago, and an impressive number of biological processes are now reported to be influenced by ingestion of whole algae or polysaccharide extracts as food or supplements (Pulz and Gross 2004; Plaza et al. 2009; Chacón-Lee and González-Mariño 2010; Lordan et al. 2011; Vo et al. 2011). Microalgal genera (Fig. 2) commonly considered as beneficial dietary supplements include *Chlorella*, *Arthrospira* (spirulina), *Dunaliella*, *Haematococcus*, *Scenedesmus*, *Aphanizomenon*, *Odontella*, and *Porphyridium*, with species of *Chlorella* being recognized as particularly rich in polysaccharides (Chacón-Lee and González-Mariño 2010). This putative bioactivity includes anticancer properties, cytokine modulation, anti-inflammatory effects, macrophage activation, and inhibition of protein tyrosine phosphatase (Hasegawa et al. 1997; Cheng et al. 2004; Kralovec et al. 2005; Sheng et al. 2007; Hsu et al. 2010). Algal polysaccharide extracts can possess strong immunomodulating effects both in vitro and in vivo (Watanabe and Seto 1989; Pasco and Pugh 2010; Suárez et al. 2010). Kwak et al. (2012) observed an immunostimulatory effect in 30 Korean volunteers fed 5 g day^{-1} *Chlorella* vs. placebo in a double-blinded 8-week trial. Acidic polysaccharide extracts from *Chlorella pyrenoidosa* have been patented (Chlon A and Respondin™) as potentially useful anti-tumor and immunostimulating supplements (Umezawa and Komiyama

1985; Komiyama et al. 1986; Kralovec 2005; Kralovec et al. 2005). Even so, the molecular structures responsible for such observed physiological functions are poorly understood because of fragmentary and sometimes conflicting information on the chemistry of these large, highly complex cell wall polymers (Řezanaka and Sigler 2007). Research also has focused on strikingly few algal species, leaving a broad window of opportunity for more expansive assessment of potential sources of bioactive compounds (Admassu et al. 2015).

The study of extracted polymer sub-fractions of structural polysaccharides provides a useful exploratory tactic for assessing the potential functional benefits of consuming macroalgal foods, and it establishes a quantitative means to determine the seasonal or environmental effects on food quality (Stengel et al. 2011; Mak et al. 2013). The predominant algal polysaccharides are the alginates in brown macroalgae, and the sulfate-esterified polysaccharides of macro- and microalgae that are widespread in red, brown, and green seaweeds (Aquino et al. 2005; Popper et al. 2011). The cellular quantities and compositions of these polysaccharides vary among species and with seasonal and environmental changes (Bourgougnon and Stiger-Pouvreau 2011; Mak et al. 2013).

Alginate

Alginate is the major polysaccharide of brown algae, comprising 14–40 % of its dry mass (cf. Ramberg et al. 2010), and was first isolated in 1881 as algin from kelp (*Laminaria* sp.) by E. C. C. Stanford. The direct consumption of brown algae as human food is of long standing (Tseng 1981; Druehl 1988; Dharmananda 2002; McHugh 2003). The purported beneficial effects of alginate include its ability to absorb toxins, decrease cholesterol uptake, alter the colonic bacterial profiles, and generate SCFAs (Brownlee et al. 2005). The metal chelating abilities of alginates makes them valuable scavengers of toxic elements in the human gut, but this scavenging also may lead to nutritional deficiencies of essential di- or polyvalent trace metals (Hollriegl et al. 2004; Brownlee et al. 2005). Most studies have investigated the effects of polysaccharide extracts rather than consumption of intact seaweeds. Although the extent of alginate dissociation from algal cell walls after ingestion is not well studied, there is little or no digestion of sodium alginate from *Ascophyllum nodosum* in humans (Percival and McDowell 1967; Painter 1983; Aarstad et al. 2012). Dietary alginates also provide a sense of satiety and so have been explored as a weight control measure, although there remains uncertainty about its efficacy in this role (Yavorska 2012).

Sulfated Heteroglycans—Ulvans

The abundant, heavily sulfated ulvans are extracted from members of the Ulvales. They are the best studied of the green seaweed polysaccharides, in part because the high production of *Ulva*spp. in eutrophic coastal waters has sparked research for new uses of these algae (Alves et al. 2013). Ulvans owe their bioactive properties to their unusual hydrophilic polyanionic features and structural analogies with animal glycosaminoglycan regulators (dermatan sulfate, heparin/heparin sulfates) and L-rhamnose specific lectins in humans. The reported bioactivities of ulvan extracts in vitro include antibacterial, anticoagulant, antioxidant, antiviral, anti-tumor, anti-hyperlipidemic, and immunoregulatory effects (Kaeffer et al. 1999; Yu et al. 2003; Mao et al. 2006, 2008; Leiro et al. 2007; Zhang et al. 2008, 2010; Lee et al. 2010; Holdt and Kraan 2011; Matloub et al. 2013).

Although the ingestion of green macroalgae by humans is rather widespread, the potential health benefits of food supplements of native ulvans or their chemically modified derivatives, let alone the direct consumption of the whole algae, are not well understood (Taboada et al. 2010; Wijesekara et al. 2011). Fermentation of *Ulva* and ulvan by human colonic bacteria was slight (16.6 and 8.9 % of organic matter, respectively) (Durand et al. 1997), indicating that they would be poor sources of SCFA production in the colon (Bobin-Dubigeon et al. 1997). However, these results cannot be generalized because only two individuals provided the bacterial inocula, and their prior dietary history relating to algal foods was unknown. A cautionary note here though is that *Ulva* can be rich in free sulfate which is readily converted to sulfide during fermentation, so consumption of more than 20 g day-*1* of the dry, unprocessed seaweed may have adverse (and odiferous) health effects (Durand et al. 1997).

Sulfated Galactans—Carrageenans

Red algal polysaccharides include the nutritionally important floridean starch, and their sulfated galactans are known to initiate or modulate a large number of biological activities of significance to human health (Prajapati et al. 2014). The most studied are the sulfated agarocolloids and the carrageenans derived from macroalgae in the orders Gelidiales, Gigartinales, and Gracilariales. Anti-viral activities include those against herpes simplex, herpes zoster, dengue-2, vaccinia, rabies, and vesicular stomatitis virus with patents and some commercial projects resulting (Richards et al. 1978; Baba et al. 1988; Vedros 1993; Bourgougnon 2003; Eccles et al. 2010; Talarico et al. 2011; Levendosky et al. 2015; Luo et al. 2015). Whether consumption of the relevant red algae or their extracts in foods is protective against viruses does not appear to be known and deserves study. Carrageenan extracts that are depolymerized to various degrees have potential as tumor inhibitors and as immunostimulators in cancer therapy. Oligomers from acid hydrolyzed κ-carrageenan injected into mice increased macrophage phagocytosis and stimulated several immune-related markers while significantly inhibiting the growth of sarcoma S180 cells (Yuan et al. 2006). Phosphorylation or further sulfation of these oligomers increased the activity of natural killer cells, the cytotoxic lymphocytes critical to immune system function (Yuan et al. 2011). Similarly, transplanted human sarcoma S180 tumors were inhibited significantly in mice fed fractionated λ-carrageenan extracts of *Chondrus ocellatus* (200 mg kg^{-1} daily) (Zhou et al. 2004). Although seaweeds containing carrageenans act as prebiotics when supplied as supplements in both poultry and rat diets (Kulshreshtha et al. 2014; Liu et al. 2015), the potential for sulfated galactans from algae to benefit human health remains to be established.

Carrageenans have the potential to be harmful (Tobacman 2001). Carrageenan extracts generate proinflammatory agents in mice (Hansra et al. 2000), and the resulting public health concerns have led to several actions regarding carrageenans in food products (Watson 2008). Carrageenan is prohibited in the EU for use in infant formulas, and, although it is permitted in the USA, it must be of high molecular mass (i.e., >100 kDa with <5 % of 50 kDa fragments). High doses of low molecular mass carrageenan cause ulceration in the guinea pig colon (Watson 2008) and lead to marked increases in the chemokine interleukin-8 and B-cell CLL/lymphoma 10 activities in the normal human colonic mucosal epithelial NCM460 cell line (Bhattacharyya et al. 2010). Oral introduction of undegraded λ-κ carrageenan in drinking water of 12-week-old mice also caused significantly higher peak glucose levels in the blood, leading to concern that carrageenan-induced insulin resistance might contribute to human diabetes (Bhattacharyya et al. 2012). However, a comprehensive examination of in vivo dietary κ-carrageenan effects in rats revealed no effects on blood glucose (Weiner et al. 2007). More recent appraisals of carrageenans as food additives could find no hazards to human health as they are currently used (McKim 2014; Weiner 2014; Barlow et al. 2015; Weiner et al. 2015). The potential benefits and negative effects of including algae or their refined products in the diet require continuing research on a case-by-case basis.

Beta-(1-3)-Glucans—Laminarans

The main polysaccharides after the alginates in brown algae include β-glucans (laminarans), cellulose, and heteroglycans, the first being an energy reserve while the others are structural components of the cell wall, fitting the definitions of dietary fiber (Jones 2014). The concentrations and composition of the β-glucans vary substantially with season and growth rates (Rioux et al. 2009). The most studied β-glucans are those from cereals and fungi, but these differ significantly in structure from those of algal origin (Rioux et al. 2010). The biological responses elicited by algal β-glucans depend strongly on details of their primary structures (Bohn and BeMiller 1995; Mueller et al. 2000; Williams et al. 2005; Hofer and Pospíšil 2011). For example, brown algal M-series laminaran showed significant hepatoprotective effects when ingested orally by rats (Neyrinck et al. 2007). The protection was organ specific and appeared to act via the Kupffer cells in the liver, thereby establishing an immunoregulatory function of this orally ingested functional fiber. These and other biological effects of β-glucans have been reviewed (Novak and Vetvica 2008; Ramberg et al. 2010; Lehtovaara and Gu 2011; Kadam et al. 2015), and certain cautions have been expressed about the functional effects of soluble and particulate forms of these compounds (Young and Castranova 2005; Hofer and Pospíšil 2011).

Sulfated Fucose-Containing Polysaccharides—Fucoidans

The fucoidans are a subset of marine fucose-containing polysaccharides (FCPs) found in brown algae (Painter 1983) that are now attracting widespread interest (Shanmugam and Mody 2000; Berteau and Mulloy

2003; Kusaykin et al. 2008; Li et al. 2008; Pomin and Mourão 2008; Courtois 2009; Pomin 2009, 2012; Fitton 2011; Jiao et al. 2011; Kim and Li 2011; Kim and Wijesekara 2011; Wijesinghe et al. 2011; Wijesinghe and Jeon 2012). Double-blind clinical trials with fucoidan extracts show anti-aging effects on skin and other benefits in cosmetic applications (Fitton et al. 2015). A common source of FCPs used in experimental studies is *Fucus vesiculosus*, but fucoidans also are found in edible species such as *Cladosiphon okamuranus*, *Saccharina japonica* (*as Laminaria japonica*), and *Undaria pinnatifida* (Fitton 2011). The highly sulfated nature and molecular weights of FCPs appear to be responsible for many demonstrated biological activities in vitro (Croci et al. 2011; Ustyuzhanina et al. 2014). The FCP structures are species-dependent and can be modified by environmental variables and the developmental status of the seaweed fronds, all of which can affect their bioactivities (Honya et al. 1999; Zvyagintseva et al. 2003; Rioux et al. 2009; Pielesz and Biniaś 2010; Skriptsova et al. 2010; Stengel et al. 2011; Anastyuk et al. 2012; Mak et al. 2013). More recently, in vitro studies have provided insight into some structure-function relationships of FCPs (Cumashi et al. 2007; Ushakova et al. 2009; Ustyuzhanina et al. 2013, 2014).

It can be concluded that knowledge of the beneficial effects of algae and their extracts as food additives for humans lags far behind that on which diets have been formulated for commercially important species in aquaculture and agriculture. The number of species exhibiting benefits is wide ranging from invertebrates (nematodes, shrimp, abalone) and finfish (sea bream to salmon) to farm animals including poultry and mammals (both ruminants and monogastric species) (reviews: Craigie 2010; O'Sullivan et al. 2010; Rajauria 2015; Heuzé et al. 2016; Makkar et al. 2016). Algal-based products Tasco™ from *Ascophyllum nodosum* and Ocean Feed™ (a blend of brown, green and red macroalgae) are commercially marketed as feed additives to improve performance, stimulate immune reactions, mitigate sea lice damage in salmonids, and other benefits. Notable is the Alternative Feeds Initiative to develop alternative dietary ingredients (NOAA 2011). In addition to conventional methods of measuring animal performance, molecular techniques have been applied to buttress claims of efficacy (cf. Kulshreshtha et al. 2014; Liu et al. 2015). Bearing in mind ethical considerations, similar approaches may be adapted to facilitate the assessment of the benefits of macroalgal ingestion by humans.

VITAMINS

Vitamins are essential organic micronutrients, which an organism cannot synthesize directly in sufficient quantities and so instead must obtain from the diet. Well-known human vitamin-deficiency diseases include beriberi (deficiency in thiamine, vitamin B_1), pellagra (niacin), pernicious anemia (cobalamin, vitamin B_{12}), and scurvy (ascorbic acid, vitamin C) (Stabler and Allen 2004; Martin et al. 2011). These compounds serve as precursors for essential enzyme cofactors and are needed for essential metabolic functions (Fig. 4). Since animals have lost the capacity to synthesize these cofactors, they must obtain them from external sources. Algal foods are rich in vitamins. Several sea vegetables—laver (*Porphyra umbilicalis*), sea spaghetti (*Himanthalia elongata*), and *Gracilaria changii*—contain levels of vitamin C comparable to common vegetables such as tomatoes and lettuce (Norziah and Ching 2000; Ferraces-Casais et al. 2012), while the vitamin C content described for the brown seaweed *Eisenia arborea*(34.4 mg $(100\ g)^{-1}$ dry wt) approaches those reported for mandarin oranges (Hernandez-Carmona et al. 2009). The vitamin content of individual algal species discussed in this section, including details of sample origin and handling, is compiled in Online Resource 4.

Sea vegetables also are a good source of B-group vitamins (particularly B_1, B_{12}), as well as the lipophilic vitamin A (derived from the carotenoid β-carotene) and vitamin E (tocopherol). Kelp (*Macrocystis pyrifera*) can contain levels of α-tocopherol (the most biologically active form of vitamin E) at par with plant oils rich in this vitamin, such as palm, sunflower seed, and soybean oils (Ortiz et al. 2009; Skrovankova 2011). Moreover, values of β-carotene (pro-vitamin A) found in the seaweeds *Codium fragile* and *Gracilaria chilensis* can exceed those measured in carrots (Ortiz et al. 2009). The vitamin composition of microalgae can be equally remarkable. Fabregas and Herrero (1990) showed that *Tetraselmis suecica*, *Isochrysis galbana*, *Dunaliella tertiolecta*, and *Chlorella stigmatophora* were particularly rich in lipid-soluble (A and E) and B-group vitamins [including vitamins B_1, B_2 (riboflavin), B_6 (pyridoxal), and B_{12}]. The concentrations of several vitamins, including E, B_1, and β-carotene, exceeded those in conventional foods considered to be rich sources of these compounds (Fabregas and Herrero 1990). It is clear then that algal foods can be an excellent source for a wide range of these essential micronutrients.

Even so, variability between samples can make direct comparisons among studies difficult (e.g., Chan et al. 1997; McDermid and Stuercke 2003; Hernandez-Carmona et al. 2009). Part of the variability may lie in the sample processing methods (Skrovankova 2011) as observed for other phytochemicals (Ling et al. 2015); for example, analysis of freeze-dried and oven dried samples of *Sargassum hemiphyllum* yielded substantially different vitamin C contents (Chan et al. 1997). But differences also can be due to environmental and seasonal factors. For instance, there are notable variations in the levels of β-carotene and vitamin C between samples of *Ulva fasciata* collected from different sites (McDermid and Stuercke 2003) (Online Resource 4). Monthly quantitation of vitamins C, B_2, B_1, and A concentrations in *Eisenia arborea* over a 12-month period revealed levels were lowest in the summer months (June, July, August) and reached the highest concentrations in April/September (for vitamins C, B_2, B_1) and January (for provitamin A) (Hernandez-Carmona et al. 2009). The proximate cause for these patterns is unknown, as is the effect of growth conditions on the content and composition of vitamins in algal foods, so this is an important topic for future research.

Algal foods offer one of the few vegetarian alternatives for cobalamin (vitamin B_{12}) in the diet. Cobalamin is not required or synthesized by higher plants (Croft et al. 2005) so fruits and vegetables are poor sources of vitamin B_{12}, which explains why vitamin B_{12}-deficiency is common among people following strict vegetarian or vegan diets (Haddad et al. 1999; Waldmann et al. 2004; Allen 2008). Over half of microalgal species surveyed have a metabolic requirement for B_{12}, and contain large amounts (Online Resource 4), but they cannot synthesize it (Croft et al. 2005; Helliwell et al. 2011). Cobalamin is synthesized only by prokaryotes (Warren et al. 2002), and it has been shown that B_{12}-synthesizing bacteria are closely associated with or reside on eukaryotic algal surfaces (Croft et al. 2005; Wagner-Döbler et al. 2010). *Pyropia yezoensis* (nori) contains up to ~0.06 mg vitamin B_{12} $(100\ g)^{-1}$ algal dry wt, comparable to that found in beef liver (Watanabe et al. 1999b; Takenaka et al. 2001). Lower levels are found in other sea vegetables such as kelps (including wakame) and hijiki, although reported concentrations vary among studies, possibly reflecting differences among strains, growing conditions, or harvesting periods (Watanabe et al. 1999a; Miyamoto et al. 2009). Given that the ultimate source of vitamin B_{12} is bacteria, changes in the character and magnitude of the epiphytic prokaryotic communities related to region or algal physiological state may contribute to variation in vitamin content; these factors currently are poorly quantified.

There is uncertainty about whether the magnitude of vitamin concentration in algal foods reflects their nutritional value. Dagnelie et al. (1991) investigated how sea vegetables affected the hematological status of B_{12}-deficient children and concluded that the algal-derived vitamin B_{12} was not bioaccessible to humans. However, their very small treatment group (n=5) may have been insufficient to draw firm conclusions. Takenaka et al. (2001) showed that feeding nori to vitamin B_{12}-deficient rats yielded a 1.9-fold increase in hepatic levels of total B_{12}compared to those without nori supplementation. Similarly, increased consumption of *Chlorella* or nori by vegan participants prevented B_{12} deficiency (Rauma et al. 1995). However, there are few data on which to base quantitative estimates of the portion of algal vitamins that are absorbed during digestion.

One approach to assessing the availability of vitamins is to distinguish among their different chemical forms. The uptake of cobalamin-based compounds, referred to more broadly as corrinoids, is largely governed by the gastrointestinal absorption system rather than their chemical liberation via digestive chemical processes (Russell-Jones et al. 1999). Pseudovitamin B_{12} (or pseudocobalamin) differs from cobalamin in its lower axial ligand (Stupperich and Krautler 1988), and this affects affinity of the mammalian B_{12}-binding protein intrinsic factor (IF) for the compound, thereby limiting its absorption in the intestine (Stupperich and Nexo 1991). This difference has human health implications because commercially produced vitamin B_{12} supplements derived from the cyanobacterium *Arthrospira* sp. (spirulina) consist largely of pseudovitamin B_{12} (Watanabe et al. 1999b; Watanabe 2007a), reducing their nutritional value. In contrast, "green" (*Ulva* [formerly *Enteromorpha* sp.]) and "purple" (*Pyropia* [formerly *Porphyra*] sp.) laver contain substantial amounts of biologically available B_{12} (Watanabe et al. 1999b), and indeed, rats fed purple laver improved their B_{12} status (Watanabe et al. 1999b). A recent study has established that the vast majority of cyanobacteria synthesize pseudocobalamin, whereas eukaryotic algae that are dependent on B_{12} for growth are like animals in that they require cobalamin (Helliwell et al. 2016). Thus, sea vegetables are likely to be a more reliable source of the appropriate form of this vitamin, although again this will be determined by the prokaryotic community associated with the algae.

These findings highlight the need for rigorous care in the analytical determinations of the vitamin content of algal foods. Bioassays using B_{12}-dependent bacteria such as *Lactobacillus delbruekii* ssp. *lactis* (ATCC7830) are inadequate because, unlike humans, these bacteria do not discriminate between vitamin B_{12} and pseudovitamin B_{12}. An alternative radioisotope dilution assay (RIDA) is likely to better reflect the functional B_{12} content (Watanabe 2007a). Distinguishing among bioavailable and non-bioavailable vitamin forms will be crucial (Watanabe 2007b). Complicating these analyses further is evidence that commercial processing methods can alter the vitamin chemistry sufficiently to affect uptake. For example, Yamada et al. (1999) showed that air-drying *Pyropia tenera* (asakusa-nori) produced B_{12} analogs that are biologically inactive. Drying by lyophilization might have better nutritional outcomes (Takenaka et al. 2001), although this has yet to be rigorously demonstrated. Other factors of particular importance to preserving vitamin content include washing methods, storage temperature, light, and moisture content (Online Resource 1, Brown 1995; Jimenez-Escrig et al. 2001; Lage-Yusty et al. 2014). There is a strong need for more detailed investigations into how the nutritional quality of sea vegetables is affected by processing methods suited for commercial-scale production.

The bioavailability of other algal-derived vitamins is also underexplored. Vitamin E encompasses eight different forms (tocopherols and tocotrienols) that differ in their biological activity (α- and γ-tocopherols are the most active). Although much less is known about their relative bioavailability compared to the vitamin B_{12} analogs, it is clear that their relative contributions affect the nutritional quality of foodstuffs (Ortiz et al. 2009). An additional concern with fat-soluble vitamins is that they must be consumed with lipid-rich foodstuffs to ensure efficient intestinal absorption (Skrovankova 2011). Although this co-dependence is understood, there currently are few data on this dependence for edible-algal species.

Most studies on algae and vitamins often focus either on analysis of vitamin concentrations in algae (e.g., Ortiz et al. 2006, 2009; Hernandez-

Carmona et al. 2009; Matanjun et al. 2009; Ferraces-Casais et al. 2012) or testing the value of an algal product as a functional food (e.g., Dagnelie et al. 1991; Rauma et al. 1995; Takenaka et al. 2001), but not both. Ideally, studies combining these two approaches should be adopted to gain meaningful insights on the true quality of algal foods as vitamin sources (Takenaka et al. 2001).

Finally, there are the ecological challenges to gaining a broad picture of algal foods as a nutritional source of vitamins. Vitamin production and metabolism can vary considerably across diverse algal lineages (Croft et al. 2006; Helliwell et al. 2011, 2013). One approach that may help reveal this complexity would be a high-throughput screening of promising algal food candidates with next-generation sequencing techniques coupled with bioinformatics to search for vitamin-biosynthesis pathways. Nevertheless, there will be continued the need for careful analytical characterizations and bioavailability testing because the up- or down-regulation of gene expression almost certainly will be environmentally regulated.

ANTIOXIDANTS

It is not surprising that there is a very broad literature on marine algae as sources of antioxidant compounds for human diets. Photosynthetic energy acquisition and transformations necessarily involve continuing redox disequilibria, with the production of reactive species that can decrease lifespan and evolutionary fitness. Microalgae and macroalgae, like other life forms, contain antioxidant organic compounds and enzymes that limit this oxidative damage, which results primarily from reduced states of oxygen—the "reactive oxygen species"—including the superoxide radical anion ($O_2^{\cdot-}$; $O_2 + 1e^-$), hydrogen peroxide (H_2O_2; $O_2 + 2e^-$), the hydroxyl free radical ($HO^\cdot$; $O_2 + 3e^-$), and singlet oxygen ($^1O_2$) (Halliwell and Gutteridge 2007). Whereas the antioxidant benefits of several terrestrial plant foods are established, much less is known about whether algal foods provide similar benefits.

The reactive oxygen metabolism in marine algae is diverse and complex, given the wide range of antioxidant compounds (Cornish and Garbary 2010), but an extension to any beneficial response from human consumption of these substrates is far less certain. Antioxidant activity can have two forms: the activity of antioxidant enzymes or the production of molecules that serve as sacrificial scavengers of reactive oxygen species. There also are two broad categories of antioxidant activity: limiting reactive oxygen species within the digestive tract, thereby decreasing oxidative stress on the gut microbiome and epithelial cells, or transport into the blood for distribution throughout the body. Evidence for direct transport is very limited, as there seems to have been no systematic study of digestive uptake of these compounds. In one study, Okada et al. (2009) examined the bioaccessibility of astaxanthin extracted from the green alga *Haematococcus* (Fig. 2) as judged from the concentration in blood serum, as a function of the timing of the ingestion of astaxanthin relative to a meal, and whether the subjects were smokers or non-smokers. Astaxanthin increased more in serum when the dose was taken 10 min after a meal rather than 2 h before, evidence of complex factors affecting its bioaccessibility. The ingestion (and topical application) of polyphenols of brown algae inhibited UVB radiation-induced skin carcinogenesis in mice (Hwang et al. 2006), and while this bioactivity remains to be determined for humans, it provides evidence that algal foods have significant functional food potential.

The foremost enzymes that restrict oxidative damage in algae and terrestrial foods include the superoxide dismutases that remove superoxide radical anions, and catalases and peroxidases that convert hydrogen peroxide to water. Superoxide dismutases in cyanobacteria have Ni, or mixtures of Fe, Mn, and Ni, as the metal, whereas eukaryotic algae have Mn or Fe, or some combination of Fe, Mn, and Cu+Zn (Wolfe-Simon et al. 2005). Catalase has an Fe-containing heme cofactor while peroxidases use a reductant to convert hydrogen peroxide to water. Of these enzyme cofactors, Cu and Zn, and particularly Fe are used in numerous human metabolic pathways. Since the ingested antioxidant enzymes are digested in the intestine, the only effect the enzymes can have in the animal is through uptake of the metal cofactors across the intestinal epithelium. The possible effects on the intestinal microbiome of any undigested enzyme, or of the released metal cofactors, have not been investigated.

There is a stronger linkage between selenium in food and antioxidant capacity in metazoans such as mammals. Selenium is an essential metal in metazoans and some algae for the production of Se-requiring glutathione peroxidase, used to metabolize hydrogen peroxide and lipid hydroperoxides (Halliwell and Gutteridge 2007; Perez et al. 2007; Gobler et al. 2011). Analyses of the elemental contents of microalgae (Quigg et al. 2011) and macroalgae (Tuzen et al. 2009; Pereira 2011) rarely include Se, even though it is present in both (Fournier et al. 2005). Se readily bioaccumulates in algae (Cases et al. 2001; Fournier et al. 2005), and Se-deficiency in rats can be alleviated by oral supplementation with Se-rich *Arthrospira*(spirulina), as indicated by increased activity of (Se-containing) glutathione peroxidase in the kidneys and liver (Cases et al. 2001). However, increases in this enzyme activity were greater in rats supplied selenite or selenomethionine (more reactive species) than with the same dosage of Se-rich cyanobacterium, likely due to lower bioavailability of the cyanobacterial Se. The factors regulating Se content of algal foods and its availability are prime research topics for the future.

Under normal metabolic conditions, the production of hydroxyl radicals and singlet oxygen cause almost immediate damage, essentially reacting with the first oxidizable molecule that they encounter. In these cases, "sacrificial" scavengers (of $HO^{\cdot}$) and quenchers (of $^1O_2$) often are the only recourse for limiting damage once the free radicals are produced (Smirnoff and Cumber 1989; Telfer et al. 1994a, b; Sunda et al. 2002; Ledford and Niyogi 2005; Halliwell and Gutteridge 2007; Ledford et al. 2007). Algae contain a wide range of molecules capable of free radical scavenging activity in vitro and in vivo. These include the water-soluble ascorbate (vitamin C) and certain compatible solutes (osmoprotectants), and the lipid-soluble α-tocopherol (vitamin E) and carotenoids such as astaxanthin (Halliwell and Gutteridge 2007). Mycosporine-like amino acids, mainly considered as UV screening compounds, are also antioxidants (Oren and Gunde-Cimerman 2007) as are a range of other solutes that act as scavengers and quenchers of reactive oxygen species in algae (Cornish and Garbary 2010). $HO^{\cdot}$scavengers include glycerol (Smirnoff and Cumber 1989), mannitol (Smirnoff and Cumber 1989; Shen et al. 1997; Larson et al. 2002), L-proline (Smirnoff and Cumber 1989), dimethylsulfoniopropionate (Sunda et al. 2002), and floridoside and isofloridoside (Li et al. 2010), although glycine betaine (or betaine: trimethylglycine) does not have this property (Smirnoff and Cumber 1989; Shen et al. 1997). Given that algal osmoprotectants are necessarily present in high concentrations (≥ 0.1 mol L^{-1}) in metabolically diverse compartments (cytosol, plastid stroma, and mitochondrial matrix), there is potential for them to have functional food roles. However, preliminary experiments showed that none of these compounds interact with O_2^{-} (Smirnoff and Cumber 1989), unlike β-carotene and other carotenoids such as fucoxanthin that quench $^1O_2$ as well as scavenging HO and O_2^{-} (Halliwell and Gutteridge 2007; Sachindra et al. 2007). Other algal components that scavenge free radicals are phenolic compounds (Ragan and Globitza 1986) including halophenols (Li et al. 2011) and phlorotannins (Shibata et al. 2007) and, as noted above, alginate (Zhao et al. 2012; Zhou et al. 2012) and sulfated polysaccharides (Barahona et al. 2012).

Most studies of the bioavailability of algal antioxidant products remain at the entry level with respect to human effects: in vitro testing of extract bioactivity on cell lines. Nwosa et al. (2011) confirmed and extended previous work showing the antioxidant activities of polyphenolic extracts from four species of edible marine algae in inhibiting Caco-2 colon cancer cell proliferation and α-glucosidase activity (see below): the green alga, *Ulva lactuca*, the brown algae *Alaria esculenta* and *Ascophyllum nodosum*, and the red alga *Palmaria palmata. Ulva lactuca* had a low yield of polyphenols relative to the other algae, but the brown and red algal polyphenolic extracts performed as well as antioxidants. However, Nwosa et al. (2011) illustrated that the method of preparing the extracts from marine algae can significantly alter their antioxidant efficacy (see also Ling et al. 2015), highlighting the need for caution in comparisons of antioxidant performance among studies. With this possible caveat, most work on antioxidant activity of algal phenols has involved red algae; some bromophenols from the marine red alga *Rhodomela confervoides* have greater in vitro antioxidant activity than ascorbate (Li et al. 2011). Olsen et al. (2013) showed that bromophenols extracted from the red alga *Vertebrata lanosa* significantly inhibited oxidant effects and lipid peroxidation in cultures of human fetal lung (MTC-5) and human hepatocellular liver carcinoma (HepG2). In this case, it was shown that bromophenol can enter cells, and thus potentially can move from the gut lumen into the blood stream. Overall, there is a strong need for more work on the in vivo effects of the antioxidant properties of phenols and other algal food constituents in mammals, and humans in particular.

Instead of serving to facilitate the control of reactive oxygen species, some algal components can inhibit their production, but most studies do not adequately distinguish between the decreased production and increased removal of oxidants. For example, dietary ingestion of phycocyanin, taken up from the gut as the chromophore component phycocyanobilin, and related bile pigment metabolites inhibits the generation by NADPH oxidase of O_2^-, which has a key role in numerous disease syndromes, e.g., antigen expression, angioplasty, cancers, glycemia and lipidemia, hypertension, immunostimulation, and age-related maculopathy (reviewed by McCarty 2007). This industry-sponsored but balanced and authoritative review shows that *Spirulina* spp. (most now transferred to *Arthrospira*) are a prominent cyanobacterial source of phycocyanobilin, a dietary supplement worthy of in-depth study.

A class of compounds attracting increasing attention are the phlorotannins found in brown algae, which have extraordinary though inconsistent antioxidant properties (see in Wang et al. 2014), in part due to the methods of extraction (Nwosa et al. 2011). The vast bulk of this work was done in vitro, much of it studying the effects of phlorotannin on carbohydrate-hydrolyzing enzymes. Nwosa et al. (2011) found that extracts of *Ascophyllum* and *Alaria* inhibited Caco-2 colon cancer cell proliferation, α-amylase activity and, to a lesser extent, α-glucosidase activity, with mass spectrometric evidence indicating that the active principal(s) were phlorotannins. Kawamura-Konishi et al. (2012) also found that phlorotannin extracts of four species of *Sargassum* significantly inhibited the salivary enzyme α-amylase in vitro, and that a novel phlorotannin from *Sargassum patens* inhibited rat pancreatic α-glucosidase action on amylopectin. Iwai (2008) showed that oral administration of extracts inhibited lipid peroxidation in the plasma, red blood cells, liver, and kidney of KK-A^y mice, indicating that the antioxidant activity of phlorotannins had beneficial properties for reducing diabetic oxidative stress. Important recent work (Corona et al. 2016) investigated the effect of food grade phlorotannins from *Ascophyllum nodosum* in trials on human subjects. The work showed the in vitro gastrointestinal modification of phlorotannins, the occurrence in plasma and urine of metabolites of phlorotannins, and a significant increase in cytokine IL-8. To conclude consideration of phlorotannins, while in vitro studies on phlorotannins are valuable, more work along the lines of that of Iwai (2008) and Corona et al. (2016) is needed to understand the uptake and systemic properties of phlorotannins, and to determine whether the in vitro effects occur in vivo and relate to their antioxidant properties (Bohn et al. 2015).

There remain substantial knowledge gaps about the efficacy of antioxidant properties of macroalgal and microalgal foods at all levels, from characterization among species through effects on gut microbiota and transport across the gut lumen to their impacts on human physiology. This will be a valuable area of emerging research over the next decade.

INORGANIC ELEMENTS

Processed seaweeds are widely used as mineral and metal nutritional supplements (e.g., Kay 1991), but the efficacy of these supplements is poorly quantified. Most studies suffer from serious experimental limitations, including short duration of the study, small sample size, and inadequate documentation of active ingredients. There is a comparatively small literature describing mineral contents of macroalgal and microalgal foods (Cabrita et al. 2016), and very little information about seasonal variations for naturally harvested sea vegetables.

The best evidence of the human nutritional benefits of sea vegetable inorganic elements is for iodine and iron, which can be highly enriched in marine macroalgae. Nutritional generalization about algal mineral contents is difficult because of sometimes large seasonal, geographic, and taxonomic variations in mineral contents of marine algae (e.g., Jensen 1993). For example, Indonesian green, brown, and red algae contain high levels of potassium, calcium, and sodium, but significantly lower levels of iron and zinc than reported for Japanese *Pyropia* (as *Porphyra*) *yezoensis*, *Ulva* (*Enteromorpha*) *intestinalis*, and *Sargassum* (*Hijikia*) *fusiformis* (Takeshi et al. 2005). These findings may indicate that macroalgal harvests from warm equatorial areas have lower mineral nutritional value than higher latitude regions (e.g., Cabrita et al. 2016); however, there are remarkably few data on which to assess the validity of such generalizations.

Iodine

There is a long history linking seaweed consumption by humans and the reduced incidence of goiter and other thyroid disorders. Iodine deficiency causes hypothyroidism while excess iodine uptake can induce either hyper- or hypothyroidism (Miyai et al. 2008). Seaweeds are a good nutritional source for iodine, particularly in regions where other foods are deficient, but the iodine content of commercially available sea vegetables varies dramatically among species, the methods of preparation (many iodine compounds are water soluble), and the duration of storage (iodine may vaporize under humid conditions) (Teas et al. 2004a). Many macroalgae are washed and dried for storage. These processing steps did not significantly reduce iodine content in three common species (*Alaria esculenta*, *Palmaria palmata*, and *Ulva intestinalis*), but rehydration followed by boiling in water lowered iodine content by 14–75 % (Nitschke and Stengel 2016).

Some kelps (*Laminaria* spp., *Saccharina* spp.) have high levels of iodine, and salts that include kelp powder are available commercially as a source of this vital nutrient. Not all brown algae accumulate high levels of iodine; for example, the kelps *Undaria* (wakame) and *Alaria*("Atlantic wakame") have lower iodine levels that are comparable to *Palmaria palmata* (dulse, a red sea vegetable) (MacArtain et al. 2007; Rhatigan 2009; Holdt and Kraan 2011; Schiener et al. 2015) (Fig. 6). In contrast, high levels of iodine in other brown macroalgae (e.g., *Laminaria*; *Saccharina*; Teas et al. 2004b; Miyai et al. 2008) have led to a strong concern that overconsumption of these particular sea vegetables can be unhealthy. Michikawa et al. (2012) reported that consumption of undefined seaweed more than 2 days per week appeared to correlate with increased risk of thyroid cancer in Japanese postmenopausal women, although Wang et al. (2016) did not find this relationship to be significant. Nevertheless, there is clear evidence that algal food consumption leads to elevated iodine levels in humans. Miyai et al. (2008) measured serum levels of thyroid hormones in conjunction with well-defined ingestion rates of kombu (*Saccharina japonica*) over short (7–10 days) and longer term (~90 days) exposure. Urinary excretion of iodine increased significantly over the short term with increasing intake (15–30 g day^{-1} of kombu—a normal consumption for some Japanese) and suppressed thyroid function for at least 3 months. Thyroid hormones returned to normal levels when seaweed intake ceased. In this case, the absorbed iodine (20–50 mg) exceeded the recommended upper daily intake of iodine (0.2 mg day^{-1} by more than an order of magnitude; WHO 1989; Dawczynski et al. 2007; Miyai et al. 2008). Food preparation (e.g., cooking, pickling) can reduce the iodine intake, but, even then, high-iodine sea vegetables, which account for only a portion of those consumed, should be restricted in the diet (Teas et al. 2004a). Research is needed on how food preparation alters the bioavailability of iodine in different sea vegetables. Indeed, Lightowler and Davies (2002) found that established food tables (UK Department of Health, 1991) gave poor estimates of dietary iodine intake and recommended developing more reliable data on iodine in foods, including the variation within food groups.

Figure 6. Examples of sea vegetable use in Japanese cuisine: a "Onigiri" is a Japanese rice ball usually wrapped by nori (*Pyropia yezoensis*) with several other ingredients: *from right to left*, rice wrapped in nori, with Japanese apricot (umeboshi) and preserved kombu (tsukudani, *Saccharina japonica* and other species of the genus *Saccharina*), wrapped with shredded kombu (torero-kombu), and wrapped in nori; b Tempura soba with wakame (*Undaria pinnatifida*). Wakame is used in Japan, noodles, soups, salads, pickles, and more. c Traditional Japanese dishes with sea vegetables: preserved kombu (tsukudani) on rice, miso soup with wakame, sliced kombu with vegetables, and sliced raw fishes (sashimi) with wakame and cucumber. d "Kobu-maki" is simmered food, often salmon or herring, wrapped in kombu, which is usually prepared for the New Year's holidays. Kombu is used in several dishes and soup stock. (Courtesy of Kazuko Sato and Yoichi Sato).

Iron

Macroalgae are a potentially rich source of iron for human diets. Garcia-Casal et al. (2007, 2009) measured seasonal differences in the iron content of four seaweed species common to Venezuelan waters (*Ulva* spp., *Sargassum* spp., *Porphyra* spp., and *Gracilariopsis* spp.). The *Gracilariopsis* spp. and *Sargassum* spp. had substantially higher iron content with *Porphyra*spp. having the lowest, and there was a distinct seasonal cycle whereby iron content was highest in spring and summer and lowest in fall and winter. *Sargassum, Ulva*, and *Porphyra* spp. have high iron content, and 15 g day^{-1} of *Sargassum* provided substantially higher amounts than daily recommended intakes (≤18 mg Fe day^{-1}, Institute-of-Medicine 2001). Cabrita et al. (2016) found that Fe (and other metal) contents varied, in some cases substantially, among macroalgal species collected at the same sites and time, presumably linked to differences in metabolic requirements. Recognizing that iron consumption in macroalgae likely does not equate with uptake, the researchers quantified iron incorporation into the blood of 93 volunteers fed diets of radioactive (gamma-emitting) ^{59}Fe-labeled *Ulva* spp., *Sargassum* spp., and *Porphyra* spp. (Garcia-Casal et al. 2007, 2009). Uptake of ^{59}Fe was dose-dependent on sea vegetable added to maize or wheat-based meals (Garcia-Casal et al. 2007), but generally was greater than the total iron content of the different macroalgal species. Thus, in addition to enhancing total dietary iron content, the seaweeds appeared to act synergistically to facilitate iron uptake from wheat and maize, possibly related to high levels of ascorbic acid (vitamin C) that converts iron to the more readily absorbed ferrous form.

The Fe content of wild algae varies seasonally and geographically depending on the metal content of coastal waters, in addition to being species-specific (Garcia-Casal et al. 2007; Cabrita et al. 2016). Wild harvesting must then be optimized for each locale known to produce algae for optimal concentrations of inorganic nutrients. Similar constraints may apply to aquaculture crops unless they are fertilized, but there are very few data currently available on sea vegetable nutritional metal content relative to species, region, or season.

PHYTOCHEMICALS

The main non-acyl glyceride compounds in lipid membranes attracting nutritional and commercial interest are the carotenoids, especially because of their dietary importance for vision (Ben-Amotz and Levy 1996). Carotenoids, among the most widespread pigments in prokaryotic and eukaryotic photosynthetic organisms (Britton et al. 2004), function as light energy harvesters, as photoprotectants, and as antioxidants (Halliwell and Gutteridge 2007, and see below). One specific keto-carotenoid, siphonaxanthin, but not fucoxanthin, induced apoptosis in human leukemia cell lines, suggesting its potential as a chemopreventative or chemotherapeutic agent (Ganesan et al. 2011), although its efficacy remains to be determined in vivo. A subset of the carotenoids, the β-carotenes, are precursors for vitamin A (Minguez-Mosquera et al. 2008), and natural β-carotene may act as a lipophilic antioxidant in vivo, providing some protection in children exposed to radiation from the Chernobyl accident (Ben-Amotz et al. 1998).

Various strains of the green microalga *Dunaliella salina* can accumulate about 8 % of their dry weight as β-carotene, which has been marketed as a functional food (Ben-Amotz and Levy 1996; Borowitzka 2013b). The US FDA recently (2011) had no questions about a filing for the powdered form of the microalga *Dunaliella* (*bardawil*) *salina* as GRAS for use as an ingredient in food products (Agency Response Letter GRAS Notice No. GRN 000351, Borowitzka 2013b). Another main source of carotenoids is the green alga *Haematococcus pluvialis*, for which cultures have developed industrially in several countries (Cysewski and Lorenz 2004). Carotenoid supplements, including β-carotene, are effective in improving the carotenoid supply in breast milk at early lactation (Nagayama et al. 2014). However, the optimal levels of β-carotene remain controversial. Undefined β-carotene supplements increased lung cancer risk in smokers or people exposed to asbestos (Druesne-Pecollo et al. 2010), while earlier work showed no reduction in the incidence of lung cancer among male smokers after 5 to 8 years of dietary supplementation with alpha-tocopherol or β-carotene (Albanes et al. 1995).

Phenolic compounds such as flavonoids and phenolic acids serve as antioxidants. Brown macroalgae (*Alaria esculenta*, *Ascophyllum nodosum*, *F. vesiculosus*, *Saccharina latissima*) had 2–15 times the total phenolics (as mg gallic acid equivalents g^{-1} dry matter) as red species (*Chondrus crispus*, *Meristotheca papulosa*, *Palmaria palmata*, *Sarcodiotheca gaudichaudii*) (Tibbetts et al. 2016). Total phenolic content was inversely related ($r = -0.81$) to in vitro protein digestibility. Phenolic compounds are challenging to extract and characterize, but improvements in analytical methodology are expected to facilitate more detailed investigation of these antioxidants in algae. Drying *Kappaphycus alvarezii* in sunlight significantly reduces total phenolic, flavonoid, anthocyanin, and carotenoid content compared with samples dried in ovens or in the shade (Ling et al. 2015).

SAFETY

Consumption of any food is not without risk, so the promotion of algal consumption must also consider potential harm to consumers. Possible risks associated with algae include excess intake of toxic metals, allergenicity, cyanotoxins, and certainly secondary metabolites (e.g., prostaglandins, kainoids) as well as contamination with pathogens, radioisotopes, and toxic synthetic compounds.

Metal Toxicity

Under normal, pristine conditions, metal uptake improves the nutritional quality of algal foods, but excessive uptake can lead to toxicity. Moreover, a substantial amount of metal associated with macroalgae can exist as colloidal-sized particles sorbed to algal surfaces (Gadd 2009; Turner et al. 2009), so surface chemistry and algal physical structures can affect metal content in addition to metabolic processes. The efficacy of macroalgae as metal collectors is why they are effective sentinel organisms for detecting anthropogenic signatures in coastal waters (Melhuus et al. 1978; Leal et al. 1997; Brown et al. 1998; Mardsen and De Wreede 2000); a feature in conflict with their use as a food source.

Despite these caveats, some general patterns appear in the literature. Metal content tends to differ among phylogenetic groups, with brown algae typically possessing higher levels of most metals in comparison with red or green algal species (Foster 1976; Munda and Hudnik 1991; Stengel and Dring 2000; Al-Masri et al. 2003; Benkdad et al. 2011). But there are exceptions. Phaneuf et al. (1999) observed that the green alga *Ulva lactuca* and *Ulva* (as *Enteromorpha*) spp. in the St. Lawrence

River estuary had greater concentrations of Co, Cr, and Cu, while the brown algae *F. vesiculosus*, *Laminaria longicruris* and *Fucus distichus* had higher As and Cd concentrations (though all below regulatory levels). These species-specific differences among algae from the same environment may reflect differing metabolic affinities for these metals. Algal uptake of Cd, Cu, Co, Zn as well as Pb occurs via transporter-mediated processes, as indicated by their uptake kinetics (e.g., Garnham et al. 1992; Knauer et al. 1997; Mehta et al. 2002; Francois et al. 2007; Costa et al. 2011). As noted, however, colloidal sorption to algal surfaces can be a contributing factor, and there are few data on colloidal metal concentrations in coastal waters (Wells 2002). Regardless, macroalgae clearly can be a vector for toxic metal transfer to humans, especially if harvested from contaminated habitats.

Perez et al. (2007) studied the concentrations of many elements in *Pyropia* (as *Porphyra*) *columbina* and *Ulva* sp. in two Argentinian regions having different exposure to human activities. They found a wide range of metal content between these species with *Pyropia columbina* being a stronger accumulator of As, Cd, Mo, and Se, while *Ulva* spp. tended to accumulate more Cr, Pb, and Ni. There also were significant seasonal variations in metal loading with *Pyropia columbina* having maximum Cd concentrations during winter while *Ulva*sp. showed highest Pb concentrations during summer. This seasonality may result not only from metabolic controls but also oceanographic influences. For example, Riosmena-Rodríguez et al. (2010) found elevated concentrations of Cd in several algal species during April that appeared to be related to local upwelling events. It is unclear to what extent the presence of these metals could affect human health, as their bioavailability was not assessed.

Metal accumulation will be influenced by both the geochemical conditions and also by metabolic control as a function of ecological growth strategies or seasonally determined productivity (Stengel et al. 2004). Schiener et al. (2015) found the concentrations of major metal salts of Ca and K and the trace element Fe to more than double during spring and summer months in the brown algae *Laminaria digitata*, *Laminaria hyperborea*, and *Saccharina latissima*, but concentrations of other trace metals did not vary in a seasonal pattern. Although concentrations of Cd, Co, Cr, Cu, Fe, Hg, Mn, Ni, Pb, or Zn differed greatly among *F. vesiculosus*, *Ascophyllum nodosum*, *Laminaria longicruris*, *Palmaria palmata*, *Ulva lactuca*, and *F. distichus* collected from the St. Lawrence River estuary, Canada, all were at very low levels and there were no metal-related health risks (Phaneuf et al. 1999). Similarly, a study of the metal content of seaweeds washed up on the Brazilian coast also showed that some toxic metals were at levels considered harmful, although with remediation of nearby industrial contamination sources these macroalgae represent a potential food alternative for humans (de Oliveira et al. 2009). In one of the more comprehensive recent analyses, Dawczynski et al. (2007) measured concentrations of six trace elements (Fe, Mn, Zn, Cu, Se, and I) and four ultra-trace elements (As, Pb, Cd, and Hg) in 34 commercially available red and brown macroalgal products originating from China, Japan, and Korea. They found that for normal consumption, daily intakes of Fe, Mn, Cu, and Se are well below daily intake recommendations of the German Society of Nutrition (DGE) and the provisional tolerable weekly intake (PTWI) values of the World Health Organization (WHO), and the ultra-trace elements were present at low, harmless concentrations. Turner et al. (2008) measured the uptake of the trace metals, Pd, Cd, Hg, and Pb, by *Ulva lactuca* along the salinity gradient ($S = 15–35$) under well-controlled laboratory conditions. Only Cd displayed salinity-dependent uptake rates, with rates decreasing at higher salinity. Their findings also showed that the presence of environmentally relevant concentrations of dissolved humic substances (3 mg L^{-1}) suppressed slightly the sorption of Pd and Hg, while moderately enhancing Pb sorption via adsorption to the algal surfaces; the uptake (internalization) of Pb was inhibited by humic substances. It is clear that processing methods that help to eliminate metals bound to algal cell walls/surfaces would be advantageous, but there is little information available on this topic.

There is thus potential for some metals to reach harmful concentrations in edible seaweeds, but there is no information on how bioaccessible or bioactive most algal metals are in human digestion. Worse, there is no consensus on a uniform or even optimal approach to quantify the bioavailability of metals. Recent advances using bio-digestive reactor approaches (e.g., Moreda-Pineiro et al. 2012) likely provide the way forward, but this remains to be established.

Arsenic

Chronic exposure to inorganic arsenic (iAs: arsenite, As(III), arsenate, As(V)) leads to a higher incidence of several cancers including skin, lung, and urinary tract cancers. The International Agency for Research on Cancer (IARC) classifies iAs as human carcinogens (Group I), while the biological metabolites dimethylarsinic acid (DMA) and monomethylarsonic acid (MMA) are classified as possibly carcinogenic (Group 2B) in humans (IARC 2012). Toxic effects of As include disruption of oxidative phosphorylation, generation of reactive oxygen stress (ROS), enzyme inhibition, and epigenetic changes (IARC 2012; NRC 2013). Exposure to iAs occurs as humans drink water enriched in As by natural geochemical processes or poor agricultural/manufacturing/mining practices and through the diet including from foods (e.g., rice) grown in contaminated soils (NRC 2013; Zhu et al. 2014; Li et al. 2016). All seafoods contain arsenic, which enters cells through phosphate transporters and aquaglyceroporins (Bhattacharjee et al. 2008; Cooney et al. 1978; Zhao et al. 2010). The World Health Organization (WHO)'s provisional maximum level of iAs in drinking water is 10 $\mu g\ L^{-1}$ (WHO 2016).

There are more than 50 arsenic species in seafood, but the absolute arsenic content of a seafood does not predict health risk because marine organisms have evolved detoxification strategies in which iAs is converted to methylated (organic) forms. Fish and crustaceans convert most iAs into arsenobetaine; humans excrete arsenobetaine, and it is not considered to be toxic (Francesconi 2010, Molin et al. 2015 [see their Table 1 for structures]). Algae, and mollusks that eat algae (e.g., oysters, clams), convert most iAs to arsenosugars, and there is some evidence that when phosphate levels are low, iAs may be converted to As-phospholipids that have a role in algal membrane function (Cooney et al. 1978; Garcia-Salgado et al. 2012). It is reassuring that fractionation studies show most macroalgae contain very little inorganic As in comparison to arsenosugars, although measurement of iAs in algae may be less reliable than in plants (de la Calle et al. 2012; Diaz et al. 2012; Hansen et al. 2003).

One of the major, excreted metabolites from algal arsenosugars in humans is dimethylarsinic acid (DMA), which the IARC (2012) considers to be possibly carcinogenic to humans (Group 2B) (IARC 2012). In vitro trials with human HepG2 cells showed that DMA toxicity occurred only at levels that were 400× the maximum DMA found in the urine of a human volunteer during clearance of arsenosugar (Raml et al. 2005). However, the rate of clearance of DMA and other metabolites of an arsenosugar that was experimentally ingested appears to vary in different individuals; for example, although 4-day urinary excretion removed ≥85 % of the As represented in an ingested arsenosugar in four individuals, two other individuals excreted only 4–15 %. This difference might reflect either retention of As, or perhaps that they absorbed less arsenosugar from the digestive tract (Raml et al. 2009).

Although most algae naturally synthesize arsenosugars from the iAs they take up from seawater, a few brown macroalgae contain a significant fraction of total As as iAs (*Laminaria digitata, Laminaria hyperborea*: Hansen et al. 2003, Taylor and Jackson 2016; *Sargassum*spp.: Nakamura et al. 2008, Magura et al. 2016). The sea vegetable hijiki (*Sargassum fusiforme*) contains unusually high levels of iAs (e.g., 60 $\mu g\ g^{-1}$ dry wt hijiki, 0.4–2.8 $\mu g\ g^{-1}$cooked hijiki) in comparison to its arsenosugar content (Francesconi 2010; Nakamura et al. 2008; Rose et al. 2007). Nevertheless, there are at least three issues that bear upon As toxicity to humans: the chemical speciation of As in the food item, the bioaccessibility after cooking (Devesa et al. 2008; Ichikawa et al. 2006), and the metabolism of As in the individual (Choi et al. 2010; Raml et al. 2009). Ichikawa et al. (2006) reported that 88.7–91.5% (w/w) of As in hijiki was removed by cooking. Nakamura et al. (2008) determined inorganic As (III+V) extracted during a simulated gastric digestion (pepsin) from cooked hijiki donated by 14 families who also provided information on their monthly consumption of hijiki. Nakamura et al. (2008) estimated that this would result in an iAs intake of 1.1 $\mu g\ kg^{-1}$ human body weight per week and could cause a non-negligible increase in skin cancer cases by their model. Currently, the WHO does not have a Provisional Tolerable Weekly Intake level of iAs, after withdrawing an earlier PTWI of 15 $\mu g\ kg^{-1}$ human body weight per week (WHO 2010). More research is needed on health risks from lower dose iAs intake following consumption of the few brown algae that store iAs. The National Research Council (USA) recommends data-driven statistical approaches vs. linear extrapolation to estimate low-dose As effects from studies where higher doses are used (NRC 2013). As Nakamura et al.

(2008) pointed out, calculations of excess cancer risk assume that the mechanism of carcinogenesis has no threshold dose; i.e., the incidence of cancer is linearly related to As intake in a low dose range, and this may not be the case. Inorganic As and arsenosugar content of hijiki varies geographically and with manufacturing method (Shimoda et al. 2010). Canada and the UK advise consumers to avoid eating hijiki (CFIA 2012; UK FSA 2016).

By estimating As bioaccessibility through the human GI tract using modifications of methods developed by nutritionists for estimating Fe uptake, Garcia-Sartal et al. (2012) concluded that only a fraction (7–15%) of the inorganic As and arsenosugars in cooked algae such as kombu, wakame, nori, and sea lettuce is bioaccessible. More research is needed to understand how different cooking processes, the particular algal food matrix, and the gut microbiome modify arsenosugar bioaccessibility and then better definition of the interaction of other As species produced in the body with cellular metabolites and macromolecules before their excretion (NRC 2013; Molin et al 2015; Carlin et al 2016). Experiments at all levels are important but it is particularly important to move beyond tests of toxicity using in vitro cell cultures in order to understand whether there is any risk of consuming any sea vegetable. The quantities of bioaccessible arsenic as iAs and/or arsenosugars in most sea vegetables appear too low to pose risk to individuals unless there is co-exposure to much higher levels of arsenic through high iAs-drinking water ($\geq 10\ \mu g\ L^{-1}$) or foods or environments contaminated by As pollutants.

Bromine

Less well-recognized toxic effects can arise from excess intake of sea salt minerals, such as bromine, which can cause nerve, DNA, and organ damage in mammals (e.g., Boyer et al. 2002). Bromine concentrations in the urine of human female subjects in China, Japan, and Korea have been shown to correlate with their seafood intake, with macroalgae apparently being a major source (Kawai et al. 2002). However, this correlation was skewed by the consumption of terrestrial crops treated with methyl bromide, which can significantly increase total bromine intake. It is an important reminder for the need to quantify total exposure of an individual to metal intake to evaluate the effects of algal consumption.

Allergenicity and Macroalgal Toxins

Relatively little has been published on the subject of possible allergenicity of algae and their products. A young man developed anaphylaxis after the first-time consumption of a spirulina tablet (Le et al. 2014), but the source of the spirulina in this tablet, or its purity, was not given. Well-known episodes of human poisoning events have occurred after consumption of wild-harvested spirulina that contained *Microcystis* and other freshwater cyanobacteria that produce neurotoxins and hepatotoxins. This emphasizes the importance of developing large-scale, controlled cultivation and daily testing of future supplement crops (e.g., Gellenbeck 2012), similar to those used now by companies that supply large quantities of GRAS-spirulina (e.g., Cyanotech, Earthrise Farms) to the food and supplement markets (Belay 2008). A dried, milled protein preparation, Whole Algalin Protein, from *Chlorella protothecoides* produced by Solazyme, Inc. was demonstrated to be unlikely to cause food allergies in nutritional studies with rats (Szabo et al. 2013).

The amino acid kainic acid, which is found in dulse and some other red algae (e.g., *Digenia simplex*), is structurally similar to glutamate, a neurotransmitter in the brain. At high doses, kainic acid is neurotoxic and used experimentally to produce disease models in mice and other animals. Concentrations of kainic acid that damage neurons are much higher than those consumed by eating dulse, but Mouritsen et al. (2013) report that no human safety standard has been established. This is both a dose and a bioavailability issue, and it needs more study because a few dwarf individuals of *Palmaria palmata* had high levels of kainic acid (>10 mg g^{-1}dry wt, Ramsey et al. 1994), even though it was undetectable or at very low level in most Atlantic dulse (Ramsey et al. 1994; Higa and Kuniyoshi 2000; Mouritsen et al. 2013). Another amino acid in the kainoid family is domoic acid (DA), which is also found at low levels in some red algae (e.g., *Chondria armata*, in the same Rhodomelaceae as *Digenia simplex*, the first discovered kainic acid-producing red alga). DA is a strong health risk during blooms of a few diatoms (e.g., *Pseudo-nitzschia*) that are bioaccumulated by filter-feeding mollusks and become a human health risk, but DA poisoning from red algae is unknown (Higa and Kuniyoshi 2000). Traditionally in both Europe and Asia, worms were eliminated from humans and animals with red algae that contain KA or DA (Mouritsen 2013).

Many *Gracilaria* (Rhodophyta) and *Caulerpa* (Chlorophyta) species are eaten as sea vegetables, especially in the western Pacific (de Gaillande et al. 2016), but illness and some deaths have occurred when a few toxic species of these genera were sold or collected by mistake (Higa and Kuniyoshi 2000; Cheney 2016). Toxic prostaglandins (PGE_2) are found in the Asian *Gracilaria vermiculophylla*, which invaded North American and European shores in recent decades (Hammann et al. 2016). Wounding of *G. vermiculophylla* increases synthesis of $PGE_2$from arachidonic acid within minutes (Noguchi et al. 1994; Nylund et al. 2011). Hammann et al. (2016) showed that this effect was enhanced in invasive populations, finding that this appears to be a defense against naïve herbivores in the non-native habitat. Their study characterized many additional metabolites present in *G. vermiculophylla*. Fatalities have resulted from consumption of fresh *Gracilaria edulis* (*Polycavernosa tsudai* in Navarro et al. 2015), due to toxic polycavernosides (Yotsu-Yamashita et al. 2007). Navarro et al. (2015)'s recent analysis suggests, however, that the polycavernoside is produced by filamentous cyanobacteria (*Okeania* sp.) that are sometimes associated with the *Gracilaria*.

SUMMARY

The considerable breadth and depth of literature on algae as nutritional and functional foods frustrates attempts for a fully comprehensive assessment of the field. It is clear that there is substantial evidence for algae as nutritional and functional foods, yet there remain considerable challenges in quantifying these benefits, and in assessing potential adverse effects. The limits to our understanding fall broadly into three areas. First is the variation of nutritional and functional composition of algae across species, seasons, and different coastal environments. The scant evidence to date suggests this variability can be substantial but it is only possible to speculate about the scope of this inconsistency. That is true also for quantification of toxic, or potentially harmful constituents present within, or adsorbed to algae. Assessments should also consider the effect of processing methods, which can increase or decrease the nutritional quality. Of the challenges ahead, these issues are the most tractable to address, given that the analytical methods are well developed.

The second, and perhaps most pressing limitation, is quantifying the bioavailability, or fraction of nutritional or functional components that actually have effect in relation to their residence time in the digestive system. These effects can manifest via translocation across the small intestinal epithelial cells into the blood, by direct interactions with the digestive epithelia, by altering uptake of other substances, by regulating the microbial consortium, or by direct contact with colonic epithelial cells in the large intestine (e.g., colon cancer). There is an increasing literature on digestive "reactor" analytical methods, and much more effort in this direction would be beneficial, but it must be recognized that bioaccessibility will be a complex function of the chemistry of the substance, the processing methods used to prepare the alga as food, the specific algal matrix containing it, the consortium of bacteria and their enzymatic competency, and the presence of other foods that may interfere or enhance uptake (or within-gut effects). Also necessary is more rigor in semantics; bioavailable, bioactive, activity, digestible, gastrointestinal absorption, and utilization are not equivalent terms but are used interchangeably in the literature. Advances in understanding bioavailability of foods in general will continue piecemeal until analytical methods and studies encompassing all of these factors become routine.

The third limitation lies in understanding how algal nutritional, and particularly functional, constituents interact in human metabolism and intermediary metabolic processes. Most investigators studying these questions report in vitro experiments or in vivo experiments employing direct introduction of purified algal constituents. These methods are useful probes to identify and mechanistically understand *potential* effects of the consumption of algal-based supplements, but they are inadequate to assess nutritional and functional *foods*. Their shortcomings most often include unrealistic or uncertain doses relative to normal algal consumption, a focus on a single or narrow range of metabolic responses, and inability to assess concurrent, or synergistic effects arising from co-occurring constituents during digestion. Coupling in vivo studies with more holistic digestive research will provide for a better assessment of algae as nutritional and functional food sources.

We envisage a rare opportunity to develop a rich and rewarding collaboration among phycological, nutritional, medical, analytical, and industrial groups investigating algae as nutritional and functional foods. Part of the challenge ahead for algal scientists is understanding the complexity of merging of basic research through clinical trials and regulatory requirements to create marine food products. The recent comprehensive reviews by Finley et al. (2014) and Borowitzka (2013b) help to clarify this pathway. But the most dramatic advances will require a rethinking of experimental and collaborative approaches, and the impetus for this research will only increase as human pressures on the climate system lead us to turn more to the oceans for food that we can harvest and grow sustainably.

ACKNOWLEDGMENTS

We thank Eva Bradford for clerical assistance in the initial stages of our literature review, and to Dr. Yoichi Sato (Riken Food Co., Ltd.) and Dr. Keiko Yoshinaga (Riken Vitamine Co, Ltd.) for providing us the most recent data on Japanese macroalgal food consumption from The National Health and Nutrition Survey in Japan, 2004–2014 (MHLW 2014). We appreciate the valuable insights Dr. Santosh P. Lall, National Research Council of Canada, provided to this work. We also thank the anonymous reviewers and Editor for their helpful comments. AGS & KEH thank the Biotechnology and Biological Sciences Research Council (BBSRC BB/1013164/1) of the UK for funding. The University of Dundee is a registered Scottish charity, No. SC015096. PP is supported by IDEALG in the frame of the stimuli program entitled "Investissements d'avenir, Biotechnologies-Bioressources" (ANR-10-BTBR-04-02). The open access fee was supported by NSF-OCE-1435021 (MLW), DIC project 1823-06 (MEC), Maine Sea Grant (NOAA) 5405971 (SHB), NSF #11A-1355457 to Maine EPSCoR at the University of Maine (SHB), and the listed funding to AGS and PP.

REFERENCES

1. Aaronson S (1986) A role for algae as human food in antiquity. Food Foodways 1:311–315
2. Aarstad OA, Tondervik A, Sletta H, Skjak-Braek G (2012) Alginate sequencing: an analysis of block distribution in alginates using specific alginate degrading enzymes. Biomacromolecules 13:106–116
3. Abdul QA, Choi RJ, Jung HA, Choi JS (2016) Health benefit of fucosterol from marine algae: a review. J Sci Food Agric 96:1856–1866

4. Admassu H, Zhao W, Yang R, Gasmalla M, Alsir E (2015) Development of functional foods: seaweeds (algae) untouched potential and alternative resource—a review. Int J Sci Technol Res 4:108–115
5. Albanes D et al (1995) Effects of alpha-tocopherol and beta-carotene supplements on cancer incidence in the alpha-tocopherol beta-carotene cancer prevention study. Am J Clin Nutr 62:S1427–S1430
6. Allen LH (2008) Causes of vitamin B_{12} and folate deficiency. Food Nutr Bull 29:S20–S34
7. Al-Masri MS, Mamish S, Budier Y (2003) Radionuclides and trace metals in eastern Mediterranean sea algae. J Environ Radioact 67:157–168
8. Almela C, Clemente MJ, Velez D, Montoro R (2006) Total arsenic, inorganic arsenic, lead and cadmium contents in edible seaweed sold in Spain. Food Chem Toxicol 44:1901–1908
9. Alves A, Sousa RA, Reis RL (2013) A practical perspective on ulvan extracted from green algae. J Appl Phycol 25:407–424
10. Anastyuk SD, Shevchenko NM, Ermakova SP, Vischuk OS, Nazarenko EL, Dmitrenok PS, Zvyagintseva TN (2012) Anticancer activity in vitro of a fucoidan from the brown alga *Fucus evanescens* and its low-molecular fragments, structurally characterized by tandem mass-spectrometry. Carbohydr Polym 87:186–194
11. Angell AR, Mata L, de Nys R, Paul NA (2016) The protein content of seaweeds: a universal nitrogen-to-protein conversion factor of five. J Appl Phycol 28:511–524
12. Aquino RS, Landeira-Fernandez AM, Valente AP, Andrade LR, Mourão PAS (2005) Occurrence of sulfated galactans in marine angiosperms: evolutionary implications. Glycobiology 15:11–20
13. Arterburn LM, Oken HA, Hoffman JP, Bailey-Hall E, Chung G, Rom D, Hamersley J, McCarthy D (2007) Bioequivalence of docosahexaenoic acid from different algal oils in capsules and in a DHA-fortified food. Lipids 42:1011–1024
14. Arterburn LM, Oken HA, Hall EB, Hamersley J, Kuratko CN, Hoffman JP (2008) Algal-oil capsules and cooked salmon: nutritionally equivalent sources of docosahexaenoic acid. J Am Diet Assoc 108:1204–1209

15. Arul AB, Al Numair K, Al Saif M, Savarimuthu I (2012) Effect of dietary beta-sitosterol on fecal bacterial and colonic biotransformation enzymes in 1,2-dimethylhydrazine-induced colon carcinogenesis. Turk J Med Sci 42:1307–1313

16. Baba M, Snoeck R, Pauwels R, deClerq E (1988) Sulfated polysaccharides are potent and selective inhibitors of various enveloped viruses, including herpes simplex virus, cytomegalovirus, vesicular stomatitis virus and human immunodeficiency virus. Antimicrob Agents Chemother 32:1742–1745

17. Bäckhed F, Ley RE, Sonnenburg JL, Peterson DA, Gordon JI (2005) Host-bacterial mutualism in the human intestine. Science 307:1915–1920

18. Bagchi D (2006) Nutraceuticals and functional foods regulations in the United States and around the world. Toxicology 221:1–3

19. Barahona T, Encinas MV, Mansilla A, Matsuhiro B, Zuniga EA (2012) A sulfated galactan with antioxidant capacity from the green variant of tetrasporic *Gigartina skottsbergii* (Gigartinales, Rhodophyta). Carbohydr Res 347:114–120

20. Barclay W, Weaver C, Metz J, Hansen J (2010) Development of docosahexaenoic acid production technology using *Schizochytrium*. In: Cohen Z, Ratledg C (eds) Single cell oils. AOCS Press, Urbana, pp 75–96

21. Barclay W, Apt K, Dong X (2013) Commercial production of microalgae via fermentation. In: Richmond A, Hu Q (eds) Handbook of microalgal culture: applied phycology and biotechnology. Blackwell, Chichester, pp 134–145

22. Barlow S, Pascal G, Cressey P, Folmer D, Smith J (2015) Carrageenan (addendum). In: Safety evaluation of certain food additives, WHO food additives series, vol 70. World Health Organization, Geneva

23. Belay A (1997) Mass culture of Spirulina outdoors—The Earthrise Farms experience. In: Vonshak A (ed) *Spirulina platensis* (*Arthrospira*): physiology, cell-biology and biotechnology. Taylor & Francis Ltd., London, pp 131–158

24. Belay A (2008) *Spirulina* (*Arthrospira*): production and quality assurance. In: Gershwin ME, Belay A (eds) *Spirulina* in human nutrition and health. CRC Press, Boca Raton, pp 1–25

25. Ben-Amotz A, Levy Y (1996) Bioavailability of a natural isomer mixture compared with synthetic all-trans beta-carotene in human serum. Am J Clin Nutr 63:729–734

26. Ben-Amotz A, Yatziv S, Sela M, Greenberg S, Rachmilevich B, Shwarzman M, Weshler Z (1998) Effect of natural beta-carotene supplementation in children exposed to radiation from the Chernobyl accident. Radiat Environ Biophys 37:187–193

27. Benjama O, Masniyom P (2012) Biochemical composition and physicochemical properties of two red seaweeds (*Gracilaria fisheri* and *G. tenuistipitata*) from the Pattani Bay in southern Thailand. Songklanakarin J Sci Technol 34:223–230

28. Benkdad A, Laissaoui A, Tornero MV, Benmansour M, Chakir E, Garrido IM, Moreno JB (2011) Trace metals and radionuclides in macroalgae from Moroccan coastal waters. Environ Monit Assess 182:317–324

29. Bernstein AM, Ding EL, Willett WC, Rimm EB (2012) A meta-analysis shows that docosahexanaenoic acid from algal oil reduces serum triglycerides and increases HDL-cholesterol and LDL-cholesterol in persons without coronary heart disease. J Nutr 142:99–104

30. Berteau O, Mulloy B (2003) Sulfated fucans, fresh perspectives: structures, functions, and biological properties of sulfated fucans and an overview of enzymes active toward this class of polysaccharide. Glycobiology 13:29R–40R

31. Bhattacharjee H, Mukhopadhyay R, Thiyagarajan S, Rosen BP (2008) Aquaglyceroporins: ancient channels for metalloids. J Biol 7:33.31–33.36

32. Bhattacharyya S, Liu H, Zhang Z, Jam M, Dudeja PK, Michel G, Linhardt RJ, Tobacman JK (2010) Carrageenan-induced innate immune response is modified by enzymes that hydrolyze distinct galactosidic bonds. J Nutr Biochem 21:906–913

33. Bhattacharyya S, O-Sullivan I, Katyal S, Unterman T, Tobacman JK (2012) Exposure to the common food additive carrageenan leads to glucose intolerance, insulin resistance and inhibition of insulin signaling in HepG2 cells and C57BL/6J mice. Diabetologia 55:194–203

34. Black WAP (1950) The seasonal variation in weight and chemical composition of the common British Laminariaceae. J Mar Biol Assoc UK 29:45–72
35. Blouin N, Calder BL, Perkins B, Brawley SH (2006) Sensory and fatty acid analyses of two Atlantic species of *Porphyra* (Rhodophyta). J Appl Phycol 18:79–85
36. Bobin-Dubigeon C, Lahaye M, Barry J-L (1997) Human colonic bacterial degradability of dietary fibres from sea-lettuce (*Ulva* sp). J Sci Food Agric 73:149–159
37. Bocanegra A, Bastida S, Benedí J, Ródenas S, Sánchez-Muniz FJ (2009) Characteristics and nutritional and cardiovascular-health properties of seaweeds. J Med Food 12:236–258
38. Boehlke C, Zierau O, Hannig C (2015) Salivary amylase—the enzyme of unspecialized euryphagous animals. Arch Oral Biol 60:1162–1176
39. Bohn JA, BeMiller JN (1995) (1→3)-β-D-Glucans as biological response modifiers: a review of structure-functional activity relationships. Carbohydr Polym 28:3–14
40. Bohn T, McDougall GJ, Alegría A, Alminger M, Arrigoni E, Aura A-M, Brito C, Cilla A, El SN, Karakaya S, Martínez-Cuesta MC, Santos CN (2015) Mind the gap—deficits in our knowledge of aspects impacting the bioavailability of phytochemicals and their metabolites—a position paper focusing on carotenoids and polyphenols. Mol Nutr Food Res 59:1323
41. Borowitzka MA (2013a) High-value products from microalgae—their development and commercialisation. J Appl Phycol 25:743–756
42. Borowitzka M (2013b) *Dunaliella*: biology, production, and markets. In: Richmond A, Hu Q (eds) Handbook of microalgal culture. John Wiley & Sons, Ltd, Chichester, pp 359–368
43. Bourgougnon N (2003) Anti-HIV compounds from red seaweeds. Biomaterials and bioprocessing, Vol. 9. Recent advances in marine biotechnology. Science Publishers, Enfield, pp 165–206
44. Bourgougnon N, Stiger-Pouvreau V (2011) Chemodiversity and bioactivity within red and brown macroalgae along the French coasts, metropol and overseas departments and territories. In: Kim

S-E (ed) Handbook of marine macroalgae: biotechnology and applied phycology. John Wiley & Sons, Chichester, pp 58–105

45. Boyer EW, Kearney S, Shannon MW, Quang L, Woolf A, Kemper K (2002) Poisoning from a dietary supplement administered during hospitalization. Pediatrics 109:e49
46. Branger B, Cadudal JL, Delobel M, Ouoba H, Yameogo P, Ouedraogo D, Guerin D, Valea A, Les Personnels des C, Zombre C, Ancel P (2003) Spiruline as a food supplement in case of infant malnutrition in Burkina-Faso (La spiruline comme complement alimentaire dans la malnutrition du nourrisson au Burkina-Faso.). Arch Pediatr 10:424–431
47. Braud J (2006) Continuous seaweed tank culture in France: from *Chondrus crispus* to co-culture of macroalge and the diatom *Odontella aurita*. In: Critchley A, Ohno M, Largo D (eds) Seaweed resources of the world. Int. Cooperation Agency, Japan
48. Britton G, Liaaen-Jensen S, Pfander PH (2004) Carotenoids: handbook. Birkhauser Verlag AG, Switzerland
49. Brown MR (1995) Effects of storage and processing on the ascorbic-acid content of concentrates prepared from *Chaetoceros calcitrans*. J Appl Phycol 7:495–500
50. Brown MT, Hodgkinson WM, Hurd CL (1998) Spatial and temporal variations in the copper and zinc concentrations of two green seaweeds from Otago Harbour, New Zealand. Mar Environ Res 47:1–10
51. Brownlee IA, Allen A, Pearson JP, Dettmar PW, Havler ME, Atherton MR, Onsoyen E (2005) Alginate as a source of dietary fiber. Crit Rev Food Sci Nutr 45:497–510
52. Butler MR (1936) Seasonal variations in *Chondrus crispus*. Biochem J 30:1338–1344
53. Cabrita ARJ, Maia MRG, Oliveira HM, Sousa-Pinto I, Almeida AA, Pinto E, Fonseca AJM (2016) Tracing seaweeds as mineral sources for farm-animals. J Appl Phycol 28:3135–3150
54. Canadian Food Inspection Agency (2012) Inorganic arsenic and hijiki seaweed consumption. http://www.inspection.gc.ca/food/information-for-consumers/fact-sheets-and-infographics/specific-products-andrisks/chemical-hazards/inorganic-arsenic/eng/133226

8146718/1332268231124. (accessed 11/10/16)

55. Cantarel BL, Lombard V, Henrissat B (2012) Complex carbohydrate utilization by the healthy human microbiome. PLoS One 7:e28742
56. Carbonell-Capella JM, Buniowska M, Barba FJ, Esteve MJ, Frigola A (2014) Analytical methods for determining bioavailability and bioaccessibility of bioactive compounds from fruits and vegetables: a review. Compr Rev Food Sci Food Saf 13:155–171
57. Carlin DJ, Naujokas MF, Bradham KD, Cowden J, Heacock M, Henry HF, Lee JS, Thomas DJ, Thompson C, Tokar EJ, Waalkes MP, Birnbaum LS, Suk WA (2016) Arsenic and environmental health: state of the science and future research opportunities. Environ Health Perspect 124:890–899
58. Cases J, Vacchina V, Napolitano A, Caporiccio B, Besançon P, Lobinski P, Rouanet J-M (2001) Selenium from selenium-rich *Spirulina* is less bioavailable than selenium from sodium selenite and selenomethionine in selenium-deficient rats. J Nutr 131:2343–2350
59. Chacón-Lee TL, González-Mariño GE (2010) Microalgae for "healthy" foods—possibilities and challenges. Comp Rev Food Sci Food Safety 9:655–675
60. Chan JCC, Cheung PCK, Ang PO (1997) Comparative studies on the effect of three drying methods on the nutritional composition of seaweed *Sargassum hemiphyllum* (Turn) C Ag. J Agric Food Chem 45:3056–3059
61. Cheney D (2016) Toxic and harmful seaweeds. In: Fleurence J, Levine I (eds) Seaweeds in health and disease. Elsevier, Amsterdam, pp 407–421
62. Cheng F-C, Lin A, Feng J-J, Mizoguchi T, Takekoshi H, Kubota H, Kato Y, Naoki Y (2004) Effects of *Chlorella* on activities of protein tyrosine phosphatases, matrix metalloproteinases, caspases, cytokine release, B and T cell proliferations, and phorbol ester receptor binding. J Med Food 7:146–152
63. Choi BS, Choi SJ, Kim DW, Huang M, Kim NY, Park KS, Kim CY, Lee HM, Yum YN, Han ES, Kang TS, Yu IJ, Park JD (2010) Effects of repeated seafood consumption on urinary excretion of arsenic species by volunteers. Arch Environ Contam Toxicol 58:222–229

64. Chrismadha T, Borowitzka MA (1994) Effect of cell-density and irradiance on growth, proximate composition and eicosapentaenoic acid production of *Phaeodactylum tricornutum*grown in a tubular photobioreactor. J Appl Phycol 6:67–74

65. Cian RE, Drago SR, de Medina FS, Martinez-Augustin O (2015) Proteins and carbohydrates from red seaweeds: evidence for beneficial effects on gut function and microbiota. Mar Drugs 13:5358–5383

66. Ciferri O (1983) *Spirulina*, the edible microorganism. Microbiol Rev 47:551–578

67. Cofrades S, López-López I, Solas MT, Bravo L, Jiménez-Colmenero F (2008) Influence of different types and proportions of added edible seaweeds on characteristics of low-salt gel/emulsion meat systems. Meat Sci 79:767–776

68. Conquer JA, Holub BJ (1996) Supplementation with an algae source of docosahexaenoic acid increases (n-3) fatty acid status and alters selected risk factors for heart disease in vegetarian subjects. J Nutr 126:3032–3039

69. Consumer-Reports (2012) Arsenic in your juice. Consumer reports magazine, January. ConsumerReports.org

70. Cooney RV, Mumma RO, Benson AA (1978) Arsonium phospholipid in algae. Proc Natl Acad Sci U S A 75:4262–4264

71. Cornish ML, Garbary DJ (2010) Antioxidants from macroalgae: potential applications to human health and nutrition. Algae 25:155–171

72. Cornish ML, Critchley AT, Mouritsen OG (2015) A role for dietary macroalgae in the amelioration of certain risk factors associated with cardiovascular disease. Phycologia 54:649–666

73. Corona G, Ji Y, Anegboonlap P, Hotchkiss S, Gill C, Yaqoob P, Spencer JPE, Rowland I (2016) Gastrointestinal modifications and bioavailability of brown seaweed phlorotannins and effects on inflammatory markers. Brit J Nutr 115:1240–1253

74. Costa S, Crespo D, Henriques BMG, Pereira E, Duarte AC, Pardal AC (2011) Kinetics of mercury accumulation and its effects on *Ulva lactuca* growth rate at two salinities and exposure conditions. Water Air Soil Pollut 217:689–699

75. Costello EK, Stagaman K, Dethlefsen L, Bohannan BJM, Relman DA (2012) The application of ecological theory toward an understanding of the human microbiome. Science 336:1255–1261

76. Cottin SC, Sanders TA, Hall WL (2011) The differential effects of EPA and DHA on cardiovascular risk factors. Proc Nutr Soc 70:215–231

77. Courtois J (2009) Oligosaccharides from plants and algae: production and applications in therapeutics and biotechnology. Opin Microbiol 12:261–273

78. Craigie JS (2010) Seaweed extract stimuli in plant science and agriculture. J Appl Phycol 23:371–393

79. Croci DO, Cumashi A, Ushakova NA, Preobrazhenskaya ME, Piccoli A, Totani L, Ustyuzhanina NE, Bilan MI, Usov AI, Grachev AA, Morozevich GE, Berman AE, Sanderson CJ, Kelly M, Gregorio PD, Rossi C, Tinari N, Iacobelli S, Rabinovich GA, Nifantiev NE (2011) Fucans, but not fucomannanoglucuronans, determine the biological activities of sulfated polysaccharides from *Laminaria saccharina* brown seaweed. PLoS One 6:e17283

80. Croft MT, Lawrence AD, Raux-Deery E, Warren MJ, Smith AG (2005) Algae acquire vitamin B_{12} through a symbiotic relationship with bacteria. Nature 438:90–93

81. Croft MT, Warren MJ, Smith AG (2006) Algae need their vitamins. Eukaryot Cell 5:1175–1183

82. Cumashi A, Ushakova NA, Preobrazhenskaya ME, D'Incecco A, Piccoli A, Totani L, Tinari N, Morozevich GE, Berman AE, Bilan MI, Usov AI, Ustyuzhanina NE, Grachev AA, Sanderson CJ, Kelly M, Rabinovich GA, Iacobelli S, Nifantiev NE (2007) A comparative study of the anti-inflammatory, anticoagulant, antiangiogenic, and antiadhesive activities of nine different fucoidans from brown seaweeds. Glycobiology 17:541–552

83. Cysewski G, Lorenz R (2004) Industrial production of microalgal cell-mass and secondary products-species of high potential: *Haematococcus*. In: Richmond A (ed) Microalgal culture: biotechnology and applied phycology. Blackwell Science, Oxford, pp 281–288

84. Dagnelie PC, Vanstaveren WA, Vandenberg H (1991) Vitamin B-12

from algae appears not to be bioavailable. Am J Clin Nutr 53:695–697

85. Dawczynski C, Schaefer U, Leiterer M, Jahreis G (2007) Nutritional and toxicological importance of macro, trace, and ultra-trace elements in algae food products. J Agric Food Chem 55:10470–10475
86. de Gaillande C, Payri C, Remoissenet G, Zubia M (2016) *Caulerpa* consumption, nutritional value and farming in the Indo-Pacific region. J Appl Phycol. doi:10.1007/s10811-016-0912-6:1-18
87. de Jesus Raposo FM, Bernado de Morais AM, Santos Costa de Morais RM (2015) Marine polysaccharides from algae with potential biomedical applications. Mar Drugs 13:2967–3028
88. de la Calle MB, Baer I, Robouch P, Cordeiro F, Emteborg H, Baxter MJ, Brereton N, Raber G, Velez D, Devesa V, Rubio R, Llorente-Mirandes T, Raab A, Feldmann J, Sloth JJ, Rasmussen RR, D'Amato M, Cubadda F (2012) Is it possible to agree on a value for inorganic arsenic in food? The outcome of IMEP-112. Anal Bioanal Chem 404:2475–2488
89. De Marco ER, Steffolani ME, Martinez CS, Leon AE (2014) Effects of spirulina biomass on the technological and nutritional quality of bread wheat pasta. LWT-Food Sci Technol 58:102–108
90. de Oliveira MN, Freitas ALP, Carvalho AFU, Sarnpaio TMT, Farias DF, Teixeira DIA, Gouveia ST, Pereira JG, de Sena M (2009) Nutritive and non-nutritive attributes of washed-up seaweeds from the coast of Ceara, Brazil. Food Chem 115:254–259
91. Delgado-Noguera MF, Calvache JA, Cosp XB, Kotanidou EP, Galli-Tsinopoulou A (2015) Supplementation with long chain polyunsaturated fatty acids (LCPUFA) to breastfeeding mothers for improving child growth and development. Cochrane Database Syst Rev. doi:10.1002/14651858.CD007901.pub3:72
92. Devesa V, Velez D, Montoro R (2008) Effect of thermal treatments on arsenic species contents in food. Food Chem Toxicol 46:1–8
93. Dhargalkar V (2015) Uses of seaweeds in the Indian diet for sustenance and well-being. Sci Cult 80:192–202
94. Dharmananda (2002) The nutritional and medicinal value of seaweeds used in Chinese medicine (2002) Institute for Traditional

Medicine. http://www.imtonline.org/. Accessed 14 Oct 2016

95. Diaz O, Tapia Y, Munoz O, Montoro R, Velez D, Almela C (2012) Total and inorganic arsenic concentrations in different species of economically important algae harvested from coastal zones of Chile. Food Chem Toxicol 50:744–749
96. Dillehay TD, Ramirez C, Pino M, Collins MB, Rossen J, Pinot-Navarro JD (2008) Monte Verde: seaweed, food, medicine and the peopling of South America. Science 320:784–789
97. Druehl LD (1988) Cultivated edible kelp. In: Lembi CA, Waaland JR (eds) Algae and human affairs. Cambridge University Press, Cambridge, pp 119–134
98. Druesne-Pecollo N, Latino-Martel P, Norat T, Barrandon E, Bertrais S, Galan P, Hercberg S (2010) Beta-carotene supplementation and cancer risk: a systematic review and metaanalysis of randomized controlled trials. Int J Cancer 127:172–184
99. Duffy LC, Raiten DJ, Hubbard VS, Starke-Reed P (2015) Progress and challenges in developing metabolic footprints from diet in human gut microbial cometabolism. J Nutr 145:1123S–1130S
100. Durand M, Beaumatin P, Bulman B, Bernalier A, Grivet JP, Serezat M, Gramet G, Lahaye M (1997) Fermentation of green alga sea-lettuce (*Ulva* sp.) and metabolism of its sulphate by human colonic microbiota in a semi-continuous culture system. Reprod Nutr Dev 37:267–283
101. Eccles R, Meier C, Jawad M, Weinmüllner R, Grassauer A, Prieschl-Grassauer E (2010) Efficacy and safety of an antiviral iotacarrageenan nasal spray: a randomized, double-blind placebo-controlled exploratory study in volunteers with early symptoms of the common cold. Respir Res 11:108. doi:10.1186/1465-9921-11-108
102. EFSA (2010) Scientific opinion on dietary reference values for carbohydrates and dietary fibre. Eur Food Safety Auth J 8:1–77
103. Eyster KM (2007) The membrane and lipids as integral participants in signal transduction: lipid signal transduction for the non-lipid biochemist. Adv Physiol Educ 31:5–16
104. Fabregas J, Herrero C (1990) Vitamin content of 4 marine microalgae—potential use as a source of vitamins in nutrition. J

Indust Microbiol 5:259–264

105. Fahy E, Subramaniam S, Brown HA, Glass CK, Merrill AH, Murphy RC, Raetz CRH, Russell DW, Seyama Y, Shaw W, Shimizu T, Spener F, van Meer G, VanNieuwenhze MS, White SH, Witztum JL, Dennis EA (2005) A comprehensive classification system for lipids. J Lipid Res 46:839–861
106. FAO (2014) The state of the world fisheries and aquaculture 2014. FAO, Rome, p 223
107. FAO (2015) FAO Global Aquaculture Production database updated to 2013—Summary information. FAO, Rome
108. FAO (2016) The State of the World Fisheries and Aquaculture 2016. Contribution to Food Security and Nutrition for All. Rome. 200 pp
109. FDA (2016) Summary: Substances Generally Regarded As Safe (Final Rule). U.S. Food and Drug Administration (HHS). http://www.fda.gov/AboutFDA/ReportsManualsForms/Reports/EconomicAnalyses/ucm517103.htm. Accessed 24 Oct 2016
110. Fernando WMADB, Ranaweera KKDS, Bamunuarachchi A, Brennan CS (2008) The influence of rice fibre fractions on the in vitro fermentation production of short chain fatty acids using human faecal micro flora. Int J Food Sci Technol 43:2237–2244
111. Ferraces-Casais P, Lage-Yusty MA, de Quiros ARB, Lopez-Hernandez J (2012) Evaluation of bioactive compounds in fresh edible seaweeds. Food Anal Methods 5:828–834
112. Finley JW, Finley JW, Ellwood K, Hoadley J (2014) Launching a new food product or dietary supplement in the United States: industrial, regulatory, and nutritional considerations. Annu Rev Nutr 34:421–447
113. Fitton JH (2011) Therapies from fucoidan; multifunctional marine polymers. Mar Drugs 9:1731–1760
114. Fitton J, Dell'Acqua G, Gardiner V-A, Karpiniec S, Stringer D, Davis E (2015) Topical benefits of two fucoidan-rich extracts from marine macroalgae. Cosmetics 2:66–81
115. Fleurence J (1999a) The enzymatic degradation of algal cell walls: a useful approach for improving protein accessibility? J Appl Phycol 11:313–314

116. Fleurence J (1999b) Seaweed proteins: biochemical, nutritional aspects and potential uses. Trends Food Sci Technol 10:25–28

117. Fleurence J, Levine I (eds) (2016) Seaweed in health and disease prevention. Elsevier, Amsterdam, pp 476

118. Foster P (1976) Concentrations and concentration factors of heavy metals in brown algae. Environ Pollut 10:45–54

119. Fournier E, Adam C, Massabuau J-C, Garnier-Laplace J (2005) Selenium bioaccumulation in *Chlamydomonas reinhardtii* and subsequent transfer to *Corbicula fluminea*: role of selenium speciation and bivalve ventilation. Environ Toxicol Chem 25:2692–2699

120. Francesconi KA (2010) Arsenic species in seafood: origin and human health implications. Pure Appl Chem 82:373–381

121. Francois L, Fortin C, Campbell PGC (2007) pH modulates transport rates of manganese and cadmium in the green alga *Chlamydomonas reinhardtii* through non-competitive interactions: implications for an algal BLM. Aquat Toxicol 84:123–132

122. Fujiwara-Arasaki T, Mino N, Kuroda M (1984) The protein value in human nutrition of edible marine algae in Japan. Hydrobiologia 116/117:513–516

123. Gadd GM (2009) Biosorption: critical review of scientific rationale, environmental importance and significance for pollution treatment. J Chem Technol Biotechnol 84:13–28

124. Galland-Irmouli A-V, Fleurence J, Lamghari R, Lucon M, Rouxel C, Barbaroux O, Bronowicki J-P, Villaume C, Gueant J-L (1999) Nutritional value of proteins from edible seaweed *Palmaria palmata* (Dulse). J Nutr Biochem 10:353–359

125. Ganesan P et al (2011) Siphonaxanthin, a marine carotenoid from green algae, effectively induces apoptosis in human leukemia (HL-60) cells. Biochim Biophys Acta - Gen Subj 1810:497–503

126. Gantar M, Svircev Z (2008) Microalgae and cyanobacteria: food for thought. J Phycol 44:260–268

127. Garcia-Casal MN, Pereira AC, Leets I, Ramirez J, Quiroga ME (2007) High iron content and bioavailability in humans from four species of marine algae. J Nutr 137:2691–2695

128. Garcia-Casal MN, Ramirez J, Leets I, Pereira AC, Quiroga MF (2009) Antioxidant capacity, polyphenol content and iron bioavailability from algae (*Ulva* sp., *Sargassum* sp and *Porphyra*sp.) in human subjects. Br J Nutr 101:79–85

129. Garcia-Salgado S, Raber G, Raml R, Magnes C, Francesconi KA (2012) Arsenosugar phospholipids and arsenic hydrocarbons in two species of brown macroalgae. Environ Chem 9:63–66

130. Garcia-Sartal C, Taebunpakul S, Stokes E, Barciela-Alonso M, Bermejo-Barrera P, Goenaga-Infante H (2012) Two-dimensional HPLC coupled to ICP-MS and electrospray ionisation (ESI)-MS/MS for investigating the bioavailability in vitro of arsenic species from edible seaweed. Anal Bioanal Chem 402:3359–3369

131. Garnham GW, Codd GA, Gadd GM (1992) Kinetics of uptake and intracellular location of cobalt, manganese and zinc in the estuarine green-alga *Chlorella salina*. Appl Microbiol Biotechnol 37:270–276

132. Gellenbeck KW (2012) Utilization of algal materials for nutraceutical and cosmeceutical applications—what do manufacturers need to know? J Appl Phycol 24:309–313

133. Georgianna DR, Mayfield SP (2012) Exploiting diversity and synthetic biology for the production of algal biofuels. Nature 488:329–335

134. Geppert J, Kraft V, Demmelmair H, Koletzko B (2005) Docosahexaenoic acid supplementation in vegetarians effectively increases omega-3 index: a randomized trial. Lipids 40:807–814

135. Gobler CJ, Berry DL, Dyhrman ST, Wilhelm SW, Salamov A, Lobanov AV, Zhang Y, Collier JL, Wurch LL, Kustka AB, Dill BD, Shah M, VerBerkmoes NC, Kuo A, Terry A, Pangilinan J, Lindquist EA, Lucas S, Paulsen IT, Hattenrath-Lehmann TK, Talmage SC, Walker EA, Koch F, Burson AM, Alejandra-Marcoval Y-ZT M, LeCleir GR, Coyne KJ, Berg GM, Bertrand EM, Saito MA, Gladyshev VN, Grigoriev IV (2011) Niche of the harmful alga *Aureococcus anophagefferens* revealed through ecogenomics. Proc Natl Acad Sci U S A 108:4352–4357

136. Gordon JI (2012) Honor thy gut symbionts redux. Science 336:1251–1252

137. Graeve M, Kattner G, Wiencke C, Karsten U (2002) Fatty acid

composition of Arctic and Antarctic macroalgae: indicator of phylogenetic and trophic relationships. Mar Ecol Prog Ser 231:67–74

138. Griffin J (2015) An investigative study into the beneficial use of seaweed in bread and the broader food industry. Dublin Institute of Technology, School of Culinary Arts and Food Technology, http://arrow.dit.ie/tfschcafdis. Accessed 05 Sept 2016
139. Gropper SS, Smith JL (2013) Advanced human nutrition and metabolism, 6th edn. Wadsworth, Belmont
140. Gubelit YI, Makhutova ON, Sushchik NN, Kolmakova AA, Kalachova GS, Gladyshev MI (2015) Fatty acid and elemental composition of littoral "green tide" algae from the Gulf of Finland, the Baltic Sea. J Appl Phycol 27:375–386
141. Guil-Guerrero JL (2007) Stearidonic acid (18:4 n-3): metabolism, nutritional importance, medical uses and natural sources. Eur J Lipid Sci Technol 109:1226–1236
142. Gupta S, Abu-Ghannam N (2011a) Bioactive potential and possible health effects of edible brown seaweeds. Trends Food Sci Technol 22:315–326
143. Gupta S, Abu-Ghannam N (2011b) Recent developments in the application of seaweeds or seaweed extracts as a means for enhancing the safety and quality of foods. Food Sci Emerg Technol 12:600–609
144. Haddad EH, Berk LS, Kettering JD, Hubbard RW, Peters WR (1999) Dietary intake and biochemical, hematologic, and immune status of vegans compared with nonvegetarians. Am J Clin Nutr 70:586S–593S
145. Hafting JT, Critchley AT, Cornish ML, Hubley SA, Archibald AF (2012) On-land cultivation of functional seaweed products for human usage. J Appl Phycol 24:385–392
146. Hafting JT, Craigie JS, Stengel DB, Loureiro RR, Buschmann AH, Yarish C, Edwards MD, Critchley AT (2015) Prospects and challenges for industrial production of seaweed bioactives. J Phycol 51:821–837
147. Halliwell B, Gutteridge JMC (2007) Free radicals in biology and medicine, 4th edn. Clarendon, Oxford

148. Hammann M, Rempt M, Pohnert G, Wang G, Boo SM, Weinberger F (2016) Increased potential for wound activated production of Prostaglandin E-2 and related toxic compounds in non-native populations of *Gracilaria vermiculophylla*. Harmful Algae 51:81–88

149. Hansen HR, Raab A, Francesconi KA, Feldmann J (2003) Metabolism of arsenic by sheep chronically exposed to arsenosugars as a normal part of their diet. 1. Quantitative intake, uptake, and excretion. Environ Sci Technol 37:845–851

150. Hansra P, Moran EL, Fornasier VL, Bogoch ER (2000) Carrageenan-induced arthritis in the rat. Inflammation 24:141–155

151. Hardy K, Brand-Miller J, Brown KD, Thomas MG, Copeland L (2015) The importance of dietary carbohydrate in human evolution. Q Rev Biol 90:251–268

152. Harnedy PA, FitzGerald RJ (2011) Bioactive proteins, peptides, and amino acids from macroalgae. J Phycol 47:218–232

153. Harnedy PA, FitzGerald RJ (2013) Extraction of protein from the macroalga *Palmaria palmata*. LWT-Food Sci Technol 51:375–382

154. Hasegawa T, Kimura Y, Hiromatsu K, Kobayashi N, Yamada A, Makino M, Okuda M, Sano T, Nomoto K, Yoshikai Y (1997) Effect of hot water extract of *Chlorella vulgaris* on cytokine expression patterns in mice with murine acquired immunodeficiency syndrome after infection with *Listeria monocytogenes*. Immunopharmacology 35:273–282

155. Haug IJ, Sagmo LB, Zeiss D, Olsen IC, Draget KI, Seternes T (2011) Bioavailability of EPA and DHA delivered by gelled emulsions and soft gel capsules. Eur J Lipid Sci Technol 113:137–145

156. Hehemann J-H, Correc G, Barbeyron T, Helbert W, Czjzek M, Michel G (2010) Transfer of carbohydrate-active enzymes from marine bacteria to Japanese gut microbiota. Nature 464:908

157. Hehemann JH, Kelly AG, Pudlo NA, Martens EC, Boraston AB (2012) Bacteria of the human gut microbiome catabolize red seaweed glycans with carbohydrate-active enzyme updates from extrinsic microbes. Proc Natl Acad Sci U S A 109:19786–19791

158. Helliwell KE, Wheeler GL, Leptos KC, Goldstein RE, Smith AG (2011) Insights into the evolution of vitamin B_{12} auxotrophy from

sequenced algal genomes. Mol Biol Evol 28:2921–2933

159. Helliwell KE, Wheeler GL, Smith AG (2013) Widespread decay of vitamin-related pathways: coincidence or consequence? Trends Genet 29:469–478

160. Helliwell KE, Lawrence AD, Holzer A, Kudahl UJ, Sasso S, Krautler B, Scanlan DJ, Warren MJ, Smith AG (2016) Cyanobacteria and eukaryotic algae use different chemical variants of vitamin B_{12}. Curr Biol 26:999–1008

161. Hernandez-Carmona G, Carrillo-Dominguez S, Arvizu-Higuera DL, Rodriguez-Montesinos YE, Murillo-Alvarez JI, Munoz-Ochoa M, Castillo-Dominguez RM (2009) Monthly variation in the chemical composition of *Eisenia arborea* JE Areschoug. J Appl Phycol 21:607–616

162. Heuzé V, Tran G, Giger-Reverdin S, Lessire M, Lebas F (2016) Seaweeds (marine macroalgae). Feedipedia, a program by INRA, CIRAD, AFZ and FAO. http://www.feedipedia.org/node/78. Accessed 04 Sept 2016

163. Higa T, Kuniyoshi M (2000) Toxins associated with medicinal and edible seaweeds. J Toxicol Toxin Rev 19:119–137

164. Hofer M, Pospíšil M (2011) Modulation of animal and human hematopoiesis by β-glucans. Molecules 16:7969–7979

165. Holdt SL, Kraan S (2011) Bioactive compounds in seaweed: functional food applications and legislation. J Appl Phycol 23:543–597

166. Holst B, Williamson G (2008) Nutrients and phytochemicals: from bioavailability to bioefficacy beyond antioxidants. Curr Opin Biotechnol 19:73–82

167. Hollriegl V, Rohmuss M, Oeh U, Roth P (2004) Strontium biokinetics in humans: influence of alginate on the uptake of ingested strontium. Health Phys 86:193–196

168. Holub BJ (2009) Docosahexaenoic acid (DHA) and cardiovascular disease risk factors. Prostagl Leukotr Essent Fatty Acids 81:199–204

169. Honya M, Mori H, Anzai M, Araki Y, Nisizawa K (1999) Monthly changes in the content of fucans, their constituent sugars and sulphate

in cultured *Laminaria japonica*. Hydrobiologia 398/399:411–416

170. Hsu H-Y, Jeyashoke N, Yeh C-H, Song Y-J, Hua KF, Chao LK (2010) Immunostimulatory bioactivity of algal polysaccharides from *Chlorella pyrenoidosa* activates macrophages via toll-like receptor-4. J Agric Food Chem 58:927–936
171. Hwang H, Chen T, Nines RG, Shin HC, Stoner GD (2006) Photochemoprevention of UVB-induced skin carcinogenesis in SKH-1 mice by brown algae polyphenols. Int J Cancer 119:2742–2749
172. IARC (2012) Arsenic, metals, fibres, and dusts Vol. 100C a review of human carcinogens. IARC, Lyon, pp 1–99
173. Ichikawa S, Kamoshida M, Hanaoka K, Hamano M, Maitani T, Kaise T (2006) Decrease of arsenic in edible brown algae *Hijikia fusiforme* by the cooking process. Appl Organomet Chem 20:585–590
174. Imhoff-Kunsch B, Stein AD, Villalpando S, Martorell R, Ramakrishnan U (2011) Docosahexaenoic acid supplementation from mid-pregnancy to parturition influenced breast milk fatty acid concentrations at 1 month postpartum in Mexican women. J Nutr 141:1762–1762
175. Institute-of-Medicine (2001) Iron. In: Dietary reference intakes for vitamin A, vitamin K, arsenic, boron, chromium, copper, iodine, iron, manganese, molybdenum, nickel, silicon, vanadium, and zinc. National Academies Press, Washington, DC, pp 290–378
176. Institute-of-Medicine (2005) Dietary, functional and total fiber. In: Dietary reference intakes for energy, carbohydrates, fiber, fat, fatty acids, cholesterol, protein, and amino acids. National Academies Press, Washington, DC, pp 339–421
177. Iwai K (2008) Antidiabetic and antioxidant effects of polyphenols in brown alga *Ecklonia stolonifera* in genetically diabetic kk-a(y) mice. Plant Foods Human Nutr 63:163–169
178. Jassby A (1988) *Spirulina*: a model for microalgae as food. In: Lembi CA, Waaland JR (eds) Algae and human affairs. Cambridge University Press, Cambridge, pp 149–179
179. Jensen A (1993) Present and future needs for algae and algal products. Hydrobiologia 261:15–23

180. Jensen CL, Voigt RG, Prager TC, Zou YLL, Fraley JK, Rozelle JC, Turcich MR, Llorente AM, Anderson RE, Heird WC (2005) Effects of maternal docosahexaenoic acid intake on visual function and neurodevelopment in breastfed term infants. Am J Clin Nutr 82:125–132

181. Jensen CL, Voigt RG, Llorente AM, Peters SU, Prager TC, Zou YLL, Rozelle JC, Turcich MR, Fraley JK, Anderson RE, Heird WC (2010) Effects of early maternal docosahexaenoic acid intake on neuropsychological status and visual acuity at five years of age of breast-fed term infants. J Pediatrics 157:900–905

182. Jiang HM, Chen F (2000) Effects of temperature and temperature shift on docosahexanenoic acid production by the marine microalga *Crypthecodinium cohnii*. JAOCS 77:613–617

183. Jiang HM, Gao KS (2004) Effects of lowering temperature during culture on the production of polyunsaturated fatty acids in the marine diatom *Phaeodactylum tricornutum*(Bacillariophyceae). J Phycol 40:651–654

184. Jiao G, Yu G, Zhang J, Ewart HS (2011) Chemical structures and bioactivities of sulfated polysaccharides from marine algae. Mar Drugs 9:196–123

185. Jimenez-Escrig A, Jimenez-Jimenez I, Pulido R, Saura-Calixto F (2001) Antioxidant activity of fresh and processed edible seaweeds. J Sci Food Agric 81:530–534

186. Johnson RB, Kim JK, Armbruster LC, Yarish C (2014) Nitrogen allocation of *Gracilaria tikvahiae* grown in urbanized estuaries of Long Island Sound and New York City, USA: a preliminary evaluation of ocean farmed *Gracilaria* for alternative fish feeds. Algae 29:227–235

187. Jones JM (2014) CODEX-aligned dietary fiber definitions help to bridge the ‘fiber gap’. Nutr J 13:14

188. Kadam SU, Tiwari BK, O’Donnell CP (2015) Extraction, structure and biofunctional activities of laminarin from brown algae. Int J Food Sci Technol 50:24–31

189. Kaeffer B, Bénard C, Lahaye M, Blottière HM, Cherbut C (1999) Biological properties of ulvan, a new source of green seaweed sulfated polysaccharides, on cultured normal and cancerous colonic

epithelial cells. Planta Med 65:527–531

190. Kawai T, Zhang ZW, Moon CS, Shimbo S, Watanabe T, Matsuda-Inoguchi N, Higashikawa K, Ikeda M (2002) Comparison of urinary bromide levels among people in East Asia, and the effects of dietary intakes of cereals and marine products. Toxicol Lett 134:285–293
191. Kawamura-Konishi Y, Watanabe N, Saito M, Nakajima N, Sakaki T, Katayama T, Enomoto T (2012) Isolation of a new phlorotannin, a potent inhibitor of carbohydrate-hydrolyzing enzymes, from the brown alga *Sargassum patens*. J Agric Food Chem 60:5565–5570
192. Kay RA (1991) Microalgae as food and supplement. Crit Rev Food Sci Nutr 30:555–573
193. Kim S-K, Li Y-X (2011) Medicinal benefits of sulfated polysaccharides from sea vegetables. Adv Food Nutr Res 64:391–402
194. Kim S-K, Wijesekara I (2011) Anticoagulant effect of marine algae. Adv Food Nutr Res 64:235–244
195. Knauer K, Behra R, Sigg L (1997) Adsorption and uptake of copper by the green alga *Scenedesmus subspicatus* (Chlorophyta). J Phycol 33:596–601
196. Komiyama K, Hirokawa Y, Morota T, Umezawa T (1986) An acidic polysaccharide Chlon A, from *Chlorella pyrenoidosa*. 2. Anti-tumor activity and immunological responses. Chemotherapy (Tokyo) 34:302–307
197. Kralovec JA (2005) *Chlorella* composition having high molecular weight polysaccharides and polysaccharide complexes. US Patent 6974576 B2
198. Kralovec JA, Power MR, Liu F, Maydanski E, Ewart HS, Watson LV, Barrow CJ, Lin TJ (2005) An aqueous *Chlorella* extract inhibits IL-5 production by mast cells in vitro and reduces ovalbumin-induced eosinophil infiltration in the airways of mice in vivo. Int Immunopharmacol 5:689–698
199. Kulshreshtha A, Zacharia AJ, Jarouliya U, Bhadauriya P, Prasad GBKS, Bisen PS (2008) *Spirulina* in health care management. Curr Pharm Biotechnol 9:400–405
200. Kulshreshtha G, Rathgeber B, Stratton G, Thomas N, Evans F,

Critchley A, Hafting J, Prithiviraj B (2014) Feed supplementation with red seaweeds, *Chondrus crispus* and *Sarcodiotheca gaudichaudii*, affects performance, egg quality, and gut microbiota of layer hens. Poult Sci 93:2991–3001

201. Kumar M, Kumari P, Trivedi N, Shukla MK, Gupta V, Reddy CRK, Jha B (2011) Minerals, PUFAs and antioxidant properties of some tropical seaweeds from Saurashtra coast of India. J Appl Phycol 23:797–810

202. Kumari P, Kumar M, Gupta V, Reddy CRK, Jha B (2010) Tropical marine macroalgae as potential sources of nutritionally important PUFAs. Food Chem 120:749–757

203. Kusaykin M, Bakunina I, Sova V, Ermakova S, Kuznetsova T, Besednova N, Zaporozhets T, Zvyagintseva T (2008) Structure, biological activity, and enzymatic transformation of fucoidans from the brown seaweeds. Biotechnol J 3:904–915

204. Kwak JH, Baek SH, Woo Y, Han JK, Kim BG, Kim OY, Lee JH (2012) Beneficial immunostimulatory effect of short-term *Chlorella* supplementation: enhancement of natural killer cell activity and early inflammatory response (randomized, double-blinded, placebo-controlled trial). Nutr J 11:8

205. Lage-Yusty MA, Alvarado G, Ferraces-Casais P, Lopez-Hernandez J (2014) Modification of bioactive compounds in dried edible seaweeds. Int J Food Sci Technol 49:298–304

206. Lahaye M (1991) Marine algae as sources of fibres: determination of soluble and insoluble dietary fibre contents in some 'sea vegetables'. J Sci Food Agric 54:587–594

207. Lane K, Derbyshire E, Li WL, Brennan C (2014) Bioavailability and potential uses of vegetarian sources of omega-3 fatty acids: a review of the literature. Crit Rev Food Sci Nutr 54:572–579

208. Larson M, Webb G, Kennington S, Kellerher N, Sheppard J, Kuo J, Enisworth-White J (2002) Mannitol in cardioplegia as an oxygen free radical scavenger measured by malondialdehyde. Perfusion 17:51–55

209. Le TM, Knulst AC, Rockmann H (2014) Anaphylaxis to *Spirulina* confirmed by skin prick test with ingredients of *Spirulina* tablets. Food Chem Toxicol 74:309–310

210. Leal MCF, Vasconcelos MT, Sousa-Pinto I, Cabral JPS (1997) Biomonitoring with benthic macroalgae and direct assay of heavy metals in seawater of the Oporto Coast (Northwest Portugal). Mar Pollut Bull 34:1006–1015
211. Ledford HK, Niyogi KK (2005) Singlet oxygen and photo-oxidation stress management in plants and algae. Plant Cell Environ 28:1037–1045
212. Ledford HK, Chin BL, Niyogi KK (2007) Acclimation to singlet oxygen stress in *Chlamydomonas reinhardtii*. Eukaryot Cell 6:919–930
213. Lee HJ, Kim HC, Vitek L, Nam CM (2010) Algae consumption and risk of type 2 diabetes: Korean national health and nutrition examination survey in 2005. J Nutr Sci Vitaminol 56:13–18
214. Legezynska J, Kedra M, Walkusz W (2014) Identifying trophic relationships within the high Arctic benthic community: how much can fatty acids tell? Mar Biol 161:821–836
215. Lehtovaara BC, Gu FX (2011) Pharmacological, structural, and drug delivery properties and applications of 1,3-β-glucans. J Agric Food Chem 59:6813–6682
216. Leiro JM, Castro R, Arranz JA, Lamas J (2007) Immunomodulating activities of acidic sulphated polysaccharides obtained from the seaweed *Ulva rigida* C. Agardh. Int Immunopharmacol 7:879–888
217. Levendosky K, Mizenina O, Martinelli E, Jean-Pierre N, Kizima L, Rodriguez A, Kleinbeck K, Bonnaire T, Robbiani M, Zydowsky TM, O'Keefe BR, Fernández-Romero JA (2015) Griffithsin and carrageenan combination to target herpes simplex virus 2 and human papillomavirus. Antimicrob Agents Chemother 59:7290–7298
218. Li B, Lu F, Wei X, Zhao R (2008) Fucoidan: structure and bioactivity. Molecules 13:1671–1695
219. Li J, Pawitwar SS, Rosen BP (2016) The organoarsenical biocycle and the primordial antibiotic methylarsenite. Metallomics 8:1047–1055
220. Li MHH, Robinson EH, Tucker CS, Manning BB, Khoo L (2009) Effects of dried algae *Schizochytrium* sp., a rich source of docosahexaenoic acid, on growth, fatty acid composition, and sensory quality of channel catfish *Ictalurus punctatus*. Aquaculture

292:232–236

221. Li YX, Lee SH, Qiam ZJ, SKKS K (2010) Inhibitors of oxidation and matrix metalloproteinases, floridoside and D-isofloridoside from marine red alga *Laurencia undulata*. J Agric Food Chem 38:578–586

222. Li C, Gao Y, Xing Y, Zhu H, Shen J, Tian J (2011) Fucoidan, a sulfated polysaccharide from brown algae, against myocardial ischemia-reperfusion injury in rats via regulating the inflammation response. Food Chem Toxicol 49:2090–2095

223. Lightowler HJ, Davies GJ (2002) Assessment of iodine intake in vegans: weighed dietary record vs duplicate portion technique. Eur J Clin Nutr 56:765–770

224. Ling ALM, Yasir S, Matanjun P, Abu Bakar MF (2015) Effect of different drying techniques on the phytochemical content and antioxidant activity of *Kappaphycus alvarezii*. J Appl Phycol 27:1717–1723

225. Liu JH, Kandasamy S, Zhang JZ, Kirby CW, Karakach T, Hafting J, Critchley AT, Evans F, Prithiviraj B (2015) Prebiotic effects of diet supplemented with the cultivated red seaweed *Chondrus crispus* or with fructo-oligo-saccharide on host immunity, colonic microbiota and gut microbial metabolites. BMC Complement Altern Med 15:279. doi:10.1186/s12906-015-0802-5

226. Lordan S, Ross RP, Stanton C (2011) Marine bioactives as functional food ingredients: potential to reduce the incidence of chronic diseases. Mar Drugs 9:1056–1100

227. Lourenço SO, Barbarino E, De-Paula JC, Pereira LODS, Marquez UML (2002) Amino acid composition, protein content and calculation of nitrogen-to-protein conversion factors for 19 tropical seaweeds. Phycol Res 50:233–241

228. Lovegrove A, Edwards C, De Noni I, Patel H, El S, Grassby T, Zielke C, Ulmius M, Nilsson L, Butterworth P, Ellis P, Shewry P (2015) Role of polysaccharides in food, digestion and health. Crit Rev Food Sci Nutr. doi:10.1080/10408398.2014.939263

229. Lubitz J (1963) The protein quality, digestibililty and composition of algae, *Chlorella* 71105. J Food Sci 28:229–241

230. Luo Z, Tian D, Zhou M, Xiao W, Zhang Y, Li M, Sui B, Wang

W, Guan H, Chen H, Fu ZF, Zhao L (2015) Lambda-Carrageenan P32 is a potent inhibitor of rabies virus infection. PLoS One 10(10):e0140586

231. Mabeau S, Fleurence J (1993) Seaweed in food products: biochemical and nutritional aspects. Trends Food Sci Technol 4:103–107
232. MacArtain P, Gill CIR, Brooks M, Campbell R, Rowland IR (2007) Nutritional value of edible seaweeds. Nutr Rev 65:535–543
233. MacFarlane GT, MacFarlane S (2012) Bacteria, colonic fermentation, and gastrointestinal health. J AOAC Int 95:50–60
234. Maehre HK, Malde MK, Eilertsen KE, Elvevoll EO (2014) Characterization of protein, lipid and mineral contents in common Norwegian seaweeds and evaluation of their potential as food and feed. J Sci Food Agric 94:3281–3290
235. Maehre HK, Edvinsen GK, Eilertsen KE, Elvevoll EO (2016) Heat treatment increases the protein bioaccessibility in the red seaweed dulse (*Palmaria palmata*), but not in the brown seaweed winged kelp (*Alaria esculenta*). J Appl Phycol 28:581–590
236. Magura J, Moodley R, Jonnalagadda SB (2016) Chemical composition of selected seaweeds from the Indian Ocean, KwaZulu-Natal coast, South Africa. J Environ Sci Health, Part B 51:525–533
237. Mak W, Hamid N, Liu T, Lu J, White WL (2013) Fucoidan from New Zealand *Undaria pinnatifida*: monthly variations and determination of antioxidant activities. Carbohydr Polym 95:606–614
238. Makkar HPS, Tran G, Heuze V, Giger-Reverdin S, Lessire M, Lebas F, Ankers P (2016) Seaweeds for livestock diets: a review. Anim Feed Sci Technol 212:1–17
239. Mao W, Zang X, Li Y, Zhang H (2006) Sulfated polysaccharides from marine green algae *Ulva conglobata* and their anticoagulant activity. J Appl Phycol 18:9–14
240. Mao W-J, Fang F, Li H-Y, Qi X-H, Sun H-H, Chen Y, Guo S-D (2008) Heparinoid-active two sulfated polysaccharides isolated from marine green algae *Monostroma nitidum*. Carbohydr Polym 74:834–839
241. Mardsen AD, De Wreede RE (2000) Marine macroalgal community structure, metal content and reproductive function near an acid mine

drainage overflow. Environ Pollut 110:431–440

242. Marrion O, Schwertz A, Fleurence J, Gueant JL, Villaume C (2003) Improvement of the digestibility of the proteins of the red alga *Palmaria palmata* by physical processes and fermentation. Nahrung 47:339–344
243. Marrion O, Fleurence J, Schwertz A, Guéant J-L, Mamelouk L, Ksouri J, Villaume C (2005) Evaluation of protein in vitro digestibility of *Palmaria palmata* and *Gracilaria verrucosa*. J Appl Phycol 17:99–102
244. Martin C, Butelli E, Petroni K, Tonelli C (2011) How can research on plants contribute to promoting human health? Plant Cell 23:1685–1699
245. Matanjun P, Mohamed S, Mustapha NM, Muhammad K (2009) Nutrient content of tropical edible seaweeds, *Eucheuma cottonii, Caulerpa lentillifera* and *Sargassum polycystum*. J Appl Phycol 21:75–80
246. Matloub AA, El-Sherbini M, Borai IH, Ezz MK, Rizk MZ, Aly HF, Fouad GI (2013) Assessment of anti-hyperlipidemic effect and physico-chemical characterization of water soluble polysaccharides from *Ulva fasciata* Delile. J Appl Sci Res 9:2983–2993
247. McCarty MF (2007) Clinical potential of *Spirulina* as a source of phycocyanobilin. J Med Food 10:566–570
248. McCauley JI, Meyer BJ, Winberg PC, Ranson M, Skropeta D (2015) Selecting Australian marine macroalgae based on the fatty acid composition and anti-inflammatory activity. J Appl Phycol 27:2111–2121
249. McDermid KJ, Stuercke B (2003) Nutritional composition of edible Hawaiian seaweeds. J Appl Phycol 15:513–524
250. McDermid KJ, Stuercke B, Haleakala OJ (2005) Total dietary fiber content in Hawaiian marine algae. Bot Mar 48:437–440
251. McHugh DJ (2003) Seaweeds as human food. In: McHugh DJ (ed) A guide to the seaweed industry, FAO Fisheries Technical Paper 411. pp 73–90
252. McKim JM (2014) Food additive carrageenan: part I: a critical review of carrageenan in vitro studies, potential pitfalls, and implications

for human health and safety. Crit Rev Toxicol 44:211–243

253. Medeiros DM, Wildman REC (2015) Advanced human nutrition. Jones & Bartlett, Burlington

254. Meharg AA, Sun GX, Williams PN, Adomako E, Deacon C, Zhu YG, Feldmann J, Raab A (2008) Inorganic arsenic levels in baby rice are of concern. Environ Pollut 152:746–749

255. Mehta SK, Singh A, Gaur JP (2002) Kinetics of adsorption and uptake of Cu^{2+} by *Chlorella vulgaris*: influence of pH, temperature, culture age, and cations. J Environ Sci Health A 37:399–414

256. Melhuus A, Seip KL, Seip HM (1978) A preliminary study of the use of benthic algae as biological indicators of heavy metal pollution in Sørfjorden, Norway. Environ Pollut 15:101–107

257. Mendes A, Reis A, Vasconcelos R, Guerra P, da Silva TL (2009) *Crypthecodinium cohnii* with emphasis on DHA production: a review. J Appl Phycol 21:199–214

258. MHLW (2014) The National Health and Nutrition Survey in Japan, 2004–2014. The Ministry of Health, Labour and Welfare. http://www.mhlw.go.jp/bunya/kenkou/kenkou_eiyou_chousa.html

259. Michel C, MacFarlane GT (1996) Digestive fates of soluble polysaccharides from marine macroalgae: involvement of the colonic microflora and physiological consequences for the host. J Appl Bacteriol 80:349–369

260. Michikawa T, Inoue M, Shimazu T, Sawada N, Iwasaki M, Sasazuki S, Yamaji T, Tsugane S, Japan Public Hlth Ctr B (2012) Seaweed consumption and the risk of thyroid cancer in women: the Japan Public Health Center-based Prospective Study. Eur J Cancer Prev 21:254–260

261. Minguez-Mosquera MI, Hornero-Mendez D, Perez-Galvez A (2008) Carotenoids and provitamin a in functional foods. In: Hurst WJ (ed) Methods of analysis for functional foods and nutraceuticals, 2nd edn. CCR Press, Boca Raton, pp 277–335

262. Mishra VK, Temelli F, Ooraikul B, Shacklock PF, Craigie JS (1993) Lipids of the red alga, *Palmaria palmata*. Bot Mar 36:169–174

263. Miyai K, Tokushige T, Kondo M, Iodine Res G (2008) Suppression of thyroid function during ingestion of seaweed "kombu" (*Laminaria*

japonica) in normal Japanese adults. Endocr J 55:1103–1108

264. Miyamoto E, Yabuta Y, Kwak CS, Enomoto T, Watanabe F (2009) Characterization of vitamin B_{12} compounds from Korean purple laver (*Porphyra* sp.) products. J Agric Food Chem 57:2793–2796

265. Molin M, Ulven SM, Meltzer HM, Alexander J (2015) Arsenic in the human food chain, biotransformation and toxicology—Review focusing on seafood arsenic. J Trace Elem Med Biol 31:249–259

266. Moreda-Pineiro J, Moreda-Pineiro A, Romaris-Hortas V, Moscoso-Perez C, Lopez-Mahia P, Muniategui-Lorenzo S, Bermejo-Barrera P, Prada-Rodriguez D (2011) In-vivo and in-vitro testing to assess the bioaccessibility and the bioavailability of arsenic, selenium and mercury species in food samples. Trends Anal Chem 30:324–345

267. Moreda-Pineiro J, Moreda-Pineiro A, Romaris-Hortas V, Dominguez-Gonzalez R, Alonso-Rodriguez E, Lopez-Mahia P, Muniategui-Lorenzo S, Prada-Rodriguez D, Bermejo-Barrera P (2012) Trace metals in marine foodstuff: bioavailability estimation and effect of major food constituents. Food Chem 134:339–345

268. Mouritsen OG (2013) Seaweeds: edible, available, and sustainable. University of Chicago Press, Chicago

269. Mouritsen OG, Dawczynski C, Duelund L, Jahreis G, Vetter W, Schroder M (2013) On the human consumption of the red seaweed dulse (*Palmaria palmata* (L.) Weber & Mohr). J Appl Phycol 25:1777–1791

270. Mueller A, Raptis J, Rice PJ, Kalbfleisch JH, Stout RD, Ensley HE, Browder W, Williams DL (2000) The influence of glucan polymer structure and solution conformation on binding to (1→3)-β-D-glucan receptors in a human monocyte-like cell line. Glycobiology 10:339–346

271. Mulvaney WJ, Jahangard S, Ingram BA, Turchini GM, Winberg PC (2015) Recovery of omega-3 profiles of cultivated abalone by dietary macroalgae supplementation. J Appl Phycol 27:2163–2171

272. Munda IM, Hudnik V (1991) Trace metal content in some seaweeds from the northern Adriatic. Bot Mar 34:241–249

273. Nagayama J, Noda K, Uchikawa T, Maruyama I, Shimomura H, Miyahara M (2014) Effect of maternal *Chlorella* supplementation on carotenoid concentration in breast milk at early lactation. Int J

Food Sci Nutr 65:573–576

274. Nakamura Y, Narukawa T, Yoshinaga J (2008) Cancer risk to Japanese population from the consumption of inorganic arsenic in cooked hijiki. J Agric Food Chem 56:2536–2540
275. Nauroth JM, Liu YC, Van Elswyk M, Bell R, Hall EB, Chung G, Arterburn LM (2010) Docosahexaenoic acid (DHA) and docosapentaenoic acid (DPA n-6) algal oils reduce inflammatory mediators in human peripheral mononuclear cells in vitro and paw edema in vivo. Lipids 45:375–384
276. Navarro G, Cummings S, Lee J, Moss N, Glukhov E, Valeriote FA, Gerwick L, Gerwick WH (2015) Isolation of polycavernoside d from a marine cyanobacterium. Environ Sci Technol Lett 2:166–170
277. Neff LM, Culiner J, Cunningham-Rundles S, Seidman C, Meehan D, Maturi J, Wittkowski KM, Levine B, Breslow JL (2011) Algal docosahexaenoic acid affects plasma lipoprotein particle size distribution in overweight and obese adults. J Nutr 141:207–213
278. Newton L (1951) Seaweed utilization. Sampson Low, London
279. Neyrinck AM, Mouson A, Delzenne NM (2007) Dietary supplementation with laminaran, a fermentable marine ß(1–3) glucan, protects against hepatoxicity induced by LPS in rat by modulating immune response in the hepatic tissue. Int Immunopharmacol 7:1497–1506
280. Nicholson JK, Holmes E, Kinross J, Burcelin R, Gibson G, Jia W, Pettersson S (2012) Host-gut microbiota metabolic interactions. Science 336:1262–1267
281. Ninomiya K (2002) Umami: a universal taste. Food Rev Int 18:23–38
282. Nitschke U, Stengel DB (2016) Quantification of iodine loss in edible Irish seaweeds during processing. J Appl Phycol. doi:10.1007/s10811-016-0868-6
283. Niwa K, Furuita H, Yamamoto T (2008) Changes of growth characteristics and free amino acid content of cultivated *Porphyra yezoensis* Ueda (Bangiales Rhodophyta) blades with the progression of the number of harvests in a nori farm. J Appl Phycol 20:687–693
284. NOAA (2011) The future of aquafeeds. Office of Aquaculture—

NOAA Fisheries—National Oceanographic and Atmospheric Administration Technical Memorandum NMFS F/SPO-124. December 2011. www.nmfs.noaa.gov/Aquaculture. Accessed 05 Sept 2016

285. Noguchi T, Matsui T, Miyazawa K, Asakawa M, Iijima N, Shida Y, Fuse M, Hosaka Y, Kirigaya C, Watabe K, Usui S, Fukagawa A (1994) Poisoning by the red-alga-ogonori (*Gracilaria-verrucosa*) on the Nojima coast, Yokohama, Kanagawa Prefecture, Japan. Toxicon 32:1533–1538
286. Norziah MH, Ching CY (2000) Nutritional composition of edible seaweed *Gracilaria changgi*. Food Chem 68:69–76
287. Novak M, Vetvica V (2008) β-Glucans, history, and the present: immunomodulatory aspects and mechanisms of action. J Immunotoxicol 5:47–57
288. NRC (2013) Critical aspects of EPA's IRIS assessment of inorganic arsenic: interim report. National Academy of Sciences, Washington, 117 pp
289. Nwosa F, Morris J, Lund JA, Stewart D, Ross HA, McDougall GJ (2011) Anti-proliferatice and potential anti-diabetic effects of phenolic-rich extracts from ebible marine algae. Food Chem 126:1006–1012
290. Nylund GM, Weinberger F, Rempt M, Pohnert G (2011) Metabolomic assessment of induced and activated chemical defence in the invasive red alga *Gracilaria vermiculophylla*. PLoS One 6(12):e29359
291. O'Sullivan L, Murphy B, McLoughlin P, Duggan P, Lawlor PG, Hughes H, Gardiner GE (2010) Prebiotics from marine macroalgae for human and animal health applications. Mar Drugs 8:2038–2064
292. Okada Y, Ishikura M, Maoka T (2009) Bioavailability of astaxanthin in *Haematococcua* algal extract: the effects of timing of diet and smoking habits. Biosci Biotechnol Biochem 73:1928–1932
293. Olsen EK, Hansen E, Isaksson J, Andersen JH (2013) Cellular antioxidant effect of four bromophenols from the red algae *Vertebrata lanosa*. Mar Drugs 11:2769–2784
294. Oren A, Gunde-Cimerman N (2007) Mycosporines and mycosporine-like amino acids: UV protectants or multipurpose secondary metabolites? FEMS Microbiol Lett 269:1–10

295. Ortiz J, Romero N, Robert P, Araya J, Lopez-Hernández J, Bozzo C, Navarrete E, Osorio A, Rios A (2006) Dietary fiber, amino acid, fatty acid and tocopherol contents of the edible seaweeds *Ulva lactuca* and *Durvillaea antarctica*. Food Chem 99:98–104

296. Ortiz J, Uquiche E, Robert P, Romero N, Quitral V, Llanten C (2009) Functional and nutritional value of the Chilean seaweeds *Codium fragile*, *Gracilaria chilensis* and *Macrocystis pyrifera*. Eur J Lipid Sci Technol 111:320–327

297. Painter TJ (1983) Algal polysaccharides. In: Aspinall GO (ed) The polysaccharides, vol 2. Academic, New York, pp 195–285

298. Pal A, Kamthania M, Kumar A (2014) Bioactive compounds and properties of seaweeds—a review. Open Access Libr J 1:e752. doi:10.4236/oalib1100752, accessed 09 Sept 2016

299. Pangestuti R, Kim SK (2011) Biological activities and health benefit effects of natural pigments derived from marine algae. J Funct Foods 3:255–266

300. Pasco D, Pugh N (2010) Potent immunostimulatory extracts from microalgae. US Patent 7846452 B2

301. Pasquet V, Ulmann L, Mimouni V, Guiheneuf F, Jacquette B, Morant-Manceau A, Tremblin G (2014) Fatty acids profile and temperature in the cultured marine diatom *Odontella aurita*. J Appl Phycol 26:2265–2271

302. Paulsen BS, Barsett H (2005) Bioactive pectic polysaccharides. In: Heinze T (ed) Polysaccharides 1: structure, characterization and use. Springer, Berlin, pp 69–101

303. Pell A, Kokkinis G, Malea P, Pergantis SA, Rubio R, Lopez-Sanchez JF (2013) LC-ICP-MS analysis of arsenic compounds in dominant seaweeds from the Thermaikos Gulf (Northern Aegean Sea, Greece). Chemosphere 93:2187–2194

304. Peng J, Yuan JP, Wu CF, Wang JH (2011) Fucoxanthin, a marine carotenoid present in brown seaweeds and diatoms: metabolism and bioactivities relevant to human health. Mar Drugs 9:1806–1828

305. Percival E, McDowell RH (1967) Chemistry and enzymology of marine algal polysaccharides. Academic, London

306. Pereira L (2011) A review of the nutrient composition of selected

edible seaweeds. In: Ponin VH (ed) Seaweeds: ecology, nutrient composition and medicinal uses. Nova Science Publishers, Hauppauge, p 30

307. Pereira CMP, Nunes CFP, Zambotti-Villela L, Streit NM, Dias D, Pinto E, Gomes CB, Colepicolo P (2016) Extraction of sterols in brown macroalgae from Antarctica and their identification by liquid chromatography coupled with tandem mass spectrometry. J Appl Phycol. doi:10.1007/s10811-016-0905-5:1-7
308. Perez AA, Farias SS, Strobl AM, Perez LB, Lopez CM, Pineiro A, Roses O, Fajardo MA (2007) Levels of essential and toxic elements in *Porphyra columbina* and *Ulva* sp from San Jorge Gulf, Patagonia Argentina. Sci Total Environ 376:51–59
309. Peters BA, Hall MN, Liu X, Parvez F, Shanchez TR, van Geen A, Mey JL, Siddique AB, Shahriar H, Uddin MN, Islam T, Balac O, Ilieveki V, Factor-Litvak P, Graziano JH, Gamble MV (2015) Folic acid and creatine as therapeutic approaches to lower blood arsenic: a randomized controlled trial. Environ Health Perspect 123:1294–1301
310. Phaneuf D, Cote I, Dumas P, Ferron LA, LeBlanc A (1999) Evaluation of the contamination of marine algae (seaweed) from the St. Lawrence River and likely to be consumed by humans. Environ Res 80:S175–S182
311. Pielesz A, Biniaś W (2010) Cellulose acetate membrane electrophoresis and FTIR spectroscopy as methods of identifying a fucoidan in *Fucus vesiculosus* Linnaeus. Carbohydr Res 345:2676–2682
312. Plaza M, Herrero M, Cifuentes A, Ibáñez E (2009) Innovative natural functional ingredients from microalgae. J Agric Food Chem 57:7159–7170
313. Pomin VH (2009) An overview about the structure-function relationship of marine sulfated homopolysaccharides with regular chemical structures. Biopolymers 91:601–609
314. Pomin VH (2012) Structure and use of algal sulfated fucans and galactans. In: Kim S-E (ed) Handbook of marine macroalgae: biotechnology and applied phycology. John Wiley & Sons, Chichester, pp 229–261

315. Pomin VH, Mourão PA (2008) Structure, biology, evolution, and medical importance of sulfated fucans and galactose. Glycobiology 18:1016–1027

316. Popper ZA, Michel G, Hervé C, Domozych DS, Willats WGT, Tuohy MG, Kloareg B, Stengel DB (2011) Evolution and diversity of plant cell walls: from algae to flowering plants. Annu Rev Plant Biol 62:8.1–8.24

317. Prajapati VD, Maheriya PM, Jani GK, Solanki HK (2014) Carrageenan: a natural seaweed polysaccharide and its applications. Carbohydr Polym 105:97–112

318. Pulz O, Gross W (2004) Valuable products from biotechnology of microalgae. Appl Microbiol Biotechnol 65:635–648

319. Quigg A, Irwin AJ, Finkel ZV (2011) Evolutionary inheritance of elemental stoichiometry in phytoplankton. Proc Roy Soc London B 278:526–536

320. Ragan MA, Globitza KW (1986) Phlorotannins, brown algal polyphenols. Progr Phycol Res 4:129–241

321. Rajauria G (2015) Seaweeds: a sustainable feed source for livestock and aquaculture. In: Tiwari B, Troy D (eds) Seaweed sustainability: food and non-food applications. Academic, Amsterdam, pp 389–420

322. Ramberg JE, Nelson ED, Sinnott RA (2010) Immunomodulatory dietary polysaccharides: a systematic review of the literature. Nutr J 9:54

323. Raml R, Goessler W, Traar P, Ochi T, Francesconi KA (2005) Novel thioarsenic metabolites in human urine after ingestion of an arsenosugar, 2′,3′-dihydroxypropyl 5-deoxy-5-dimethylarsinoyl-β-D-riboside. Chem Res Toxicol 18:1444–1450

324. Raml R, Raber G, Rumpler A, Bauernhofer T, Goessler W, Francesconi KA (2009) Individual Variability in the human metabolism of an arsenic-containing carbohydrate, 2′,3′-dihydroxypropyl 5-deoxy-5-dimethylarsinoyl-β-D-riboside, a naturally occurring arsenical in seafood. Chem Res Toxicol 22:1534–1540

325. Ramsey UP, Bird CJ, Shacklock PF, Laycock MV, Wright JLC (1994) Kainic acid and 1’hydroxykainic acid from *Palmariales*. Nat Toxins 2:286–292

326. Rauma AL, Torronen R, Hanninen O, Mykkanen H (1995) Vitamin B-12 status of long-term adherents of a strict uncooked vegan diet ("living food diet") is compromised. J Nutr 125:2511–2515

327. Read BA, Kegel J, Klute MJ, Kuo A et al (2013) Pan genome of the phytoplankton *Emiliania*underpins its global distribution. Nature 499:209–213

328. Řezanaka T, Sigler K (2007) Structural analysis of a polysaccharide from *Chlorella kessleri* by means of gas chromatography–mass spectrometry of its saccharide alditols. Folia Microbiol 52:246–252

329. Řezanka T, Petránková M, Cepák V, Přibyl P, Sigler K, Cajthaml T (2010) *Trachydiscus minutus*, a new biotechnological source of eicosapentaenoic acid. Folia Microbiol 55:265–269

330. Rhatigan P (2009) Irish seaweed kitchen; the comprehensive guide to healthy everyday cooking with seaweeds. Booklink

331. Riosmena-Rodríguez R, Talavera-Sáenz A, Acosta-Vargas B, Gardner SC (2010) Heavy metals dynamics in seaweeds and seagrasses in Bahía Magdalena, B.C.S., México. J Appl Phycol 22:283–291

332. Rioux L-E, Turgeon S (2015) Seaweed carbohydrates. In: Tiwari B, Troy D (eds) Seaweed sustainability: food and non-food applications. Elsevier, Amsterdam, pp 141–192

333. Rioux L-E, Turgeon SL, Beaulieu M (2009) Effect of season on the composition of bioactive polysaccharides from the brown seaweed *Saccharina longicruris*. Phytochemistry 70:1069–1075

334. Richards JT, Kern ER, Glasgow LA, Overall JC Jr, Deigh EF, Hatch MT (1978) Antiviral activity of extracts from marine algae. Antimicrob Agents Chemother 14:24–30

335. Rioux L-E, Turgeon SL, Beaulieu M (2010) Structural characterization of laminaran and galactofucan extracted from the brown seaweed *Saccharina longicruris*. Phytochemistry 71:1586–1595

336. Rose M, Lewis J, Langford N, Baxter M, Origgi S, Barber M, MacBain H, Thomas K (2007) Arsenic in seaweed—forms, concentration and dietary exposure. Food Chem Toxicol 45:1263–1267

337. Rossoll D, Bermudez R, Hauss H, Schulz KG, Riebesell U, Sommer U, Winder M (2012) Ocean acidification-induced food quality deterioration constrains trophic transfer. PLoS One 7(4):e34737

338. Ruperez P, Saura-Calixto F (2001) Dietary fibre and physicochemical properties of edible Spanish seaweeds. Eur Food Res Technol 212:349–354

339. Russell-Jones GJ, Arthur L, Walker H (1999) Vitamin B_{12}-mediated transport of nanoparticles across Caco-2 cells. Int J Pharm 179:247–255

340. Rutherfurd SM, Fanning AC, Miller BJ, Moughan PJ (2015) Protein digestibility-corrected amino acid scores and digestible indispensable amino acid scores differentially describe protein quality in growing male rats. J Nutr 145:372–379

341. Sachindra NM, Sato E, Maeda H, Hosokawa M, Niwano Y, Kohno M, Miyashita K (2007) Radical scavenging and singlet oxygen quenching activity of marine carotenoid fucoxanthin and its metabolites. J Agric Food Chem 55:8516–8522

342. Safi C, Charton M, Pignolet O, Silvestre F, Vaca-Garcia C, Pontalier PY (2013) Influence of microalgae cell wall characteristics on protein extractability and determination of nitrogen-to-protein conversion factors. J Appl Phycol 25:523–529

343. Safi C, Zebib B, Merah O, Pontalier PY, Vaca-Garcia C (2014) Morphology, composition, production, processing and applications of *Chlorella vulgaris*: a review. Renew Sust Energy Rev 35:265–278

344. Saiki T (1906) The digestibility and utilization of some polysaccharide carbohydrates derived from lichens and marine algae. J Biochem, Tokyo 2:251–265

345. Sayanova O, Haslam RP, Caleron MV, Ruiz-Lopez N, Worthy C, Rooks P, Allen MJ, Napier JA (2011) Identification and functional characterisation of genes encoding the omega-3 polyunsaturated fatty acid biosynthetic pathway from the coccolithophore *Emiliania huxleyi*. Phytochemistry 72:594–600

346. Schiener P, Black KD, Stanley MS, Green DH (2015) The seasonal variation in the chemical composition of the kelp species *Laminaria digitata*, *Laminaria hyperborea*, *Saccharina latissima* and *Alaria esculenta*. J Appl Phycol 27:363–373

347. Schmid M, Guiheneuf F, Stengel DB (2014) Fatty acid contents and profiles of 16 macroalgae collected from the Irish Coast at two seasons. J Appl Phycol 26:451–463
348. Schuchardt JP, Schneider I, Meyer H, Neubronner J, von Schacky C, Hahn A (2011) Incorporation of EPA and DHA into plasma phospholipids in response to different omega-3 fatty acid formulations—a comparative bioavailability study of fish oil vs. krill oil. Lipids Health Dis 10:7
349. Sensoy I (2014) A review on the relationship between food structure, processing, and bioavailability. Crit Rev Food Sci Nutr 54:902–909
350. Shanmugam M, Mody KH (2000) Heparinoid-active sulphated polysaccharides from marine algae as potential blood anticoagulant agents. Curr Sci 79:1672–1683
351. Shen B, Jones RG, Bohnert HJ (1997) Mannitol protects against oxidation by hydroxyl radicals. Plant Physiol 115:527–532
352. Sheng J, Yu F, Xin Z, Zhao L, Zhu X, Hu Q (2007) Preparation, identification and their antitumor activities in vitro of polysaccharides from *Chlorella pyrenoidosa*. Food Chem 105:533–539
353. Shibata T, Ishimaru K, Kawaguchi S, Yoshikawa H, Hama Y (2007) Antioxidant activities of phlorotannins isolated from Japanese Laminariaceae. J Appl Phycol 20:705–711
354. Shimoda Y, Suzuki Y, Endo Y, Kato K, Tachikawa M, Endo G, Yamanaka K (2010) Speciation analysis of arsenics in commercial hijiki by high performance liquid chromatography-tandem-mass spectrometry and high performance liquid chromatography-inductively coupled plasma mass spectrometry. J Health Sci 56:47–56
355. Shuuluka D, Bolton JJ, Anderson RJ (2013) Protein content, amino acid composition and nitrogen-to-protein conversion factors of *Ulva rigida* and *Ulva capensis* from natural populations and *Ulva lactuca* from an aquaculture system, in South Africa. J Appl Phycol 25:677–685
356. Signes-Pastor AJ, Deacon C, Jenkins RO, Haris PI, Carbonell-Barrachina AA, Meharg AA (2009) Arsenic speciation in Japanese rice drinks and condiments. J Environ Monit 11:1930–1934
357. Skriptsova AV, Shevchenko NM, Zvyagintseva TN, Imbs TI (2010)

Monthly changes in the content and monosaccharide composition of fucoidan from *Undaria pinnatifida* (Laminariales, Phaeophyta). J Appl Phycol 22:79–86

358. Skrovankova S (2011) Seaweed vitamins as nutraceuticals. In: Kim SK (ed) Marine medicinal foods: implications and applications, macro and microalgae. Elsevier, San Diego, pp 357–369
359. Smirnoff N, Cumber QJ (1989) Hydroxyl radical scavenging activity of compatible solutes. Phytochemistry 28:1057–1060
360. Spindler SR, Mote PL, Flegal JM (2014) Dietary supplementation with Lovaza and krill oil shortens the life span of long-lived F1 mice. Age 36:1345–1352
361. Stabler SP, Allen RH (2004) Vitamin B_{12} eficiency as a worldwide problem. Annu Rev Nutr 24:299–326
362. Stengel DB, Dring MJ (2000) Copper and iron concentrations in *Ascophyllum nodosum*(Fucales, Phaeophyta) from different sites in Ireland and after culture experiments in relation to thallus age and epiphytism. J Exp Mar Biol Ecol 246:145–161
363. Stengel DB, Macken A, Morrison L, Morley N (2004) Zinc concentrations in marine macroalgae and a lichen from western Ireland in relation to phylogenetic grouping, habitat and morphology. Mar Pollut Bull 48:902–909
364. Stengel DB, Connan S, Popper ZA (2011) Algal chemodiversity and bioactivity: sources of natural variability and implications for commercial application. Biotechnol Adv 29:483–501
365. Stiger-Pouvreau V, Bourgougnon N, Deslandes E (2016) Carbohydrates from seaweeds. In: Fleurence J, Levine I (eds) Health and disease prevention. Academic, San Diego, pp 223–274
366. Stupperich E, Krautler B (1988) Pseudo vitamin B_{12} or 5-hydroxybenzimidazolyl-cobamide are the corrinoids found in methanogenic bacteria. Arch Microbiol 149:268–271
367. Stupperich E, Nexo E (1991) Effect of the cobalt-N coordination on the cobamide recognition by the human vitamin B_{12} binding proteins intrinsic factor, transcobalamin and haptocorrin. Eur J Biochem 199:299–303
368. Suárez ER, Kralovec JA, Grindley TB (2010) Isolation of

phosphorylated polysaccharides from algae: the immunostimulatory principle of *Chlorella pyrenoidosa*. Carbohydr Res 345:1190–1204

369. Sunda W, Kieber DJ, Kiene RP, Huntsman S (2002) An antioxidant function for DMSP and DMS in marine algae. Nature 418:317–320
370. Switzer L (1980) *Spirulina*: the whole food revolution, 2nd edn. Proteus Corp, Berkeley
371. Szabo NJ, Matulka RA, Chan T (2013) Safety evaluation of whole algalin protein (wap) from *Chlorella protothecoides*. Food Chem Toxicol 59:34–45
372. Taboada C, Millan R, Miguez I (2010) Composition, nutritional aspects and effect on serum parameters of marine algae *Ulva rigida*. J Sci Food Agric 90:445–449
373. Taboada MC, Millán R, Miguez MI (2013) Nutritional value of the marine algae wakame (*Undaria pinnatifida*) and nori (*Porphyra purpurea*) as food supplements. J Appl Phycol 25:1271–1276
374. Takenaka S, Sugiyama S, Ebara S, Miyamoto E, Abe K, Tamura Y, Watanabe F, Tsuyama S, Nakano Y (2001) Feeding dried purple laver (nori) to vitamin B_{12}-deficient rats significantly improves vitamin B_{12} status. Br J Nutr 85:699–703
375. Takeshi S, Yumiko Y-S, Joko S (2005) Mineral components and anti-oxidant activities of tropical seaweeds. J Ocean Univ China 4:205–208
376. Talarico LB, Noseda MD, Ducatti DRB, Duarte MER, Damonte EB (2011) Differential inhibition of dengue virus infection in mammalian and mosquito cells by iota-carrageenan. J Gen Virol 92:1332–1342
377. Taylor VF, Jackson BP (2016) Concentrations and speciation of arsenic in New England seaweed species harvested for food and agriculture. Chemosphere 163:6–13
378. Teas J, Hebert JR, Fitton JH, Zimba PV (2004a) Algae—a poor man's HAART? Med Hypotheses 62:507–510
379. Teas J, Pino S, Critchley A, Braverman LE (2004b) Variability of iodine content in common commercially available edible seaweeds. Thyroid 14:836–841
380. Telfer A, Bishop SM, Phillips D, Barber J (1994a) Isolated

photosystem II reaction centers of photosystem II as a sensitizer for the formation of singlet oxygen. Detection and quantum yield determination using a chemical trapping technique. J Biol Chem 269:13244–13253

381. Telfer A, Dhami S, Bishop SM, Phillips D, Barber J (1994b) β-carotene quenches singlet oxygen formed by isolated photosystem II reaction centres. Biochemistry 33:14469–14474
382. Terada A, Hara H, Mitsuoka T (1995) Effect of dietary alginate on the faecal microbiota and faecal metabolic activity in humans. Microb Ecol Health Dis 8:259–266
383. The nutritional and medicinal value of seaweeds used in Chinese medicine (2002) Institute for Traditional Medicine. www.imtonline.org
384. Thomas F, Barbeyron T, Tonon T, Genicot S, Czjzek M, Michel G (2012) Characterization of the first alginolytic operons in a marine bacterium: from their emergence in marine Flavobacteria to their independent transfers to marine Proteobacteria and human gut Bacteroides. Environ Microbiol 14:2379–2394
385. Tibbetts SM, Milley JE, Lall SP (2016) Nutritional quality of some wild and cultivated seaweeds: nutrient composition, total phenolic content and in vitro digestibility. J Appl Phycol. doi:10.1007/s10811-016-0863-y:1-11
386. Tiwari B, Troy D (eds) (2015) Seaweed sustainability: food and non-food applications. Academic, London
387. Tobacman JK (2001) Review of harmful gastrointestinal effects of carrageenan in animal experiments. Environ Health Perspect 109:983–994
388. Tseng CK (1981) Commercial cultivation. In: Lobban CS, Wynne MJ (eds) The biology of seaweeds. Blackwell Science Publications, Oxford, pp 680–725
389. Turner NJ (2003) The ethnobotany of edible seaweed (*Porphyra abbottae* and related species; Rhodophyta: Bangiales) and its use by First Nations on the Pacific Coast of Canada. Can J Bot 81:283–293
390. Turner A, Pedroso SS, Brown MT (2008) Influence of salinity and humic substances on the uptake of trace metals by the marine macroalga *Ulva lactuca*: experimental observations and modelling

using WHAM. Mar Chem 110:176–184

391. Turner A, Pollock H, Brown MT (2009) Accumulation of Cu and Zn from antifouling paint particles by the marine macroalga, *Ulva lactuca*. Environ Pollut 157:2314–2319

392. Tuzen M, Verep B, Ogretmen AO, Soylak M (2009) Trace element content in marine algae species from the Black Sea, Turkey. Environ Monit Assess 151:363–368

393. UK Food Standards Agency (2016) Trade Information Sheet 8. https://www.food.gov.uk/sites/default/files/tradeinfo08.pdf accessed 11 October 2016

394. Umezawa I, Komiyama K (1985) A new acidic polysaccharide CH-1 isolated from a culture of cells of the algae *Chlorella pyrenoidosa*. US Patent 4533548 A

395. Urbano MG, Goni I (2002) Bioavailability of nutrients in rats fed on edible seaweeds, Nori (*Porphyra tenera*) and Wakame (*Undaria pinnatifida*), as a source of dietary fibre. Food Chem 76:281–286

396. Ursu AV, Marcati A, Sayd T, Sante-Lhoutellier V, Djelveh G, Michaud P (2014) Extraction, fractionation and functional properties of proteins from the microalgae *Chlorella vulgaris*. Bioresour Technol 157:134–139

397. Ushakova NA, Morozevich GE, Ustyuzhanina NE, Bilan MI, Usov AI, Preobrazhenskaya NE (2009) Anticoagulant activity of fucoidans from brown algae. Biochemistry (Moscow) Suppl Ser B 3:77–83

398. Ustyuzhanina NE, Ushakova NA, Zyuzina KA, Bilan MI, Elizarova AL, Somonova OV, Madzhuga AV, Krylov VB, Preobrazhenskaya ME, Usov AI, Kiselevskiy MV, Nifantiev NE (2013) Influence of fucoidans on hemostatic system. Mar Drugs 11:2444–2458

399. Ustyuzhanina NE, Bilan MI, Ushakova NA, Usov AI, Kiselevskiy MV, Nifantiev NE (2014) Fucoidans: pro- or antiangiogenic agents? Glycobiology 24:1265–1274

400. Valera B, Suhas E, Counil E, Poirier P, Dewailly E (2014) Influence of polyunsaturated fatty acids on blood pressure, resting heart rate and heart rate variability among french polynesians. J Am Coll Nutr 33:288–296

401. Van Blervliet S, Devos M, Delhaye T, Van Biervliet JP, Robberecht E, Christophe A (2008) Oral DHA supplementation in ΔF508 homozygous cystic fibrosis patients. Prostaglandins Leukot Essent Fat Acids 78:109–115

402. van Ginneken VJT, Helsper J, de Visser W, van Keulen H, Brandenburg WA (2011) Polyunsaturated fatty acids in various macroalgal species from north Atlantic and tropical seas. Lipids Health Dis 10:104

403. Vedros NA (1993) Topical demulcent for viral and inflammatory diseases of the skin. US Patent No. 5198217 A

404. Viso AC, Marty JC (1993) Fatty-acids from 28 marine microalgae. Phytochemistry 34:1521–1533

405. Vo T-S, Ngo D-H, Ta QV, Kim S-K (2011) Marine organisms as a therapeutic source against herpes simplex virus infection. Eur J Pharm Sci 44:11–20

406. Wagner-Döbler I, Ballhausen B, Berger M, Brinkhoff T, Buchholz I, Bunk B, Cypionka H, Daniel R, Drepper T, Gerdts G, Hahnke S, Han C, Jahn D, Kalhoefer D, Kiss H, Klenk HP, Kyrpides N, Liebl W, Liesegang H, Meincke L, Pati A, Petersen J, Piekarski T, Pommerenke C, Pradella S, Pukall R, Rabus R, Stackebrandt E, Thole S, Thompson L, Tielen P, Tomasch J, von Jan M, Wanphrut N, Wichels A, Zech H, Simon M (2010) The complete genome sequence of the algal symbiont *Dinoroseobacter shibae*: a hitchhiker's guide to life in the sea. Isme J 4:61–77

407. Waldmann A, Koschizke JW, Leitzmann C, Hahn A (2004) Homocysteine and cobalamin status in German vegans. Public Health Nutr 7:467–472

408. Wang CC, Harris WS, Chung M, Lichtenstein AH, Balk EM, Kupelnick B, Jordan HS, Lau J (2006) N-3 fatty acids from fish or fish-oil supplements, but not alpha-linolenic acid, benefit cardiovascular disease outcomes in primary- and secondary-prevention studies: a systematic review. Am J Clin Nutr 84:5–17

409. Wang T, Jonsdottir R, Olafsdottir G, Kristinsson HG (2014) Antioxidant properties of marine macroalgae. In: Kristinsson HG (ed) Antioxidants and functional components in aquatic foods. Wiley-Blackwell, Chichester

410. Wang C, Yatsuya H, Li Y, Ota A, Tamakoshi K, Fujino Y, Mikami H, Iso H, Tamakoshi A (2016) Prospective study of seaweed consumption and thyroid cancer incidence in women: the Japan collaborative cohort study. Eur J Cancer Prev 25:239–245

411. Warrand J (2006) Healthy polysaccharides. The next chapter in food products. Food Technol Biotechnol 44:355–370

412. Warren MJ, Raux E, Schubert HL, Escalante-Semerena JC (2002) The biosynthesis of adenosylcobalamin (vitamin B_{12}). Nat Prod Rep 19:390–412

413. Watanabe F (2007a) Vitamin B_{12} from edible algae—from food science to molecular biology. Vitamins (Kyoto) 81:49–55

414. Watanabe F (2007b) Vitamin B_{12} sources and bioavailability. Exp Biol Med 232:1266–1274

415. Watanabe S, Seto A (1989) Ingredient effective for activating immunity obtained from *Chlorella minutissima*. US Patent 4831020 A

416. Watanabe F, Katsura H, Takenaka S, Fujita T, Abe K, Tamura Y, Nakatsuka T, Nakano Y (1999a) Pseudovitamin B_{12} is the predominant cobamide of an algal health food, *Spirulina*tablets. J Agric Food Chem 47:4736–4741

417. Watanabe F, Takenaka S, Katsura H, Hussain Masumder SAMZ, Abe K, Tamura Y, Nakano Y (1999b) Dried green and purple lavers (nori) contain substantial amounts of biologically active vitamin B_{12} but less of dietary iodine relative to other edible seaweeds. J Agric Food Chem 47:2341–2343

418. Watson DB (2008) Public health and carrageenan regulation: a review and analysis. J Appl Phycol 20:505–513

419. Weiner ML (2014) Food additive carrageenan: part II: a critical review of carrageenan in vivo safety studies. Crit Rev Toxicol 44:244–269

420. Weiner ML, Nuber D, Blakemore WR, Harriman JF, Cohen SM (2007) A 90-day dietary study on kappa carrageenan with emphasis on the gastrointestinal tract. Food Chem Toxicol 45:98–106

421. Weiner ML, Ferguson HE, Thorsrud BA, Nelson KG, Blakemore WR, Zeigler B, Cameron MJ, Brant A, Cochrane L, Pellerin M,

Mahadevan B (2015) An infant formula toxicity and toxicokinetic feeding study on carrageenan in preweaning piglets with special attention to the immune system and gastrointestinal tract. Food Chem Toxicol 77:120–131

422. Weintraub HS (2014) Overview of prescription omega-3 fatty acid products for hypertriglyceridemia. Postgrad Med 126:12
423. Wells ML (2002) Marine colloids and trace metals. In: Dennis AH, Carlson CA (eds) Biogeochemistry of marine dissolved organic matter. Academic, San Diego, pp 367–404
424. Wen X, Peng C, Zhou H, Lin Z, Lin G (2000) Nutritional composition and assessment of *Gracilaria lemaneiformis* Bory. J Integr Plant Biol 48:1047–1053
425. WHO (1989) Evaluation of certain food additives and contaminants. WHO, Geneva
426. WHO (2010) Exposure to Arsenic: a major public health concern. www.who.int/ipcs/features/arsenic.pdf accessed 14 October 2016
427. WHO (2016) Arsenic Fact Sheet. http://www.who.int/mediacentre/factsheets/fs372/en/accessed 11 October 2016
428. Wijesekara I, Pangestuti R, Kim SK (2011) Biological activities and potential health benefits of sulfated polysaccharides derived from marine algae. Carbohydr Polym 84:14–21
429. Wijesinghe W, Jeon YJ (2012) Biological activities and potential industrial applications of fucose rich sulfated polysaccharides and fucoidans isolated from brown seaweeds: a review. Carbohydr Polym 88:13–20
430. Wijesinghe WAJP, Athukorala Y, Jeon Y-J (2011) Effect of anticoagulative sulfated polysaccharide purified from enzyme-assistant extract of a brown seaweed *Ecklonia cava* on Wistar rats. Carbohydr Polym 86:917–921
431. Wilke T, Faulkner S, Murphy L, Kealy L, Kraan S, Brouns F (2015) Seaweed enrichment of feed supplied to farm-raised Atlantic salmon (*Salmo salar*) is associated with higher total fatty acid and LC n-3 PUFA concentrations in fish flesh. Eur J Lipid Sci Technol 117:767–772
432. Williams DL, Lowman DW, Ensley HE (2005) Introduction to

the chemistry and immunobiology of β-glucans. In: Young S-H, Castranova V (eds) Toxicology of 1→3-beta glucans: glucans as a marker for fungal exposure. CRC Press, Boca Raton, pp 65–93

433. Wolfe-Simon F, Grzebyk D, Schofield O, Falkowski PG (2005) The role and evolution of superoxide dismutases in algae. J Phycol 41:453–465
434. Wong KH, Cheung PCK (2003) Effect of fiber-rich brown seaweeds on protein bioavailability of casein in growing rats. Int J Food Sci Nutr 54:269–279
435. Woolf PJ, Fu LL, Basu A (2011) vProtein: identifying optimal amino acid complements from plant-based foods. PLoS One 6:e18836
436. Yamada K, Yamada Y, Fukuda M, Yamada S (1999) Bioavailability of dried asakusanori (*Porphyra tenera*) as a source of cobalamin (vitamin B_{12}). Int J Vitam Nutr Res 69:412–418
437. Yavorska N (2012) Sodium alginate—a potential tool for weight management: effect on subjective appetite, food intake, and glycemic and insulin regulation. J Undergrad Life Sci 6:66–69
438. Yotsu-Yamashita M, Abe K, Seki T, Fujiwara K, Yasumoto T (2007) Polycavernoside C and C2, the new analogs of the human lethal toxin polycavernoside A, from the red alga, *Gracilaria edulis*. Tetrahedron Lett 48:2255–2259
439. Young S-H, Castranova V (2005) Animal model of (1→3)-β-D-glucan-induced pulmonary inflammation in rats. In: Young S-H, Castranova V (eds) Toxicology of 1→3-beta glucans: glucans as a marker for fungal exposure. CRC Press, Boca Raton, pp 65–93
440. Yu PZ, Li N, Liu XG, Zhou GF, Zhang QB, Li PC (2003) Antihyperlipidemic effects of different molecular weight sulfated polysaccharides from *Ulva pertusa* (Chlorophyta). Pharmacol Res 48:543–549
441. Yuan H, Song J, Li X, Li N, Dai J (2006) Immunomodulation and antitumor activity of κ-carrageenan oligosaccharides. Cancer Lett 243:228–234
442. Yuan H, Song J, Li X, Li N, Liu S (2011) Enhanced immunostimulatory and antitumor activity of different derivatives of κ-carrageenan oligosaccharides from *Kappaphycus striatum*. J Appl Phycol 23:59–65

443. Zeller S (2005) Safety evaluation of single cell oils and the regulatory requirements for use as food ingredients. In: Cohen Z, Ratledge C (eds) Single cell oils. AOCS Press, Urbana, pp 161–181

444. Zhang H-J, Mao W-J, Fang F, Li H-Y, Sun H-H, Chen Y, Qi X-H (2008) Chemical characteristics and anticoagulant activities of a sulfated polysaccharide and its fragments from *Monostroma latissimum*. Carbohydr Polym 71:428–434

445. Zhang Z, Wang F, Wang X, Liu X, Hou Y, Zhang Q (2010) Extraction of the polysaccharides from five algae and their potential antioxidant activity in vitro. Carbohydr Polym 82:118–121

446. Zhao FJ, McGrath SP, Meharg AA (2010) Arsenic as a food chain contaminant: mechanisms of plant uptake and metabolism and mitigation strategies. Annu Rev Plant Biol 61:535–559

447. Zhao X, Li B, Xue C, Sun L (2012) Effect of molecular weight on the antioxidant property of low molecular weight alginate from *Laminaria japonica*. J Appl Phycol 24:295–300

448. Zhou G, Sun Y, Xin H, Zhang Y, Li Z, Xu Z (2004) In vivo antitumor and immunomodulation activities of different molecular weight lambda-carrageenans from *Chondrus ocellatus*. Pharmacol Res 50:47–53

449. Zhou C, Xu X, Zhang Y, He R, Ma H (2012) Ultrasonic degradation, purification and analysis of structure and antioxidant activity of polysaccharides from *Porphyra yexoensis* Udea. Carbohydr Polym 87:2046–2051

450. Zhu Y-G, Yoshinga MY, Zhao F-J, Rosen BP (2014) Earth abides arsenic transformations. Annu Rev Earth Planet Sci 42:443–467

451. Zvyagintseva TN, Shevchenko NM, Chizhov AO, Krupnova TN, Sundukova EV, Isakov VV (2003) Water-soluble polysaccharides of some far-eastern brown seaweeds. Distribution, structure, and their dependence on the developmental conditions. J Exp Mar Biol Ecol 294:1–13

CHAPTER

3

SEAWEED AQUACULTURE IN NORWAY: RECENT INDUSTRIAL DEVELOPMENTS AND FUTURE PERSPECTIVES

Pierrick Stévant[1,2], Céline Rebours[1,3], and Annelise Chapman[1]

[1]Møreforsking Ålesund AS, Ålesund, Norway

[2]Department of Biotechnology, Norwegian University of Science and Technology (NTNU), Trondheim, Norway

[3]Norwegian Institute for Bioeconomy Research (NIBIO), Ås, Norway

ABSTRACT

The use of cultivated seaweeds as a feedstock for multiple industrial applications has gained increasing interest in the Western World over the past decades. Norway has an extensive coastline and a well-established aquaculture sector offering suitable preconditions for developing large-scale cultivation of seaweed biomass both in monoculture and in Integrated Multi-Trophic Aquaculture (IMTA) systems. Recent efforts from research, industry and public authorities have been committed to develop a Norwegian bio-economy based on cultivated seaweed, focusing on cultivation and processing of the biomass. This review reports on the status of seaweed aquaculture in Norway, supported by production data collected since the delivery of the first commercial cultivation permits at sea in 2014. Although novel product developments are currently limited, future industrial perspectives based on cultivated biomass are being discussed. Upscaling from experimental cultivation schemes to commercial production requires a thorough assessment of the risks and benefits associated with seaweed aquaculture, as well as the development of a regulative framework adapted to this industry. Issues associated with upscaling the macroalgal production that needs to be addressed includes (i) genetic interactions between cultivated and wild crops, (ii) impacts of seaweed cultivation on surrounding ecosystems, (iii) epiphytes and diseases, (iv) area utilization and (v) threats from climate change. Addressing these issues and adapting production practices will ensure the environmental and economic sustainability of an emerging industry based on cultivated seaweed biomass in Norway.

Keywords: Bioeconomy, Integrated multitrophic aquaculture (IMTA), Regulation, Macroalgae, Seaweed, Sustainability

INTRODUCTION

Aquaculture has a great potential for meeting challenges of increasing demand for food. It is the fastest growing food production sector and accounts for almost 50% of aquatic resources globally (FAO 2016a). The world's total aquaculture production in 2014 was in excess of 100 million t, of which 27 million t were aquatic plants, which are mostly seaweeds with a commercial value of 5.6 billion US$ (FAO 2016b). Global seaweed cultivation is largely dominated by Asian countries, i.e. China (producing more than 13 million t), Indonesia, The Philippines, North and South Korea and Japan. In these countries, the biomass is either used for direct human consumption (both fresh and dried), or it is further processed for other food applications (mainly phycocolloids such as alginates, agar and carrageenan). In addition, non-food applications include fertilizers, pharmaceuticals and cosmetics. Aside from having multiple industrial applications, brown, red and green seaweeds have other assets which make them suitable alternatives to terrestrial biomass production—as autotrophic organisms, marine macroalgae utilize dissolved carbon dioxide and nutrients (e.g. dissolved inorganic nitrogen and phosphorus) in combination with light for growth and do not require feed nor fertilizer. They can be cultivated on large scales in coastal areas with little or no demand on fresh water resources in production cycles. Moreover, the fact that macroalgal growth rates exceed those of terrestrial plants (Brinkhuis et al. 1987), have spurred the interest for cultivation of seaweed biomass, illustrated by a doubling of global production during the past decade (FAO 2016b).

The rapid development of fed aquaculture (e.g. finfish or shrimps) is often associated with environmental concern regarding the discharge of nutrients to coastal areas, potentially leading to deterioration of local marine environments (Buschmann et al. 2006). Integrated multitrophic aquaculture (IMTA) systems represent a practical solution for mitigating the negative effects of fish farming wastes by utilising excess nutrients as a valuable resource for marine plant production (Chopin et al. 2008). Seaweeds in such integrated cultivation systems, e.g. in proximity to fish farms, function as extractive components within a cultivation food web. In addition to reducing the environmental impact of intensive fish aquaculture, IMTA systems add value to the investment in finfish aquaculture by increasing the yield of total biomass produced on a single site (Chopin et al. 2001; Neori et al. 2004; Troell et al. 2009; Chopin et al. 2012).

For more than five decades, seaweeds have been the subject of considerable industrial utilization in Norway based on harvesting of natural biomass. Harvested wild seaweed biomass of primarily *Laminaria hyperborea* (commonly known as 'forest kelp', Northern kelp' or 'cuvie') but also *Ascophyllum nodosum* (knotted wrack) amounted to over 154,000 t fresh weight in 2014 (FAO 2016b). Harvesting of wild *L. hyperborea*, using specifically designed trawling equipment, remains somewhat controversial as the removal of and interference with natural habitats (specifically kelp forests) has the potential to affect local biodiversity and ecosystem integrity, and may contribute to coastal erosion (Mork 1996; Christie et al. 1998; Christie et al. 2003; Lorentsen et al. 2010; Smale et al. 2013). Although Norwegian management regimes for seaweed harvesting are considered to be among the best in the world (Smale et al. 2013), regional conflicts persist between the seaweed trawling industry and other coastal zone users (representing for instance coastal fisheries, marine conservation groups and ornithologists). Even with positive developments in knowledge-based integrated coastal zone management with broad stakeholder involvement, and the consideration of a prospective geographic expansion of kelp trawling areas northward, it is not foreseen that future harvesting of wild kelp biomass will be able to meet the growing demands of an industry that is mainly producing alginate from *L. hyperborea*.

Norway's long and complex coastline extending over 100,000 km and characterized by fjords, islands and skerries is highly suited for aquaculture. The cold temperate waters of the Northeast Atlantic are home to more than 400 species of brown, red and green seaweeds (Rueness 1998), many of which have recognized commercial value. Norway is the world's second largest exporter of seafood and has natural advantages which create opportunities to increase seafood production through aquaculture. In 2015, 1.38 million t of salmon and trout were farmed in Norway (Statistisk Sentralbyrå 2016). As salmon and trout production in Norway is predicted to increase by approximately 3 to 5% per year (Olafsen et al. 2012) over the coming decades, measures must be taken to limit the negative effects of excess nutrient discharge, especially environmental eutrophication and resulting ecological degradation and biodiversity loss. In addition, finite global phosphorus resources (Cordell and White 2011) provide a powerful argument for recovering excess nutrients, specifically phosphorus, from all sources including aquaculture and in view of their high value. The benefit of employing IMTA in this context, i.e. recovering organic and inorganic nutrients discharged from fish farms, is emphasized in a study by Wang et al. (2012) which highlights the potential for seaweed production in Norway, based on the estimated total amount of dissolved inorganic nutrients released by salmon farms. In this context, the development of seaweed aquaculture in Norway is in line with the European strategy driving 'Blue Growth' initiatives (European Commission 2012).

STATUS OF SEAWEED AQUACULTURE IN NORWAY

Experimental cultivation of kelps (Laminariales) at sea in Norway started around 2005 following successful trials conducted in France, Germany, Ireland and Scotland. Joint efforts of key actors from both research and private sectors primarily focused on developing cultivation technology in order to upscale the biomass production. Current efforts regarding the industrial scale production of seaweeds, mainly sugar kelp (*Saccharina latissima*), are now being directed at developing efficient farming strategies that reduce the need for technical maintenance, along with marine technology, including mechanisation of seedling deployment, biomass harvest and crop handling logistics (MACROSEA, www.sintef.no/projectweb/macrosea/). Efficient cultivation technology along with access to sufficient area at sea is a prerequisite for the sustainable development of an industry based on the utilization of seaweed biomass. Although activities to date have been characterised by research and pilot scale production, the first commercial permits for cultivation of seaweeds were granted in 2014 following the creation of a specific–interim-licensing system for macroalgae by public authorities. The Ministry of Trade, Industry and Fisheries coordinates the processing of seaweed farming applications and considers them according to the Aquaculture Act. The evaluation of applications involves several authorities, including the Directorate of Fisheries, The Norwegian Coastal Administration, the Norwegian Food Safety Authority, The County Governor's Environmental Department and The Norwegian Water Resources and Energy Directorate, each one considering potential conflicts of the application within their area of responsibility. Whether the seaweed farm is in concordance with coastal zone spatial plans, is considered by the concerned municipal authorities, as coastal zone management lies within the municipalities' responsibility.

Since 2014, several permits for sea cultivation of seaweeds, including in IMTA systems, have been granted following the rapid increase in the number of private actors involved in this activity (Table 1). The surface allocated to seaweed cultivation has more than tripled between 2014 and 2016, currently reaching a total of ca. 277 ha along the entire Norwegian coast and corresponding to a production potential of approximately 16,000 t (as estimated from the total maximum production by applicants on each cultivation sites). However, this potential by far exceeds the real production output, given that most companies involved in this activity are at a start-up phase characterized by experimental production. In fact, only 10 of 16 companies holding a cultivation permit were operating in 2015 (Directorate of Fisheries 2016a, 2016b). In Norway as in the rest of Europe, efforts for large-scale cultivation of seaweed biomass have been focussing largely on kelp species, particularly *S. latissima*, because of the species' potential for high biomass yields (Kraan et al. 2000; Handå et al. 2013) and valuable nutritional content (Schiener et al. 2015). The cultivation of *S. latissima* represented 96% (49 t wet weight) of the total production output, while cultivation of winged kelp (*Alaria esculenta*) represented the remaining 4% (2 t wet weight). Although licences for cultivation of a wide range of species (e.g. *Palmaria palmata, Ulva* spp., *Laminaria digitata, Porphyra* spp.) were granted, none of these are currently cultivated at sea. Given the data reported by the Directorate of Fisheries (2016b) for the year 2015, the retail value of *S. latissima* and *A. esculenta* was 399 US\$ t^{-1} and 1099 US\$ t^{-1}, respectively. Although no information is available regarding the retail markets of the produced biomass, *A. esculenta* is often sold in dried or fresh form as a high value food ingredient or sea vegetable whereas *S. latissima* has a broader range of market outputs, e.g. dried for human consumption or as meal in animal feed, which may explain the difference in market value. A higher value for *A. esculenta* may also simply result from a lower biomass availability on the market. As a comparison, retail value of the global seaweed production output is estimated to 206 US\$ t^{-1} in 2014 based on figures mentioned earlier (FAO 2016b).

Table 1. Statistics from seaweed cultivation in Norway between 2014 and 2016. Source: Directorate of Fisheries (2016a, 2016b). Production data for 2016 have not yet been registered (as per January 2017)

	2014	**2015**	**2016**
Total seaweed cultivation sites at sea	8	24	36
Total surface area for seaweed cultivation at sea (ha)	76	194	277
Seaweed production output (t wet weight)	na	51	na
Commercial value (US$)[a]	na	21,760	na
Number of companies involved in seaweed cultivation	6	16	20

[a]Values initially expressed in Norwegian krone (NOK). The exchange rate of 1NOK = 0.1221 US$ was used

Despite a long coastline and an increase in the number of permits at sea, seaweed cultivation sites are still scarce and not evenly distributed along the Norwegian coast (Fig. 1). A very large part of the allocated surface area (87%) shared by 28 cultivation sites is concentrated on ten municipalities, localized in few counties, namely Hordaland and Sogn and Fjordane in the west, Sør- and Nord-Trøndelag in the mid-part and

Nordland in the northern part of the country. Hence, there is a large potential to establish seaweed cultivation in other areas, especially the Møre and Romsdal county in the west, as well as Troms and Finnmark counties in the north, both of which are characterized by extensive areas and the presence of salmon aquaculture (see possibilities for integration of seaweed and salmon farming above). At present, only a few of the existing cultivation sites are in the vicinity of fish farms, hence characterized as IMTA. Most of the so-called IMTA sites were developed as 'add-on' to a pre-existing fish farm and to date, few have been fully designed for the purpose of fish and seaweed co-culture. However, many sites of seaweed monoculture are located in areas of intense salmon production (Directorate of Fisheries' interactive map: http://kart.fiskeridir.no/default.aspx?gui=1&lang=2) thus, are likely to directly benefit from nutrient releases originating from fish farming activities.

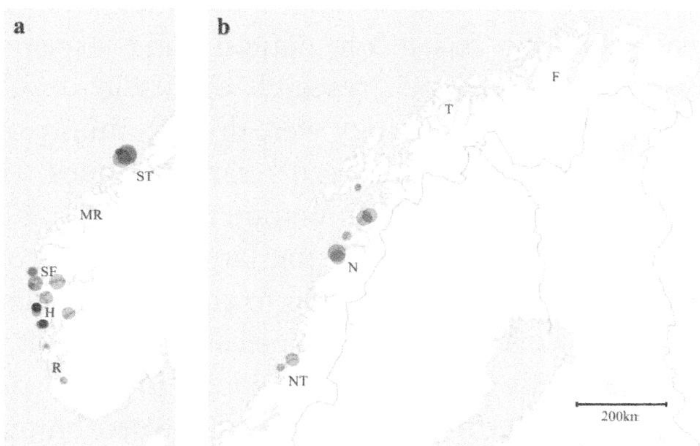

Figure 1. Surface area allocated to seaweed cultivation sites along the coast of southern Norway (a) and Northern Norway (b) as of January 2017. Counties are listed from south to north: Rogaland (*R*), Hordaland (*H*), Sogn and Fjordane (*SF*), Møre and Romsdal (*MR*), Sør-Trøndelag (*ST*), Nord-Trøndelag (*NT*), Nordland (*N*), Troms (*T*) and Finnmark (*F*). Darker circles indicate several cultivation sites close to each other. Data from the Directorate of Fisheries (2016a).

The cultivation cycle of the kelp species is based on their life cycle (see Bartsch et al. 2008 for more details on the life cycle of the Laminariales). Sporelings are grown from microscopic stages i.e. zoospores or gametophytes (derived from zoospores) under controlled conditions and directly seeded onto appropriate substrata (e.g. twine, nets, ribbons, sheets) which are later on deployed at sea for further growth under natural conditions. Although the production of spores in natural kelp populations is seasonal and species-specific (Bartsch et al. 2008), sorus (reproductive structure) formation can be induced artificially by, for example, controlling the light regimes provided to adult sporophytes. In this way, viable sporelings can be produced at all seasons (Forbord et al. 2012). However, biofouling of the biomass from epiphytic organisms limits the cultivation period at sea from autumn to spring (see Risk assessment 1—epiphytes and diseases).

PRODUCT APPLICATIONS—INDUSTRIAL PERSPECTIVES

Seaweed cultivation has been subject to increasing interest in Europe over the past decades, and activities have benefited from recent trends towards a bio-economy based on natural, and especially marine, biological resources. In Norway, research efforts to develop seaweed cultivation and processing technology were driven initially by the interest for producing bioenergy from seaweed's carbohydrates. However, the production of biofuel and biogas on commercial scales is impaired by the difficulties associated with producing large biomass volumes, since neither cultivation technology nor access to aquaculture sites at sea have yet been sufficiently developed for large-scale production. In addition, the market value for seaweed biomass as biofuel is low compared with other product types (Skjermo et al. 2014). Seaweed biomass is a rich source of a wide range of bioactive products of potentially high market value (Holdt and Kraan 2011). Relevant applications include (i) human food, (ii) domestic animal and fish feed products, (iii) fertilizer, (iv) prebiotics, (v) cosmetics, (vi) bioactive peptides and (vii) pharmaceuticals and nutraceuticals, and these are predicted to play an important role in the emerging Norwegian bio-economy based on seaweed cultivation (Skjermo et al. 2014).

In the past, mainly brown algae have been utilized in Norway. Primary applications have been alginates (used as food additives, health products and bio-medicinal ingredients, to name just a few) or agricultural products (i.e. animal feed supplement and soil enhancers/ fertilizers). In most cases, the seaweeds have been used for the functional properties of their polysaccharides or their mineral content. Although the seaweed chemical composition is variable depending on species, collection time and habitat, the nutritional value of several species is well characterized. Most species share a low lipid content on the one hand, while being rich in polysaccharides (both structural and storage, Dawczynski et al. 2007) and minerals on the other (Mabeau and Fleurence 1993; Rupérez 2002; MacArtain et al. 2007; Pereira 2011). Other nutritional properties, such as a relatively high protein content and high quality profiles of amino-acids (Fleurence 2004; Mæhre et al. 2014), lipids (Sánchez-Machado et al. 2004) and minerals (Cofrades et al. 2008; López-López et al. 2009) are highly relevant towards food and feed applications. Positive effects of including seaweeds in feed meals have been reported for both, marine and terrestrial organisms (Soler-Vila et al. 2009; Kandasamy et al. 2012; Evans and Critchley 2014; Wilke et al. 2015), and several feed products based on powdered dried seaweed are commercially available.

Seaweeds have traditionally been a major element of the diet in Asian countries and are used extensively as food in China, Japan, Korea and the Philippines where they are recognized for their nutritional value as well as for rich and unique flavors. In Europe, traditions of eating seaweeds have been limited to coastal communities, and over the past centuries, few regions have had an explicit culture for utilizing seaweeds as a significant component of the people's diet. The use of seaweeds as food in Norway goes back to the Vikings who likely brought dried dulse on long expeditions at sea for its virtues in protecting against scurvy, following its high content of vitamins (Mouritsen et al. 2013). In recent decades, Asian dishes such as sushi, have enjoyed increasing popularity and contributed to introducing seaweeds to European palates. In Scandinavia, the renewal of the Nordic cuisine initiated by high-end restaurants and based on locally available natural ingredients has aroused interest for including seaweeds in the gastronomy (Mouritsen et al. 2012). Previous studies have revealed the potential of seaweed species native to the Atlantic coast of Europe (e.g. the red algae *P. palmata* and the kelps *S. latissima* and *A. esculenta*) to be used as food ingredients in a wide range of foodstuffs both as vegetables and flavour enhancers (Mouritsen et al. 2012; Chapman et al. 2015). In Western Europe, i.e. France, Ireland, Spain and UK, a number of small and medium-sized enterprises are using wild or cultivated seaweeds in food products. This new type of businesses has developed in recent years following the rising demand from European consumers (NETALGAE 2012). In France, edible seaweeds e.g. *S. latissima*, *P. palmata*, *Undaria pinnatifida* and *Himanthalia elongata* are sold in fresh or dried form, as "sea vegetables" or condiments, or are incorporated in various product types (e.g. spreads, dressings, soups) as identified by Le Bras et al. (2015a, 2015b) in a comprehensive study among a wide range of retailers across the country. The retail price of these products, mainly sold in organic or health food shops, varies greatly with product category, seaweed species and the amount of seaweed used, but the identified market value is generally high i.e. over 100 € kg^{-1} (ca. 105 US\$ kg^{-1}) on average for dried and flaked seaweeds (Le Bras et al. 2015b) . Although there is no data available regarding seaweed consumption in Norway, it is truly marginal, and few Nordic seaweed food products are currently available. The development of such 'niche' food products, based on market and consumer insight, can contribute to creating value from Norwegian produced seaweed biomass on a relative short-term. Thoroughly investigating the food safety of commercialized edible seaweeds (see Risk assessment 2—product related) is, however, essential.

Moreover, seaweeds contain a large variety of phytochemical constituents that can be used in the prevention and treatment of health diseases by various mechanisms (Holdt and Kraan 2011; Brown et al. 2014). Bioactive compounds of interest include polysaccharides (e.g. alginates, fucoidans), polyphenols and pigments, to name just a few. Besides being used mainly in the food industry as thickening or emulsifying agent, alginates are regarded as a source of dietary fibre with beneficial effects both on colonic and cardiovascular health (Brownlee et al. 2005). Alginate polymers can be artificially modified to achieve tailored properties suited in various medical applications including cell micro-encapsulation, tissue engineering and regeneration, and dental impressions (Ruvinov and Cohen 2016). The company leading the alginate production in Norway (FMC BioPolymer) manufactures ultrapure alginate from wild *Laminaria hyperborea* for research and high-end commercial applications. The same firm recently obtained the authorization to commercialize high purity extracts of fucoidan as a dietary ingredient. Fucoidan represents a group of polysaccharides from brown seaweeds, mainly composed of sulfated fucose with documented bioactivity (Berteau and Mulloy 2003). Other compounds which are currently not commercially exploited in Norway may offer industrial opportunities in the future. Carotenoid pigments from brown algae, namely fucoxanthin, are recognized for their antioxidant activity as well as positive health effects against obesity and type-2 diabetes (Maeda 2015), and can be found in dietary supplements and nutraceuticals. Phycoerythrin, a pigment from red algae, is used extensively in fluorescent applications in clinical and immunological analysis as well as natural dye in the food and cosmetic industry (Sekar and Chandramohan 2008). Extraction procedures of this pigment using common species along the Atlantic coast of Europe, e.g. *P. palmata,* have been suggested (Galland-Irmouli et al. 2000; Dumay et al. 2013) together with methodology for industrial upscaling (Denis et al. 2009). Polyphenols are secondary metabolites which are abundant in brown seaweeds such as *A. nodosum, Fucus* spp., and widely recognized for their antioxidant properties (Wang et al. 2009; Tierney et al. 2013). Polyphenol-rich extracts from brown algae have inhibitory effects against lipid oxidation in fish (Wang et al. 2010) and fish oil (Xiaojun et al. 1996). Hence, polyphenols may provide a safer alternative to synthetic commercial antioxidants such as butylated hydroxytoluene (BHT), butylated hydroxyanisole (BHA) and tert-butyl hydroxyquinone (TBHQ) which are currently used in the food and feed industry, and criticized for their potential toxicity and carcinogenic effects (Gupta and Abu-Ghannam 2011). The commercial exploitation of such high value-added products will require relatively low biomass volumes in comparison to bulk products (e.g. biofuels) and is predicted to generate economic value from cultivated macroalgal biomass (Holdt and Kraan 2011; Skjermo et al. 2014). However, the extraction of these substances is technology demanding thus, associated with high investment costs. More knowledge is also required regarding the market potential of seaweed bioactive compounds to identify commercial opportunities.

In contrast to current processes in the Norwegian alginate industry, in which only a fraction of the harvested biomass is utilized and by-products are discarded, recent trends towards bio-economy favour complete utilization of produced biomass, recycling of nutrients in complete cycles and hence, reducing negative environmental impacts. Integrated processes maximizing biomass utilization by recovering a stream of products from seaweed biomass have been studied with promising results (Hou et al. 2015; Baghel et al. 2016). An increasing number of initiatives involving both research and industrial partners are aiming to develop biorefinery processes of marine macroalgae in Norway, in order to optimize feedstock utilization towards a wide range of products. However, the high water content of seaweeds (70 to 90%, Jensen 1993) represents a challenge for conserving and transporting large amounts of biomass (as in industrial production) from harvesting to processing sites. Seaweeds are characterized by a rapid microbial decomposition once harvested (Enríquez et al. 1993), and thus require appropriate preservation methods to maintain biomass quality and ensure product safety. Primary processes such as drying can effectively stabilize the biomass but require technology and may be difficult to implement close to harvesting sites for large biomass volumes. Drying also affects the chemical content (Chan et al. 1997; Gupta et al. 2011) with consequences for the product's nutritional value as well as extraction yields of bioactive compounds during further processing. Post-harvest treatments that will contribute to maintaining biomass quality include seawater storage (Paull and Chen 2008), cold storage (Liot et al. 1993), silage (Herrmann et al. 2015) and freezing (Choi et al. 2012). A recent study by Broch et al. (2016) stressed that large-scale seaweed cultivation sets standards for efficient logistic strategies and suggests the use of energy surplus from existing industrial facilities along the coast for processing of large biomass volumes. Another national research project investigates the possibilities for using energy surplus from a waste incinerator located on the west coast of Norway for seaweed processing, including its technical, economic and logistic potential (PROMAC, www.promac.no/).

RISK ASSESSMENT 1—MANAGERIAL TOOLS ASSOCIATED WITH SEAWEED CULTIVATION AND IMTA

Despite a growing interest for seaweed cultivation in the western world and a wide range of industrial applications globally, the tools and methods for establishing a seaweed industry in Norway—whether they are technical, economic, environmental, social or regulatory—still need to be developed or adapted from Asian models to fit European frameworks. This will help aquaculture enterprises and policy makers gain a better understanding of the risks and benefits associated with seaweed cultivation in both monoculture and IMTA systems, and provide conditions suitable for the development of a sustainable bioeconomy based on macroalgal biomass.

By cultivating extractive species (e.g. seaweeds) near fish farms, IMTA systems increase the biomass production of an aquaculture site associated with a certain area license. Whereas seaweed cultivation integrated with fish farms is currently being viewed as an experimental 'add-on' to existing aquaculture sites of salmonid fishes, the potential of seaweed production for improving environmental and economic performance of fish aquaculture and consequently increasing biomass production, is significant. The regulatory frameworks largely differ among species (i.e. fish, seaweed, shellfish) and are non-existent for species combinations (particularly concerning the integration of finfish, shellfish and seaweed production). The current interim regulatory process in use includes seaweed cultivations and IMTA systems, but needs to be specifically adapted for the various seaweed farming scenarios. An adaptation of the legal framework needs to achieve the transition from fed-monoculture (fish or shellfish) to poly- and integrated aquaculture (Alexander et al. 2015). In addition, the adaptation of existing regulations and policies for IMTA will demand an assessment of the risks and benefits of developing new aquaculture systems or using new aquaculture species.

The "Strategy for an Environmentally Sustainable Norwegian Aquaculture Industry" (Ministry of Fisheries and Coastal affairs 2009) provides a guideline for current Norwegian aquaculture policies, focusing on five key problem areas, i.e. (i) genetic interaction and escape, (ii) pollution and emissions, (iii) disease, (iv) area utilization and (v) feed and feed resources. Whereas these policies were designed for fish aquaculture, they are equally relevant for seaweed cultivation, and are discussed below in this context.

Genetic Interactions

There are several infamous cases of accidental introductions of non-native seaweed species to new coastal environments, e.g. *Undaria pinnatifida* and *Sargassum muticum* to Western Europe and *Caulerpa taxifolia* to the Mediterranean Sea, which have resulted in massive population explosions of the non-native species and consequently significant ecosystem alterations. Because spreading of non-native species can have large environmental and economic impacts (Schaffelke et al. 2006), their intentional use in mariculture must be prevented.

Though seaweeds have been exploited in Asia for centuries, selective breeding of crops is a relatively recent practice, carried out for the purpose of increasing yields and/or value by optimizing seaweed growth, improving the biomass' general quality or increasing the content of one or several desirable compounds, as well as improving performance towards pathogens and competitors. Commercial varieties of *Saccharina japonica* (Zhang et al. 2007) as well as eucheumoid species like *Kappaphycus alvarezii* and *Eucheuma denticulatum* (Ask and Azanza 2002) illustrate successful efforts towards seaweed crop improvement. However, crop selection and the use of the selected varieties on a large scale may lead to gene swamping, where genetic material from cultivated crops is introduced to wild populations (Snow et al. 2010).

The genetic structure and degrees of isolation between populations of species that are relevant for aquaculture in Norway are still poorly understood. Recent studies conducted on kelp species showed increasing genetic isolation correlated with distance among populations. Water current patterns were identified as the primary force responsible for spreading of dispersal stages, resulting in gene flow among populations (Alberto et al. 2011; Robuchon et al. 2014). High extent of isolation may lead to phenotypic differences among populations, which can be affected by selection. Accordingly, Futsæter and Rueness (1985) reported morphological differences between sugar kelps from Northern and Southern Norway, as well as differences in growth under certain conditions. In this regard, the use and spreading of crop material along the coast of Norway must be regulated to limit the effects of "crop-to-wild" gene flows. Fredriksen and Sjøtun (2015) recommended using local ecotypes for crop breeding and restricting the produced crops within fjord systems and coastal areas defined by the eco-region based map catalogue from the Norwegian Environment Agency (Miljødirektoratet, https://kartkatalog.miljodirektoratet.no/Dataset/Details/501). Sterilization technologies for cultivated seaweeds have also been suggested as a measure to mitigate gene swamping (Loureiro et al. 2015), but examples of supposedly sterile aquaculture organisms adapting over time and eventually being able to reproduce after all (Jouaux et al. 2010) advise caution in such cases.

Impact on Ecosystems

As organisms which take up dissolved nutrients from surrounding seawater, seaweeds function as extractive components in culture systems. Thus, they assimilate dissolved inorganic nitrogen and phosphorus, which, in many coastal ecosystems, originate from terrestrial runoffs or from fish farming activities. In this way, seaweed cultivations can mitigate potential eutrophication of coastal environments (Kim et al. 2015). In addition to nutrients, seaweeds assimilate dissolved carbon in form of CO_2 as part of the photosynthetic processes. Pedersen et al. (2012) estimated the annual production of a Norwegian kelp forest (*L. hyperborea*) as between 14 and 25 t ha^{-1} dry weight, corresponding to 5.6 to 10 t ha^{-1} of carbon-bound biomass growth, ranking these kelp forests among the most productive vegetation systems on earth. Chung et al. (2013) estimated the potential of the brown alga *Ecklonia cava* cultivated on ropes to extract approximately 10 t CO_2 ha^{-1} $year^{-1}$. Large-scale seaweed cultivation is justifiably regarded as a potential temporary CO_2 sink and may contribute substantially to the management of human impacts on atmospheric and oceanic carbon cycle interactions.

Seaweed cultivations create a temporary habitat for invertebrates and fishes, increasing habitat complexity, possibly functioning as artificial reefs and contributing to ecological interactions within nearby ecosystems. Although such an effect has not been systematically investigated, seaweed cultivation fields have been observed to house associated biota, i.e. mobile drifting or swimming organisms and attached fauna and flora (Stamenic 2015). These observations indicate an overall positive effect on local biodiversity, but environmental monitoring over time will provide evidence. Similarly, benthic environments associated with seaweed cultivations are bound to be altered, and the effects need to be documented systematically. An ongoing research project in collaboration with Norwegian fish farmers (TARELAKS, contact Céline Rebours) is addressing ecosystem effects of co-cultivation of seaweeds and salmonids at pilot IMTA-sites. In this context, the effects of seaweeds on water quality (e.g. current speed, dissolved nutrient content, chlorophyll *a* content) and fish health parameters, as well as on biological diversity near the seaweed cultivation sites are being assessed. Preliminary observations obtained from a Norwegian farm in a European project (IDREEM, www.idreem.eu/) indicate high abundances of juvenile fishes associated with macroalgae cultivations (Hughes et al. 2016). Especially lumpfish, *Cyclopterus lumpus,* appear to be common in seaweed farms (Stamenic 2015). This is a potentially valuable association, as lump suckers are being used as cleaning fish in salmon farms to remove parasites (salmon lice) from the farmed fish. Salmon lice infestations represent one of the major challenges facing the salmon aquaculture industry in Norway, and developing biological controls is a priority.

Nutrient transfers form seaweeds and kelp forests to other ecosystems are key processes in nature, increasing food availability to otherwise nutrient-poor environments (Vetter 1995; Krumhansl and Scheibling 2012). However, besides acting as a temporary nutrient sink and having potentially positive effects on surrounding pelagic environments, large-scale seaweed cultivations may also cause high material deposits on the seafloor which could have adverse consequences on benthic ecosystems and biodiversity through decomposition processes and associated oxygen depletion (Viaroli et al. 2008). In addition, physical shading of both pelagic and benthic ecosystem by large-scale seaweed cultivation may affect primary production in the case of shallow inshore areas, although these effects have not yet been studied.

Epiphytes and Diseases

Problems with pathogens and epiphytes of seaweeds may arise in large-scale and high-density cultivations. In Norway, the presence of epiphytic invertebrates, especially bryozoans (e.g. *Membranipora membranacea* and *Electra pilosa*), but also blue mussels (*Mytilus edulis*), hydroids and fouling macroalgae has been reported for sugar kelp cultivated at sea during summer months (Forbord et al. 2012; Handå et al. 2013; Lüning and Mortensen 2015; Førde et al. 2016). In worst case scenarios, such fouling may lead to extensive blade damage and loss of biomass or considerable quality deterioration. Whereas, the diversity of associated organisms in natural kelps beds carries important ecological functions in coastal areas (Christie et al. 2009), biofouling by epiphytes is a major constraint for the year-round cultivation of seaweed biomass, forcing producers to harvest in May–June, hence limiting the year-round seaweed cultivation at sea (Forbord et al. 2012; Peteiro and Freire 2013; Førde et al. 2016). Outbreaks of encrusting colonies of the bryozoan *M. membranacea* are the result of settlement of cyphonaut larvae, which are present in the water column of coastal area throughout the year (Ryland 1965; Saier and Chapman 2004) with a peak of abundance during the summer (Ryland 1965; Førde et al. 2016). Larval recruitment and the extent of biofouling which may vary among cultivation sites, are linked to zooplankton population dynamics, influenced by a wide range of both biotic and abiotic factors including the temperature history and wave action (Saunders and Metaxas 2008). Seawater temperature seems to play an important role since the onset of fouling is somewhat delayed at higher latitudes (Fieler and Matsson 2015). However, the same time delay is observed for growth of kelps, with growth seasons extended into late summer in Northern Norway, compared to more southern locations. Hence, the growing season for cultivated sugar kelp is limited from October to May in Southern Norway, and possibly to August in Northern Norway (Fieler and Matsson 2015), and the crops must be harvested before the onset of fouling. In the case of IMTA, the peak nutrient release from fish farms occurs during the summer months, due to high metabolic activity and growth of the fish (Wang et al. 2012). Developing crops resistant to fouling or alternative seaweed species as summer crops, is key to increasing the production efficiency in IMTA systems.

Seaweed production in IMTA systems will also alter the surrounding water quality and the fish-rearing environment in general, e.g. by changing the oxygen content resulting from photosynthesis and respiration, as well as hydrographic regimes to the fish cages. Large-scale seaweed cultivation may also influence fish health indirectly either by acting as a physical and/or biological barrier for pathogens, or in the opposite case, as a reservoir for fish disease agents.

Whereas, evidence from IMTA sites elsewhere indicates many benefits, both on conceptual, practical, environmental and economic levels (Barrington et al. 2008; Chopin et al. 2012), potential risks associated with integrated fish and seaweed production need to be evaluated and assessed in the context of specific geography, production systems (regarding both fish and seaweed), scales of production and associated technology. Such issues are also addressed in on-going research projects in Norway.

Area Utilization

A special commission mandated by the Norwegian Ministry of Trade, Industry and Fisheries stressed that a sufficient number of locations must be assigned to aquaculture with focus on efficient production and minimal environmental impact (Gullestad 2011). An economically viable seaweed aquaculture sector requires large areas for cultivation which may lead to conflicts with other users of the coastal zone. Integrated management tools applied across municipal boundaries are therefore necessary to avoid conflicts between seaweed cultivation on the one hand and other uses, including fisheries, fairways, nature protection, renewable energy production, kelp trawling areas as well as oil and gas installations on the other (Broch et al. 2016).

Although the current demand for seaweed biomass can be met by cultivation sites in the Norwegian coastal zone, model simulations including environmental variables critical to seaweed growth (e.g. light intensity, nutrient, temperature and current speed) showed better and more stable conditions off the continental shelf than closer to the coast (Broch et al. 2016). The modelling study by Broch et al. (2016) was carried out for the coast of Møre and Romsdal county with an overall conclusion of a prospective upscaling of seaweed biomass production placed offshore.

However, offshore working environments entail technical challenges including mooring of the cultivation structures as well as logistic aspects from deployment of seedlings at sea to harvesting and processing of the biomass. Knowledge and competence transfer from existing maritime industries (e.g. oil and gas) is pointed out by the authors as a key factor for the success of the future offshore seaweed production.

Threats from Climate Change

Climate change is likely to impose threats to marine ecosystems and consequently aquaculture. Rising sea temperatures could eventually lead to significant changes both in terms of aquaculture species, optimal range for the production and localization patterns. Higher sea temperatures will involve a relocation of the boundaries of the distribution of organisms in the ocean, including seaweeds, and the general northern displacement of farmed organisms (NOU 2010). Higher summer sea surface temperatures may be problematic for farmed species that are adapted for life in cold water. As a result, the productivity of aquaculture species can be affected, and Southern Norway may be less suitable for species such as salmon, with socio-economic impacts (Hermansen and Heen 2012). Such climatic changes may eventually entail that northern areas will be better suited for mariculture than southern regions (Bergh et al. 2007). A study by Callaway et al. (2012) assessing the consequences of climate change on aquaculture species in the UK and Ireland, suggested that elevated sea temperatures and changes in hydrodynamic regimes will impact macroalgal cultivation; however, impacts will likely vary according to species and geographic locations. It is therefore particularly important to investigate the framework conditions for seaweed monoculture and IMTA along the entire Norwegian coast.

RISK ASSESSMENT 2—PRODUCT RELATED

In contrast to the many features of seaweeds beneficial for human health and nutrition, seaweeds may also accumulate undesired compounds such as heavy metals (lead, mercury, cadmium, arsenic) as well as minerals and trace elements which may be toxic above a certain limit (iodine, manganese, zinc) (Almela et al. 2006; Rose et al. 2007; Besada et al. 2009).

Variations in content and concentrations of undesirable and toxic compounds are dependent on species (Phaneuf et al. 1999), seasons and locations (Ar Gall et al. 2004), and in general, levels of toxic compounds are often related to the quality of the surrounding environment of harvested or cultivated seaweeds. However, few studies about negative consequences of consuming seaweeds have been carried out, but among the best documented are effects of high iodine content in some brown algae (Lüning and Mortensen 2015; Desideri et al. 2016). The potentially mitigating effects of primary processing of edible seaweeds on their content of toxins as well as antinutrients (i.e. compounds reducing digestibility and bio-availability of health-advancing ingredients) and on their microbial stability require further investigation.

Current research efforts in collaboration with the Norwegian Food Safety Authority (Duincker et al. 2016) are addressing these relevant questions in order to establish a regulatory framework for the use of macroalgae as human food and animal feed. Such a regulatory framework is a prerequisite for further industrial development of seaweeds as raw material for human consumption and animal production.

CONCLUSIONS

The seaweed industry is a growing economic sector in Norway. Both research communities, public and private stakeholders are collaborating to further the sustainable development of the Norwegian bio-economy based on cultivated seaweed biomass. Although cultivation technology has come a long way, it still requires optimization to upscale the biomass production and reduce the need for technical operators as well as maintenance. Current needs from seaweed producers are directed towards energy efficient processing standards that will maintain or increase product quality, ensure consumer safety and maximize biomass utilization. Integrated models using industrial surplus heat as energy source for seaweed processing are suggested, and their feasibility are currently under study.

Despite a large body of scientific literature reporting on the quality of macroalgal phytochemical constituents, commercial outputs from Norwegian cultivated seaweeds are currently limited. High-value products to be used in food, cosmetics as well as medical and pharmaceutical applications, are predicted to play an important role in creating value from Norwegian cultivated seaweed biomass. Nevertheless, expert knowledge regarding markets and prices for different seaweed products is required in order to establish market strategies. Successful product developments along with efficient logistic solutions throughout the entire value chain will build competitiveness of the Norwegian seaweed sector. The future production of macroalgae on large-scale and at relatively low cost will open perspectives for using the biomass as feedstock for bulk products (high volume and low price) e.g. bio-fuels and -chemical, and animal feed.

Upscaling seaweed cultivation requires specific solutions regarding the technical, spatial and practical integration of the production and will be specific for each production site. Although the current demand for seaweed raw material can be met by cultivating biomass within the coastal zone, large-scale production is foreseen to occur outside the continental shelf of Norway, due to more stable growth conditions and avoidance of conflicts arising with other users of the coastal zone. The Norwegian regulations regarding seaweed cultivation (including monoculture and IMTA) are not yet fully developed. Hence, the cooperation between public authorities, the industry and research institutions is needed in order to develop a knowledge-based legal framework for a sustainable industry based on cultivated seaweed biomass in Norway.

REFERENCES

1. Alberto F, Raimondi PT, Reed DC et al (2011) Isolation by oceanographic distance explains genetic structure for *Macrocystis pyrifera* in the Santa Barbara Channel. Mol Ecol 20(12):2543–2554
2. Alexander KA, Potts TP, Freeman S et al (2015) The implications of aquaculture policy and regulation for the development of integrated multi-trophic aquaculture in Europe. Aquaculture 443:16–23
3. Almela C, Clemente MJ, Velez D et al (2006) Total arsenic, inorganic arsenic, lead and cadmium contents in edible seaweed sold in Spain. Food Chem Toxicol 44(11):1901–1908
4. Ar Gall E, Küpper FC, Kloareg B (2004) A survey of iodine content in *Laminaria digitata*. Bot Mar 47(1):30
5. Ask EI, Azanza RV (2002) Advances in cultivation technology of commercial eucheumatoid species: a review with suggestions for future research. Aquaculture 206(3–4):257–277
6. Baghel RS, Trivedi N, Reddy CRK (2016) A simple process for recovery of a stream of products from marine macroalgal biomass. Bioresource Technol 203:160–165
7. Barrington K, Ridler N, Chopin T et al (2008) Social aspects of the sustainability of integrated multi-trophic aquaculture. Aquacult Int 18(2):201–211
8. Bartsch I, Wiencke C, Bischof K et al (2008) The genus *Laminaria* Sensu Lato: recent insights and developments. Eur J Phycol 43(1):1–86
9. Bergh Ø, Asplin L, Boxaspen K et al. (2007). Climate change - its consequences for Norwegian aquaculture. Marine Research, Topics 2–2007. Institute of Marine Research. https://brage.bibsys.no/xmlui/bitstream/handle/11250/116175/No.2_2007.pdf?sequence=2
10. Berteau O, Mulloy B (2003) Sulfated fucans, fresh perspectives structures, functions and biological properties of sulfated fucans and an overview of enzymes active toward this class of polysaccharide. Glycobiology 13(6):29–40

11. Besada V, Andrade JM, Schultze F et al (2009) Heavy metals in edible seaweeds commercialised for human consumption. J Marine Syst 75(1–2):305–313
12. Brinkhuis BH, Levine HG, Schlenk CG et al. (1987). *Laminaria* cultivation in the far east and North America. In: Bird and Benson (ed) Seaweed cultivation for renewable resources. Development in aquaculture and fisheries science. Elsevier, New York: 107–146.
13. Broch OJ, Skjermo J and Handå A (2016). Potensialet for storskala dyrking av makroalger i Møre og Romsdal. SINTEF Fisheries and Aquaculture - Report A27869. https://www.sintef.no/publikasjoner/publikasjon/?pubid=SINTEF+A27869
14. Brown EM, Allsopp PJ, Magee PJ et al (2014) Seaweed and human health. Nutr Rev 72(3):205–216
15. Brownlee IA, Allen A, Pearson JP et al (2005) Alginate as a source of dietary fiber. Crit Rev Food Sci Nutr 45(6):497–510
16. Buschmann A, Riquelme V, Hernandez-Gonzalez M et al (2006) A review of the impacts of salmonid farming on marine coastal ecosystems in the Southeast Pacific. ICES J Mar Sci 63(7):1338–1345
17. Callaway R, Shinn AP, Grenfell SE et al (2012) Review of climate change impacts on marine aquaculture in the UK and Ireland. Aquat Conserv 22(3):389–421
18. Chan JCC, Cheung PCK, Ang PO (1997) Comparative studies on the effect of three drying methods on the nutritional composition of seaweed *Sargassum hemiphyllum* (turn.) C. Ag.†. J Agr Food Chem 45(8):3056–3059
19. Chapman AS, Stévant P, Emblem Larssen W (2015) Food or fad? Challenges and opportunities for including seaweeds in a Nordic diet. Bot Mar 58(6):423–433
20. Choi J-S, Lee B-B, An S et al (2012) Simple freezing and thawing protocol for long-term storage of harvested fresh *Undaria pinnatifida*. Fisheries Sci 78(5):1117–1123
21. Chopin T, Buschmann A, Halling C et al (2001) Integrating seaweeds into marine aquaculture systems: a key toward sustainability. J Appl Phycol 37:975–986

22. Chopin T, Robinson SMC, Troell M et al. (2008). Multitrophic Integration for Sustainable Marine Aquaculture. In: Jorgensen SE and Fath BD (ed) The Encyclopedia of Ecology, Ecological Engineering. Elsevier, Oxford: 2463–2475.

23. Chopin T, Cooper JA, Cross S et al (2012) Open-water integrated multi-trophic aquaculture: environmental biomitigation and economic diversification of fed aquaculture by extractive aquaculture. Rev Aquacult 4(4):209–220

24. Christie H, Fredriksen S, Rinde E (1998) Regrowth of kelp and colonization of epiphyte and fauna community after kelp trawling at the coast of Norway. Hydrobiologia 375(376):49–58

25. Christie H, Jørgensen NM, Norderhaug KM et al (2003) Species distribution and habitat exploitation of fauna associated with kelp (*Laminaria hyperborea*) along the Norwegian coast. J Mar Biol Assoc UK 83(4181):1–13

26. Christie H, Norderhaug KM, Fredriksen S (2009) Macrophytes as habitat for fauna. Mar Ecol-Prog Ser 396:221–233

27. Chung IK, Oak JH, Lee JA et al (2013) Installing kelp forests/ seaweed beds for mitigation and adaptation against global warming: Korean project overview. ICES J Mar Sci 70(5):1038–1044

28. Cofrades S, López-López I, Solas MT et al (2008) Influence of different types and proportions of added edible seaweeds on characteristics of low-salt gel/emulsion meat systems. Meat Sci 79(4):767–776

29. Cordell D, White S (2011) Peak phosphorus: clarifying the key issues of a vigorous debate about long-term phosphorus security. Sustainability 3(10):2027

30. Dawczynski C, Schubert R, Jahreis G (2007) Amino acids, fatty acids, and dietary fibre in edible seaweed products. Food Chem 103(3):891–899

31. Denis C, Massé A, Fleurence J et al (2009) Concentration and pre-purification with ultrafiltration of a R-phycoerythrin solution extracted from macro-algae *Grateloupia turuturu*: process definition and up-scaling. Sep Purif Technol 69(1):37–42

32. Desideri D, Cantaluppi C, Ceccotto F et al (2016) Essential and

toxic elements in seaweeds for human consumption. J Toxicol and Env Heal, Part A 79(3):112–122

33. Directorate of Fisheries (2016a). Akvakulturregisteret. Cited 11 Jan 2017. http://www.fiskeridir.no/Akvakultur/Registre-og-skjema/Akvakulturregisteret
34. Directorate of Fisheries (2016b). Akvakulturstatistikk (tidsserier)-Alger. Cited 02 Nov 2016. http://www.fiskeridir.no/Akvakultur/Statistikk-akvakultur/Akvakulturstatistikk-tidsserier/Alger
35. Duincker A, Sunde Roiha I, Amlund H et al. (2016). Potential risks posed by macroalgae for application as feed and food - a Norwegian perspective. National Institute of Nutrition and Seafood Research (NIFES). http://www.mattilsynet.no/mat_og_vann/produksjon_av_mat/fisk_og_sjomat/rapport_makroalger_2016_nifes.23097/binary/Rapport%20makroalger%202016%20Nifes
36. Dumay J, Clément N, Morançais M et al (2013) Optimization of hydrolysis conditions of *Palmaria palmata* to enhance R-phycoerythrin extraction. Bioresource Technol 131(0):21–27
37. Enríquez S, Duarte CM, Sand-Jensen K (1993) Patterns in decomposition rates among photosynthetic organisms: the importance of detritus C:N:P content. Oecologia 94(4):457–471
38. European Commission, (2012). Communication from the commission to the european parliament, the council, the european economic and social committee and the committee of the regions: Blue Growth opportunities for marine and maritime sustainable growth, COM/2012/0494 final. http://eur-lex.europa.eu/legal-content/EN/ALL/?uri=CELEX:52012DC0494
39. Evans FD, Critchley AT (2014) Seaweeds for animal production use. J Appl Phycol 26:891–899
40. FAO, (2016a). The state of world fisheries and aquaculture: Food and Agriculture Organization of the United Nations (FAO), Fisheries and Aquaculture Department. Rome, 2012. http://www.fao.org/3/a-i5555e.pdf
41. FAO (2016b). FAO Yearbook - Fishery and Aquaculture Statistics Summary tables. Cited 01 Nov 2016. ftp://ftp.fao.org/FI/STAT/summary/default.htm
42. Fieler R and Matsson S, (2015). FoU aktiviteter ved Akvaplan-

Niva. Results presented at the SIG (Special Interest Group) seaweed workshop - Cultivation technology. Industrial Biotech Network Norway, 28th of May, Trondheim, Norway. http://www.indbiotech.no/sites/default/files/Akvaplan_Niva.pdf

43. Fleurence J (2004). Seaweed proteins. In: Yada RY (ed) Proteins in food processing. Woodhead publishing, Cambridge: 197–213
44. Forbord S, Skjermo J, Arff J et al (2012) Development of *Saccharina latissima* (Phaeophyceae) kelp hatcheries with year-round production of zoospores and juvenile sporophytes on culture ropes for kelp aquaculture. J Appl Phycol 24(3):393–399
45. Førde H, Forbord S, Handå A et al (2016) Development of bryozoan fouling on cultivated kelp (*Saccharina latissima*) in Norway. J Appl Phycol 28(2):1225–1234
46. Fredriksen S and Sjøtun K (2015). Risikovurdering ved utsetting av ikke–stedegen tare. Miljødirektoratet (Norwegian Environment Agency). Rapport M-299. http://www.miljodirektoratet.no/Documents/publikasjoner/M299/M299.pdf
47. Futsæter G and Rueness J, (1985). Metode for kontroll av livssyklus, krysningsforsøk og masseproduksjon av sporofytter i kultur av *Laminaria saccharina* (L) Lamour. University of Oslo. 23 pp.
48. Galland-Irmouli AV, Pons L, Luçon M et al (2000) One-step purification of R-phycoerythrin from the red macroalga *Palmaria palmata* using preparative polyacrylamide gel electrophoresis. J Chromatogr B 739:117–123
49. Gullestad P, (2011). Effektiv og bærekraftig arealbruk i havbruksnæringen - areal til begjær. Rapport Fra Et Ekspertutvalg Oppnevnt Av Fiskeri- Og Kystdepartementet, Oslo, 4. feb 2011. https://www.regjeringen.no/globalassets/upload/fkd/vedlegg/rapporter/2011/effektiv_og_baerekraftig_arealbruk_i_havbruksnaeringen.pdf
50. Gupta S, Abu-Ghannam N (2011) Recent developments in the application of seaweeds or seaweed extracts as a means for enhancing the safety and quality attributes of foods. Innov Food Sci Emerg 12(4):600–609
51. Gupta S, Cox S, Abu-Ghannam N (2011) Effect of different drying temperatures on the moisture and phytochemical constituents of

edible Irish brown seaweed. Food Sci Technol-LEB 44(5):1266–1272

52. Handå A, Forbord S, Wang X et al (2013) Seasonal- and depth-dependent growth of cultivated kelp (*Saccharina latissima*) in close proximity to salmon (*Salmo salar*) aquaculture in Norway. Aquaculture 414-415:191–201
53. Hermansen Ø, Heen K (2012) Norwegian salmonid farming and global warming: socioeconomic impacts. Aquacult Econ Manage 16(3):202–221
54. Herrmann C, FitzGerald J, O'Shea R et al (2015) Ensiling of seaweed for a seaweed biofuel industry. Bioresource Technol 196:301–313
55. Holdt SL, Kraan S (2011) Bioactive compounds in seaweed: functional food applications and legislation. J Appl Phycol 23(3):543–597
56. Hou X, Hansen JH, Bjerre A-B (2015) Integrated bioethanol and protein production from brown seaweed *Laminaria digitata*. Bioresource Technol 197:310–317
57. Hughes AD, Corner RA, Cocchi M et al. (2016). Beyond fish monoculture: Developing Integrated Multi-trophic Aquaculture in Europe. IDREEM (Increasing Resource Efficiency in European Mariculture) project report. Project funded by the European Union's Seventh Framework Programme for research, technological development and demonstration under grant agreement no. 308571. http://www.idreem.eu/cms/wp-content/uploads/2016/10/IDREEM_FINALREPORT_PRINT_710_web_2.pdf
58. Jensen A (1993). Present and future needs for algae and algal products. In: Chapman ARO, Brown MT and Lahaye M (ed) Fourteenth International Seaweed Symposium. Springer Netherlands. 85: 15–23
59. Jouaux A, Heude-Berthelin C, Sourdaine P et al (2010) Gametogenic stages in triploid oysters *Crassostrea gigas*: irregular locking of gonial proliferation and subsequent reproductive effort. J Exp Mar Biol Ecol 395(1–2):162–170
60. Kandasamy S, Khan W, Evans F et al (2012) Tasco(R): a product of *Ascophyllum nodosum*enhances immune response of *Caenorhabditis*

elegans against *Pseudomonas aeruginosa*infection. Mar Drugs 10(1):84–105

61. Kim JK, Kraemer GP, Yarish C (2015) Use of sugar kelp aquaculture in Long Island sound and the Bronx River estuary for nutrient extraction. Mar Ecol-Prog Ser 531:155–166
62. Kraan S, Verges Tramullas A, Guiry M (2000) The edible brown seaweed*Alaria esculenta*(Pheophyceae, Laminariales) hybridization growth and genetic comparisons of six Irish populations. J Appl Phycol 12:577–583
63. Krumhansl KA, Scheibling RE (2012) Production and fate of kelp detritus. Mar Ecol-Prog Ser 467:281–302
64. Le Bras Q, Lesueur M, Lucas S et al. (2015a). Etude du marché français des algues alimentaires. Catalogue et analyse des produits existants. Programme IDEALG Phase 2: Les publications du Pôle halieutique AGROCAMPUS OUEST, vol. 37. 41 pp. http://halieutique.agrocampus-ouest.fr/pdf/4661.pdf
65. Le Bras Q, Lesueur M, Lucas S et al. (2015b). Etude du marché français des algues alimentaires. Panorama de la distribution en magasins. Programme IDEALG Phase 2: Les publications du Pôle halieutique AGROCAMPUS OUEST, vol. 36. 42 pp. http://halieutique.agrocampus-ouest.fr/pdf/4662.pdf
66. Liot F, Colin A, Mabeau S (1993) Microbiology and storage life of fresh edible seaweeds. J Appl Phycol 5(2):243–247
67. López-López I, Cofrades S, Ruiz-Capillas C et al (2009) Design and nutritional properties of potential functional frankfurters based on lipid formulation, added seaweed and low salt content. Meat Sci 83(2):255–262
68. Lorentsen S-H, Sjøtun K, Grémillet D (2010) Multi-trophic consequences of kelp harvest. Biol Conserv 143(9):2054–2062
69. Loureiro R, Gachon CMM, Rebours C (2015) Seaweed cultivation: potential and challenges of crop domestication at an unprecedented pace. New Phytol 206(2):489–492
70. Lüning K, Mortensen LM (2015) European aquaculture of sugar kelp (*Saccharina latissima*) for food industries: iodine content and epiphytic animals as major problems. Bot Mar 58(6):449–455

71. Mabeau S, Fleurence J (1993) Seaweed in food products: biochemical and nutritional aspects. Trends Food Sci Tech 4:103–107

72. MacArtain P, Gill CIR, Brooks M et al (2007) Nutritional value of edible seaweeds. Nutr Rev 65(12):535–543

73. Maeda H (2015) Nutraceutical effects of Fucoxanthin for obesity and diabetes therapy: a review. J Oleo Sci 64(2):125–132

74. Mæhre HK, Malde MK, Eilertsen KE et al (2014) Characterization of protein, lipid and mineral contents in common Norwegian seaweeds and evaluation of their potential as food and feed. J Sci Food Agric 94(15):3281–3290

75. Ministry of Fisheries and Coastal affairs, (2009). Strategy for an Environmentally Sustainable Norwegian Aquaculture Industry. https://www.regjeringen.no/globalassets/upload/fkd/vedlegg/diverse/2009/strategy-for-an-sustainable-aquaculture.pdf

76. Mork M (1996). Wave Attenuation due to Bottom Vegetation. In: Grue J, Gjevik B and Weber JE (ed) Waves and Nonlinear Processes in Hydrodynamics. Springer Netherlands, Dordrecht: 371–382

77. Mouritsen OG, Williams L, Bjerregaard R et al (2012) Seaweeds for umami flavour in the new Nordic cuisine. Flavour 1(1):4

78. Mouritsen OG, Dawczynski C, Duelund L et al (2013) On the human consumption of the red seaweed dulse (*Palmaria palmata* (L.) Weber & Mohr). J Appl Phycol 25:1777–1791

79. Neori A, Chopin T, Troell M et al (2004) Integrated aquaculture: rationale, evolution and state of the art emphasizing seaweed biofiltration in modern mariculture. Aquaculture 231(1–4):361–391

80. NETALGAE, (2012). Overview of the seaweed industry by country: Ireland, France, Norway, Portugal, Spain, United Kingdom. A summary of the initial findings of the Netalgae project regarding the current status of the seaweed industries in the participant countries. Netalgae project financed by the Atlantic Area Operational Programme of the European Union. http://www.netalgae.eu/uploadedfiles/posters_final_UK_1.pdf

81. NOU, (2010). Adapting to a changing climate. Norway's vulnerability and the need to adapt to the impacts of climate change. Recommendation by a committee appointed by Royal Decree of 5 December 2008, submitted to the Ministry of the Environment on

15 November 2010. Official Norwegian Reports, Noregs Offentlege Utgreiingar (NOU) https://www.regjeringen.no/contentassets/00f70698362f4f889cbe30c75bca4a48/pdfs/nou201020100010000en_pdfs.pdf

82. Olafsen T, Winther U, Olsen Y et al. (2012). Value created from productive oceans in 2050. Report of a working group appointed by the Royal Norwegian Society of Sciences (DKNVS) and the Norwegian Academy of Technological Sciences (NTVA). https://www.sintef.no/contentassets/f025260af6b8435394eced5e03939e11/value-created-from-productive-oceans-in-2050.pdf
83. Paull RE, Chen NJ (2008) Postharvest handling and storage of the edible red seaweed *Gracilaria*. Postharvest Biol Techn 48(2):302–308
84. Pedersen MF, Nejrup LB, Fredriksen S et al (2012) Effects of wave exposure on population structure, demography, biomass and productivity of the kelp *Laminaria hyperborea*. Mar Ecol-Prog Ser 451:45–60
85. Pereira L (2011). A review of the nutrient composition of selected edible seaweeds. In: Pomin VH (ed) Seaweed: Ecology, Nutrient Composition and Medicinal Uses. Nova Science Publishers, Inc.: 15–50.
86. Peteiro C, Freire Ó (2013) Epiphytism on blades of the edible kelps *Undaria pinnatifida* and *Saccharina latissima* farmed under different abiotic conditions. J World Aquacult Soc 44(5):706–715
87. Phaneuf D, Côté I, Dumas P et al (1999) Evaluation of the contamination of marine algae (seaweed) from the St. Lawrence River and likely to be consumed by humans. Environ Res 80(2):S175–S182
88. Robuchon M, Le Gall L, Mauger S et al (2014) Contrasting genetic diversity patterns in two sister kelp species co-distributed along the coast of Brittany, France. Mol Ecol 23(11):2669–2685
89. Rose M, Lewis J, Langford N et al (2007) Arsenic in seaweed—forms, concentration and dietary exposure. Food Chem Toxicol 45(7):1263–1267
90. Rueness J (1998). Alger i farger. En felthåndbok om kystensmakroalger. Almater Forlag, Oslo.

91. Rupérez P (2002) Mineral content of edible marine seaweeds. Food Chem 79:23–26

92. Ruvinov E, Cohen S (2016) Alginate biomaterial for the treatment of myocardial infarction: progress, translational strategies, and clinical outlook: from ocean algae to patient bedside. Adv Drug Deliver Rev 96:54–76

93. Ryland JS (1965). Polyzoa (Bryozoa). Order Cheilostomata, cyphonautes larvae. In: Fiches d'identification du zooplancton, vol. 107. Conseil International pour l'Exploration de la Mer, ICES Copenhagen, p 6.

94. Saier B, Chapman AS (2004) Crusts of the alien bryozoan *Membranipora membranacea* can negatively impact spore output from native kelps (*Laminaria longicruris*). Bot Mar 47:265

95. Sánchez-Machado DI, López-Cervantes J, López-Hernández J et al (2004) Fatty acids, total lipid, protein and ash contents of processed edible seaweeds. Food Chem 85(3):439–444

96. Saunders M, Metaxas A (2008) High recruitment of the introduced bryozoan *Membranipora membranacea* is associated with kelp bed defoliation in Nova Scotia, Canada. Mar Ecol-Prog Ser 369:139–151

97. Schaffelke B, Smith JE, Hewitt CL (2006) Introduced Macroalgae—a growing concern. J Appl Phycol 18(3):529–541

98. Schiener P, Black KD, Stanley MS (2015) The seasonal variation in the chemical composition of the kelp species *Laminaria digitata*, *Laminaria hyperborea*, *Saccharina latissima* and *Alaria esculenta*. J Appl Phycol 27(1):363–373

99. Sekar S, Chandramohan M (2008) Phycobiliproteins as a commodity: trends in applied research, patents and commercialization. J Appl Phycol 20(2):113–136

100. Statistisk Sentralbyrå (2016). Akvakultur, 2013, endelige tall. Cited 16 Dec 2016. http://www.ssb.no/jord-skog-jakt-og-fiskeri/statistikker/fiskeoppdrett

101. Skjermo J, Aasen IM, Arff J et al. (2014). A new Norwegian bioeconomy based on cultivation and processing of seaweeds: opportunities and R&D needs. SINTEF Fisheries and Aquaculture - Report A25981. https://www.sintef.no/contentassets/04aa6cd738b2

45979b334715163c0506/seaweed-bioeconomy-2014-revised-2.pdf

102. Smale DA, Burrows MT, Moore P et al (2013) Threats and knowledge gaps for ecosystem services provided by kelp forests: a Northeast Atlantic perspective. Ecol Evol 3(11):4016–4038
103. Snow AA, Culley TM, Campbell LG et al (2010) Long-term persistence of crop alleles in weedy populations of wild radish (*Raphanus raphanistrum*). New Phytol 186(2):537–548
104. Soler-Vila A, Coughlan S, Guiry M et al (2009) The red alga *Porphyra dioica* as a fish-feed ingredient for rainbow trout effects on growth. Feed efficiency and carcass composition. J Appl Phycol 21:617–624
105. Stamenic J (2015). From monoculture to Integrated Multi-Trophic Aquaculture (IMTA): mapping and valuation of ecosystem services after the integration of seaweed cultivation on a Norwegian salmon farm. Thesis submitted in partial fulfillment of the requirements for the Degree of Msc in Environmental Management for Business, Cranfield University.
106. Tierney MS, Smyth TJ, Rai DK et al (2013) Enrichment of polyphenol contents and antioxidant activities of Irish brown macroalgae using food-friendly techniques based on polarity and molecular size. Food Chem 139(1–4):753–761
107. Troell M, Joyce A, Chopin T et al (2009) Ecological engineering in aquaculture—potential for integrated multi-trophic aquaculture (IMTA) in marine offshore systems. Aquaculture 297(1–4):1–9
108. Vetter EW (1995) Detritus-based patches of high secondary production in the nearshore benthos. Mar Ecol-Prog Ser 120(1):251–262
109. Viaroli P, Bartoli M, Giordani G et al (2008) Community shifts, alternative stable states, biogeochemical controls and feedbacks in eutrophic coastal lagoons: a brief overview. Aquat Conserv 18(S1):S105–S117
110. Wang T, Jónsdóttir R, Ólafsdóttir G (2009) Total phenolic compounds, radical scavenging and metal chelation of extracts from Icelandic seaweeds. Food Chem 116(1):240–248
111. Wang T, Jónsdóttir R, Kristinsson HG et al (2010) Inhibition of haemoglobin-mediated lipid oxidation in washed cod muscle and

cod protein isolates by *Fucus vesiculosus* extract and fractions. Food Chem 123(2):321–330

112. Wang X, Olsen LM, Reitan KI et al (2012) Discharge of nutrient wastes from salmon farms: environmental effects, and potential for integrated multi-trophic aquaculture. Aquac Environ Interact 2(3):267–283
113. Wilke T, Faulkner S, Murphy L et al (2015) Seaweed enrichment of feed supplied to farm-raised Atlantic salmon (*Salmo salar*) is associated with higher total fatty acid and LC n-3 PUFA concentrations in fish flesh. Eur J Lipid Sci Tech 117(6):767–772
114. Xiaojun Y, Xiancui L, Chengsu Z et al (1996) Prevention of fish oil rancidity by phlorotannins from *Sargassum kjellmanianum*. J Appl Phycol 8(3):201–203
115. Zhang Q-S, Tang X-X, Cong Y-Z et al (2007) Breeding of an elite *Laminaria* variety 90-1 through inter-specific gametophyte crossing. J Appl Phycol 19(4):303–311

CHAPTER

4

SEAWEEDS: AN OPPORTUNITY FOR WEALTH AND SUSTAINABLE LIVELIHOOD FOR COASTAL COMMUNITIES

Céline Rebours[1,12], Eliane Marinho-Soriano[2], José A. Zertuche-González[3], Leila Hayashi[4], Julio A. Vásquez[5], Paul Kradolfer[6], Gonzalo Soriano[7], Raul Ugarte[8], Maria Helena Abreu[9], Ingrid Bay-Larsen[10], Grete Hovelsrud[10], Rolf Rødven[1], and Daniel Robledo[11]

[1]Bioforsk, Norwegian Institute for Agricultural and Environmental Research, Ås, Norway

[2]Department of Oceanography and Limnology, Federal University of Rio Grande do Norte, Natal, Brazil

[3]Instituto de Investigaciones Oceanológicas, Universidad Autónoma de Baja California, Ensenada, Mexico

[4]Aquaculture Department, Federal University of Santa Catarina, Florianópolis, Brazil

[5]Facultad de Ciencias del Mar, Centro de Estudios Avanzados en Zonas Aridas, Universidad Católica del Norte, Coquimbo, Chile

[6]PSW SA, Lima 16, Peru

[7]Soriano SA, Gaiman, Argentina

[8]Acadian Seaplants Limited, Dartmouth, Canada

[9]ALGAPlus Lda, Ílhavo, Portugal

[10]Nordland Research Institute, Bodø, Norway

[11]Department of Marine Resources, CINVESTAV-IPN, Mérida, Mexico

[12]Arctic Agriculture and Land Use Division, Bioforsk, Bodø, Norway

ABSTRACT

The European, Canadian, and Latin American seaweed industries rely on the sustainable harvesting of natural resources. As several countries wish to increase their activity, the harvest should be managed according to integrated and participatory governance regimes to ensure production within a long-term perspective. Development of regulations and directives enabling the sustainable exploitation of natural resources must therefore be brought to the national and international political agenda in order to ensure environmental, social, and economic values in the coastal areas around the world. In Europe, Portugal requires an appraisal of seaweed management plans while Norway and Canada have developed and implemented coastal management plans including well-established and sustainable exploitation of their natural seaweed resources. Whereas, in Latin America, different scenarios of seaweed exploitation can be observed; each country is however in need of long-term and ecosystem-based management plans to ensure that exploitation is sustainable. These plans are required particularly in Peru and Brazil, while Chile has succeeded in establishing a sustainable seaweed-harvesting plan for most of the economically important seaweeds. Furthermore, in both Europe and Latin America, seaweed aquaculture is at its infancy and development will have to overcome numerous challenges at different levels (i.e., technology, biology, policy). Thus, there is a need for regulations and establishment of "best practices" for seaweed harvesting, management, and cultivation. Trained human resources will also be required to provide information and education to the communities involved, to enable seaweed utilization to become a profitable business and provide better income opportunities to coastal communities.

Keywords: Coastal management, Latin America, Marine resources, Seaweed industry

INTRODUCTION

The worldwide seaweed industry provides a wide variety of products for direct or indirect human uses that have an estimated total value of US$10 billion per year (Bixler and Porse 2011; FAO 2013). Sea vegetables for human consumption constitute about 83 % of production (Craigie 2011), while the remainder is used as fertilizers and animal feed additives, medical applications (Zimmermann et al. 2005; Ehrhart et al. 2013), and biotechnological applications (McHugh 2003). Worldwide, macroalgal production increases 5.7 % every year and more than 18 million tons of macroalgae were produced from global capture and aquaculture in 2011 (FAO 2014). In 2011, 96 % of the global total production of macroalgae came from aquaculture, with Asian countries dominating seaweed culture production (99.05 % by quantity and 99.36 % by value, FAO 2014). Five genera (e.g., *Saccharina*, *Undaria*, *Porphyra*, *Eucheuma*/*Kappaphycus*, and *Gracilaria*) represented around 98 % of the world's cultivated seaweed production (Suo and Wang 1992; Pereira and Yarish 2008). *Saccharina japonica* was the most cultivated algae in the world until 2010 when the production of *Eucheuma*/*Kappaphycus* reached over 5.5 million tons for a value over US$1.3 billion (Suo and Wang 1992; McHugh 2003; FAO 2014). *Saccharina* and *Eucheuma*/*Kappaphycus* are mostly produced as raw materials for the food and food polymer industries. Aquaculture of seaweed is scarce outside of Asia, which triggered a worldwide search for hitherto unexploited natural seaweed resources. In 2011, 786,466 t of seaweeds was commercially harvested in 28 countries, ranging from cold to tropical coastlines in both hemispheres, with over 55 % of the biomass harvested in Latin America and almost 32 % in Europe (FAO 2014). The top producers are Chile and Norway respectively accounting for 51.3 and 19.2 % of the global catches of natural seaweed (FAO 2014).

The Pacific coast of South America is naturally rich in marine resources (Thiel et al. 2007; Ortiz 2008; Taylor et al. 2008; Vásquez et al. 2012; Vega et al. 2013). In Latin America, macroalgae are an important group for species richness of all regions, varying from 4.9 to 8.7 % of total marine species biodiversity (O'Dor et al. 2010). The highest biodiversity of macroalgal species is reported for the Brazilian region (10.6 species per 100 km of coast), followed by the Humboldt Current system (7.3 species per 100 km of coast), the Tropical West Atlantic (7.1 species per 100 km of coast), and the Tropical East Pacific (6.0 species per 100 km of coast), with the lowest diversity attributed to the Patagonian Shelf (4.7 species per 100 km of coast; Miloslavich et al. 2011). The trends reported so far, however, may not truly reflect real patterns, as sampling has not been equal throughout the continent, and taxonomic capacity is very uneven from one country to another, as is the case in the Caribbean

(Robledo and Townsend 2006; Miloslavich et al. 2010). In this context, there are still some efforts to be made on basic research to describe and evaluate some of the macroalgal resources in the region (Critchley et al. 2006).

Macroalgae or seaweeds are one of major components of primary biomass production in coastal maritime ecosystems and play an essential ecological role as habitat and substrata for invertebrates, fish, mammals, and birds (Vásquez 1992; Graham et al. 2007). Drastic reduction of any macroalgal community directly influences marine biodiversity, as well as reproduction, recruitment, and growth rates of marine fauna (Vásquez and Santelices 1984; Vásquez 1993). Furthermore, macroalgae may also protect coastlines against erosion (Dayton et al. 1984) and contribute significantly to the marine carbon cycle (Thiel et al. 2007; Ugarte et al. 2010; Vásquez et al. 2013). It is possible that overexploitation of natural seaweed resources could lead to significant ecological, economic, and social consequences at local, regional, and even global scales (Graham et al. 2007; Rebours and Karlsen 2007). In this regard, Alvarez and Vodden (2009) examined the relationships between human actors and *Chondracanthus chamissoi*harvesting in the community of Pisco in Peru. By using local ecological knowledge (LEK), these authors found that a disorderly extraction of *C. chamissoi* generates reductions in the resource that could be exacerbated by climatic change ("El Niño" and "La Niña" phenomena) and the rupture of ecological cycles due to pollution of the marine space and decline of marine species.

Accurate data about seaweed aquaculture or the exploitation of the natural resource stock are difficult to obtain through the international channels in most of the Latin American countries (except in Chile). The necessary information was extracted from database queries at the Harvest data from Instituto de Fomento Pesquero (IFOP) in Chile; Secretaría de Agricultura, Ganadería, Desarrollo Rural, Pesca y Alimentación (SAGARPA) in Mexico; numerous Dirección Regional de Produce (DIREPRO) in Peru; and the FAO Fisheries and Aquaculture Information and Statistics Service to discuss the trends for some Latin

American countries, especially for those harvesting seaweeds. The potential for seaweed aquaculture, particularly for red seaweeds, has been addressed in Hayashi et al. (2013). In Latin America, exploitation of seaweed happened or has happened in Argentina, Brazil, Chile, Mexico, Peru, Uruguay, and Venezuela (Critchley et al. 2006; FAO 2013). The main seaweed exploitation has been reported in Argentina, Brazil, Chile, Mexico, and Peru. In these countries, traditional harvesting and/or aquaculture is presently active or has been attempted and has generated an estimated biomass volume around 554,585 t in 2009 (equivalent to US$115 million in value), with an estimated decline in total volume to 444,686 t in 2011 (equivalent to US$22 million; Table 1). The main objective of this study is to present a short overview of the management and regulations in place for the exploitation and harvesting of seaweeds in the five above-mentioned Latin American countries. Some of the environmental, social, economic, and political obstacles to sustainable development for seaweed utilization in Latin America will be identified through multidisciplinary analyses, combining natural, ecological, and social sciences. The European Environmental Agency (EEA) also stressed the need for an analytical approach for coastal areas and to place this in an ecosystem-based management approach combining integrated spatial planning and management. Included in this is the need for a consolidated knowledge base and widespread information sharing to support informed policy development and management actions (EEA 2013). Examples for long-term management from Europe (Norway, Portugal) and North America (Canada) are then shortly presented, and successful practices in these Northern regions are then identified and presented as possible as "best practices" model for long-term sustainable seaweed exploitation in Latin America. Particular emphasis is given to identify the possible actions in particular to overcome social, economic, and political obstacles for using Latin American seaweeds as an economic resource for wealth and sustainable livelihoods for coastal communities.

Table 1. Harvesting of seaweeds in Argentina, Brazil, Chile, Mexico, and Peru and commercial aquaculture of seaweeds attempted in only four of these countries. Harvest and aquaculture volumes of seaweed biomass in tons

	2000	2001	2002	2003	2004	2005	2006	2007	2008	2009	2010	2011
Seaweed harvest[a]												
Argentina	3											
Brazil												
Chile	280,844	299,759	315,660	349,128	410,850	425,343	334,674	337,206	406,223	456,184	380,742	417,965
México	33,555	46,927	30,124	28,996	27,418	5,277	5,751	5,093	4,900	5,814	6,009	5,721
Peru	1,323	5,517	6,176	7,864	7,418	5,000	3,118	9,213	12,107	3,874	3,836	5,801
Seaweed aquaculture[b]												
Argentina	3											
Brazil									320[c]	520[c]	730[c]	730[c]
Chile	33,471	65,538	71,648	40,079	20,273	15,493	33,586	23,668	21,687	88,193	12,179	14,469
México												
Peru	11	12										

IFOP Instituto de Fomento Pesquero (www.ifop.cl), *SAGARPA* Secretaría de Agricultura, Ganadería, Desarrollo Rural, Pesca y Alimentación (http://www.sagarpa.gob.mx), *DIREPRO* Dirección Regional de Produce (Regional Direction from Production Ministry)

[a]Harvest data from Chile (IFOP), Mexico (SAGARPA), Peru (DIREPRO)

[b]Aquaculture data from FAO (18/12/2013)

[c]FAO estimates

STATE-OF-THE ART ON THE MANAGEMENT AND REGULATION IN LATIN AMERICA SEAWEED INDUSTRY

In Latin America, several species of red and brown seaweed are harvested as a raw material for phycocolloid extraction. Chile was one of the first agar producers in the world and supplies 10 % of the global biomass for alginates (Bixler and Porse 2011; Vásquez et al. 2012). In Chile, coastal communities have harvested primarily *Gracilaria* and *Lessonia* species since the 1980s, while in more recent years, *Macrocystis* species has been cultivated, first to feed abalone and most recently for bioenergy production (Buschmann et al. 2014). In Chile, several other benthic fisheries like sea urchins and molluscs depend on existence of seaweed beds. The exploitation of natural seaweed resources is therefore regulated with aims to protect the target species as well as the associated biodiversity. Several management strategies have been implemented successfully during the last 10 years, considering comanagement between fishermen unions and the state, using strategies with biological and ecological bases like morphological constraints, quotas by fishing areas, biological bans, rotation of harvesting, and also using experimental areas for harvesting and collection (Vásquez 2008; Vásquez et al. 2012). Lately, Vega et al. (2013) suggested an ecosystem approach and community indicators to manage wild brown seaweed stocks along the country.

In Argentina, different commercial species can be found such as *Undaria pinnatifida, Gracilaria gracilis, Gigartina skottsbergii, Lessonia vadosa, Macrocystis pyrifera, Porphyra columbina, Ulva lactuca*, and *Codium fragile*. In 1958, *G. gracilis* and *G. skottsbergii* were already harvested for the agar and carrageenan industries, respectively. Since 1980, *L. vadosa*and *M. pyrifera* are exported to the USA and China to supply the alginate industry, and since 1999, the uses of the Argentinian seaweeds have expended to new markets, such as for human consumption, nutraceuticals, and cosmetic including the fucoidan industries. All seaweed is harvested in Patagonia, mostly in the provinces of Chubut (Fig. 1a, b) and Santa Cruz. Local farmers directly sell the seaweeds to the processing companies or companies with concessions which directly employ their own workers for harvesting during the year and contracted divers in the summer (Fig. 1 c–i). Harvest is regulated since 1970 by local government through special licenses for 3, 10, or 30 years. The duration of the license is decided based on the investment of the companies.

Today, there is only one company (Soriano SA) that has worked steadily through the years producing agar and carrageenan. The National Center of Patagonia (CENPAT) guarantees that the harvesting methods are performed in a sustainable way. Regulations for the management of brown seaweeds and marine concessions are particularly well developed, and the supply in brown seaweed to the alginate industry is well managed and organized (Zaixso et al. 2006; Soriano SA unpublished report).

Figure 1. Argentina. Harvesting of seaweed in Patagonia. (a) Beach-casted seaweed is harvested. (b) Seabed of *U. pinnatifida* is harvested at low tide. (c) Transport in truck of the harvest biomass to the drying site. (d) Algae are harvested in nets. e Suspended net for drying the *U. pinnatifida*. (f) Selection of high-quality algae (processing of high-quality algae). (g) Preparation of high-quality *U. pinnatifida* for the Japanese market. (h) Storage of *Lessonia* sp. in bulk for the Chinese market. (i) Quality control on final product (photos: G Soriano).

In Mexico, the federal government manages all fisheries. However, under a law published in 2009, individual states can also manage sessile marine resources through an agreement to the federation. Presently, Mexico's seaweed biomass is sold to the phycocolloid industry, abalone farming, and seaweed extracts for agriculture (DOF 2012; Zertuche-González et al. 2013). None of the four commercial seaweed species harvested in Mexico (*M. pyrifera*, *Gelidium robustum*, *Chondracanthus canaliculatus*, *Gracilariopsis lemaneiformis*) are endangered, thanks to the application of proper harvesting methods (Hernández-Garibay et al. 2006; Robledo 2006; DOF 2012). The highest reduction of the seaweed resource was observed along the Pacific Mexican coast during the years when El Niño occurred, but even then, the resources recovered successfully (Ladah et al. 1999; Casas-Valdez et al. 2003). In Mexico, seaweed harvesting may be classified into three broad categories of activity: by hand (at low tide), by diving, and by mechanized methods (specialized boats). Men and women participate both in harvesting and drying. Women and other family members are however mostly involved in the manual gathering, while men dominate the diving and mechanical harvesting, due to the hardship of the work. *C. canaliculatus* has been harvested for carrageenan, by hand, during low tide since 1966, and it is a sustainable harvest up to today (Fig. 2 a–d). The economic outcome from sustainable exploitation of natural resources does not yet necessarily assure a good livelihood to fishermen. Fishermen involved in seaweed harvesting by boat and diving (such as in the case of *G. robustum* and *M. pyrifera*) get better income than those harvesting by hand in low tide. *C. canaliculatus*, for instance, has been harvested for carrageenan, by hand, during low tide since 1966, and it is a sustainable harvest up to today (Fig. 2 a–d); however, harvesting and drying methods are artisanal and seasonal and therefore generate low income. Fishermen that harvest seaweeds from boats and diving gear complement their income by participating in other fisheries such as abalone or sea urchin. Although the seaweed fishery is economically or environmentally sustainable, the livelihood of the fishermen is not always attractive. It is therefore important to look at the fishery in the larger context, and the cultivation of seaweeds (as opposed to their harvesting) may actually offer a better alternative to coastal communities, as it can give them the opportunity to increase production and improve productivity and quality.

In this regard, a socio-economic analysis of seaweed farming involving coastal communities was performed in the Gulf of Mexico coast using assumptions and data from the pilot cultivation of *Kappaphycus alvarezii* carried out in Dzilam, Yucatan Peninsula (Robledo et al. 2013). Based on this experience, various social and institutional factors in seaweed farming were discussed, which indicated a good potential for seaweed farming to become an integrated part of the local livelihood strategies for community development (Fig. 2e–g). One of the major conclusions of this study was that any integration of seaweed growing and industrialization in Mexico would require an interest from industry as well as local investors and government authorities in order to ensure successful implementation and development of the activity.

Figure 2. Mexico. (a–d) Harvesting of *Chondracanthus canaliculatus* in San Quintin, Baja California. *C. canaliculatus* is harvested by hand in low tide. In the harvest, men and women participate together. (Photos: J Zertuche). (e–g) *Kappaphycus alvarezii*aquaculture and harvesting in Dzilam de Bravo, Yucatan (photos: D Robledo).

Peru is just starting to establish regulations for management or exploitation of the seaweed resources including some harvesting restrictions associated with the brown seaweed populations (Vásquez et al. 2012). Southern Peru has approximately 1,000 km of exploitable coastline, with several major, natural populations of brown seaweed species, such as *Lessonia nigrescens*, *Lessonia trabeculata*, *Macrocystis integrifolia*, and *M. pyrifera* (Acleto 2006). The Peruvian production was previously based on the harvesting of *Gracilaria* species. Until 2001, drift brown seaweeds were collected along the shoreline and directly sold to one processing company. McHugh (2002) reported that *L. nigrescens* and *L. trabeculata* were available from natural beds in the south of Peru, although the biomass was strongly affected by El Niño events (Taylor et al. 2008). In the years when the natural marine resources are not negatively affected by El Niño events, about 3,000 t of dry *L. nigrescens* and dry *L. trabeculata* is exported to Asia (Peruvian seaweed exports 2012, PSW SA unpublished report). *M. pyrifera* is also widely available, and prior to 2001, it was exported only in small quantities. In 2012, 95.5 % of the Peruvian seaweed exports was done by the four main Peruvian-based companies (one Chilean and three Chinese). Today, there are four companies purchasing seaweed from the local communities: two Peruvian companies, one Chilean, and one Chinese-owned company. The occurrence of Chinese-owned companies in the Peruvian market has triggered an intensive competition for obtaining the seaweed biomass from the local fishers. As captured fisheries (including seaweed) decline because of overharvesting, the prices of target species often increase dramatically. Likewise, the most commonly harvested fish in Peru, anchoveta (Engraulis ringens), has had a dynamic history of overharvest and fluctuating production (Diana 2009). As a result, beach prices of dried seaweeds increased from US$60 to US$400 t^{-1}between 2006 and 2008 (Table 2, PSW SA unpublished report), creating in the short term an uncontrolled intensification of the harvesting and resulting in a depletion of the natural resources that could have long-term consequences for the Peruvian fishers and Peruvian seaweed industry. Thus, while anchoveta indeed is the fundamental fish species in the Peruvian ecosystem, there are other fisheries to be considered for management (including seaweed). There are trade-offs in managing fisheries, and ideally, such trade-offs should be known when setting fisheries policies (Christensen et al. 2014). This area is well known for experiencing large changes in the abundance and species composition of its main fish resources. Three decades ago, the Peruvian fishery also experienced an economic and social crisis due to an absence of adequate management actions in relation to overexploitation of the anchovy resource (Schreiber 2013). Particular favorable environmental conditions, good recruitment coupled with careful management, and a surveillance scheme have apparently contributed to the fast recovery of the anchovy resource at the end of the 1990s, provided by the political regulatory organism for fisheries resource PRODUCE (Ministry of Production) with the information given by their scientific and technical entity Marine Institute of Peru (IMARPE).

Table 2. Prices in US dollars per ton of dried seaweed paid to the Peruvian fishermen excluding the tax (IGV 19 %; PSW SA unpublished data)

Species	Early 2007	Early 2008	At present
Macrocystis pyrifera	74	170	400
Chondracanthus chamissoi	440	–	840

In the case of Brazil, historically, the seaweed industry is based on harvesting of seaweed from wild populations, a practice maintained until today. Industries based on *Kappaphycus* farming have also been established in the last 20 years but still need to address the issues relative to the acceptance of this new activity from the authorities (Hayashi et al. 2013). Despite the large number of economically commercial species, only few species of *Gracilaria* and *Hypnea* are exploited commercially in Brazil (Marinho-Soriano et al. 2006). Red seaweed exploitation began in the 1960s along the northeast coast and was undertaken without regulations. In 2006, the Brazilian Institute of Environment and Renewable Natural Resources (IBAMA) implemented a regulation on the harvest of natural resources along the coastline. However, the lack of taxation caused overexploitation of natural beds and most of activities that were then dependent on the harvesting of these species have nowadays stopped (Hayashi et al. 2013). Seaweeds are still a significant source of income and support the livelihoods of many people within coastal communities. Harvest is considered a secondary activity, undertaken without the participation of the family heads, who are generally fishermen (E Marinho-Soriano unpublished). On the other hand, for many women, harvesting is considered their main occupation, even if it is performed only part time, during low tide (Fig. 3a–c). The harvest is done on the shores or by diving from a small boat ("jangadas," Fig. 3d, e), when the seaweed is further out. This activity is mostly carried out by women and their children equipped with a net bag (Fig. 3b, c). After the harvest, the biomass is brought back to the shore and then sun-dried for 2 to 3 days on land (Fig. 3f). Finally, the dried seaweed is sold to an intermediate collector and will be used for agar or carrageenan production (E Marinho-Soriano unpublished).

Figure 3. Brazil. Harvesting and aquaculture in the community of Rio do Fogo, Rio Grande do Norte, Brazil. (a–b) Women and their children harvesting the natural bed of seaweeds at low tide. (c–d) Gathering of the seaweed in bags.

(d–e) Harvesting of the seaweed by boat. (f) Sun-drying of the seaweeds. (g) Preparation of the seaweed seedling net by the women of the community. (h) Culture line of *Gracilaria birdie* (photos: E Marinho-Soriano).

To improve the situation of poor coastal fishing communities of Brazil, in 2001, the Brazilian government started an FAO Technical Cooperation Project (TCP/BRA/0065) with the objective of assisting the establishment of a sustainable seaweed farming sector that could benefit poor coastal communities in the northeast states (Rio Grande do Norte, Paraiba e Ceará). The scope of the project was testing farming techniques in pilot communities and verifying the technical and financial feasibility of the packages, promoting the associative work among the producers, monitoring the social impact of the introduction of these new techniques, and facilitating the establishment of an institutional framework for this new production sector (Freddi and Aguilar-Manjarrez 2003). Moreover, the project showed that community members easily learned techniques for seaweed cultivation, and red seaweed farming (*Gracilaria birdiae*) became a significant source of income, particularly for the poorer segments of the coastal fishing population (Fig. 3g, h). A fivefold increase in the prices of dry seaweed collected from natural beds or from the beaches was experienced by the seaweed-farming industries (Freddi and Aguilar-Manjarrez 2003). Nowadays, the kilogram of dried *Gracilaria* is sold for US$3.50; seaweeds can therefore bring to the coastal communities an additional income as described in the last data published by the Brazilian Institute of Geography and Statistics (IBGE 2014).

In Brazil, the size of the coastal zone and the large ecosystems and diversity produce a false idea of an inexhaustible exploitation potential, causing adoption of development policies that do not take into account the sustainable use of the resources (MMA/REVIZEE 2006). As a result, over 80 % of the resource is currently overexploited (Miloslavich et al. 2011). The importance of the fishing industry in Brazil is, however, incontestable. Marine fisheries contribute with 63 % of the national fish production and involve approximately 800,000 professionals creating 3.5 million jobs directly or indirectly on the sector (ACEB 2014). According to the IBGE (2014), approximately one quarter of the Brazilian population lives in the coastal area, accounting 50.7 million residents. The process of urbanization in which population and economic activities are concentrated particularly along the coastline has been one of the main processes responsible for the modification of habitat and communities, as well as increasing pressures on water resources (Marques et al. 2004). Only a small portion of the enormous Brazilian coastline is under some form of protection or management, and large areas are subject to anthropogenic pressures, encroachments, and overuse (Amaral and Jablonski 2005).

While Mexico, Chile, and Argentina have developed coastal management plans and appropriate regulations for harvesting of their natural resources, Peru and Brazil do not have well-established regulations or sustainable practices in regard to the exploitation of their natural seaweed resources to avoid depletion of their natural beds (Santelices and Doty 1989; Vásquez et al. 2012). Due to the significance of the seaweed forest for coastal ecosystems and its place in the maritime food chain, a widespread reduction in South American seaweed resources may have an impact on the overall productivity and stability of the coastal communities, with only moderate capacity to adopt to societal and environmental changes. In Peru, the Humboldt Current Large Marine Ecosystem (HCLME) Project therefore attempts to develop terms of reference and comprehensive plan of action (PIA), for the repopulation of natural seaweed beds along the Peruvian coast. In the Bay of Paracas ecosystem (Ica Region, Peru), through laboratory cultivation, transplanting, and harvesting, the PIA enables the recovery of the kelp populations, which were negatively affected by indiscriminate harvesting. The PIA also allows for the sustainability of the economic activity for the community of artisanal fishermen in the area in order to start a sustainable production that includes the transfer of appropriate technologies to them. The understanding of the synergy between ecology, economy, and sociology involved in the exploitation of selected seaweeds provides a sustainable platform for seaweed exploitation in Latin America (Clausen and York 2008).

EUROPEAN AND AMERICAN EXPERIENCES: WHAT ARE THE "LESSONS TO LEARN" ON REGULATIONS AND MANAGEMENT FOR A LONG-TERM SUSTAINABLE SEAWEED INDUSTRY IN LATIN AMERICA?

In Europe, the Norwegian seaweed industry is producing over 60 % of the biomass and is almost completely reliant on natural beds of *Ascophyllum nodosum* and *Laminaria hyperborea. L. hyperborea* accounts for about 90 % of the national harvest (Meland and Rebours 2012a), and the alginate industry is by far the most important sector. The remaining biomass is used directly for food, fodder, cosmetics, and the health sectors. Regulations for harvesting seaweed (Meland and Rebours 2012b) are applied to seabed algae such as *L. hyperborea*, and private owner rights only regulate the harvest of foreshore algae such as *A. nodosum*. However, environmental protection laws and other regulations can restrict areas for harvesting. Even a strong interest for seaweed aquaculture, particularly as part of integrated multitrophic aquaculture (Leonczek 2013), the development of any new industrial sector in Norway follows the principle of precaution and is for now highly restricted and controlled by the existing regulation (Alexander et al. 2014) as regulations are under development (Meland and Rebours 2012b).

Portugal seaweed activities are now reduced to the seasonal and low-volume harvest of agarophyte and carrageenophyte species, whereas in the 1980s it was one of the world's largest agar producer (Sousa-Pinto and Araújo 2006). Despite the existence of several economical valuable species, regulations only exist for residual (once important) activities such as Gelidium harvesting and beach cast seaweed activities. A revision of the existing regulation that allows the exploitation of new species is required not only in order to establish sustainable harvesting activities but also to prevent the uncontrolled harvest of economically important sensitive species (Araújo et al. 2009). On the other hand, the country's recent focus on aquaculture development opens the door for seaweed cultivation in monoculture or integrated multitrophic aquaculture (IMTA) systems (Abreu et al. 2011a).

Commercial exploitation of seaweeds in eastern Canada began in the early 1940s with the harvest of the carrageenophyte *Chondrus crispus*.

From 1965 to 1980, Canada was the world's main supplier of raw material for the carrageenan industry, with a peak landing over 50,000 wet t in 1974. By the 1990s, the landings sharply declined to less than 10,000 wet t due to a combination of international competition and biological fluctuations in the abundance of the resource (Chopin and Ugarte 2006). Today, *A. nodosum* is the main economic seaweed resource in Canada, with over 40,000 t landed in 2010 (Ugarte and Sharp 2011). This seaweed is being used for among other applications as a biostimulant extract for crops and animal feed supplements, incorporating more than US$40 million to the local economy. *A. nodosum* has been managed since 1995 under an ecosystem approach, following strict regulations established in the Ocean Act (Ugarte and Sharp 2001). To solve problems mainly of fluctuating supply and avoid the risks of overexploitation, the aquaculture of *C. crispus* was developed in the 1970s. *C. crispus* production is the main commercial Canadian seaweed aquaculture, with the culture of *Porphyra* species and *Saccharina latissima* as organic extractive components of integrated multitrophic aquaculture system (Levine 2004; Chopin and Ugarte 2006; Chopin 2008).

Like other European and American countries, the Brazilian and Peruvian dependency on traditional coastal fisheries and eventual growth in aquaculture is followed by an urgent need for robust institutions for planning and managing the coastal zone with focus on developing an integrated and sustainable management plan for the seaweed resources (Prates 2007; Castello 2010; Johnsen et al. 2014). The steep growth in the use of marine biological resources represents a fundamental change in the way humans derive benefits from the oceans. Marine aquaculture production is growing at 7 % per year and the sector of domestication of marine species is growing at about 3 % per year, while the number of natural marine products of commercial interest is growing at a rate of 4 % per year (Duarte et al. 2007). Large-scale integrated conservation and management plans are thus urgently required in order to address the sustainable development of various economic and societal sectors. In addition to developing and implementing management plans for sustainable use, these countries also face huge tasks connected to mapping, investigating, and monitoring their marine biological diversity. Information and "lessons learned" may be provided by the scientific communities and management plans in Latin American countries such as

Chile or Mexico (Vásquez et al. 2006; 2012; Ávila et al. 2013; Vega et al. 2013) and European directives such as those from Norway (Meland and Rebours 2012a) or the ecosystem-based management regulated by the Ocean Act in Canada (Ugarte and Sharp 2001). National regulations should be established on best practices for seaweed harvesting, management, and cultivation, and the knowledge should be passed on to the coastal communities, which could then make seaweed a sustainable opportunity for income.

Alternatively, the seaweed biomass could be obtained through cultivation (McHugh 2002). The seaweed aquaculture industry still requires technological and management improvements, institutional changes, and appropriate environmental and social frameworks (Valenti 2008; Oliveira 2009; Abreu et al. 2011b; Marroni and Asmus 2013). As seaweed aquaculture is at its infancy both in Europe and America, its development will have to overcome numerous challenges by introducing innovations at different levels, i.e., technology, biology, and policy. However, cultivation techniques are standardized and economically sustainable especially in Asia (Miura 1980) and cultivated macroalgae now supply 96 % of the global demand (FAO 2014). The aquaculture of macroalgae is also strongly recommended by the EU regulation (EC No 710/2009) for the organic production of animals in aquaculture systems. The development of regulations for supporting the aquaculture of seaweed is a necessary action in recent years, and these regulations should be guided by best practices. The combined effect of rapidly increasing domestication and production with increasing demand for seaweed products is likely to promote innovation in seaweed biotechnology, and these efforts must be accompanied by substantial scientific effort (Mazarrasa et al. 2013). In addition, revision of the Common Agricultural Policy (CAP) intends to include management with High Nature Value in organic certification. Nordic countries such as Denmark, Sweden, and Norway are also aiming to increase organic products to 15 % of the home market. Several food and food supplement products directly prepared from algae are already certified in Europe and in Norway as organic food and given opportunities for producers to access to a niche market (Friis Pedersen et al. 2013).

STEPS TOWARDS WEALTH AND SUSTAINABLE LIVELIHOODS FOR LATIN AMERICA COASTAL COMMUNITIES

The seascapes are increasingly managed for multiple functions and services in addition to provision of food, and this requires the integration of ecological and socio-economic research, policy innovation, and public education. The multiuse dilemma has driven many researchers, experts, and policy makers to try and address issues related to the sustainability of coastal development from disciplinary/sectorial perspectives addressing the interactions and functioning within the wider ecosystem, social, economic, and political contexts (Buchholz et al. 2012). A review by Harley et al. (2012) addressed the significant gaps in understanding, which hamper ability to predict the outcomes of global change in seaweed-dominated systems. In particular, it addressed the lack of general or even basic understanding of (a) the importance of rates, timing, magnitude, and duration of environmental change; (b) nonadditive effects of multiple stressors; (c) population-level implications of variable environmental impacts among life history stages; (d) the scope for population- or species-level adaptation to environmental change; and (e) ecological responses at the level of communities and ecosystems, including tipping points and sudden phase shifts. In this regard, biological responses (i.e., ecophysiological) to key environmental drivers or combinations of drivers can be incorporated into demographic models to better describe and predict changes in population growth or decline. The expansion of seaweed cultivation particularly in tropical regions contributes significantly to carbon sequestration since the rapid turnover in seaweed culture, approximately 3 months per crop (in the tropics) with yields of over 2,500 wet t ha^{-1} (De Silva and Soto 2009; Vásquez et al. 2013). Nevertheless, some authors have pointed out that a significant proportion (estimates range up to 60 %) of the carbon they fix photosynthetically is released into the water, and a proportion of this released dissolved organic carbon (DOC) is highly labile, entering in the bacterial loop and rapidly remineralizing back to CO_2 (Hughes et al. 2012). Environmental impacts of seaweed farming in the tropics have been reviewed by Zemke-White and Smith (2006). Some authors have also pointed out other environmental impacts of algal farming both positive (i.e., increase on fish assemblages; Bergman et al. 2001) and/or negative (i.e., effect on the meiobenthos;

Olafsson et al. 1995). All of these impacts should be considered when the environmental effects of seaweed aquaculture are taken into account. The concept of ecosystem-based management approaches based on an integrated approach of entire ecosystem including humans should also be considered to develop the coastal spatial planning and the best practices guidelines for the exploitation of seaweed (both harvesting and aquaculture) in order to avoid spatial and temporal mismatches of the governance (Crowder and Norse 2008).

The benefits for the well-being of coastal communities as increase of direct permanent employment in previously disadvantaged coastal communities, where unemployment is not only an economic issue but also a socio-political concern, have been exemplified in an IMTA farm of abalone and seaweed (Nobre et al. 2010). Sustainable management of coastal resources creates new economic activities based on the exploitation of a raw material and could participate in the local socio-economic development in coastal areas and communities in Latin America. Developing long-term management plans will also produce fundamental long-term results of interest for the international research community. Socio-economic benefits derived from seaweeds have been already exemplified in the Philippines where approximately 116,000 families consisting of one million individuals were farming more than 58,000 ha of seaweed, making seaweed farming the largest and most productive form of livelihood among the coastal population. In Zanzibar (Tanzania), more than 90 % of seaweed farmers are women. Like in Latin America, changing the life in the villages by women gaining independent economic power will contribute in reducing childhood malnutrition (as an indicator that the health of their mothers has improved), increasing the numbers of children attending schools regularly, and reversing the trend of rural depopulation by self-employment of the village youths (Msuya 2006; Msuya et al. 2007).

Innovation should be promoted when trying to integrate seaweed harvesting or aquaculture as part of the wealth for coastal communities. In this regard, Castellacci (2010) pointed out that the technology dynamics of a country depend on three main factors: its innovative intensity, its human capital, and its technological infrastructures. Nowadays, Latin American countries show a much lower innovative strength than Norway or Canada, where the innovation gap has been quantified with a ratio of 16:1 for patents and 10:1 for scientific articles (Castellacci 2010). In order to get closer and eventually jump to the innovation development stage, developing economies such as in some Latin American countries should implement an appropriate combination of policies that takes into account the need to simultaneously develop R & D activities, traditional infrastructures, information and communication technologies, and advanced human skills. Human capital education explains differences in economic performance across countries; education is therefore a necessity to promote social inclusion and cohesion as well as employment. By focusing on marine resources involving low-cost technology requirements, such as the production of seaweed, developing countries are provided an opportunity to access an emerging market, propelled by a diversification of the demand for seaweed products from traditional uses to bioenergy, cosmetics, and biomedicine applications.

CONCLUSIONS

The European, Canadian, and Latin American seaweed industries rely on the harvesting of their natural resources—activity limited by the biomass availability and the potential for the harvested species to recover. Different scenarios associated to seaweed resource exploitation can be observed, and several countries are in need of long-term management plans for the sustainable exploitation of their natural seaweed (Hersoug and Revold 2012). As this activity increases, there is also an urgent need to develop and implement ecosystem-based management models and integrated coastal zone planning. Policy makers must develop regulations and directives that enable a sustainable exploitation of the natural resource, not only to preserve marine and coastal ecosystems but also to ensure social stability and economic income of local communities.

Both European and Latin American countries need to address capacity building and adaptive governance towards seaweed resources. Norway and Canada have well-established management regimes for the sustainable exploitation of their seaweed resources based on the development and application of a sound knowledge and cross-sectorial management and spatial plans. Portugal, on the contrary, needs to review the national and regional management plans concerning the exploitation of the seaweed resources. In Latin America, it is important to stress that Chile has succeeded in establishing a sustainable harvesting in the south of the country. The further expansion of the seaweed industry in this region thus depends on reliable access to raw material, the development of value-added products, and the transfer of expertise between developed and less developed regions. An alternative solution to produce seaweed biomass for this growing sector is offered by aquaculture, which will also require guidelines for evolving into a responsible and well-managed farming.

Finally, trained human resources are required in order to provide education to coastal communities, based on best practices for harvesting and cultivation, in order to establish profitable businesses which could provide socio-economic development leading to better living conditions to the coastal rural communities. The current weak participation of developing countries in the seaweed market could then be reversed by strengthening cooperation to transfer technological knowledge and investment to developing nations active in seaweed management and aquaculture, thereby promoting sustainable development based on their own natural resources, as encouraged by the Convention on Biological Diversity (http://www.cbd.int/doc/legal/cbd-en.pdf). In this context, South American, European, and North American experts should therefore collaborate to develop guidelines to plan an integrated coastal management for Latin American seaweed resources. It is also of great importance for the legitimacy of these guidelines that these originate from transnational and cross-sectorial cooperation, including political, cultural, commercial and industrial actors, NGOs, and research communities. As a first action of knowledge transfer, research and technology development (RTD) and small and medium enterprises (SME) representatives from Norway, Argentina, Brazil, Canada, Chile, Mexico, Peru, and Portugal have established a common working framework in order to support the development of the Latin American seaweed sector. These industry/academia networks encourage cooperation among the seaweed stakeholders; across the project areas; in all aspects related to seaweed production, research, ecosystem services, and management of artisanal and small-scale aquaculture; and traditional and alternative market and economics.

ACKNOWLEDGMENTS

The NorLARNET program (Norway), the County Council of Nordland (Norway), Bioforsk (Norway), and Ministry of Fisheries and Aquaculture (Brazil) financially supported this work. The LA Seaweed Consortium met in Brazil at the Latin American Congress of Algal Biotechnology (IV CLABA).

REFERENCES

1. Abreu MH, Pereira R, Yarish C, Buschmann A, Sousa-Pinto I (2011a) IMTA with *Gracilaria vermiculophylla*: productivity and nutrient removal performance of the seaweed in a land-based pilot scale system. Aquaculture 312:77–87
2. Abreu M, De Mattos P, Lima P, Padula A (2011b) Shrimp farming in coastal Brazil: reasons for market failure and sustainability challenges. Ocean Coast Manag 54:658
3. ACEB (2014) Anuário da Pesca e Aquicultura Brasileira 2014. 1st Brazilian Fishery and Aquaculture Yearbook 2014: http://www.mpa.gov.br/index.php/publicidade/publicacoes
4. Acleto CO (2006) Seaweed resources in Peru. In: Critchley AT, Ohno M, Largo DB (eds) CD-ROM World Seaweed Resources—an authoritative reference system. Version:1.0 Margraf Publishers GmbH
5. Alexander K, Potts TP, Freeman S, Israel D, Johansen J, Kletou D, Meland M, Pecorino D, Rebours C, Shorten M, Angel DL (2014) The implications of aquaculture policy and regulation for the development of integrated multi-trophic aquaculture in Europe. Aquaculture (manuscript number: AQUA-D-14-00294, Submitted 17.02.2014)
6. Alvarez JC, Vodden K (2009) Local ecological knowledge and the impacts of global climatic change on the community of seaweed extractors in Pisco-Perú. PICMET - Proceedings, August 2–6, Portland, Oregon USA
7. Amaral ACZ, Jablonski S (2005) Conservation of marine and coastal biodiversity in Brazil. Conserv Biol 19:625–631
8. Araújo R, Vaselli S, Almeida M, Serrão EA, Sousa-Pinto I (2009)

Disturbance on marginal populations: effects of human trampling on *Ascophyllum nodosum* assemblages at its southern distribution limit. Mar Ecol Prog Ser 378:81–92

9. Ávila M, Piel MI, Alcapan A (2013) Indoor and outdoor culture of *Callophyllis variegata* (Bory) Kützing (Gigartinales, Rhodophyta) in southern Chile. J Appl Phycol. doi:10.1007/s10811-013-0109-1
10. Bergman KC, Svensson S, Öhman MC (2001) Influences of algal farming on fish communities. Mar Pollut Bull 42:1379–1389
11. Bixler HJ, Porse H (2011) A decade of change in the seaweed hydrocolloids industry. J Appl Phycol 23:321–335
12. Buchholz CM, Krause G, Buck BH (2012) Seaweed and man. In: Wiencke C, Bischof K (eds) Seaweed biology. Springer, Berlin, pp 471–494
13. Buschmann AH, Prescott S, Potin P, Faugeron S, Vásquez JA, Camus C, Infante J, Hernández-González MC, Gutierrez A, Varela DA (2014) The status of kelp exploitation and marine agronomy, with emphasis on Macrocystis pyrifera, in Chile. Adv Bot Res 71
14. Casas-Valdez M, Serviere Zaragoza E, Lluch Belda D, Marcos R, Aguila Ramírez R (2003) Effect of climatic change on the harvest of the kelp *Macrocystis pyrifera* on the Mexican Pacific Coast. Bull Mar Sci 73:545–556
15. Castellacci F (2010) Closing the technology gap? MPRA Paper No. 27586, posted Online at http://mpra.ub.uni-muenchen.de/27586/
16. Castello JP (2010) O futuro da pesca e da aquicultura marinha no Brasil: A pesca costeira. Cienc Cult 62(3):32–35
17. Chopin T (2008) Integrated multi-trophic aquaculture (IMTA) will also have its place when aquaculture moves to the open ocean. Fish Farmer 31(2):40–41
18. Chopin T, Ugarte R (2006) Seaweed resources of Eastern Canada. In: Critchley AT, Ohno M, Largo DB (eds) CD-ROM World Seaweed Resources—an authoritative reference system. Version:1.0 Margraf Publishers GmbH
19. Christensen V, de la Puente S, Sueiro JC, Steenbeek J, Majluf P (2014) Valuing seafood: the Peruvian fisheries sector. Marine Policy 44:302–311

20. Clausen R, York R (2008) Global biodiversity decline of marine and freshwater fish: a cross-national analysis of economic, demographic, and ecological influences. Soc Sci Res 37:1310–1320
21. Craigie JS (2011) Seaweed extract stimuli in plant science and agriculture. J Appl Phycol 23:321–335
22. Critchley AT, Ohno M, Largo DB (2006) World Seaweed Resources: an authoritative reference system. CD-ROM World Seaweed Resources—an authoritative reference system. Version:1.0 Margraf Publishers GmbH, Germany
23. Crowder L, Norse E (2008) Essential ecological insights for marine ecosystem-based management and marine spatial planning. Mar Policy 32:772–778
24. Dayton PK, Currie V, Gerrodette T, Keller B, Rosenthal R, Ven Tresca D (1984) Patch dynamics and stability of some California kelp communities. Ecol Monogr 54:253–289
25. De Silva SS, Soto D (2009) Climate change and aquaculture: potential impacts, adaptation and mitigation. In: Cochrane K, De Young C, Soto D, Bahri T (eds) Climate change implications for fisheries and aquaculture: overview of current scientific knowledge. FAO Fisheries and Aquaculture Technical Paper. No. 530. FAO, Rome, pp 151–212
26. Diana J S (2009) Aquaculture Production and Biodiversity Conservation. Bioscience 59(1):27–38
27. DOF (2012) Diario Oficial de la Federación: Plan de Manejo para la Pesquería de Macroalgas en Baja California, México. Secretaría de Agricultura, Ganadería, Desarrollo Rural, Pesca y Alimentación. http://dof.gob.mx/nota_detalle.php?codigo=5280840&fecha=30/11/2012
28. Duarte CM, Marbá N, Holmer M (2007) Rapid domestication of marine species. Science 316:382–383
29. EEA (2013) Balancing the future of Europe's coasts—knowledge base for integrated management. European Environmental Agency, Denmark, p 67, Report 12
30. Ehrhart F, Mettler E, Böse T, Weber MM, Vásquez JA et al (2013) Biocompatible coating of encapsulated cells using ionotropic gelation. PLoS ONE 8(9):e73498

31. FAO (2013) Fisheries and Aquaculture Information and Statistics Service – 18/12/2013
32. FAO (2014) Fisheries and Aquaculture Information and Statistics Service - 16/03/2014
33. Freddi A, Aguilar-Manjarrez J (2003) Small-scale seaweed farming in North East Brazil. TCP/BRA/0065 35 FAO Aquaculture Newsletter 32
34. Friis Pedersen S, Meland M, Rebours C (2013) Macroalgae for an increasing organic market. Bioforsk Fokus 8(2):337–338
35. Graham MH, Vásquez JA, Buschmann AH (2007) Global ecology of the giant kelp *Macrocystis*: from ecotypes to ecosystems. Oceanogr Mar Biol Annu Rev 45:39–88
36. Harley CDG, Anderson KM, Demes KW, Jorve JP, Kordas RL, Coyle TA, Graham MH (2012) Effects of climate change on global seaweed communities. J Phycol 48:1064–1078
37. Hayashi L, Bulboa C, Kradolfer P, Soriano G, Robledo D (2013) Cultivation of red seaweeds: a Latin American perspective. J Appl Phycol. doi: 10.1007/s10811-013-0143-z
38. Hernández-Garibay E, Guardado-Puentes J, Bautista-Alcantar J, Reyes-Tisnado R (2006) Macroalgas del Oceáno Pacífico. In: Sustentabilidad y Pesca Responsable en Mexico. Ed J Cuellar. Instituto Nacional de la Pesca. SAGARPA: 235–244
39. Hersoug B, Revold J (2012) Assessment criteria for Norwegian assistance to aquaculture in developing countries. A draft report to Norad. The Norwegian College of Fishery Science, University of Tromsø, Tromsø, p 47
40. Hughes AD, Black KD, Campbell I, Davidson K, Kelly MS, Stanley MS (2012) Does seaweed offer a solution for bioenergy with biological carbon capture and storage? Greenh Gas Sci Technol 2:402–407
41. IBGE (2014) Instituto Brasileiro de Geografia e Estatística. Available: saladeimprensa.ibge.gov.br/en/noticias
42. Johnsen JP, Hersoug B, Solås A-M (2014) The creation of coastal space—how local ecological knowledge becomes relevant. Mar Stud 13:2. doi:10.1186/2212-9790-13-2

43. Ladah LB, Zertuche-González JA, Hernández-Carmona G (1999) Giant kelp (*Macrocystis pyrifera*, Phaeophyceae) recruitment near its southern limit in Baja California after mass disappearance during ENSO 1997–1998. J Phycol 35:1106–1112

44. Leonczek A (2013) Traditional and integrated aquaculture. Today's environmental challenges and solutions of tomorrow. The Bellona Foundation, report, 117pp

45. Levine IA (2004) *Porphyra* Cultivation in Maine, USA and New Brunswick, Canada Aquaculture Compendium 1–13

46. Marinho-Soriano E, Moreira WSC, Carneiro MAA (2006) Some aspects of the growth of *Gracilaria birdiae* (Gracilariales, Rhodophyta) in an estuary in northeast Brazil. Aquacult Int 14:327–336

47. Marques M, Costa MF, Mayorga MIO, Pinheiro PRC (2004) Water environments: anthropogenic pressures and ecosystem changes in the Atlantic drainage basins of Brazil. Ambio 33:68–77

48. Marroni EV, Asmus ML (2013) Historical antecedents and local governance in the process of public policies building for coastal zone of Brazil. Coast Ocean Manag 76:30–37

49. Mazarrasa I, Olsen YS, Mayol E, Marbà N, Duarte CM (2013) Rapid growth of seaweed biotechnology provides opportunities for developing nations. Nat Biotechnol 31:591–592

50. McHugh DJ (2002) Prospects for seaweed production in developing countries. FAO Fisheries Circular No. 968 FIIU/C968

51. McHugh DJ (2003) A guide to the seaweed industry. FAO Fisheries Technical paper, No 441: 105 p

52. Meland M, Rebours C (2012a) Short description of the Norwegian seaweed industry. Bioforsk Fokus 7(2):278–279

53. Meland M, Rebours C (2012b) Introduction to the management and regulation of Norwegian seaweed industry. Bioforsk Fokus 7(2):275–277

54. Miloslavich P, Díaz JM, Klein E, Alvarado JJ, Díaz C et al (2010) Marine biodiversity in the Caribbean: regional estimates and distribution patterns. PLoS ONE 5(8):e11916

55. Miloslavich P, Klein E, Díaz JM et al (2011) Marine biodiversity

in the Atlantic and Pacific coasts of South America: knowledge and gaps. PLoS ONE 6(1):e14631

56. Miura A (1980) Seaweed cultivation: present practices and potentials. In: Carter A (ed), Living resources, University of Chicago, pp 57–68
57. MMA/REVIZEE (2006) Avaliação do potencial sustentável de recursos vivos na zona econômica exclusiva: relatório executivo/ MMA, Secretaria de Qualidade Ambiental. Brasília: Brazil
58. Msuya FE (2006) The impact of seaweed farming on the social and economic structure of seaweed farming communities in Zanzibar, Tanzania. In: Critchley AT, Ohno M, Largo DB (eds) CD-ROM World Seaweed Resources—an authoritative reference system. Version:1.0 Margraf Publishers GmbH Germany
59. Msuya FE, Shalli MS, Sullivan K, Crawford B, Tobey J, Mmochi AJ (2007) A comparative economic analysis of two seaweed farming methods in Tanzania. The sustainable coastal communities and ecosystems program. Coastal Resources Center, University of Rhode Island and the Western Indian Ocean Marine Science Association, pp 27
60. Nobre AM, Robertson-Andersson D, Neori A, Sankar K (2010) Ecological–economic assessment of aquaculture options: comparison between abalone monoculture and integrated multi-trophic aquaculture of abalone and seaweeds. Aquaculture 306:116–126
61. O'Dor R, Miloslavich P, Yarincik K (2010) Marine biodiversity and biogeography—regional comparisons of global issues, an introduction. PLoS ONE 5(8):e11871
62. Ólafsson E, Johnstone RW, Ndaro SGM (1995) Effects of intensive seaweed farming on the meiobenthos in a tropical lagoon. J Exp Mar Biol Ecol 191:101–117
63. Oliveira RC (2009) O panorama da aquacultura no Brasil: a prática com foco na sustentabilidade. Rev Intertox Toxicol Risco Ambient Soc 2(1):71–89
64. Ortiz M (2008) Mass balanced and dynamic simulations of trophic models of kelp ecosystems near the Mejillones Peninsula of northern Chile (SE Pacific): comparative network structure and assessment of harvest strategies. Ecol Model 216:31–46

65. Pereira R, Yarish C (2008) Mass production of marine macroalgae. Encyclopedia of Ecology, 2236–2247
66. Prates APL (2007). National Plan for Protected Areas Marine and Coastal Areas Context. In: MMA/SBF. 2007. Aquatic Protected Areas as Fisheries Management tools. 18–24 p
67. Rebours C, Karlsen Å (2007) Seaweeds in the North: new scopes for coastal farming. Bioforsk Fokus 2(13):73–76
68. Robledo D (2006) Seaweed resources of Mexico. In: Critchley AT, Ohno M, Largo DB (eds) CD-ROM World Seaweed Resources—an authoritative reference system. Version:1.0 Margraf Publishers GmbH Germany
69. Robledo D, Townsend W (2006) Seaweed and mangroves: improving environmental practices in coastal communities to secure sustainable livelihoods. In: Breton Y, Brown D, Davy B, Haughton M, Ovares L (eds), Coastal resource management in the wider Caribbean: resilience, adaptation, and community diversity. Ian Randle Publishers and International Development Research Center
70. Robledo D, Gasca-Leyva E, Fraga J (2013) Social and economic dimensions of carrageenan seaweed farming in Mexico. In: Valderrama D, Cai J, Hishamunda N, Ridler N (eds) Social and economic dimensions of carrageenan seaweed farming. FAO, Rome, pp 185–204, Fisheries and Aquaculture Technical Paper No. 580
71. Santelices B, Doty MS (1989) A review of *Gracilaria* farming. Aquaculture 78:95–133
72. Schreiber MA (2013) Institutions for sustainable fisheries governance—the case of the commercial Peruvian anchovy fishery. PhD Thesis, Universität Bremen, 132 p
73. Sousa-Pinto I, Araújo R (2006) Seaweed resources of Portugal. In: Critchley AT, Ohno M, Largo DB (eds) CD-ROM World Seaweed Resources—an authoritative reference system. Version:1.0 Margraf Publishers GmbH, Germany
74. Suo R, Wang Q (1992) *Laminaria* culture in China. INFO Fish Int 1(92):40–42
75. Taylor MH, Wolff M, Mendo J, Yamashiro C (2008) Changes in trophic flow structure of Independence Bay (Peru) over an ENSO cycle. Prog Oceanogr 79:336–351

76. Thiel M, Macaya EC, Acuña E et al (2007) The Humboldt Current System of northern-central Chile: oceanographic processes, ecological interactions and socioeconomic feedback. Oceanogr Mar Biol Annu Rev 45:195–344

77. Ugarte R, Sharp G (2001) A new approach to seaweed management in Eastern Canada: the case of *Ascophyllum nodosum*. Cah Biol Mar 42:63–70

78. Ugarte R, Sharp G (2011) Management and production of the brown algae *Ascophyllum nodosum* in the Canadian maritimes. J Appl Phycol 24:409–416

79. Ugarte RA, Craigie JS, Critchley AT (2010) Fucoid flora of the rocky intertidal of the Canadian Maritimes: implications for the future with rapid climate change. In: Israel A, Einav R, Seckbach J (eds) Seaweeds and their roles in globally changing environments. Springer, Dordrecht, pp 69–90

80. Valenti WC (2008) Aqüicultura Brasileira é sustentável? Palestra apresentada durante o IV Seminário Internacional de Aqüicultura, Maricultura e Pesca, Aquafair 2008, Florianópolis, 13–15 de maio de 2008. pp 1–11 (www.avesui.com/anais)

81. Vásquez JA (1992) *Lessonia trabeculata*, a subtidal bottom kelp in northern Chile: a case of study for a structural and geographical comparison. In: Seeliger U (ed) Coastal plant communities of Latin America. Academic, San Diego, pp 77–89

82. Vásquez JA (1993) Natural mortality of giant kelp *Macrocystis pyrifera* affecting the fauna associated with its holdfasts. Pac Sci 47:180–184

83. Vásquez JA (2008) Production, use and fate of Chilean brown seaweeds: resources for a sustainable fishery. J Appl Phycol 20:457–467

84. Vásquez JA, Santelices B (1984) Comunidades de macroinvertebrados en discos de adhesión de *Lessonia nigrescens* en Chile central. Rev Chil Hist Nat 57:131–154

85. Vásquez JA, Vega JMA, Buschmann AH (2006) Long term studies on El Niño-La Niña in northern Chile: effects on the structure and organization of subtidal kelp assemblages. J Appl Phycol 18:505–519

86. Vásquez JA, Piaget N, Alonso Vega JM (2012) The *Lessonia nigrescens* fishery in northern Chile: "how you harvest is more important than how much you harvest". J Appl Phycol 24:417–426

87. Vásquez JA, Zuñiga S, Tala F, Piaget N, Rodriguez DC, Vega JMA (2013) Economic evaluation of kelp forest in northern Chile: values of good and service of the ecosystem. J Appl Phycol. doi: 10.1007/s.10811-013-0173-6

88. Vega JMA, Broitman B, Vásquez JA (2013) Monitoring the sustainability of *Lessonia nigrescens* complex (Laminariales, Phaeophyta) in northern Chile under strong harvest pressure. J Appl Phycol. doi: 10.1007/s.10811-013-0167-4

89. Zaixso AB, Ciancia M, Cerezo AS (2006) Seaweed resources in Argentina. In: Critchley AT, Ohno M, Largo DB (eds) CD-ROM World Seaweed Resources—an authoritative reference system. Version:1.0 Margraf Publishers GmbH, Germany

90. Zemke-White WL, Smith JE (2006) Environmental impacts of seaweed farming in the tropics. In: Critchley AT, Ohno M, Largo DB (eds) CD-ROM World Seaweed Resources—an authoritative reference system. Version:1.0 Margraf Publishers GmbH, Germany

91. Zertuche-González JA, M Sánchez-Barredo, JM Guzmán-Calderón, Z Altamirano-Gómez (2013) *Eisenia arborea* J.E. Areschough as abalone diet on an IMTA farm in, Baja California, México. J Appl Phycol. doi: 10.1007/s10811-013-0138-9

92. Zimmermann H, Zimmermann D, Reuss R et al (2005) Towards a medically approved technology for alginate-based microcapsules allowing long-term immune isolated transplantation. J Mater Sci Mater Med 16:491–501

CHAPTER

5

IS A COOPERATIVE APPROACH TO SEAWEED FARMING EFFECTUAL? AN ANALYSIS OF THE SEAWEED CLUSTER PROJECT (SCP), MALAYSIA

Adibi M. Nor[1], Tim S. Gray[2], Gary S. Caldwell[1], and Selina M. Stead[1]

[1]School of Marine Science and Technology, Newcastle University, Ridley Building, Claremont Road, Newcastle upon Tyne, England NE1 7RU, UK

[2]School of Geography, Politics and Sociology, Newcastle University, Newcastle upon Tyne, England NE1 7RU, UK

ABSTRACT

Seaweed (*Kappaphycus* spp.) farming has been practiced in Malaysia since the late 1970s following government policy incentives (training and farming inputs). However, numerous governance, economic, environmental, technological and sociocultural challenges have limited the industry from achieving its full potential. The Seaweed Cluster Project (SCP) was introduced in 2012 to address some of these challenges. We sought to evaluate the effectiveness of the SCP in delivering its central objectives of increasing seaweed production, optimizing the farming area, improving seaweed quality and farming efficiency, raising farmers' income, and reducing the environmental impact of seaweed farming. Community and industry perceptions of the SCP were obtained from seven communities using a mixed-methods approach based on face-to-face semi-structured interviews, focus group discussions, household surveys, observation and secondary data. Views on the SCP outcomes were generally negative, including low take-up rates by indigenous people, poor stakeholder participation in decision-making, limited acceptance of new technologies, economic vulnerability, a complex marketing system, and low social cohesion of seaweed farming communities. Positive perceptions included recognition that the SCP confers high social status upon a community, reduces operating costs, and facilitates the production of certified seaweed. The SCP's problems are linked to poor multi-level governance, weak market mechanisms and unintegrated community development. The study concludes with five recommendations to improve the SCP: promote the participation of indigenous people; legalize existing migrant farmers; strengthen local seaweed cooperative organizations; provide entrepreneurship skills to farmers; and fully integrate stakeholders into decision-making.

Keywords: Aquaculture, Seaweed farming, Governance, Community, Policy, Malaysia

INTRODUCTION

Demand for carrageenan from the processed food industry has increased substantially since the 1960s (Hurtado et al. 2012), fuelling strong growth in the seaweed aquaculture sector. *Kappaphycus* farming expanded into Eastern Malaysia (Semporna district, Sabah) in the late 1970s (Sade et al. 2006). However, ecological problems and low seaweed pricing resulted in poor uptake from the indigenous people. Civil war in the southern Philippines precipitated the migration of Filipinos to Sabah in the 1970s and 1980s (Allerton 2014; Razali et al. 2015) where they worked in the fishing, palm oil and cocoa industries, and subsequently adopted seaweed farming. From the late 1980s, the indigenous communities began to accept seaweed farming as an alternative livelihood to fishing, having learned to farm from the Filipino community. The Sabah Fisheries Department (SFD) subsequently provided assistance to 10 local fisher families occupying 10 ha of farming space (Dalli 1988). This action prompted additional coastal communities to engage in seaweed farming. A further stimulus stemmed from strong demand from foreign processors, driving prices upwards from MYR 0.50 (US$0.15) per kg in 1991 to MYR 1.70 (US$0.51) per kg in 1996.

In 2012, the Malaysian government (through the National Seaweed Technical Committee, NSTC) introduced the Seaweed Cluster Project (SCP) into Semporna, focusing on the villages of Lok Butun and Gelam-gelam. The SCP was designed to aggregate the existing farmers under a system of cooperative-owned enterprises, managed by the farmers but controlled by the government. The goals of the SCP were to increase national production, raise farmers' income, optimize the farming area, improve farming efficiency and seaweed quality, and reduce the environmental impact of seaweed farming activities. The goals were to be met by adopting farming approaches that were new to Malaysia, specifically: the replacement of raffia tie-tie with polyethylene tie-tie; using poly-floats rather than recycled plastic bottles; switching to detachable eco-friendly longlines; use of cement anchoring systems; the supply of seedling (propagule) tying tables, wooden drying platforms, and fiberglass boats; and the establishment of the Malaysian Good Aquaculture Practices (MyGAP) certification scheme. The remaining sections in this paper explain and evaluate for the first time the SCP's attempts to meet these goals. Lessons learned have wider implications for other seaweed producing nations facing similar problems.

METHODS

Ethics Statement

Ethical approval for this study was obtained from the Newcastle University Ethics Committee. Additional permissions were granted by the Malaysian government (Economic Planning Unit, Prime Minister's Department).

Study Area

Situated in Southeast Asia, Malaysia (population 28.3 million) comprises two non-contiguous land areas separated by the South China Sea: 'Peninsular Malaysia' bordering Thailand on the northern side and 'East Malaysia' on the northern side of Borneo. East Malaysia comprises two states, Sabah and Sarawak. Sabah has the third highest state population (3.21 million) in Malaysia, but the second lowest population density (44 people km^{-2}) (DOS 2010). Sabah was the major contributor to the national Gross Domestic Product (GDP) for the agriculture sector in 2013; 16.8% of MYR 56.1 billion (US$ 17.1 billion). At the state level, agriculture contributed 20.6% of GDP, and the fishing sector 2.4% (DOS 2013).

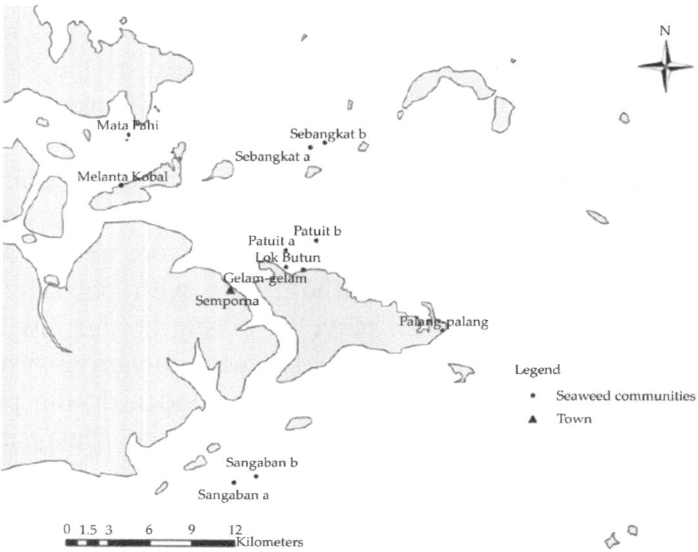

Figure 1. Map of the study area showing the locations of the seaweed communities and their nearby towns.

Sabah was chosen as the research site as it is the only seaweed producing state in Malaysia operating at a commercial scale (26,076 t in 2015). Semporna (population 137,868) spans 114,477 ha, comprising the mainland and 50 islands. Bum-bum, the second largest island in Semporna, contains 7.3% (>10,000 inhabitants) of the total Semporna population (Semporna District Office, unpublished data). Semporna comprises different ethnicities including Bajau (the largest), Bugis, Kadazan, Malay and Tausug (Suluk) (DOS 2010). With a total budget of MYR 8.03 million (US$2.64 million) the SCP was introduced in 2012 to four areas in Sabah: three cluster groups in Semporna district (Lok Butun, Gelam-gelam and Silungun villages) and one cluster group in Tawau district (Merotai village). The current study was, by necessity, limited to the neighbouring villages of Lok Butun (population 618) and

Gelam-gelam (population 1085) (Semporna District Office, unpublished data) in response to the following three factors; 1) both villages were SCP pioneers; 2) fieldwork resources prevented wider inclusion of other SCP communities; and, 3) safety concerns as a direct consequence of travel restrictions imposed in respose to conflict between insurgents and Malaysian security forces in Lahad Datu and Semporna, Sabah. Seven communities were chosen to evaluate the SCP: two SCP communities (Lok Butun and Gelam-gelam), four individual farmer communities, i.e., individual households not part of a cooperative (Sebangkat a and b, Sangaban a and b, Patuit a and b and Palang-palang), and one cooperative farming community (Mata Pahi/Melanta Kobal) (Fig. 1).

Data Collection

Fieldwork began with a reconnaissance and fact-finding activity (April 2015) as an initial exploration of a seaweed farming community. Furthermore, this activity was conducted to gain awareness of community leaders, key authorities, government departments, and other stakeholders related to seaweed farming and management of the seaweed industry. The full survey (April to June, 2015) utilised four research instruments: key informant (KI) interviews, focus group discussions (FGD), household surveys (HS), and observation and secondary data. Face-to-face semi-structured interviews (n = 42) were conducted with 42 key informants, including government officials at local, state, and national levels; a seaweed pioneer (the first local person to introduce seaweed farming to Malaysia); seaweed buyers; carrageenan processors; community representatives; SCP leaders; and a non-governmental organization (NGO) representative (World Wide Fund for Nature, WWF). The sample comprised 7 females and 35 males. The KI were selected using non-probability sampling through purposive sampling. The snowball method (by asking people to suggest the names of potential KI) was used to identify the KI in managing the seaweed industry. The selection of KI for semi-structured face-to-face interviews and FGD was based on their expertise and decision-making roles. Three FGD were carried out with groups of seaweed farmers, processors and buyers (n = 40).

The HS were administered to 144 households in the seven study communities. The SCP participants were selected from a list of names provided by the Sabah Fisheries Department (SFD). The selection reflected the impacts of the introduction of the SCP at the household level. In total, 25 and 15 participant households were selected in Lok Butun and Gelam-gelam respectively. Using purposive sampling and a list provided by the Semporna Area Farmers' Association (SAFA), the selection represented individual farmers (n = 87) and cooperative members (n = 17). There was not a list of individual farmers, so the household heads were selected through purposive sampling due to their status as migrants. A boat was hired on a daily basis to survey these farmers. Using non-probability sampling methods through purposive sampling, three farmer categories were selected: (1) SCP participants; (2) individual participants or migrant farmers living in stilt houses (*pondohan*); and (3) local cooperative members.

The KI interviewees were asked questions in face-to-face in-depth interviews using questionnaires with open-ended questions. The survey questionnaires were developed in English, translated into 'Bahasa Melayu', and focused on the following topics: (1) farmers' demographic information, (2) household economy, (3) perceptions of community development, (4) social participation, (5) farm characteristics and farming activities, (6) material style of life, (7) perceptions of access to resources, (8) perceptions of problems encountered in seaweed farming, (9) perceptions of seaweed industry management, and (10) perceptions of the SCP. Interviewees were asked whether they agreed, disagreed or were neutral on questions related to (3) (4) (7) (8) (9) using a Likert-type scale (Thurstone 1928). The SCP participants were asked (10) whether or not the SCP had changed their income or production since they joined the cooperative. Using pictorial guides as part of the questionnaires, the participants were also asked to rank the eight types of assistance provided as part of the SCP in terms of usefulness and practicality.

Permission letters were sent to KI prior to interviews, and invitation letters were sent to the respondents before the FGD. All participants were contacted by telephone, email or met in person to obtain assurances of their attendance. The interviews took place at various locations including at sea, at their homes, business premises and offices, in restaurants, community halls and public library meeting rooms.

Data Analysis

The selection of themes for analysis was decided using the method proposed by Ryan and Bernard (2003). The data were textual, and transcribed from an audio recorder and interview guide. The transcripts from the KI interviews, FGD, fieldwork notes, observation and HS were coded. Different codes were chosen to allow thematization of underlying patterns. Themes were analysed for connections between data, concepts and theories. Interview data were transcribed, uploaded into NVivo 11 qualitative research software (QSR International Pty Ltd.), and organised into four broad categories: governance, economic, environmental, technological and sociocultural. Survey data were analysed as a descriptive analysis using SPSS version 21 quantitative research software (IBM Corp.) and plotted as a principal component analysis (PCA) using Plymouth Routines in Multivariate Ecological Research (PRIMER) software (Plymouth Marine Laboratory) based on the variables and questions shown in Table 1. The assessment of social network strength and leadership followed Bodin and Crona (2008); however, in the present study PCA was incorporated to explore possible relationships between social capital, leadership and the need for government intervention.

Table 1. Examples of the variables and questions asked to assess social network strength in seaweed communities in Semporna

Variables	Questions
Individualistic	Do you think people are only interested in their own welfare?
Conflicts	Are there unresolved issues or problems between people in the community?
Intervention	Do you think the government should solve major problems that affect the community?
Trusting	Do you think most people in the community are honest and can be trusted?
Harmonious	Are relationships in the community generally good?
Accepted	Do you feel accepted as part of the community?
Leadership	Do you think there are strong leaders in the community who are respected?
Problem solving	Does everyone work together to solve problems that affect the community?

RESULTS

Demographic Composition of Semporna Seaweed Farmers

The demographic composition of seaweed communities in Semporna is presented in Table 2. Although the majority of farmers were married (90%), there was limited female participation in farming activities (less than 30%), with women mainly assisting males in longline cleaning, seeding or drying.

Table 2. Demographic data of seaweed communities in Semporna, taking into account gender, age, ethnicity, religion, marital status, education, extent of experience of seaweed farming, place of origin, and time living within the farming community. SCP = Seaweed Cluster Project, SAFA = Semporna Area Farmers' Association

	Type of farmers		
Variables	SCP (n = 40)	Non-SCP (n = 87)	SAFA (n = 17)
Gender (%)			
Male	77.5	75.9	70.6
Female	22.5	24.1	29.4
Age (%)			
Below 50 years old	77.5	80.5	76.5
Above 50 years old	22.5	19.5	23.5
Ethnicity (%)			
Bajau	97.5	5.7	64.7
Suluk	2.5	77	5.9
Other	0	17.2	29.4
Religion (%)			
Islam	100	100	100
Marital status (%)			
Married	90	92	100
Single	2.5	3.4	0
Widowed	2.5	4.6	0
Divorced	5	0	0
Level of education (%)			

Higher education	0	1.2	0
Secondary	42.5	10.5	23.5
Primary	37.5	45.3	52.9
No formal education	20	43	23.5
Experience in seaweed farming (%)			
Less than 5 years	10	16.1	52.9
6–10 years	10	28.7	23.5
11–15 years	15	24.1	0
16–20 years	22.5	18.4	11.8
More than 20 years	42.5	12.6	11.8
Place of origin (%)			
This community	97.5	3.5	30.1
This island	2.5	15.1	14.7
This country	0	12.8	9.8
Other country	0	68.6	45.5
Time living within community (%)			
Less than 5 years	0	12.6	41.2
6–10 years	0	23	0
11–15 years	0	12.6	5.9
16–20 years	10	14.9	17.6
More than 20 years	90	36.8	35.3

Despite there being an age limit criterion to becoming an SCP beneficiary (under 50 years of age), 22.5% were older than 50 years old.

"We only selected participants below 50 years old. Moreover, priority went to unemployed youths with a family. Previously, we had some participants that were over 50. We revised the age factor because, in the seaweed certification program (MyGAP), participants were required to undergo a medical health check-up. This resulted in elderly participants not qualifying for MyGAP. The Sabah Fisheries Department instructed us to substitute elderly participants with younger ones" (SCP Leader LB KI interview, 13 Apr 15).

On the other hand, non-SCP farmers (migrant and SAFA) had the highest proportion of farmers younger than 50 years (80.5%), most commonly in the 30–39 age category (migrant =34.5%, SAFA =52.9%).

However, a Kruskal-Wallis test showed that age was not significantly different between the seaweed communities.

All farmers were Muslim and the majority of the SCP and SAFA farmers (97.5 and 64.7% respectively) were Bajau, while migrant farmers — either war refugees or illegal Filipino migrants — were Suluk. The Suluks occupied most of the productive farming areas that were unoccupied by indigenous farmers, i.e., reef areas or remote islands. A KI SFD official stated that *"the majority of people involved with seaweed farming are illegal immigrants*." More than 45% of non-SCP farmers originated from another country, whereas 97.5% of the SCP farmers originated from the same community, with 90% having resided in the community for more than 20 years. SAFA farmers had the highest proportion of farmers (41.2%) having lived in the area for less than five years. Most migrant farmers (68.6%) originated from the southern Philippines, and 45.5% of SAFA farmers with less than five years residency were of Filipino extraction. As a respondent noted:

"Seaweed farming is dominated by the Suluk ethnicity (originated from the Southern Philippines). I think their contribution to seaweed farming is important. Only a few indigenous people are interested in seaweed farming. One of the challenges in this industry is how to increase participation amongst indigenous people (Sabahan) and how to solve issues with immigrant farmers from the Southern Philippines – i.e., legalize their existence as seaweed farmers or take over their farming sites and give them to the indigenous people" (PPK Manager KI interview, 14 Apr 15).

The KI reported that the highest seaweed production levels were achieved by the migrant farmers and *"they are the key people in this industry that provide a large proportion of the labor force"* (Processor LF KI interview, 13 Apr 15).

More than 50% of farmers in all communities had obtained some level of formal education, either in Malaysia or their country of origin. Migrant farmers had a higher proportion of no formal education and only 1.2% had obtained higher education. The SCP farmers had the highest percentage (42.5%) of farmers with more than 20 years of farming experience.

The Institutional Arrangements in the SCP Decision-Making Process

The federal government at ministry level has the power to formulate policies and regulations, although some powers are delegated to state and local government agencies. Management of the seaweed industry at the national level is led by the Ministry of Agriculture and Agro-Based Industry with the Department of Fisheries Malaysia (DOFM) as the responsible agency. The DOFM, which is based in the Federal Territory of Putrajaya, is responsible for governing fisheries, using the Fisheries Act 1985 (Amended 2012) as a legal framework. The Sabah Ministry of Agriculture and Food Industry and the SFD are the responsible agencies for managing the seaweed industry at the state level, while the district office (i.e., Semporna) is responsible for managing seaweed projects at the local level.

The federal, state and local levels were all involved in the SCP decision-making process: the federal government was responsible for planning, monitoring, and providing funds to the state government; the state government, through the SFD, selected the sites; while the local community leader selected the SCP participants from within the local communities (SCP Manager KI interview, 06 May 15). The SFD contacted the community leaders, who provided names of local people that were interested in joining the SCP. Ahead of project approval, the SCP committee at the local level (comprising SFD officials and community leaders) visited each village and interviewed all participants to ensure that the participant list was legitimate and contained genuine candidates, i.e., Malaysian citizens employed full-time as seaweed farmers. The finalized SFD participants list was submitted to the federal level for approval. The DOFM (the federal level) then released funding to the SFD to implement the project after NSTC approval.

This study found that 76% of KI (n = 23) were not members of the NSTC and not involved in decision-making for seaweed industry management. However, 100% felt they were informed of seaweed interventions and knew of at least one seaweed program implemented in Sabah. Although they were informed about seaweed programs, more than half were unsure about the impact of the programs. Overall, 48% of the KI were involved passively in seaweed industry management, and 71% had communicated with more than ten organizations in their social network relationships. As shown in Fig. 2a, 65.2% of KI felt that seaweed farming was very important for the country, particularly coastal communities, whereas 26.1% and 8.7% perceived seaweed farming to be important or less important, respectively. However, there were split views within that 65.2% over whether the government had prioritised seaweed industry management issues less than other government issues (less important: 21.7%, unimportant: 17.4%, very unimportant: 26.1%) (Fig. 2b).

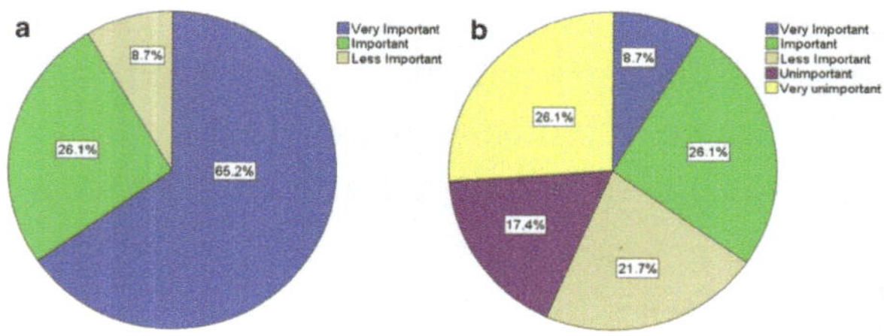

Figure 2. The perceptions of the importance of; a) seaweed farming to the country; and, b) seaweed industry management issues in comparison to other government priorities.

Perceptions of the SCP Implementation and Its Community Impact

User Reactions to the SCP

Negative reactions to the SCP were more evident than positive. For instance, positive reactions included the following. A KI claimed that the SCP concept would offer solutions to seaweed problems by increasing efficiency, productivity, and incomes for farmers through the adoption of innovative farming techniques, and the formation of farmer organizations (DOFM KI interview, 14 Jun 15). The community leaders and members agreed to the project as it symbolized high social status as a successful seaweed community. As one SCP leader stated, *"the government has chosen our community because we have the most productive seaweed area and we produce good quality seaweed. There are seaweed farmers in other villages on this island, but our community has the most active farmers"* (SCP leader GG KI interview, 10 May 15). Similarly, another

SCP leader felt that the SCP implementation demonstrated that the government acknowledged the importance of seaweed farmers: *"By joining the SCP we felt that the government appreciated us as seaweed farmers and it symbolized that our participation in this activity is important"* (SCP leader LB KI interview, 13 Apr 15).

Negative reactions included criticism that while most interview respondents (73.9%) felt they were informed of the SCP implementation, they were unsure about the programs impact. They said they were not involved in the seaweed industry management decision-making at a national/federal level, and although the SFD engaged with community leaders and members before the SCP implementation, their views were not taken as part of the planning. Rather,

"The SFD conducted a meeting with the communities in March 2012 to inform them of the SCP and we provided an explanation of the SCP concept and the benefits of this project. We also explained that it will take some time to see the SCP impacts at that time, the government had the funding and we wanted to select individuals to participate in this project and provide them with assistance"

(SCP Manager KI interview, 06 May 15).

Feelings of a lack of participation in the SCP decision-making process were strongest in Gelam-gelam. During an SCP farmers' focus group in 2015, the participants claimed that:

"The SFD did not think carefully when they introduced the SCP. The SFD did not consult us before implementation. We only knew about the project when they wanted to build the infrastructure at our village. The system is unsuitable for us. We accepted because it was the governments' project"

(Gelam-gelam SCP farmers" Focus Group, 02 Jun 15).

Moreover, there was confusion over the purpose of the SCP. One KI at the local level stated that *"the SCP cooperative was formed as a formality to receive assistance from the government"* (SFD Local KI interview, 09 Apr 15). During a focus group in Gelam-gelam in 2015, the SCP participants voiced dissatisfaction that although they belonged to the SCP cooperative the SFD official only used their name to gain government assistance (Gelam-gelam SCP farmers' Focus Group, 02 Jun 15). A SAFA manager claimed that the federal level wanted to transform seaweed farming to a commercial scale, but local people felt the seaweed project was another poverty eradication program (SAFA Manager KI interview, 14 Apr 15).

More serious criticism was voiced that the government failed to choose genuine farmers or companies to participate, and that the SCP project was largely a publicity stunt: "*the government's aim for seaweed programs such as the SCP, the mini estate or seaweed National Key Economic Area (NKEA), are that they are all just for publicity and to look good on paper. The participants they chose might not be genuine players and the government did not know what was happening on the ground and had no idea what will happen to this industry in the future*" (Processor LF KI interview, 13 Apr 15). The government rejected this criticism, insisting that it took seriously its responsibilities to rigorously select appropriate farmers for the SCP. The government claimed that not only did it enforce the strict criteria to be met by individual farmers (such as Malaysian citizenship, residency in the local community, full-time seaweed farmer, and entitlement to the farming site), but it also required additional criteria to be met, and the SFD gave a full mandate to community leaders in choosing the SCP participants, because "*the villagers only deal with people they know and that is why we can only meet with them through their leader*" (SCP Manager KI interview, 06 May 15).

A further source of resentment among the SCP participants was over the subsidized aid given to them by the government. Participants at both sites considered the number of boats provided to be neither sufficient nor practical. Some participants quarreled over boat usage (LOCAL SFD KI interview, 09 Apr 15), for example, participants complained that '*the boat given is not enough and we have to take turns. If everybody wants to go to the farm we will face problems*' (Lok Butun SCP farmers' Focus Group, 03 Jun 15). Lack of boat ownership/access is a major constraint in seaweed farming in Malaysia. The SCP farmers perceived that they could not farm without a boat unless they borrow from relatives or friends (Gelam-gelam SCP farmers' Focus Group, 02 Jun 15). This study found that 42.5% of the SCP farmers did not own a boat, whereas all (100%) of non-SCP and SAFA farmers owned at least one boat.

The Impact of the SCP in Increasing Seaweed Production and Farmers' Income

The SFD provided both individual and collective farming inputs to the SCP participants (SCP Manager KI interview, 06 May 15). Individual assistance ensured that each participant received 600 kg of seedlings, a free assessment for seaweed farm certification, and 50 EFL (Figs 3 and 4). Excluding the cost of farm certification, the government estimated the initial cost per individual to be MYR 30,000 (US$ 9870). Collective assistance ensured that the government provided two fiberglass boats, eight tying tables and built drying platforms for each SCP site (Figs 3 and 4), costing approximately MYR 1.0 million (US$ 330,000) per site. In return, the participants had to utilize the innovations as a condition of the project.

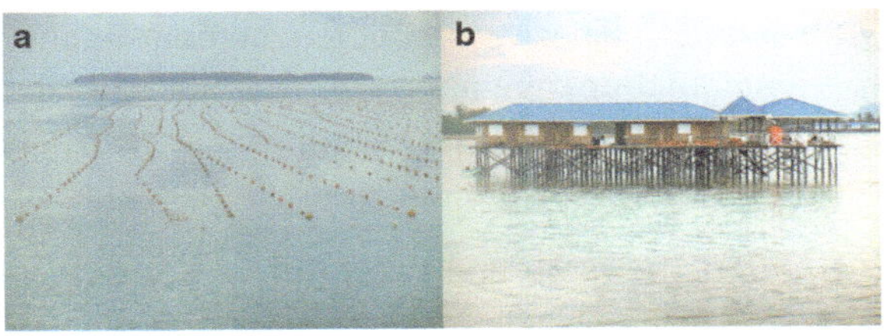

Figure 3. Individual and collective farming inputs provided to the SCP participants: a) Eco-friendly longlines installed at the SCP site at Lok Butun, Bum-bum Island; and, b) the SCP drying platform at Gelam-gelam, Bum-bum Island.

Figure 4. Selected technological interventions provided as part of the SCP: a) used plastic bottles for flotation were replaced with poly-floats; b) raffia tie-tie were replaced with polyethylene; c) a concrete/cement anchoring system replaced the conventional anchor rope approach (the anchor rope was often tied to live coral); and d) dugout canoes were replaced with fiberglass boats and seedling tying was done from a tying table at the SCP drying platform.

The SFD estimated the seaweed production per cycle for 50 EFL in a two-hectare farming area to be one tone dry weight per ha per cycle, i.e., six tones dry weight per year (based on three cycles per year). KI and seaweed farmers reported that a production cycle lasted seven to eight weeks, therefore the theoretical maximum number of cycles per year is eight. However, production is seasonally affected (Stakeholders' Focus Group, 04 Jun 15) so that in reality most farmers could only farm between two and six cycles depending on site. The Lok Butun SCP participants experienced two to four cycles per year while Gelam-gelam SCP participants, SAFA farmers in Mata Pahi/Melanta Kobal, and migrant farmers in Sebangkat and Palang-palang experienced two to six cycles; whereas migrant farmers in Patuit and Sangaban experienced three to eight cycles. Figure 5 shows the seaweed productivity from 144 farmers. All SCP farmers produced less than five tons of dried seaweed per year. However, the majority of migrant (60.9%) and SAFA (94.1%) farmers produced at least six tons per year.

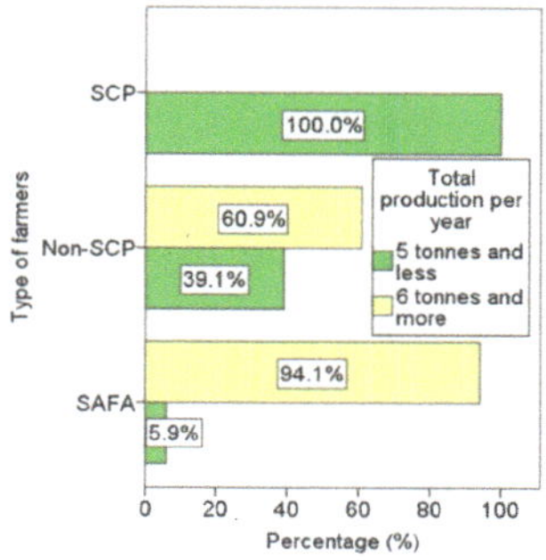

Figure 5. The proportion of farmers (SCP, non-SCP, and SAFA) achieving productivity targets of six tonnes dry weight per year.

In 2008, the Sabah state government introduced the *Projek Mini Estet Sejahtera* (MESEJ) *Rumpai Laut* (Prosperous Mini State Seaweed Project) to the Bum-bum island community as an initiative to reduce poverty. The same participants, or their children or grandchildren, were chosen to participate in the SCP. Under the MESEJ program, the Lok Butun participants were designated a farming area based on traditional customs or inheritance from parents. According to a community leader, the older generations who started seaweed farming in the 1980s chose farming sites close to their houses. When the SFD selected Lok Butun as an SCP site the Semporna District Office declined to establish the SCP at the MESEJ program sites, but instructed the SFD to build the SCP infrastructure in designated Seaweed Aquaculture Industry Zones (SAIZ) (tracts of sea that had been gazetted by the government for seaweed aquaculture activities only). When the villagers were selected to participate in the SCP, they were allocated a new farming area within the SAIZ, which was distant from their houses. The Lok Butun farmers could choose to plant seaweed at the MESEJ or SCP sites, of which 52% farmed two sites using differing farming systems – the conventional longlines system at the MESEJ sites (210 m ropes with raffia strings as tie-tie and used plastic bottles as buoys); and the SCP system (100 m ropes with double PE strings as tie-tie with poly-floats as buoys) at the SCP sites. Based on the survey, 16% farmed the MESEJ sites only, a decision based on four main reasons: a preference for raffia tie-tie (although non-SCP farmers eventually changed from raffia to PE tie-tie as they became aware of seaweed quality implications); the farm location was nearer; a perception that the SCP site was problematic due to seaweed predation; and that one person could manage the conventional farming system whereas at least two people were needed to work the SCP system. As mentioned by the SCP participants, *"if we were given a choice, we will choose raffia tie-tie. Seaweed biomass is heavier if we use raffia tie-tie...the location of the drying platform is far from our house. It's not practical"* (Lok Butun SCP farmers' Focus Group, 03 Jun 15). In contrast, all Gelam-gelam SCP farmers allowed their previous farming systems to be replaced by the SCP system (SCP Leader GG KI interview, 10 May 15). Previously, they were not involved with the MESEJ program, therefore their sites had no constraints on other forms of seaweed intervention.

Although the level of abject poverty in Malaysia decreased from 49.3% in 1970 to 0.6% in 2014, Sabah remains a relatively poor state (Hatta and Ali 2013), and Semporna one of its poorest districts:

"Seaweed farming has been used as a poverty alleviation program for districts facing poverty issues, and Semporna district is one of poorest in Sabah. Most of the poor people are fishermen and from coastal communities" (Local SFD KI interview, 09 Apr 15).

According to the SCP manager, all of the SCP farmers selected by the community leaders were on the poverty list, but had not previously received any government aid. He said *"we get the list from the e-kasih list* (poverty list) *provided by the district office. None of the SCP participants had received any government welfare schemes except individual aids"*. However, the HS found that 35.0% of the SCP farmers had obtained support through the governments' welfare house scheme through the MESEJ project, which contradicts the claim made by the SCP manager. Nevertheless, the SCP farmers were poorer than the non-SCP farmers. For example, only 40.0% of the SCP farmers owned their houses compared with 77.9% of non-SCP farmers and only 6.7% lived in welfare houses (mainly from the SAFA community, since other non-SCP farmers, particularly migrants, were ineligible for government assistance). Therefore, the SCP project did target poorer members of the community, with the exception of migrants.

Total SCP production for each farmer was estimated at 6.0 t dry weight per year with an annual income of MYR 14,000 (US$ 4600) if the price per kg dry weight was MYR 2.50 (US$ 0.82). However, SCP farmers did not see the SCP project as either alleviating their poverty or increasing their income. The HS showed that the actual production was less than 5.0 t dry weight per year per SCP farmer (Fig. 6), and the actual annual income was less than MYR 14,000 (US$ 4600) per SCP farmer. This is mainly due to the poor market price for seaweed. During the SCP farmers' focus group in 2015, the participants agreed that *"seaweed farming is not a lucrative business"* (Gelam-gelam SCP farmers' Focus Group, 02 Jun 15). The low seaweed price was a strong disincentive for the SCP participants to focus solely on seaweed farming. As mentioned by the KI, *"one of the seaweed farm owners met with me last night and expressed his worries for this industry due to the low price. At the moment, the guy bought a fishing net and gave it to his workers because of seaweed farming being slow. The workers need to do other things to get income for them and their employer"* (SFD Official KI interview, 28 May 15). One reason for the low market price for seaweed was the restricted marketing system based on intermediaries (buyers). All of the SCP participants in both villages agreed that the only market source for their produce was through intermediaries. The majority (82.5%) of participants chose intermediaries, while 15.0% and 2.5% chose good friends or relatives (who impose no quality control) as their buyers, respectively. One of the reasons that discouraged most SCP participants from selling seaweed directly to the carrageenan processor or to exporters for a better price was their lack of seaweed marketing knowledge (82.5% of total participants). Even though some (17.5%) knew about seaweed marketing, especially white seaweed or *rumpai putih* for direct consumption, they could not sell directly to the end users because of poor marketing networks (37.5% agreed, 40% unsure and 22.5% disagreed) and poor entrepreneurship (47.5% agreed, 30% unsure and 22.5 disagreed) even after they had joined the SCP.

Figure 6. Improving seaweed quality and achieving MyGAP certification by encouraging the SCP farmers to improve post-harvest handling by switching from drying seaweed at their house to the SCP drying platform.

This study found that 87.5% of the SCP farmers earned less than US$ 10 per day per farmer from seaweed farming, which is below the national Poverty Line Income (PLI) (Hatta and Ali 2013), while only 43.3% of non-SCP farmers earned less than US$10.0 per day per farmer. Only 8 and 20% of the SCP farmers in Lok Butun and Gelam-gelam, respectively, earned more than the national PLI level from seaweed farming. By contrast, seaweed farming alleviated the poverty of most of the non-SCP migrant (55.2%) and SAFA farmers (64.7%). Therefore on balance, the SCP has not alleviated poverty for most participants: we found the same results as the SFD annual report which stated that the SCP participants' incomes for 2013 and 2014 were less than US$ 10 per day per farmer (Sabah Fisheries Department, unpublished data). These findings mean the cluster project has not met its aim of increasing household income to more than US$ 10 per day per farmer. The SCP participants who perceived that their income had not improved felt that five factors were responsible – the government has not provided enough seedlings for them; seasonal changes; diseases; predation problems; and price fluctuations. However, they felt that the assistance they received from the government had reduced their operating costs on farming inputs and replaced broken farming ropes and buoys.

The Impact of Technical Innovations to Improve Seaweed Quality and Farming Efficiency

Malaysia has laid down seaweed standards on seaweed cultivation, dried seaweed and semi-refined carrageenan. The price of dried seaweed is determined by using dried seaweed from carrageenophytes to meet Specification (MS 2528:2013) standard. However, the certification scheme is voluntary, and farmers and processors can choose whether they want to participate or not. As mentioned by MARDI RO KI interview, 09 Jun 15: *"The MyGAP seaweed certification scheme helps the farmers in terms of their hygienic practices but does not increase seaweed prices"*. Even though the seaweed certification scheme has been implemented in Malaysia, this only applies to the local farmers and of those, is limited to the SCP farmers in Lok Butun. The government perceived that the Lok Butun SCP farmers were ready to participate as they have farming facilities that would comply with the certification scheme. It therefore seems that MyGAP certification is limited to farmers who operate from the government's facilities i.e., drying platforms. One KI commented that *"we have to upgrade the infrastructure provided by the federal government to get MyGAP certification. With MyGAP, seaweed produced in SCP Lok Butun can easily be exported because of hygiene and being safe to eat"* (SCP Manager KI interview, 06 May 15). But farmers were not taking full advantage of certification: "*The SCP farmers can demand a higher price if their seaweed is certified; however, at the moment they have not exported their produce and follow the normal market prices*" (SCP Manager KI interview, 06 may 15). The farmers perceived that seaweed prices remained the same regardless of whether they complied with the certification requirements because the intermediaries dictate the price, and they are the only source of marketing (Stakeholders' Focus Group, 04 Jun 15).

Technological interventions were viewed as necessary if the SCP was to increase productivity; however, the farmers' views on the practicality, economic sustainability, and environmental and social acceptability of the interventions revealed some resistance to their adoption — which may have hindered the program's success. The SFD provided both individual and collective farming inputs to the selected participants (SCP Manager KI interview, 06 May 15) (Fig. 6). The eight SCP technological interventions (PE tie-tie, poly-floats, detachable eco-friendly long lines (EFL), concrete anchoring systems, seedling tying tables, wooden drying platforms, fiberglass boats, and the seaweed certification program - received a wide variety of responses. Among the criticisms of government assistance was that while each EFL was 100 m long and could be planted with 120 kg wet weight of seedling, equating to each participant requiring 6.0 t of seedlings for 50 EFL, the government only provided 600 kg seedlings or 10% of the total seedlings required by each participant to start their farming activity. The SFD said that participants had either to produce their own future seedlings using vegetative propagation from the given 600 kg, or to obtain future seedlings by purchasing them from other farmers. But participants complained that vegetative propagation took a long time and was uncertain, and they could not afford to buy seedlings from other farmers. The EFL were simply regarded as equivalent to standard PE ropes without any improvement on current practices, apart from a shortened rope length. In contrast, the SCP participants who chose the SCP site preferred — in terms of usefulness and practicality — the poly-float buoys, followed by the EFL, the seaweed certification scheme, the boats, the anchoring system, the drying platform, the PE tie-tie, with the tying table the least liked. The design and color of the poly-floats attracted visitors to the villages; this 'SCP identity' had a positive perception. The Lok Butun SCP community held mixed views of the farming efficiency. As mentioned by their leader:

"We have two groups; A and B. Each group was allocated with one drying platform by the government. The difference between group A and B is the set-up of farming sites. Group A use the same site as their previous farms before they joined the SCP while group B use the new farming sites that were allocated by the Sabah Fisheries Department in the SAIZ area. Group A and B use either farming systems either SCP-only or in combination with conventional systems. The EFL uses double tie-tie points for seedlings while the conventional system only uses one tie-tie. I think the EFL system optimized the farming site because one line of EFL is equivalent to three lines of the conventional method" (SCP Leader LB KI interview, 13 Apr 15).

The SCP participants in Gelam-gelam had to farm at one site only. All of the Gelam-gelam SCP participants perceived the farming area was optimized with the installation of the SCP farming systems on two hectare areas, whereas 52% of the Lok Butun SCP participants felt that their farming area was optimized. Despite the higher percentage in Gelam-gelam, 80% of respondents felt that the production output was the same as with the previous farming system. The majority of the SCP participants (more than 90%) in both villages felt that seaweed quality and labor usage were the same as previously (Table 3).

Table 3. Percent distribution of perceptions at two SCP communities of increased seaweed production, successful optimization of the farming areas, and improved seaweed quality and farming efficiency

Indicators	**SCP Participants**	
	Lok Butun (n = 25)	**Gelam-gelam (n = 15)**
a) Perception of optimizing farming area by using the SCP farming systems		
i. Optimized	48%	100%
ii. Not optimized	52%	0%
b) Perception of production by using the SCP farming systems		

i. Better	0%	20%
ii. Same	100%	80%
c) Perception of seaweed quality produced by using the SCP farming systems		
i. Better	4%	5.7%
ii. Same	96%	93.3%
d) Perception of labor usage by using the SCP farming systems		
i. Reduced labor	4%	0%
ii. Same	96%	100%
Total	100%	100%

SCP Initiatives to Reduce Environmental Impacts

The survey revealed two perspectives on environmental impacts; one focusing on plastics pollution and the other on water quality. The SCP farming system uses EFL as its main feature, which uses neither raffia tie-tie nor plastic bottles. The EFL replaces the conventional hanging longlines with the intention that it would reduce the environmental impact of seaweed farming (Fig. 6). The hanging longlines system used recycled plastic bottles as buoys and raffia tie-tie to tie seaweed to the main farming ropes; however, the SFD regarded these as polluting as the farmers discarded broken bottles and raffia tie-tie into the sea. The MyGAP seaweed certification scheme prohibits any pollution from farming activities, and the SFA viewed the EFL as the best approach to fulfil the requirement (Fig. 6) as described by the KI:

"The government have improved the farming system by introducing environmentally friendly methods in the SCP. If you visit the farm you can see the differences between the SCP and individual farmers. Most of the individual farmers still use empty bottles as floats which polluted the environment. The SCP participants do not pollute the environment because they use poly-floats that are environmentally friendly. Also, the SFD monitor the SCP using the Malaysia Good Agriculture Practices (MyGAP) approach that gives certification to this system" (PEMANDU KI interview, 15 Jun 15).

His opinion was shared by the SCP leader by saying this: *"I think the difference between the traditional and the SCP system is from the environmental aspect because the SCP uses an eco-friendly long line system. When we obtained the seaweed MESEJ poverty alleviation project in 1996, the development agents complained about pollution from our farming activities, for instance, pollution from the plastic bottles that we use as a float. But we don't have a choice because this system has been practiced since the 1990s and the government is the one whom introduced it to us"* (SCP Leader LB KI interview, 13 Apr 15).

Paradoxically, the introduction of EFL caused an immediate increase in pollution. Prior to the introduction of the SCP, the communities collected all the good condition plastic bottles that floated under their houses to be used as spares for their existing floats. Anecdotal evidence showed that this practice was abandoned with the advent of the EFL. The community leader raised his concerns on this matter by saying that "*nowadays, the environmental issue is severe because of pollution from the community. You now see a lot of waste, particularly used plastic bottles in the sea during the north east season. They do not care about it anymore"* (Community Leader LB KI interview, 16 Apr 15). When questioned on the environmental impacts of seaweed farming, 97.9% thought that it does not contribute to marine pollution. KI also did not perceive a risk of environmental pollution in seaweed farming areas.

The success of seaweed farming is closely associated with the physiochemical parameters of seawater such as temperature, pH, salinity and turbidity. According to the SCP leader:

"Another issue was seaweed growth not being the same as when I started this activity in the year 2000. I think this is related to pollution caused by the increasing number of houses and people in the community. I used to harvest 2-3 kg of seaweed per plant but nowadays the most I can get is 1.5 kg, a big difference. Last time the seawater color was blue and nowadays it is green. The water quality is poor and is causing the spread of ice-ice disease in the seaweed. That is why the seaweed size is smaller nowadays" (SCP Leader LB KI interview, 13 Apr 15).

Whereas there is no evidence that the SCP has directly impacted seawater quality, should seaweed farming continue to grow and attract additional practitioners from outside the local communities, this may exacerbate any decline in water quality with knock-on consequences for seaweed production.

Social Network Strength

The migrant and SAFA farmers' social cohesion was strong and they would help each other rather than rely on the government to solve problems within their communities. Moreover, both communities believed they have strong and respected leadership within the communities. In contrast, the SCP farmers did not think highly of local leadership, had low social cohesion, and relied on the government to solve problems (Fig. 7). This study found that there was resentment between the SCP farmers because of dissatisfaction with farmers who were selected despite not being seaweed farmers and not owning seaweed farms at the Gelam-gelam SCP site. One KI said, "*they do not farm seaweed and kept the inputs given by the government at their home*" (SCP Leader GG KI interview, 10 May 15). This KI said that the previous SCP leader had chosen farmers whom he knew were neither seaweed farmers nor owners of farming sites. This created resentment among the SCP farmers in east Gelam-gelam who were genuine farmers and had to give up some of their farming areas to the new SCP farmers in west Gelam-gelam, who were not genuine farmers:

"Most of the farmers in west Gelam-gelam are new farmers and they do not own the farm, including the previous leader. Meanwhile, all the farmers in east Gelam-gelam are seaweed farmers and owned seaweed farms before joining the SCP. They had to give their existing farming areas to the SFD so that they can manage and divide to other farmers" (SCP Leader GG KI interview, 10 May 15).

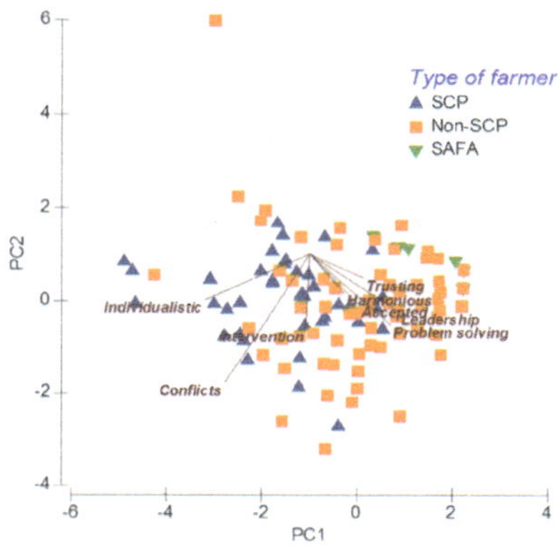

Figure 7. Principal component analysis of the social network strength within the seaweed communities (SCP, non-SCP, and SAFA) in Semporna.

There was evidence of poor governance principles which can impact the level of social cohesion as the results identified although most of the SCP farmers in Lok Butun have family relationships with each other they did not discuss their problems openly and did not interact with their leader to solve the problems within the communities.

DISCUSSION

The study has identified five main issues that may explain the success or otherwise of the SCP program; relations between indigenous and migrant farmers; stakeholder participation; human capital; livelihood strategies; and local leadership.

Relationship between Indigenous and Migrant Farmers

The tense relationship between indigenous and migrant seaweed farmers is in part linked to the split between female and male farmers. Local community involvement in seaweed farming has broadened through growing female participation, although this remains low as the women lack seafaring skills (Cooke 2004). The longlines system requires deep water (2 to 5 m), necessitating the use of boats. As outboard engines are costly, farmers use dugout canoes (*bogo-bogo*) for farm access, which are difficult to handle without the necessary seafarer skills. Also, using the conventional longlines system a farmer usually works under the hot sun when tying seedlings, whereas the women prefer to assist from home (Kronen et al. 2010). In contrast, seaweed farming in Zanzibar and India is female dominated as they received assistance through credit services and skills training (Valderrama et al. 2013; Periyasamy et al. 2014). Also, the farming system used in Zanzibar is fixed-off bottom in shallow water, which avoids boat use (Msuya 2013). In Semporna, the female SCP participants were either widows or divorcees, preferring to hire labor (mostly from migrant communities) rather than farm themselves.

The Moro conflict in the southern Philippines forced more than 50,000 people to migrate to Sabah in the 1970s (UNHCR 2014). In the late 1980s, the Malaysian government divided the Filipino migrants into three categories; illegal migrants, war refugees, and economic migrants (Kassim 2009). Seaweed farming became a safety net for migrant communities living in *pondohan* stilt huts, not only because they felt safe living at sea but because the authorities did not obstruct their way of life or their farming activities. Although official data on the number of migrant farmers are unavailable, KI claimed that they dominate the industry, generating more than 50% of total national seaweed production. The migrant farmers also had strong leaders (usually a respected person holding Malaysian citizenship), who deal with the indigenous people and authorities if required. The *pondohan* migrant communities have strong social cohesion, into which the leaders will not admit random immigrants as they are aware of governmental concerns surrounding immigrant issues. Migrants generally do not commit crimes or create social problems as they do not want to attract the attention of the authorities. Moreover, social capital was high among migrant communities, whereas it was low among the indigenous SCP farmers who were often involved in disputes over the project. Nevertheless, migrants were blamed for the lack of government support for the seaweed industry, as the industry is considered a threat to national security as it attracts more migrants from the southern Philippines. Consequently, the government is reluctant to grant sea tenure to migrants, or permit them to participate in the SCP, despite their obvious value as seaweed producers.

Stakeholder Participation

There was disconnect between farmers and government officials. Farmers felt estranged from the decision-making process concerning community development programs, and perceived the government to have absolute decision-making power. Government officials confirmed that farmers were not involved with decision-making at the national level, but were involved to some extent at state and local levels. Respondents argued that before the implementation of seaweed projects, the government should engage with all stakeholders, including farmers, buyers, processors, officials and NGOs; inclusiveness being an effective approach to gather information and input from stakeholders (Gray 2005; Lockwood et al. 2010). Interviewees complained that poor stakeholder participation leads to feelings of marginalization from decision-making. Embedding local input and knowledge in decision-making would also produce a better understanding of socio-ecological systems (Reed 2008).

Human Capital

Interviewees criticized the government for failing to provide the SCP farmers with entrepreneurial skills that would have allowed them to make the transformation from individual small-scale producers to commercial-scale production. The SCP farmers perceived that they possessed neither the economic nor human capital to manage their farms. They claimed that the seedlings provided by the government were insufficient so most of the farming lines were left empty, and that they lacked training in technical skills to make use of the new technology. Seaweed farming is a simple activity, using cheap and recycled materials to establish the farm and individuals can learn farming skills by themselves or through observation. However, work experience using the same practices does not necessarily result in higher productivity, unless the individual switches to new technology (Hanna et al. 2014). All seaweed intervention projects are about introducing new innovations, but farmers need training in such skills (Vredegoor and Pennink 2013).

The SCP farmers also complained that they were not given guidance on marketing strategies. Seaweed farming is a risky financial activity, not least because the seaweed price constantly fluctuates and often seaweed farmers receive very low prices from the buyers who come to their islands or villages. Seaweed farmers usually do not sell directly to the processors because they lack marketing information: they believed that processors hire their own agents or buyers to obtain specific quantities of seaweed. The government do not assist farmers to develop marketing skills. Carrageenan processors mentioned that when they met with government agencies to discuss seaweed marketing, they perceived the agencies do not know how to deal with it as the government lacks the capability of managing the seaweed industry and its marketing system. Although several KI perceived that the seaweed industry will develop and expand because of government interventions and continued strong global demand from the processed food industry, respondents identified many problems with marketing. For example, seaweed buyers said that if the government wanted them to export legally, they should facilitate the export procedure and build a port and appoint an export agent based in Semporna rather than in Tawau as it will be costly for them to go to Tawau for documentation preparation. Also, most seaweed buyers did not know how to register or use online marketing services. Despite this, stakeholders believed that the industry will develop and expand due to government interventions and the continued strong global demand for carrageenan.

Livelihood Strategies

Sabah is one of the poorest Malaysian states with a GDP of MYR 18,603 (US$5673) per capita in 2013, compared with the national GDP per capita of MYR 33,000 (US$10,064) (DOFM 2013; DOS 2013). The Sabah fishing communities (especially in Semporna) are among the poorest in Malaysia (Solaymani and Kari 2014) with small-scale fishers portrayed as *"the poorest of the poor"* (Allison and Ellis 2001). Part of the attraction of seaweed farming as a poverty reduction strategy is that it is a household enterprise involving farmer's wives and family members (Cooke 2004). However, farmers in Semporna perceived seaweed farming to be a low income source which drove a lack of interest among coastal communities. Seaweed farming is not perceived as the main livelihood, rather fishing remains the main income source of those who make their living from the sea (Hill et al. 2012).

Local Leadership

Local leaders provide a vital link between the community members, industry and government. In addition, strong local leadership can increase social cohesion among community members (Bodin and Crona 2008; Gutiérrez et al. 2011). The SCP leader has an important role as s/he is responsible for choosing and managing the SCP participants. KI drew attention to deficiencies of leadership in their SCP projects, claiming a lack of trust. For instance, ten Lok Butun SCP participants quit the project as they thought the leader was corrupt. They also perceived that the leader was not fulfilling his promise of the project's development: for example, he resided in the neighboring village, did not monitor the project unless there was a visit from government officials or politicians, and was head of the youth wing of the local political party, which compromised his independence.

CONCLUSIONS AND RECOMMENDATIONS

This research found there to be conflicting views over the outcomes of the SCP in Semporna. Negative perceptions included low take-up rates by indigenous people, poor stakeholder participation in decision-making processes, economic vulnerability of seaweed farmers, a complex marketing system, low social cohesion among beneficiaries, and limited acceptance of new technologies. Positive perceptions included a recognition that the SCP confers high social status upon a community, reduces operating costs of seaweed production, and facilitates the production of certified seaweed. Tellingly, the non-SCP farmers were more successful than the SCP participants in performing seaweed farming activities. This study concludes by making five recommendations for improving the SCP.

Firstly, the government should increase and promote the participation of indigenous people in seaweed farming, particularly females and youths. Although, the current seaweed projects were awarded to local people, their engagement was poor and they perceived seaweed farming as an unpredictable source of income. Secondly, the government should confer legal status upon the existing migrant farmers rather than ignore their existence. However, the government has to set the criteria for the selection and make sure the authenticity of the documents, i.e., legal status as the war refugees. Additionally, this effort will provide the government with an accurate view of the current status of migrant farmers in Malaysia, which eventually will reinforce national security. Thirdly, the local seaweed cooperative organizations should be strengthened so that these organizations could provide the platform for farmers, buyers and processors to address seaweed related issues. The cooperatives could also provide services and become a business network for members. Fourth, the government should teach entrepreneurship skills to farmers so that they may negotiate with buyers and find other market outlets, thus making farmers more independent. Finally, stakeholders should be fully integrated into decision-making processes, thereby tackling issues based on local needs. Such steps could transform the prospects for seaweed production in Malaysia.

ACKNOWLEDGEMENTS

This work was funded by the Government of Malaysia and Newcastle University, UK. The authors would like to thank the Economic Planning Unit, Prime Minister's Department Malaysia, Department of Fisheries Malaysia and Sabah Fisheries Department for granting permission to conduct this fieldwork. Thanks are due to the seaweed communities in Semporna, community leaders, government officials, NGOs, carrageenan processors and villagers in general for their kindness, cooperation and hospitality which made our work much easier. We would also like to thank Dr. James Alin, Ms. Nazmahwati Walli and Mr. Mizpal Ali of Universiti Malaysia Sabah (UMS) for providing insight and assistance during fieldwork. Also, Mr. Masri Bangkoka of Gelam-gelam village for assisting with community assimilation. Any shortcomings of the paper belong to the authors.

REFERENCES

1. Allerton C (2014) Statelessness and the lives of the children of migrants in Sabah, East Malaysia. Tilburg Law Rev: J Int European Law 19:26–34
2. Allison EH, Ellis F (2001) The livelihoods approach and management of small-scale fisheries. Mar Policy 25:377–388
3. Bodin Ö, Crona BI (2008) Management of natural resources at the community level: exploring the role of social capital and leadership in a rural fishing community. World Dev 36:2763–2779
4. Cooke MF (2004) Symbolic and social dimensions in the economic production of seaweed. Asia Pac Viewp 45:387–400
5. Dalli ASA (1988) Seaweed Farming in East Malaysia. In: ASEAN/UNDP/FAO Regional Small-scale Coastal Fisheries Development Project, RAS/84/016.
6. DOFM (2013) Portal Rasmi Jabatan Perikanan, Department of Fisheries Malaysia, http://www.dof.gov.my/index.php/pages/view/1948.
7. DOS (2010) Population distribution and basic demographic Characteriestics Putrajaya. Report of the Department of Statistics, Malaysia

8. DOS (2013) Gross domestic product (GDP) by state, 2013. Putrajaya. Report of the Department of Statistics, Malaysia
9. Gray TS (2005) Theorising about participatory fisheries governance. In: Gray TS (ed) Participation in fisheries governance. Springer, Netherlands, pp. 1–25
10. Gutiérrez NL, Hilborn R, Defeo O (2011) Leadership, social capital and incentives promote successful fisheries. Nature 470:386–389
11. Hanna R, Mullainathan S, Schwartzstein J (2014) Learning through noticing: theory and evidence from a field experiment. Q J Econ 129:1311–1353
12. Hatta ZA, Ali I (2013) Poverty reduction policies in Malaysia: trends, strategies and challenges. Asian Cult Hist 5:48–56
13. Hill NA, Rowcliffe JM, Koldewey HJ, Milner-Gulland EJ (2012) The interaction between seaweed farming as an alternative occupation and fisher numbers in the Central Philippines. Conserv Biol 26:324–334
14. Hurtado AQ, Montaño MNE, Martinez-Goss MR (2012) Commercial production of carrageenophytes in the Philippines: ensuring long-term sustainability for the industry. J Appl Phycol 25:733–742
15. Kassim A (2009) Filipino refugees in Sabah: state responses, public stereotypes and the dilemma over their future. Southeast Asian Stud 47:52–88
16. Kronen M et al (2010) Gender and seaweed farming on Wagina Island, Choiseul Province in Solomon Islands. SPC Women Fish Info Bull 21:3–10
17. Lockwood M, Davidson J, Curtis A, Stratford E, Griffith R (2010) Governance principles for natural resource management. Soc Nat Resour 23:986–1001
18. Msuya FE (2013) Social and economic dimensions of carrageenan seaweed farming. FAO Fisheries and Aquaculture Technical Paper 580:115–146
19. Periyasamy C, Anantharaman P, Balasubramanian T (2014) Social upliftment of coastal fisher women through seaweed (*Kappaphycus alvarezii* (Doty) Doty) farming in Tamil Nadu, India. J Appl Phycol 26:775–781

20. Razali DDE, Bahron A, Alin JM (2015) Sustaining seaweed farming in Malaysia. Int J Bus Manag 3:201–205
21. Reed MS (2008) Stakeholder participation for environmental management: a literature review. Biol Conserv 141:2417–2431
22. Ryan GW, Bernard HR (2003) Techniques to identify themes. Field Meth 15:85–109
23. Sade A, Ali I, Mohd Ariff MR (2006) The seaweed industry in Sabah East Malaysia. Jati 11:97–107
24. Solaymani S, Kari F (2014) Poverty evaluation in the Malaysian fishery community. Ocean Coastal Manag 95:165–175
25. Thurstone LL (1928) Attitudes can be measured. Am J Sociol 33:529-554
26. UNHCR (2014) Refugee Factsheet. United Nations High Commissioner for Refugees, Kuala Lumpur. http://www.unhcr.org/56167f6b6.pdf
27. Valderrama D, Cai J, Hishamunda N, Ridler N (2013) Social and economic dimensions of carrageenan seaweed farming. FAO Fisheries and Aquaculture Technical Paper, vol 580. Rome
28. Vredegoor MTJ, Pennink BJW (2013) Including capabilities of local actors in regional economic development: empirical results of local seaweed industries in Sulawesi. South East Asian J Manag 7:61–88

SECTION III

BIOLOGY AND MANAGEMENT TECHNIQUES

CHAPTER

6

GROWTH, MORPHOLOGY AND GROWTH RELATED HORMONE LEVEL IN KAPPAPHYCUS ALVAREZII PRODUCED BY MASS SELECTION IN GORONTALO WATERS, INDONESIA

Siti Fadilah[1], Alimuddin[2], Petrus Rani Pong-Masak[1], Joko Santoso[3], and Andi Parenrengi[4]

[1]Research and Development Institute for Seaweed Aquaculture, Boalemo, Gorontalo 96265, Indonesia

[2]Department of Aquaculture, Faculty of Fisheries and Marine Sciences (FFMS), IPB, Bogor 16680, Indonesia

[3]Department of Fish Processing Technology, FFMS, IPB, Bogor 16680, Indonesia

[4]Research and Development Institute for Brackishwater Aquaculture, Maros, Sulawesi Selatan 90512, Indonesia

ABSTRACT

The use of high quality seed can support the success of the seaweed cultivation. This study was conducted to evaluate the growth performance, morphology and growth related hormone level of brown strain seaweed *Kappaphycus alvarezii* seed produced by mass selection.

Selection was performed in the Tomini Gulf, Gorontalo, based on mass selection of seaweed seed protocol with a slight modification in cut-off 10% of the highest daily growth rate. Selection was carried out for four generations. The selected 4th generation of seed was then used in cultivation performance test in the Celebes Sea, North Gorontalo, for three production cycles. The results showed that the selected *K. alvarezii* has higher clump weight and daily growth rate, longer thallus, more number of branches, and shorter internodes compared to the unselected control and seaweed from the farmer as external control. Furthermore, total sugar content, levels of kinetin hormone and kinetin:indole-3-acetic acid ratio were higher in selected seaweeds than that of unselected control and external control. Thus, mass selection method could be used to produce high growth of seed, and kinetin and indole-3-acetic acid play an important role in growth of *K. alvarezii*.

Keywords: Growth; Hormone; Kappaphycus alvarezii; Mass selection; Morphology

INTRODUCTION

Kappaphycus alvarezii (Doty) is a seaweed species of Rhodophyceae class (red algae) producing kappa carrageenan which is widely used in the food industry, pharmaceuticals, cosmetics, and others. The need for carrageenan products and raw materials of *K. alvarezii* is predicted to increase in the future. Farming of high quality *K. alvarezii* seed can increase the production level.

Seaweed seeds are generally obtained by some of the crop through vegetative propagation. Repeated vegetative propagation causes a decrease in genetic variability and resulting in decreased growth rate, carrageenan yield, gel strength and increase of susceptibility to disease (Hurtado and Cheney 2003). Therefore, a systematic attempt to generate a high quality seed is necessary.

Improvement of genetic quality can be performed by mass selection and tissue culture techniques. Tissue culture technique of seaweed have been developed by several research groups (Reddy *et al.* 2003; Kumar *et al.* 2004; Mulyaningrum *et al.* 2012; Sulistiani *et al.* 2012; Suryati *et al.* 2013). The success of acclimatization and field cultivation of seaweed *K.*

alvarezii from somatic embryo culture has also been reported by Reddy *et al.* (2003) in India. Seaweed is monitored for seven generations and shows daily growth rate (DGR) of 1.5–1.8 times higher than that of the farmers (Reddy *et al.* 2003). In Indonesia, the success of field cultivation of *Gracilaria verrucosa* seed generated by tissue culture has been reported by Suryati *et al.* (2013). During the period of mass production, 1200 kg of seed is obtained in five cultivation cycles (Suryati *et al.* 2013).

On the other hand, mass selection methods have been found to increase yield in higher plants and improve the proportion of superior genotypes in a short time (Taran *et al.* 2004; Yang *et al.* 2013). The use of mass selection method to produce a high quality of *K. alvarezii* seed has been initiated by Pong-Masak *et al.* (2013). However, morphological and physiological characteristics of the selected seed remain to be examined. This study was performed to examine the growth, morphology and growth related hormone levels of the mass selected *K. alvarezii* brown strain with cut-off 10% of the highest DGR.

MATERIALS AND METHODS

Seaweed Source and Mass Selection Procedure

K. alvarezii brown strain was collected from local farmers at Anggrek Waters, North Gorontalo Regency. Healthy, fresh, clean, brightly colored, highly branched, lush thalli and 30-day-old seaweed were chosen for mass selection. Selection was conducted in March to July 2013 in the Tomini Gulf, Boalemo Regency, Gorontalo (Figure 1).

Figure 1. Research locations for selection in Tomini Gulf and cultivation performance test in Celebes Sea, Gorontalo.

Selection was conducted as described in seaweed selection protocols (RICA 2011) with slight modification in cut-off to 10% of highest DGR. Cultivation was performed for 30 days of each cycle, using long-line method. Cultivation construction was 50 × 35 m^2, containing 50 long ropes and each rope was 35 m in length, and the distance between the ropes was 1 m. Each rope contained 230 seed clumps with distance between clumps of 15 cm. Seaweed seed (G0) with initial clump weight of 50 g was tied to the rope. The seeds were placed at 30 cm from water surface. Seaweed clean-up activities of pests were done as needed.

Each clump was weighed at the end of each cultivation cycle. Seeds had 10% of highest DGR value were selected. The selected seed clumps were then cut into new clumps and tied to the new rope, and they were cultivated with the same methods, processes and period as previous cycles. Selected seed resulted from each cycle of mass selection was referred to a generation. In each cultivation cycle, there were internal control and external control. Internal control was unselected clumps that have DGR mean of population, while the external control was seeds from farmers around research location. Selection was carried out until 4th generation which usually had stabilized DGR and uniformity level about >90%.

Mass Production and Cultivation Performance Test

The 30-day-old selected seeds of 4th generation were used in mass cultivation to compare their growth performance, morphological and growth related hormone level with unselected internal control and seaweed from the farmers as external control. Mass production and cultivation performance test were conducted in the Celebes Sea, Anggrek, North Gorontalo Regency, in August to October 2013 (Figure 1). Cultivation performance test was performed for three cultivation cycles which 30 days for each cycle. Cultivation method was performed by the same procedures as applied in mass selection. Three ropes were prepared for each of the selected seeds, internal and external controls. Each rope contained 230 clumps of seed. Three samples were taken randomly to measure clump wet weight at the end of each cultivation cycle. Whole clump was used to analyze morphological measurements and total sugar content. Branch clumps were used to analyze hormone content.

Variables and Statistical Analysis

Morphological characteristics and physiological measurements were performed at the end of each cultivation cycle. Morphological characteristics measured were thallus length (the main thallus, branch I, II and III), internodes (primary and secondary), thallus diameter (primary, secondary and tertiary), branch numbers (I, II and III). Primary diameter is the diameter of the main thallus, secondary diameter is the diameter of the 1st branch, while the tertiary diameter is the diameter of the branches II and III (Figure 2). Physiological characteristics observed were DGR, total sugar, and hormone level of kinetin and indole-3-acetic acid (IAA). Total sugar content was estimated using the Luff Schoorl method, whereas kinetin and IAA hormone levels were estimated by spectrophotometric method (Unyayar *et al.* 1996).

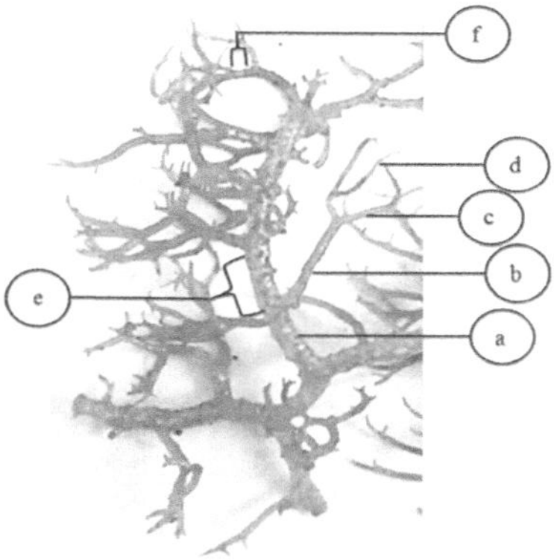

Figure 2. Morphology of seaweed: a) the main thallus, b) branch I, c) branch II, d) branch III, e) primary internodes, f) secondary internodes.

Water qualities measured were *in situ* observation of temperature, dissolved oxygen, and salinity, while *ex situ* observation were NO_3 (nitrate), PO_4 (phosphate) and total organic matter. Rainfall as secondary data was obtained from the Gorontalo Meteorological Station.

Morphological and physiological characteristics data were analyzed by analysis of variance at 5% significance level using software SPSS version 17. If there is a significant difference, analysis was continued using the Duncan multiple range test. Water qualities and rainfall data were analyzed descriptively.

RESULTS

Growth Performance

Mass selection is a simple method in selecting the best broods to produce the next generation. Seaweed mass selection can produce the best seeds for seaweed culture. In our previous study, DGR response at the 4th generation of selected seaweed was 2.22% – 5.21%. This DGR value showed improvement by 27.99% higher than the unselected internal controls, and 35.26% higher than the seaweed derived from the farmer as the external controls (RDISA 2013).

Mass production in this study aimed to expand the number of selected seeds to be cultivated at field level. The results showed that selected seaweed had higher biomass and DGR than the external and internal controls (Figure 3). Internal and external controls seeds did not grow due to a decline of the clump weight mean.

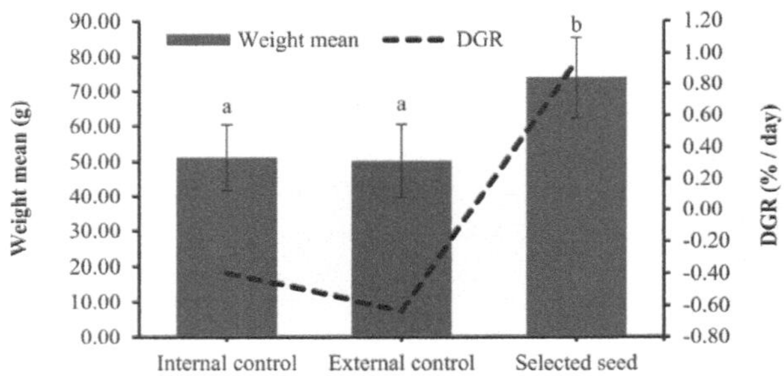

Figure 3. Clump weight mean and daily growth rate (DGR) of unselected internal control, farmer seaweed as the external control and the selected seeds.

Morphological Characteristics

The results showed that the total thallus length of the selected seaweed was 66.94% longer ($p < 0.05$) than the external control and 48.35% longer than the internal control (Table 1). Higher thallus length in the selected seaweed was significant contribution of branch II and branch III thalli length (Table 1). The total number of branches in the selected seaweed (316 branches) was 134.07% higher ($p < 0.05$) than external controls (135 branches) and 69.89% higher than internal control (186 branches). The increasing of branch number in the selected seaweed was contribution of branch II and III (Table 1). Selected and internal control seaweeds have shorter primary and secondary internodes than the external controls (Table 1; $p < 0.05$). Selected and internal control seeds showed thalli branches grow in cluster along the main thallus and branch I, that they look more lush clumps, whereas external control seaweed from the northern Gorontalo Waters has spreading branches (Figure 4). Furthermore, primary and secondary thallus diameters of three types of seaweed seed were not significantly different (Table 1; $p > 0.05$).

Table 1. Thallus length, branches number, internodes and diameter of selected, internal control and external control seaweed seeds

Variables	Seaweed seed		
	Internal control, *n* = 3	**External control, *n* = 3**	**Selected seed, *n* = 3**
Thallus length (cm)			
Main thallus	15.10 ± 2.71[a]	42.88 ± 4.47[c]	28.63 ± 4.16[b]
Branch I	102.39 ± 14.12[a]	123.76 ± 18.39[a]	177.65 ± 45.99[a]
Branch II	155.49 ± 23.90[a]	117.04 ± 15.95[a]	229.01 ± 26.68[b]
Branch III	83.60 ± 12.81[b]	33.18 ± 7.77[a]	93.70 ± 15.53[b]
Total	356.58 ± 44.60[a]	316.86 ± 32.98[a]	528.98 ± 76.53[b]
Branches number			
Branch I	34 ± 3[a]	43 ± 5[a]	57 ± 13[a]
Branch II	70 ± 7[a]	59 ± 8[a]	154 ± 45[b]
Branch III	82 ± 12[b]	33 ± 5[a]	105 ± 25[b]
Total	186 ± 21[ab]	135 ± 11[a]	316 ± 80[c]
Internodes length (cm)			
Primary	0.63 ± 0.04[a]	1.20 ± 0.09[b]	0.85 ± 0.13[a]
Secondary	0.80 ± 0.03[ab]	0.96 ± 0.07[b]	0.72 ± 0.07[a]
Thallus diameter (cm)			
Primary	0.66 ± 0.03[a]	0.64 ± 0.03[a]	0.73 ± 0.03[a]
Secondary	0.41 ± 0.03[a]	0.47 ± 0.04[a]	0.44 ± 0.03[a]
Tertiary	0.29 ± 0.04[a]	0.28 ± 0.03[a]	0.26 ± 0.01[a]

Different superscript letters in the same line indicate significantly different (Duncan test; $p < 0.05$) and *n* indicates number of sample.

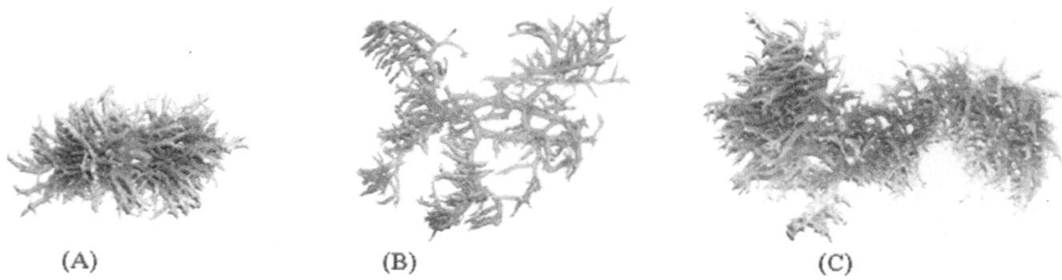

Figure 4. Performance of internal control (A), external control (B) and selected seaweed (C).

Physiological Characteristics

Total sugar content of selected seaweed was 15.52% higher than the internal control and 16.42% higher than external control ($p > 0.05$; Figure 5). Levels of kinetin and ratio of kinetin to IAA were higher in the selected seaweed (Figures 6A and B). Total sugar content, kinetin level and ratio of kinetin to IAA were in line with higher DGR in selected seed ($p < 0.05$).

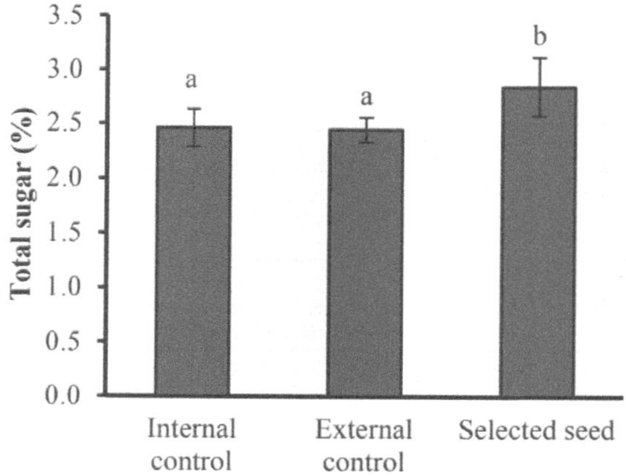

Figure 5. Total sugar content in internal control, external control and the selected seaweed. Internal control was unselected seaweed and external control was seaweed from the farmer around research location. Data with significant differences are indicated by different letters (Duncan test; $p < 0.05$).

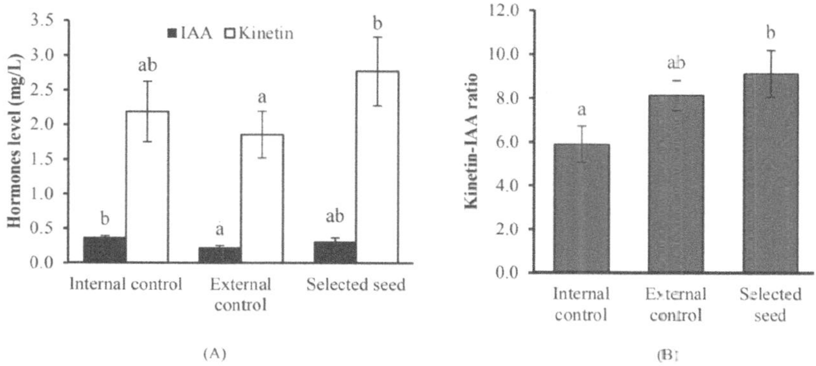

Figure 6. Hormone levels of indole acetic acid (IAA), kinetin (A), and kinetin-IAA ratio (B) in internal control, external control and selected seeds. Internal control was unselected seaweed and external control was seaweed from the farmer around research location. Different letters in the same bar type indicate significant difference (Duncan test; $p < 0.05$).

Environmental Conditions of the Study Site

Seasonal condition during mass selection conducted in March to July 2013 was the rainy season with monthly rainfall intensity range of 9.71–16.62 mm (Figure 7A). However, frequency of rain during cultivation performance test conducted from August to October 2013 began to decline. Even in September to the end of October there was no rain (Figure 7B). In cultivation performance test, some of water quality variables were not in optimum conditions for seaweed growth (Table 2). This led to the biofouling attachment of barnacles and epiphytic algae to thalli surface in August. All thalli seeds also suffered from bleaching to pale yellow in September. In October, the thallus color became light brown.

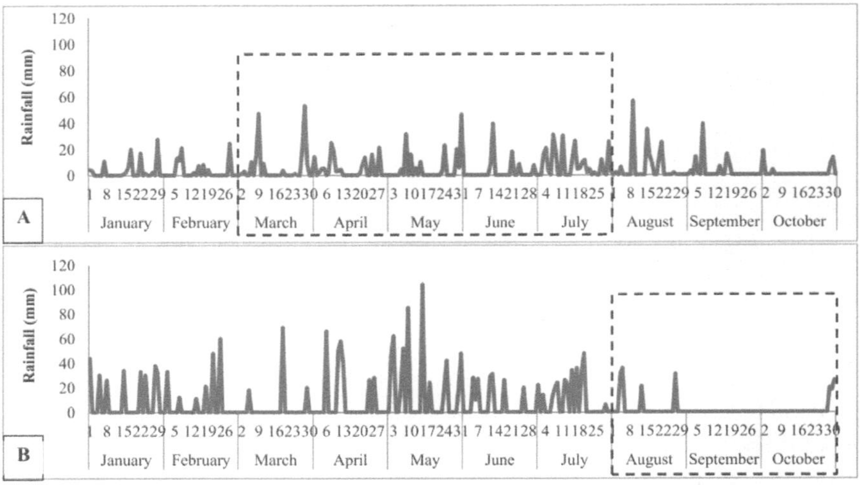

Figure 7. Rainfall level during *Kappaphycus alvarezii* mass selection (A) and cultivation performance test (B). The dashed lines indicate the study time. Image was processed from rainfall data of Tilamuta (A) and Kwandang stations (B), the Gorontalo meteorological station.

Table 2. Water qualities of North Gorontalo and *Kappaphycus alvarezii* condition at cultivation performance test

Variable	Unit	Month			Optimum level*
		August	September	October	
Water qualities					
Temperature	°C	28	29	30	24–30
Dissolved oxygen	mg/L	5.62	6.31	5.93	>5
Salinity	g/L	36	33	34	28–34
Total phosphate	mg/L	0.03	<0.002	0.15	0.10–0.20
Nitrate	mg/L	0.40	0.19	0.78	0.01–0.07
Total organic matter	mg/L	47.03	36.07	38.65	
Seaweed condition					
Thallus color		Dark brown	Pale yellow	Light brown	
Biofouling		Barnacles; epiphytic algae	–	–	

*Aslan (1998)

DISCUSSION

Biomass and DGR of all seaweed seeds show unfavorable growth in cultivation performance test. The thalli were attacked by biofouling and suffered from bleaching (whitening). This condition occurred because of water quality especially lower total phosphate and higher nitrate in August and September (Table 2). This phenomenon did not occur in mass selection activity.

The selected seeds obtained in this study had the longer thallus, more number of branches, but shorter in internodes than the controls. Selected seed and internal control look more like lush clumps, while external control has spreading branches. Presumably, selected and internal control seaweeds were affected by environmental conditions in the south Gorontalo Waters during mass selection.

Total sugar content of selected seaweed was higher than both controls. This result was in line with research of Pong-Masak *et al.* (2013) that carrageenan content in the selected seaweed was 6.07% and 12.72% higher than the internal control and external control, respectively. Carrageenans are sulfated linear polysaccharides of D-galactose and 3,6-anhydro-D-galactose extracted from certain Rhodophyceae seaweeds which is their main cell wall material (Campo *et al.* 2009). Generally, seaweed with higher DGR has higher carrageenan yield (Naguit *et al.* 2009). Carrageenan and biomass were strongly associated with the optimum harvest time. The optimum harvest time for *Kappaphycus* species is 8–9 weeks of cultivation (Villanueva *et al.* 2011). Thus, cultivation of selected seaweed seed has high potential to increase yield and carrageenan production.

Higher level of kinetin and ratio of kinetin to IAA in selected seaweed indicates that the two hormones play important role in the seaweed growth. Kinetin is a type of cytokinin, whereas IAA is a type of auxin. Root and shoot development in higher plant depends on the cytokinin:auxin ratio. Shoot formation was supported by high concentrations of cytokinin, whereas rooting was promoted by a high concentration of auxin (Brault and Maldiney 1999; Moubayidin *et al.* 2009). The presence of endogenous hormones such as auxin, cytokinin and abscisic acid in 11 species of red algae from the coast of Brazil has been reported by Yokoya (2010). Auxin and cytokinin had a regulatory role in the growth and morphogenesis *in vitro* in some species of red algae, such as *Gracilaria tenuistipitata, Gracilaria perplexa* (Yokoya *et al.* 2004), *Solieria filiformis* (Yokoya and Handro 2002).

Physiologically, cytokinin controls cytokinesis (cell division) and bud development, whereas auxin induces elongation growth and apical dominance (Tarakhovskaya *et al.*2007). Cytokinins play a role in mediating the response to extrinsic factors and response to biotic and abiotic stresses. This role together contributes to the regulation of quantitative growth in plants (Werner and Schmulling 2009). Basipetal auxin transport in the shoot actively inhibits bud development. Instead, acropetal transport of cytokinin was involved in the promotion of bud outgrowth (Ongaro and Leyser 2008).

Seaweed growth is affected by the environment (Glen and Doty 1990). Seasonal condition during mass selection was the rainy season with low rainfall intensity. Seaweed needs low rainfall. Therefore during mass selection, condition of seaweed was good, there was no disease and biofouling infected. It was caused by water quality that supported seaweed growth. However, in cultivation performance test, some of water quality variables were not supportable for seaweed growth. This led to the biofouling attachment and bleaching thalli seed in August and September, respectively. Bleaching thalli to pale yellow indicated as early of ice-ice disease. In October, the thallus color became light brown which indicated thalli began to recovery. Mizuta *et al.* (2002) investigated that seasonal changes affected pigment composition in two red algae, *Gloiopeltis furcata* and *Porphyra yezoensis*. Levels of chlorophyll a and phycoerythrin were high during the winter, but decreased in the late spring or summer. These changes were accompanied by changes in color from deep red to green or yellow, closely linked to fluctuations in phycoerythrin content. Phycoerythrin content was significantly correlated with the nitrogen content in both species (Mizuta *et al.* 2002).

Unfavorable condition in this study was consistent with Pong-Masak *et al.* (2009) that productive season of seaweed *K. alvarezii* in the North Gorontalo Waters was November to April, whereas in May to October, they were less productive. Because of the decline of water quality, biofouling attachment and ice-ice disease infection, seaweed DGR was <3% (Pong-Masak *et al.* 2009). In addition, our results also indicated that seaweed cultivation should not be performed during transition period from rainy season to dry season and in dry season. In that period, water quality declined especially higher nitrate and lower phosphate in August and September. Fried *et al.* (2003) conclude that if only one nutrient increases substantially the other nutrients will limit growth. If nitrogen is present in very high concentrations and low phosphorus levels, the phosphorus will be a limiting factor in the growth of algae (Fried *et al.* 2003). According to Correll (1998), phosphorus is an essential component of nucleic acids and many intermediary metabolites, such as sugar phosphates and adenosine phosphates, which are an integral part of the metabolism of all life forms. Although nitrogen is a component of proteins and found in many biological compounds (Rezael and Samimi 2013), Orbita (2013) found that growth and carrageenan content on thallus *K. alvarezii* positively correlated with water flow, phosphate and nitrate. Water flow during the rainy season increases the hydrodynamic thus increasing growth and carrageenan. In addition, phosphorus and nitrogen are nutrients that play an important role in the algae growth and carrageenan. Maximum carrageenan is obtained during the period of the highest nutrient supply (Orbita 2013).

REFERENCES

1. Aslan LM. 1998. Seaweed Cultivation. Yogyakarta: Kanisius.
2. Brault M, Maldiney R. 1999. Mechanisms of cytokinin action. Plant Physiol Biochem 37:403–12.
3. Campo VL, Kawano DF, Da Silva Jr DB, Carvalho I. 2009. Carrageenans: biological properties, chemical modifications and structural analysis. Carbohydr Polym 77: 167–80.
4. Correll DL. 1998. The role of phosphorus in the eutrophication of receiving waters: a review. J Environ Qual 27:261–6
5. Fried S, Mackie B, Nothwehr E. 2003. Nitrate and phosphate levels positively affect the growth of algae species found in Perry Pond. Tillers 4:21–4.
6. Glen EP, Doty MS. 1990. Growth of the seaweeds Kappaphycus alvarezii, K. striatum and Eucheuma denticulatum as affected by environment in Hawaii. Aquaculture 84:245–55.
7. Hurtado AQ, Cheney DP. 2003. Propagule production of Eucheuma denticulatum (Burman) Collins et Harvey by tissue culture. Bot Mar

46:338–41.

8. Kumar GJ, Reddy CRK, Ganesan M, Thiruppathi S, Dipakkore S, Eswaran K, Rao PVS, Jha B. 2004. Tissue culture and regeneration of thallus from callus of Geladiella acerosa (Geladiales, Rhodophyta). Phycologia 43:596–602.
9. Mizuta H, Shirakura Y, Yasui H. 2002. Relationship between phycoerythrin and nitrogen content in Gloiopeltis furcata and Porphyra yezoensis. Algae 17:89–93.
10. Moubayidin L, Di Mambro R, Sabatini S. 2009. Cytokinin-auxin crosstalk. Trends Plant Sci 14:557–62.
11. Mulyaningrum SRH, Nursyam H, Risjani Y, Parenrengi A. 2012. Regeneration filament of Kappaphycus alvarezii callus by different plant growth regulator. J Pen Perik 1:52–60.
12. Naguit MRA, Tisera WL, Lanioso A. 2009. Growth performance and carrageenan yield of Kappaphycus alvarezii (Doty) and Eucheuma denticulatum (Burman) Collins et Harvey, farmed in Bais Bay, Negros Oriental and Olingan, Dipolog City. Threshold 4:38–51.
13. Ongaro V, Leyser O. 2008. Hormonal control of shoot branching. J Exp Bot 59:67–74
14. Orbita MLS. 2013. Growth rate and carrageenan yield of Kappaphycus alvarezii (Rhodophyta, Gigartinales) cultivated in Kolambugan, Lanao del Norte, Mindanao, Philippines. AAB Bioflux 5:128–39.
15. Pong-Masak PR, Pantjara B, Rachmansyah. 2009. Seaweed Planting Season in Anggrek Waters, North Gorontalo. Yogyakarta (ID): Gajahmada University. pp. 1–10. Annual National Seminar VI Fisheries and Marine Research Report; 2009 Jul 25; Yogyakarta, Indonesia.
16. Pong-Masak PR, Parenrengi A, Tjaronge M. 2013. Producing high quality Kappaphycus alvarezii seed. In: Sukadi F, Djunaidah IS, Nurkhakim S, Sugama K, Heruwati ES, Purba M, Hanggono A (Eds.). Recommendation of Marine and Fisheries Technology. Jakarta: Agency for Marine and Fisheries Research and Development. pp. 161–75.

17. RDISA [Research and Development Institute for Seaweed Aquaculture]. 2013. Development of Fast Growth Seaweed Kappaphycus alvarezii Seed by Selection. Boalemo: Research and Development Institute for Seaweed Aquaculture. pp. 20–3.. Annual Report 2013 of RDISA.
18. Reddy CRK, Kumar GRK, Shiddhanta AK, Tewari A. 2003. In vitro somatic embryogenesis and regeneration of somatic embryos from pigmented callus of Kappaphycus alvarezii (Doty) Doty (Rhodophyta, Gigartinales). J Phycol 39:610–6.
19. Rezael R, Samimi A. 2013. Effects of phosphorus and nitrate in wastewater Shahinshahr City use for oil refinery. Int J Innov Appl Stud 2:250–8.
20. RICA [Research Institute for Coastal Aquaculture]. 2011. Mass/ Variety Selection Protocol of Superior Seaweed Seed. Marcs: RICA (Research Institute for Coastal Aquaculture).
21. Sulistiani E, Soelistyowati DT, Alimuddin Yani SA. 2012. Callus induction and filaments regeneration from callus of cottonii seaweed (Kappaphycus alvarezii (Doty)) collected from Natuna Islands, Riau Islands Province. Biotropia 19: 103–14.
22. Suryati E, Mulyaningrum SRH, Parenrengi A. 2013. Gracilaria verrucosa cultivation using seed produced by tissue culture. In: Sukadi F, Djunaidah IS, Nurkhakim S, Sugama K, Heruwati ES, Purba M, Hanggono A (Eds.). Recommendation of Marine and Fisheries Technology. Jakarta: Agency for Marine and Fisheries Research and Development. pp. 207–13.
23. Tarakhovskaya ER, Maslov YI, Shishova MF. 2007. Phytohormones in algae. Russ J Plant Physiol 54:163–70.
24. Taran SA, Nasibullah S, Mian MA, Shah H. 2004. Genetic improvement through mass selection in maize (Zea mays L.). J App Emerg Sci 1:152–7.
25. Unyayar S, Topcuoglu SF, Unyayar A. 1996. A modified method for extraction and modification of indole-3-acetic acid (IAA), gibberellic acid (GA3), abscisic acid (ABA) and zeatin produced by Phanerochate chrysosporium ME446. Bulg J Plant Physiol 22:105–10.

26. Yang Y, Chen J, Zhao M, Li C, Meng Z, Wang J, Chen Z, Zhang G, Yang S. 2013. Phenotypic and genetic uniformity in three populations of Panax notoginseng by mass selection. Mol Plant Breed 4:169–76.

27. Yokoya NS, Handro W. 2002. Effects of plant growth regulators and culture medium on morphogenesis of Solieria filiformis (Rhodophyta) cultured in vitro. J Appl Phycol 14:97–102.

28. Yokoya NS, West JA, Luchi AE. 2004. Effects of plant growth regulators on callus formation, growth and regeneration in axenic tissue cultures of Gracilaria tenuistipitata and Gracilaria perplexa (Gracilariales, Rhodophyta). Phycol Res 52: 244–54.

29. Yokoya NS. 2010. Endogenous cytokinins, auxins and abscisic acid in red algae from Brazil. J Phycol 46:1198–205.

30. Villanueva RD, Romero JB, Montano MNE, de la Pena PO. 2011. Harvest optimization of four Kappaphycus species from the Philippines. Biomass Bioenergy 35:1311–6.

31. Werner T, Schmulling T. 2009. Cytokinin action in plant development. Plant Biol 12: 527–38.

CHAPTER

7

METHODS FOR THE INDUCTION OF REPRODUCTION IN A TROPICAL SPECIES OF FILAMENTOUS ULVA

Christina Carl, Rocky de Nys, Rebecca J. Lawton, and Nicholas A. Paul

MACRO – the Centre for Macroalgal Resources and Biotechnology, and School of Marine and Tropical Biology, James Cook University, Townsville, Queensland, Australia

ABSTRACT

The green seaweed *Ulva* is a major fouling organism but also an edible aquaculture product in Asia. This study quantified for the first time the effect of key factors on the reproduction of a tropical species of filamentous *Ulva* (*Ulva* sp. 3). The controlled timing of release of swarmers (motile reproductive bodies) was achieved when experiments were initiated in the early afternoon by exposing the thalli to a temperature shock (4°C) for 10 min and subsequently placing them into autoclaved filtered seawater under a 12 h light: 12 h dark photoperiod at 25°C. The release of swarmers then peaked two days after initiation. In contrast, segmentation, dehydration, salinity or time of initiation of experiments had no effect of any magnitude on reproduction. The released swarmers were predominantly biflagellate (95%), negatively phototactic and germinated without complementary gametes. This indicates that *Ulva* sp. 3 has a simple asexual life history dominated by biflagellate zoids.

INTRODUCTION

The green seaweed *Ulva* (syn. *Enteromorpha*; [1]) is a major fouling organism on marine infrastructure [2]–[4], with the prevention of settlement and growth of this genus being a primary goal for antifouling technologies [5]–[7]. Paradoxically, *Ulva* is also an important high-value aquaculture product in Japanese and Asian cuisine, known as 'aonori' [8] with economic benefits relying on the enhanced settlement and growth of these species [9]–[12]. Finally, *Ulva* is a target for the bioremediation of nitrogen and phosphorous from aquaculture with a similar goal to aquaculture of enhanced growth [13]–[16].

The key to developing both antifouling and aquaculture technologies for *Ulva* is the manipulation of reproduction thereby inducing the release of swarmers to be used as bioassay test organisms or as seedlings. The term 'swarmer' refers to motile reproductive bodies of *Ulva*, the nature of which depends on the three life history stages from which the swarmers are sourced. This can be either sexual or one of two asexual life history stages [17]. The sexual life history is characterised by an alternation of isomorphic sporophytes and gametophytes [18], with both generations only being distinguishable by their reproductive structures [19], [20]. Sporophytes release quadriflagellate zoospores, whereas gametophytes release biflagellate female (+) or male (−) gametes which fuse and form zygotes. In contrast, *Ulva* species possessing a simple asexual life history produce either exclusively biflagellate or quadriflagellate zoids [17], [21]. In all cases, pre-existing vegetative cells of *Ulva* transform directly into reproductive cells [19], [22].

The liberation of zoospores and gametes is driven by lunar periodicity in temperate and cold waters [10], [12], [23]–[25]. In addition, the formation and release of zoospores, gametes, and asexual zoids of *Ulva* in temperate environments is induced by a range of stimuli including dehydration [26], [27], extended dark periods over several days [27], [28], fragmentation [12], [29]–[33], renewal of culture medium [28], [34], salinity [12], and change in temperature [29], [30], [35], [36]. In contrast, the identified stimuli that induce the reproductive development of *Ulva* in tropical environments are restricted to salinity, dehydration, segmentation, and temperature [37]–[40], and there is a lack of understanding of other key drivers controlling reproduction, particularly for filamentous species of *Ulva* (formerly *Enteromorpha*).

Given the economic importance of *Ulva* and its capacity for rapid and intense growth under tropical conditions [15], [41], [42], there is an imperative to understand the mechanisms that influence the reproduction of this species in tropical environments. Therefore, the aim of this study was to understand the fundamental physiological drivers of reproduction in the most common filamentous species of tropical *Ulva* (*Ulva* sp. 3) in Eastern Australia [42], and subsequently develop manipulative treatments to control reproduction. This will provide a baseline method for the reliable supply of swarmers with a direct application to laboratory fouling studies and the commercial production of filamentous *Ulva* in the tropics. The effects of six key factors on reproduction of *Ulva* sp. 3 were quantified in laboratory-based experiments. These were (1) salinity, (2) dehydration, (3) segmentation, (4) photoperiod, (5) time of initiation of experiments, and (6) temperature shock. Furthermore, the most successful combination of treatments was used to quantify the reproductive output of *Ulva* sp. 3, the number of flagella and phototactic response of the swarmers and their ability to settle and germinate.

MATERIALS AND METHODS

Study Species and Collection

The genus *Ulva* has a worldwide distribution with a broad tolerance of salinity, with species occurring in hypersaline to freshwater environments [43]–[45]. The species used in this study is characterized by flat tubular thalli. Because of the broad intraspecific variation and taxonomic difficulties associated with the *Ulva* genus [46], the species used in this study was analysed using molecular barcoding. Barcoding compares short DNA sequences from a standardized region of the genome - the ‘barcode’ - to a library of reference sequences derived from individuals of known identity [47]. The species was identified as *Ulva* sp. 3 [43] using newly generated DNA sequences from the internal transcribed spacer (ITS) region of the ribosomal cistron (Genbank accession number KF534755). Phylogenetic trees constructed using these sequences show that this species occupies a unique clade that is distinct from all other filamentous species of *Ulva* used in previous studies of reproduction (Figure 1).

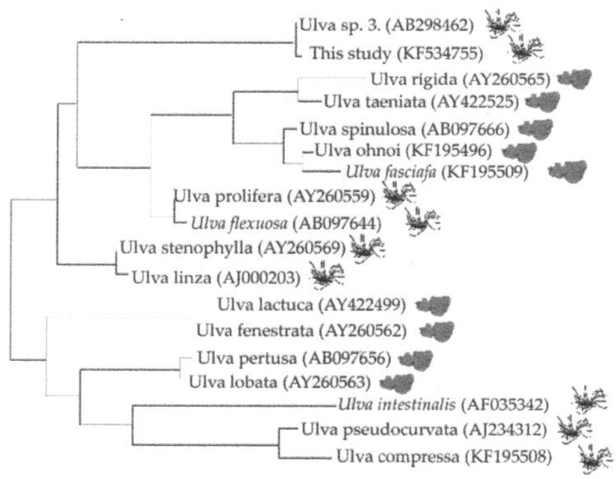

Figure 1. Reduced *Ulva* ITS phylogenetic tree. Reduced version of maximum likelihood tree of *Ulva* ITS sequences, showing the species used in this study (shown in bold) and *Ulva* species used in previous studies investigating the reproduction of this genus. Species investigated in tropical environments are shown in italic. Pictures next to species represent the morphology (flat and blade-like or tubular and filamentous thalli).

Ulva sp. 3 (hereafter *Ulva*) was collected by hand from a land-based aquaculture facility at Guthalungra (19° 55′S, 147° 50′E), Queensland, Australia. Permission was obtained from owners to collect algae from this site. Notably, reproductive patterns driven by lunar periodicity as found for temperate species of *Ulva* [25], [48] were not observed in the field populations collected over a four month period. Samples were placed in a 25 L container filled with pond water and then transported within 3 h to the laboratory at James Cook University (JCU) in Townsville, Australia. Subsequently, the samples were gently washed three times with filtered seawater (FSW; 0.2 μm and UV sterilised) to remove debris, epiphytes and invertebrates.

Laboratory Experiments

Effect of Salinity, Dehydration, and Segmentation

To determine the effect of salinity, dehydration and segmentation on the formation and release of swarmers, these factors were manipulated in a fully factorial experiment under laboratory conditions using samples collected in the morning and gently cleaned with FSW as described above. Firstly, *Ulva* was exposed to a lower salinity by placing filaments in dechlorinated tap water (DC) for 10 min with FSW being used as a salinity control. Secondly, the effect of dehydration was tested for each salinity treatment. All filaments were dried using paper towels to remove excess water and subsequently either dehydrated by exposure to air for 45 min in the dark at 25°C (dehydration treatments), or alternatively placed in FSW in the dark for 45 min (non-dehydrated control). Thirdly, the effect of segmentation was tested by cutting a section of approximately 50–70 mm from each filament (hereafter referred to as 'whole'), or alternatively cutting this section further into pieces <5 mm using a razor blade (hereafter referred to as 'cut'). Both treatments had similar biomass per unit volume. Each of the whole and cut filaments were placed in individual Petri dishes (Iwaki; 1010-060) filled with 10 mL autoclaved FSW and sealed using Parafilm to prevent evaporation. The Petri dishes were then placed in a culture cabinet at 25°C at an irradiance of 125 μmol photon m^{-2} s^{-1} under a 12 h light (L):12 h dark (D) photoperiod for three days. A total of three replicates was used for each treatment combination (n=3 for each salinity×dehydration×segmentation).

The percentage of the total area of each filament (either whole or cut) that had released swarmers (hereafter referred to as 'discharge') was visually quantified daily at 3 pm using a dissecting microscope (Olympus SZ61) according to Nielsen and Nordby [49]. Vegetative cells of *Ulva* transform directly into reproductive cells (Figure 2). The vegetative (Figure 2a), reproductive (Figure 2c) and discharged cells (Figure 2d) had a green, brown-ish and white colour, respectively [19], [50]. The discharge was quantified using only white cells that had unequivocally released swarmers (Figure 2d).

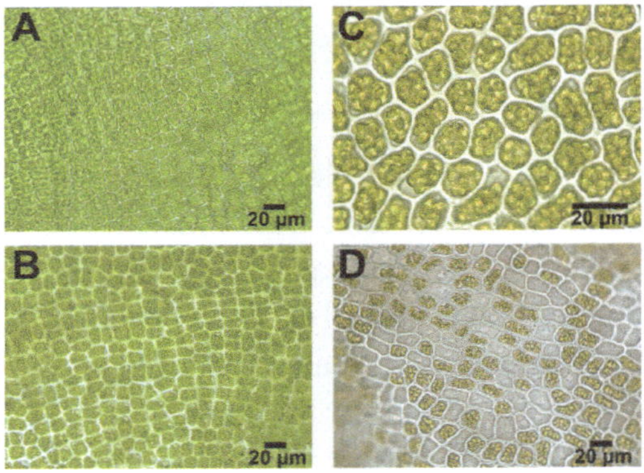

Figure 2. Light micrographs of *Ulva* sp. 3. Transformation of vegetative cells into reproductive cells of *Ulva* sp. 3. (a) Vegetative cells, (b) formation of swarmers after 26 h, (c) reproductive cells with fully formed swarmers after 44 h, (d) reproductive cells with fully formed and discharged swarmers after 46 h. Scale bar, 20 µm.

Effect of Time of Initiation of Experiments and Photoperiod

To determine the effect of time of initiation of experiments and photoperiod, on the formation and release of swarmers, samples were collected in the morning and gently cleaned with FSW as described above. Algae were subsequently maintained in an outdoor aquarium at the Marine & Aquaculture Research Facilities Unit at JCU until experiments were initiated by cutting a single section from each filament. Experiments were initiated at five times, two on the day of collection at 1 pm and 7 pm, and three on the day after collection at 7 am, 1 pm and 7 pm. The times of initiation affected the length of exposure to light and darkness at the start of the experiment (experiments initiated at 7 am were initially exposed to 12 h light in contrast to experiments initiated at 7 pm where thalli were initially exposed to 12 h darkness).

For all replicates, a single section of approximately 50–70 mm was cut from each filament (see Results) and placed in Petri dishes (Techno Plas; S6014S10) filled with 10 mL autoclaved FSW and sealed with Parafilm. A total of 1080 replicate dishes were then placed in one of three culture cabinets at 25°C at photoperiods of a normal light period (12 h L:12 h D), an extended light period (18 h L:6 h D), or constant light (24 h L). In addition, another 360 Petri dishes were wrapped in aluminium foil to provide constant darkness (24 h D) and these were split between the culture cabinets.

The discharge of swarmers was quantified at 7 pm on the day of collection and then daily at 7 am, 1 pm, and 7 pm for the following three days. At each sampling point, nine haphazardly selected dishes were destructively sampled from each treatment (n=9 for each photoperiod×time of experimental initiation) and the discharge quantified as a percentage of the total area of each filament, as described above.

Effect of Temperature Shock and Controlled Release of Swarmers

This experiment determined the effect of temperature shock on the formation of swarmers and assessed the potential to control their release by initially constraining and subsequently inducing the immediate release of swarmers to rapidly obtain a dense suspension of swarmers. Samples were collected in the morning and gently cleaned with FSW as described above. In the early afternoon, filaments of *Ulva* were placed in chilled (4°C) FSW and stored in the fridge for 10 min prior to placing a single section of approximately 50–70 mm from each filament (see Results) in Petri dishes (Techno Plas; S6014S10) filled with 10 mL autoclaved FSW. The use of a 10-min period was based on a pilot experiment where extended exposure periods to 4°C did not increase the formation and release of swarmers. As controls, filaments were either placed in 25°C FSW and stored in the dark for 10 min or immediately used in assays after the cleaning procedure without any further pre-treatment. The dishes were then placed in a culture cabinet at 25°C at an irradiance of 125 μmol photon m^{-2} s^{-1} under a 12 h L:12 h D photoperiod.

After two days in the culture cabinet, prior to the expected onset of the release of swarmers, filaments were wrapped in moist paper towels at 7 am or exposed to air for 4 h by placing each filament individually on baking paper. After 4 h (11 am), each filament was placed in a new Petri dish filled with 10 mL autoclaved FSW. These methods were used in an attempt to constrain the release of swarmers to a short period of time in order to rapidly obtain a dense suspension of swarmers. A control remained submersed over this same period of time. The time of day selected reflected the onset of release in previous experiments (see Results). The discharge of swarmers was determined at 11 am within 5 min post re-submersion, and at 4 pm. A total of ten replicates was used for each treatment combination (n=10 for temperature pre-treatment×restraining-treatment).

Reproductive Output

The reproductive output (the number of swarmers released per unit area) of *Ulva* was estimated by quantifying the surface area of individual thalli and subsequently determining the percentage area of released swarmers, and the numbers of discharged swarmers over time for each thallus. Samples of *Ulva* were collected in the morning and gently cleaned with FSW as previously described. The most successful combination of treatments tested above was used to induce sporulation and control the release of swarmers (see Results). Briefly, the samples were chilled at 4°C for 10 min to maximize the formation and release of swarmers prior to cutting a section of approximately 50–70 mm from filaments (n=50). Images of the filament pieces were captured using a camera (Olympus DP25) attached to a dissecting microscope (Olympus SZ61) for surface area measurements (Image J). Subsequently, these single filaments were placed in Petri dishes filled with 10 mL autoclaved FSW, which were then placed in a culture cabinet at 25°C at an irradiance of 125 μmol photon $m^{-2} s^{-1}$ under a 12 h L:12 h D photoperiod.

To determine the number of released swarmers over time, two days after the initiation of experiments each filament was transferred into a new Petri dish filled with 10 mL autoclaved FSW every 90 min (from 7 am until 1 pm). This time period was the window of peak release for swarmers (see Results). To obtain accurate numbers of discharged swarmers in the water column, the dishes and water were changed at each measurements point (every 90 min) to minimise settlement. The water in the dishes was preserved in 1% Lugol's solution and subsequently transferred into sealable sample tubes. To increase the concentration of swarmers, the samples were centrifuged (400 *g* for 1 min; [51]) and concentrated to a volume of 2.5 mL. Subsequently, the number of swarmers was determined using a haemocytometer. The reproductive output (RO) of each filament was calculated using the equation $RO=NoS/(SA\cdot D)$, where NoS is the number of released swarmers, SA the surface area of each filament and D is the discharge (percentage area of swarmer release). Finally, the number of flagella on released swarmers was counted using a compound microscope (Olympus BX53). The number of flagella on approximately 20 swarmers per sample was quantified in order to determine whether swarmers were bi- or quadriflagellate. Due to the low discharge of some thalli, only 59 samples could be used for counting the number of flagella on swarmers.

In addition, the viability and germling development of released swarmers was determined by inducing sporulation in a second section from each of the 50 filaments. These filaments were placed in individual Petri dishes filled with 10 mL autoclaved FSW and incubated as previously described to induce sporulation. After two days, the filaments were removed at 2 pm and nutrients were added to each dish (AlgaBoost *1000x* f/2). Subsequently, the dishes were returned to the culture cabinet at 25°C under a 12 h L:12 h D photoperiod. After a culture period of five days, the swarmers released by each filament were examined for germination using an inverted microscope (Olympus CKX41).

Experiments were conducted with three independent collections of *Ulva* over an eight day period with 50 replicates from each collection time.

Phototactic Behavior of Released Swarmers

To determine the phototactic behavior of released swarmers, sporulation was induced by chilling samples and subsequently incubating an excised section of filaments (n=40) in individual Petri dishes as previously described. After two days, the dishes were placed on a window sill at noon and any lights in the laboratory were turned off so that natural light was the primary light source. Released swarmers showed phototactic responses and consequently concentrated on the dark side of the dish facing away from the natural light when negatively phototactic or on the side with natural light source when possessing a positive phototaxis. Phototactic swarmers were removed and collected in sealable sample tubes over 90 min using a transfer pipette and preserved in 1% Lugol's solution. Subsequently, the number of flagella was counted using a microscope (Olympus BX53). In addition, images of a total of 64 released swarmers were taken for size measurements (width and length).

Statistical Analysis

Data were analyzed by permutational analysis of variance (PERMANOVA) using PRIMER 6 (v. 6.1.13) and PERMANOVA+ (v. 1.0.3) [52]. The Bray-Curtis dissimilarity measure was used for all PERMANOVAs and *p*-values were calculated using permutation of residuals under a reduced model with 9999 random permutations. If there was a significant difference, pair-wise *a posteriori* comparisons were made among the significant groups using the Bray-Curtis similarity measure (α=0.05). All data are reported as mean±1 standard error (S.E.) unless stated otherwise.

The effects of salinity shock, dehydration, and segmentation on the discharge of swarmers were considered as fixed factors in the first experiment. Because the second experiment evaluated the interactive effects of different times of initiation and photoperiod, these data had to be analyzed in two ways. A formal comparison using PERMANOVA was made between the effects of time of initiation and photoperiod by treating the day of sampling (day 3 and 4) and the sampling time of day (7 am, 1 pm, and 7 pm) as fixed factors because dishes were destructively sampled. An alternative plot of the effect of time of initiation (5 initiation times over 2 days past sampling) was then made for the key photoperiod (12 h L:12 h D, see Results) by standardising the time of initiation at 0 h but was not formally analysed. The third experiment assessing the effect of temperature shock and treatments to control the release of swarmers, PERMANOVAs were run for each sampling point (11 am and 4 pm) with pre-treatment and restraining-treatment as fixed factors. To test for the effect of batch on the release of swarmers over time in the fourth experiment, a two-factor PERMANOVA was used with time as a fixed factor and batch as a random factor.

RESULTS

Effect of Salinity, Dehydration, and Segmentation

There was no effect of salinity shock (three-factor PERMANOVA: $F_{(1, 16)}$=0.86, *p*=0.382), dehydration ($F_{(1, 16)}$=0.55, *p*=0.512), or segmentation ($F_{(1, 16)}$=0.04, *p*=0.948) on the discharge of swarmers among treatments after two and three days (Figure 3). No swarmers were released one day after initiation of experiments. After two days, the discharge of swarmers ranged from 23.3±23.3% (FSW×dehydrated×segmented) to 76.7±23.3% (FSW×non-dehydrated×whole) and was generally lower than after three days, where the discharge ranged from 23.3±23.3% (FSW×dehydrated×segmented) to 90.0±10.0% (FSW×non-dehydrated×whole, and also, DC×dehydrated×segmented). Therefore, the treatment combination of FSW×non-dehydrated×whole was chosen for the subsequent experiments on the basis of maximised discharge of swarmers with the least number of treatments.

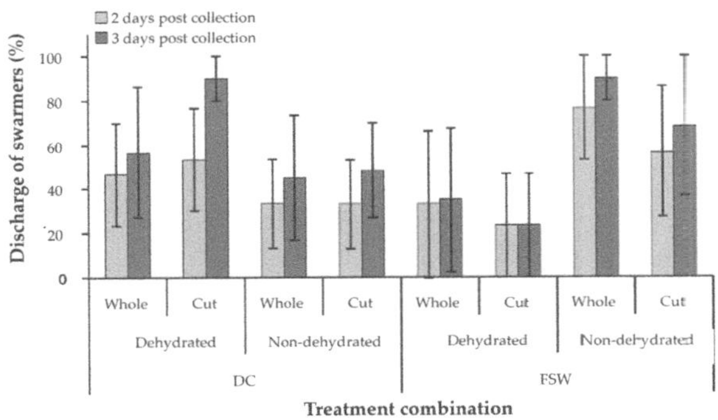

Figure 3. Discharge of swarmers in experiments testing the effect of salinity, dehydration, and segmentation. Mean (± S.E.) discharge of swarmers (%) after two and three days post collection and treatment. Samples were rinsed in dechlorinated tap water (DC) or filtered seawater (FSW) 10 min prior to dehydration for 45 min (Dehydrated). Non-dehydrated filaments (Non-dehydrated) were used as a control. Filaments were either left whole (Whole) or segmented into pieces <5 mm (Cut).

Effect of Time of Initiation of Experiments and Photoperiod

The discharge of swarmers varied significantly between photoperiods (Figure 4), with the highest discharge in the 12 h L:12 h D photoperiod (47.8±10.4%). The discharge of swarmers decreased with extended light periods and was below 31% and 23% at photoperiods of 18 h L:6 h D and 24 L, respectively. The lowest overall discharge occurred under constant darkness, with less than 12% discharge at any time (Figure 4). Photoperiod had a significant interactive effect with time of initiation ($p<0.001$; Table 1), sampling day ($p<0.001$) and time of sampling day

(p=0.001). Furthermore, there was a complex interactive effect of photoperiod, sampling day, and time of sampling day on the discharge of swarmers (p<0.047; Table 1). This effect was driven by significant differences in the release of swarmers between photoperiods and higher overall discharge on day 4 in comparison to day 3.

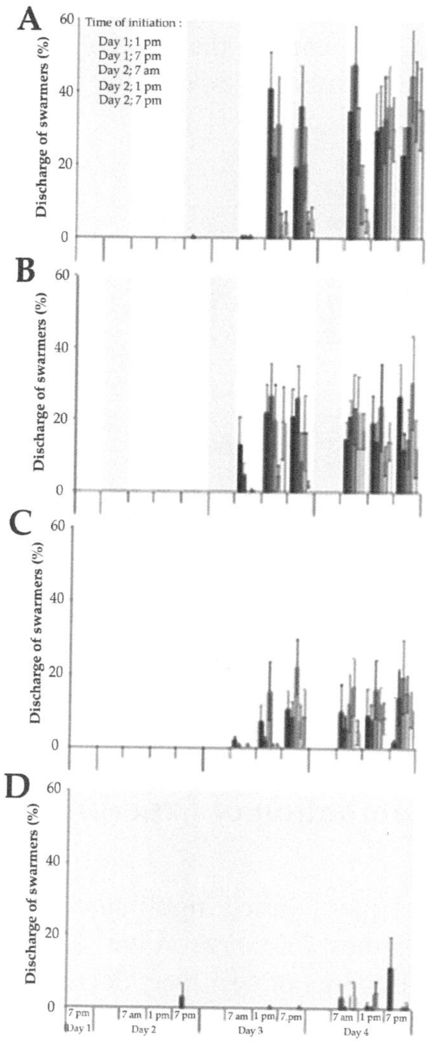

Figure 4. Discharge of swarmers testing the effect of time of initiation of experiments and photoperiod. Mean (± S.E.) discharge of swarmers (%) under

photoperiods of (a) 12 h light:12 h dark; (b) 18 h light:6 h dark; (c) constant light; and (d) constant darkness. Experiments were initiated at 1 pm and 7 pm on the day of collection (Day 1) and at 7 am, 1 pm, and 7 pm one day after collection (Day 2). Grey shaded background indicates duration of the dark period.

Table 1. PERMANOVA output testing the effect of photoperiod (Ph; 12 h light:12 h dark, 18 h light:6 h dark, 24 h light, and 24 h dark), time of initiation (In; 1 pm and 7 pm on collection day, 7 am, 1 pm, and 7 pm one day after collection), sampling day (Day; 3 and 4 days past collection), and time of sampling day (sTime; 7 am, 1 pm, and 7 pm) (all fixed factors) on the discharge of swarmers.

Source	*df*	*F*	*P*
Ph	3	**68.05**	**<0.001**
In	4	**5.01**	**<0.001**
Day	1	**68.66**	**<0.001**
sTime	2	**12.89**	**<0.001**
Ph×In	12	**2.67**	**<0.001**
Ph×Day	3	**5.33**	**<0.001**
Ph×sTime	6	**2.81**	**0.001**
In×Day	4	1.72	0.101
In×sTime	8	0.90	0.549
Day×sTime	2	**6.83**	**<0.001**
Ph×In×sTime	12	1.08	0.356
Ph×In×sTime	24	0.89	0.688
Ph×Day×sTime	6	1.43	0.140
In×Day×sTime	8	**1.70**	**0.047**
Ph×In×Day×sTime	24	0.75	0.886
Residual	960		

doi:10.1371/journal.pone.0097396.t001

Regardless of the time of initiation, the number of released swarmers were similar at the end of the experiment for each photoperiod and ranged from 22.8±7.4% to 44.4±12.2% under the normal photoperiod (12 h L:12 h D), with generally large variations in the discharge between samples of the same treatment combinations.

Under the normal photoperiod (12 h L:12 h D), the discharge generally peaked between 42 and 48 h after the initiation of experiments (two days post initiation), with the exception of the early initiation of 7 am one day past collection where the discharge peaked after only 30 h (Figure 5). However, the same trend of an onset of release in the morning after the filaments were exposed to light occurred under normal photoperiod across all initiation times.

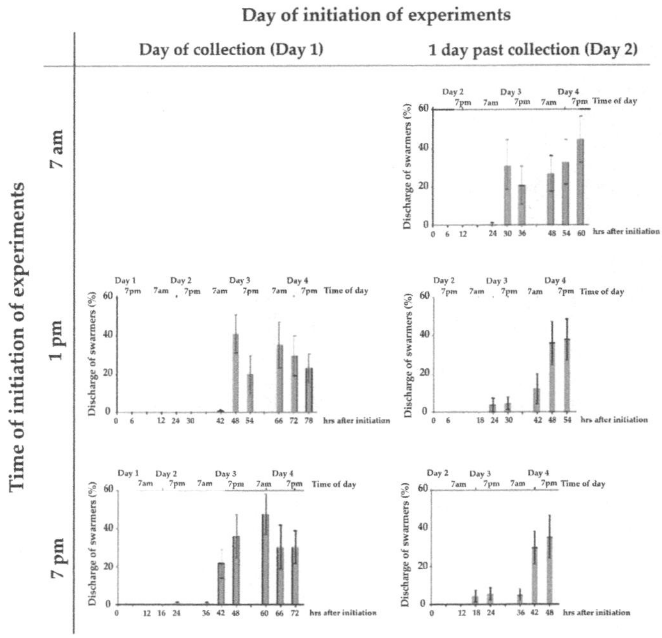

Figure 5. Discharge of swarmers under a 12;12 h D photoperiod testing the effect of time of initiation of experiments. Mean (± S.E.) discharge of swarmers (%) under a 12 h light:12 h dark photoperiod. Experiments were initiated at 1 pm and 7 pm on the day of collection (Day 1), and at 7 am, 1 pm, and 7 pm one day after collection (Day 2). Grey shaded background indicates the dark period.

Effect of Temperature Shock and Controlled Release of Swarmers

In general, the discharge of swarmers was higher for chilled filaments (4°C) than the other treatments (Figure 6), with the mean discharge being nearly double (34.4±4.6%) that of both the 25°C pre-treatment (20.9±7.9%) and without any pre-treatment (18.5±3.9%). However, the variance within the pre-treatments was relatively high (Figure 6) and pre-treatment was not a significant effect (two-factor PERMANOVA: $F_{(2,81)}=1.88$, $p=0.107$). There was no clear effect of restraining-treatment combinations to manipulate the release of swarmers ($F_{(2,81)}=0.57$, $p=0.718$).

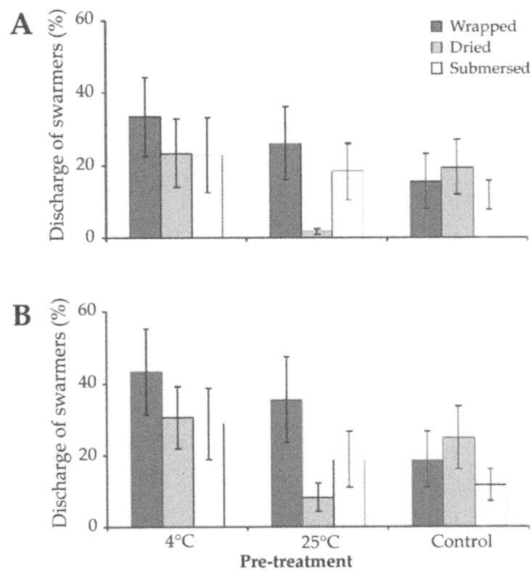

Figure 6. Discharge of swarmers testing the effect of temperature shock and controlled release of swarmers. Mean (± S.E.) discharge of swarmers (%) two days after initiation of experiments at (a) 11 am and (b) 4 pm. Filaments of *Ulva* were exposed to 4°C and 25°C FSW and without pre-treatment as a control. After two days, the filaments were wrapped in moist paper, dried, or remained submersed from 7 am for 4 h.

The discharge of swarmers was not constrained for filaments wrapped in moist paper towel or dried for 4 h (Figure 6a). In fact, treatments to constrain the discharge of swarmers resulted in slightly higher discharge at 11 am than continuously submersed filaments, with the exception of the 25°C pre-treatment (submersed: 18.3±7.8%; dried; 1.7±0.8%; wrapped: 26.1±10.0%). The unwrapping of filaments at 11 am revealed a high discharge of swarmers on the moist paper towel and these were clearly visible due to the change from white to green/brown-ish colour of the paper towel. Furthermore, constraining the discharge of swarmers by wrapping and drying was unsuccessful indicated by similar discharge between 11 am and 4 pm for those filaments (Figure 6a, b). Up to 33% of the biomass discharged swarmers while being wrapped (Figure 6a), whereas the discharge ranged from 18.9±7.7% (control) to 43.3±11.9% (4°C) after being re-submersed for 5 h (4 pm; Figure 6b). Similarly, the discharge of dried filaments was up to 23% while exposed to air (Figure 6a) and increased marginally at 4 pm, ranging from 8.3±4.0 (25°C) to 30.6±8.5% (4°C) (Figure 6b).

Reproductive Output

The time of day had a significant effect on the number of swarmers released (two-factor PERMANOVA: $F_{(4,\ 735)}=14.19$, $p=0.003$), with a peak of release at 11:30 am with 842,708±190,123 swarmers released (mean of three batches ±1 SE) (Figure 7a). The number of swarmers released was an order of a magnitude smaller at all other times and ranged from 34,375±18,311 (7 am) to 94,791±51,549 (1 pm). There was also a significant effect of batch on the number of swarmers released ($F_{(2,\ 735)}=3.16$, $p=0.025$); however, the number of swarmers released showed a similar trend among batches and the highest release consistently occurred at 11:30 am, regardless of batch (Figure 7b).

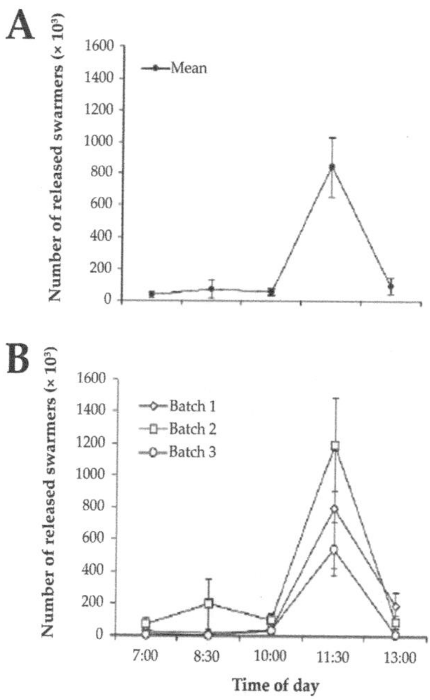

Figure 7. Number of released swarmers over time. (a) Mean (± S.E.) number of released swarmers over time (n=3). (b) Mean number (± S.E.) of released swarmers over time from three independently collected batches of algal biomass (n=50). Algal batches were collected on 6 May 2013 (Batch 1), 7 May 2013 (Batch 2), and 14 May 2013 (Batch 3).

The reproductive output differed between batches and was nearly doubled for batch 1 and 2 in comparison to batch 3 with 2.3±0.9, 2.4±0.5

and 1.3±0.4×10^6 released swarmers per cm^2, respectively (Figure 8). Biflagellate swarmers were much more common (95%; 56 out of 59 analysed samples) than quadriflagellate swarmers (5%; 3 out of 59 analysed samples).

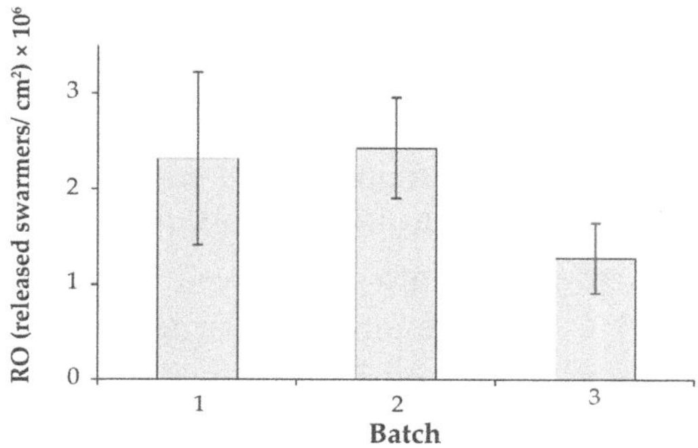

Figure 8. Reproductive output of *Ulva* sp. 3. Mean (± S.E.) reproductive output (RO) of three independently collected algal batches (n=50). Experiments were initiated on 6 May 2013 (Batch 1), 7 May 2013 (Batch 2), and 14 May 2013 (Batch 3).

Out of 150 thalli used to determine the viability and germling development of swarmers, a total of 147 thalli released swarmers. The released swarmers, both bi- and quadriflagellate, settled and germinated successfully in all 147 samples after five days.

Phototactic Behavior of Released Swarmers

A total of 22 thalli released swarmers, all of which were biflagellate with a negative phototactic response. However, on one occasion, a small number of released biflagellate swarmers showed positive phototaxis, while the vast majority of swamers released from the same thallus were negatively phototactic. The average length and width of the biflagellate swarmers was 6.55±0.85 (mean ± S.D.) and 3.75±0.52 µm, respectively.

DISCUSSION

This study provides a foundation for understanding the factors affecting the reproduction of the filamentous species of tropical *Ulva*, and identifies the best treatment combination to induce reproduction of a common and widely distributed species under controlled conditions. Photoperiod and temperature shock were successfully manipulated to enhance the formation and release of swarmers, while the effects of segmentation, dehydration, salinity, and time of initiation of experiments were negligible. The efficient manipulation of photoperiod and temperature shock is therefore the key in the reliable supply of swarmers with applications for fouling studies of tropical *Ulva* species and the seeding of nets for mass-cultivation. The proposed method–factoring in the maximum release and minimum practical effort and timing–is therefore to collect *Ulva* in the morning and initiate the experiments in the early afternoon (at 1 pm) by washing the thalli in FSW, subsequently chilling the thalli for 10 min at 4°C and then placing them into autoclaved FSW under a 12 h L:12 h D photoperiod at 25°C. Consequently, swarmers are released with peak after two days between 10:00 and 11:30 am (Figure 9).

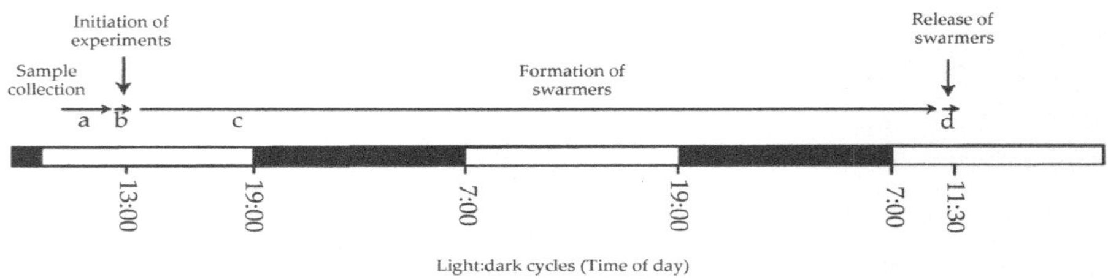

Figure 9. Time course of induced sporulation of tropical *Ulva sp 3*. (a) Collection of algal samples and subsequent transportation to the laboratory. (b) Initiation of experiments at 1 pm by washing the thalli in FSW, subsequently chill *Ulva* for 10 min and then place thalli into autoclaved FSW under a 12 h light:12 h dark photoeriod at 25°C. (c) Induction of sporulation with visible formation of swarmers after approximately 26 h. (d) Release of swarmers peaks between 10:00 and 11:30 am.

Notably, photoperiod had a significant effect on the formation and release of swarmers, with a discharge of up to 50% under normal photoperiod (12 h L:12 h D). In contrast, extended light periods resulted in lower discharge and this is in agreement with temperate environments, where the day length plays a key role in the reproduction of seaweed [22], [53]. While shorter days result in a reduced growth rate and trigger reproduction, longer days allow continuation of the vegetative growth phase for seaweeds in temperate systems [54]. However, unlike previous studies on *Ulva* in temperate and cold waters where continuous light and dark cycles restrained the discharge of swarmers [22], [55]–[57], tropical

Ulva sp. 3 continued to form and release swarmers under both extremes (up to 23% and 12%, respectively).

The present study highlights the importance of the dark phase for the formation and release of swarmers in tropical *Ulva* sp. 3, as discharge was halved when thalli were kept under continuous light. In general, the dark phase is essential for the formation of swarmers as most cells of *Ulva* divide during the dark period [29], [58], [59], forming zoospores and gametes by meiosis and mitotic division, respectively [29], [35]. A dark phase of at least 1 h is essential to induce the release of swarmers of temperate *U. pseudocurvata* [22]. Furthermore, the dark period is an important factor to trigger the release of swarmers [22], with a discharge in the morning [26] after a few minutes [22] and up to 2–5 h after the onset of the light period [31] for *U. pseudocurvata* and *U. pertusa*, respectively. For *Ulva* sp. 3, the release of swarmers peaked at 4½ h after onset of light, around 11:30 am.

A further factor affecting the formation and release of swarmers of *Ulva* sp. 3 was temperature shock. Sporulation was increased by approximately 10% when thalli were exposed to chilled (4°C) seawater for 10 min. This is consistent with a study in temperate waters identifying temperature shock as a trigger to induce reproduction of *U. lactuta* [36]. Furthermore, refrigeration of fertile thalli is also commonly used to maximise the release of swarmers in temperate environments [19], [48], [60]. The increased formation and release of swarmers following temperature shock may be a strategy to disperse under unfavourable conditions.

In general, stress treatments of segmentation [12], [61], dehydration [27], [39], [40], [62], and salinity [12] can initiate the formation and release of swarmers of *Ulva* [63]. In contrast, these factors had no effect on the sporulation of *Ulva* sp. 3. This was unexpected, particularly that segmentation did not result in an increase of sporulation as shown for *U. mutabilis*, where sporulation increased by more than 60% when thalli were fragmented [64]. Notably, the 'punching method' and other segmentation techniques are commonly used to induce reproduction of a wide range of *Ulva* species within a few days, including *U. fenestrata* [65], *U. ohnoi* [66], *U. pertusa* [31], [50], [56], [61], *U. prolifera* [12], [32], [33], *U. pseudocurvata* [22], *U. spinulosa* [21], and *U. mutabilis* [29], [59]. It is unclear at this stage whether the traits listed above are uniquely temperature or sub-tropical and do not have direct relevance to tropical species.

The dehydration of filaments was also ineffective in increasing the sporulation of *Ulva* sp. 3. Previous studies have used this method to maximise the release of swarmers and dehydration times ranged from less than 1 h [27] up to 12 h [39], [40]. Therefore, the tested dehydration times of 45 min and 4 h are well within these timeframes and provide confidence that dehydration is ineffective for tropical *Ulva* sp. 3. However, dehydration was proven to be effective for other tropical *Ulva* species [37]–[40]. In addition, neither dehydration nor wrapping in moist paper towel were effective constraining treatments for *Ulva* sp. 3 in the present study and swarmers were released during the desiccation process, indicating that once the reproductive cells are ready for release, this process might not be delayed.

The overall discharge of swarmers was generally low, at approximately 50% of previous studies, where sporulation reached up to 90–100% [12], [31], [37], [64]. In general, the discharge differs between sporophytes and gametophytes, with more than 90% in sporophytes, and only 40% for gametophytes [31]. A total of 95% of thalli in the present study released biflagellate swarmers. The possession of a majority of negative phototaxis and the ability to germinate and grow within five days without fusing with complementary gametes suggests a simple asexual life history via biflagellate zoids [17]. However, this requires confirmation through the cultivation of successive generations. In general, gametes are positively phototactic [34], [67], [68], while asexual biflagellate zoids are negatively phototactic [17], as are quadriflagellate zoospores [37] and zygotes [34]. The negative phototaxis guides swarmers to suitable surfaces for settlement and attachment [69]. Quadriflagellate swarmers were also released in the present study, although only from 5% of the thalli, and it remains unknown as to whether these are zoospores or asexual zoids [21]. In general, species of *Ulva* with a simple asexual life history produce either exclusively bi- of quadriflagellate zoids [17], [21]. It is possible that both bi- and quadriflagellate swarmers were found in this study because used samples could have been from different species of *Ulva* due to morphological similarity between filamentous species of *Ulva*; alternatively *Ulva* sp. 3 may have several life histories as reported for *U. prolifera* (formerly *Enteromorpha prolifera*; [17]). Overall, the simplicity of the reproduction of *Ulva* sp. 3 may be a unique trait of tropical species related to *Ulva* sp. 3 that has genetic basis. Based on phylogenetic trees constructed using ITS sequences, *Ulva* sp. 3 forms a unique clade separate to other species that have been investigated in previous studies.

In conclusion, a baseline method to induce the sporulation in tropical *Ulva* sp. 3 within 30 to 48 h after initiation was established. Sporulation was enhanced by temperature shocking thalli prior to incubating at a photoperiod of 12 h L:12 h D. Swarmers were released two days after initiating experiments with a maximum release between 10:00 and 11:30 am. A total of 95% of the collected field population released biflagellate swarmers with negative phototaxis. These swarmers also had the ability to settle and germinate without crossing with complementary gametes. The findings of this study have application in laboratory antifouling assays for the tropics and in the seeding of nets for the mass-cultivation of seaweed.

ACKNOWLEDGMENTS

We thank M. Martinez, A. Ricketts and B. Bao for assistance with experiments, and Coral Coast Barramundi Ltd for allowing collection of seaweed.

AUTHOR CONTRIBUTIONS

Conceived and designed the experiments: CC RdN RL NP. Performed the experiments: CC RL. Analyzed the data: CC RL NP. Contributed reagents/materials/analysis tools: CC RdN RL NP. Wrote the paper: CC RdN RL NP.

REFERENCES

1. Hayden HS, Blomster J, Maggs CA, Silva PC, Stanhope MJ, et al. (2003) Linnaeus was right all along: *Ulva* and *Enteromorpha* are not distinct genera. Eur J Phycol 38: 277–294.
2. Godwin LS (2003) Hull fouling of maritime vessels as a pathway for marine species invasions to the Hawaiian Islands. Biofouling 19 Suppl: 123–131

3. Callow ME (1986) Fouling algae from "in-service" ships. Bot Mar 29: 279–372.
4. Hodson SL, Burke CM, Bissett AP (2000) Biofouling of fish-cage netting: the efficacy of a silicone coating and the effect of netting colour. Aquaculture 184: 277–290.
5. Long CJ, Schumacher JF, Robinson PAC, Finlay JA, Callow ME, et al. (2010) A model that predicts the attachment behavior of *Ulva linza* zoospores on surface topography. Biofouling 26: 411–419.
6. Thome I, Pettitt ME, Callow ME, Callow JA, Grunze M, et al. (2012) Conditioning of surfaces by macromolecules and its implication for the settlement of zoospores of the green alga *Ulva linza*. Biofouling 28: 501–510.
7. Schumacher JF, Carman ML, Estes TG, Feinberg AW, Wilson LH, et al. (2007) Engineered antifouling microtopographies – effect of feature size, geometry, and roughness on settlement of zoospores of the green alga *Ulva*. Biofouling 23: 55–62.
8. Kawashima Y, Akasaki T, Matsumoto Y, Yamazaki Y, Shimada S (2013) Species identification of imported and Japanese commercial green algal products based on phylogenetic analyses using the nrITS2 and 5S rDNA spacer regions. Fish Sci 79: 521–529.
9. Shokita S, Kakazu K, Tomori A, Toma T (1991) Mariculture of seaweeds. In: Shokita S, Kakazu K, Tomori A, Toma T, Yamaguchi M, editors. Aquaculture in tropical areas. Tokyo: Midori Shobo. 31–95.
10. Pandey RS, Ohno M (1985) An ecological study of cultivated *Enteromorpha*. Rep Usa Mar Biol Inst 7: 21–31.
11. Ohno M (1993) Cultivation of the green algae, *Monostroma* and *Enteromorpha* "Aonori. " In: Ohno M, Critchley AT, editors. Seaweed cultivation and marine ranching. Jokosuka: Japan International Cooperation Agency. 7–16.
12. Dan A, Hiraoka M, Ohno M, Critchley AT (2002) Observations on the effect of salinity and photon fluence rate on the induction of sporulation and rhizoid formation in the green alga *Enteromorpha prolifera* (Müller) J. Agardh (Chlorophyta, Ulvales). Fish Sci 68: 1182–1188.
13. Yokoyama H, Ishihi Y (2010) Bioindicator and biofilter function of

Ulva spp. (Chlorophyta) for dissolved inorganic nitrogen discharged from a coastal fish farm – potential role in integrated multi-trophic aquaculture. Aquaculture 310: 74–83.

14. Neori A, Msuya FE, Shauli L, Schuenhoff A, Kopel F, et al. (2003) A novel three-stage seaweed (*Ulva lactuca*) biofilter design for integrated mariculture. J Appl Phycol 15: 543–553.
15. Msuya FE, Kyewalyanga MS, Salum D (2006) The performance of the seaweed *Ulva reticulata* as a biofilter in a low-tech, low-cost, gravity generated water flow regime in Zanzibar, Tanzania. Aquaculture 254: 284–292.
16. Cruz-Suárez LE, León A, Peña-Rodríguez A, Rodríguez-Peña G, Moll B, et al. (2010) Shrimp/*Ulva* co-culture: A sustainable alternative to diminish the need for artificial feed and improve shrimp quality. Aquaculture 301: 64–68.
17. Hiraoka M, Dan A, Shimada S, Hagihira M, Migita M, et al. (2003) Different life histories of *Enteromorpha prolifera* (Ulvales, Chlorophyta) from four rivers on Shikoku Island, Japan. Phycologia 42: 275–284.
18. Pickett-Heaps JD (1975) The Ulotrichales. In: Pickett-Heaps JD, editor. Green algae: structure, reproduction and evolution in selected genera. Upper Saddle River: Prentice-Hall. 167–218.
19. Fletcher RL (1989) A bioassay technique using the marine fouling green alga *Enteromorpha*. Int Biodeterior 25: 407–422.
20. Graham LE, Wilcox LW (2000) Algae. Upper Sadle River: Prentice-Hall International. 640 p.
21. Hiraoka M, Shimada S, Ohno M, Serisawa Y (2003) Asexual life history by quadriflagellate swarmers of *Ulva spinulosa* (Ulvales, Ulvophyceae). Phycol Res 51: 29–34.
22. Lüning K, Kadel P, Pang S (2008) Control of reproduction rhythmicity by environmental and endogenous signals in *Ulva pseudocurvata* (Chlorophyta). J Phycol 44: 866–873.
23. Christie AO, Evans LV (1962) Periodicity in the liberation of gametes and zoospores in *Enteromorpha intestinalis* Link. Nature 193: 193–194.
24. Evans LV, Christie AO (1970) Studies on the ship-fouling alga

Enteromorpha. Ann Bot 34: 451–466.

25. Callow ME, Callow JA, Pickett-Heaps JD, Wetherbee R (1997) Primary adhesion of *Enteromorpha* (Chlorophyta, Ulvales) propagules: quantitative settlement studies and video microscopy. J Phycol 33: 938–947.
26. Smith GM (1947) On the reproduction of some Pacific coast species of *Ulva*. Am J Bot 34: 80–87.
27. Corradi MG, Gorbi G, Zanni C (2006) Hypoxia and sulphide influence gamete production in *Ulva* sp. Aquat Bot 84: 144–150.
28. Thiadens AJH, Zeuthen E (1967) Meiosis and sporulation induced in sporophytes of *Ulva mutabilis* ("slender") with synchronous mitosis. Planta 72: 60–65.
29. Nordby Ø, Hoxmark RC (1972) Changes in cellular parameters during synchronous meiosis in *Ulva mutabilis* Føyn. Exp Cell Res 75: 321–328.
30. Nordby Ø (1974) Light microscopy of meiotic zoosporogenesis and mitotic gametogenesis in *Ulva mutabilis* Føyn. J Cell Sci 15: 443–455.
31. Hiraoka M, Enomoto S (1998) The induction of reproductive cell formation of *Ulva pertusa* Kjellman (Ulvales, Ulvophyceae). Phycol Res 46: 199–203.
32. Hiraoka M, Oka N (2008) Tank cultivation of *Ulva prolifera* in deep seawater using a new "germling cluster" method. J Appl Phycol 20: 97–102.
33. Gao S, Chen X, Yi Q, Wang G, Pan G, et al. (2010) A strategy for the proliferation of *Ulva prolifera*, main causative species of green tides, with formation of sporangia by fragmentation. PLoS One 5(1): e8571.
34. Kuwano K, Hashioka T, Nishihara GN, Iima M (2012) Durations of gamete motility and conjugation ability of *Ulva compressa* (Ulvophyceae). J Phycol 48: 394–400.
35. Bråten T, Nordby Ø (1973) Ultrastructure of meiosis and centriole behaviour in *Ulva mutabilis* Føyn. J Cell Sci 13: 69–81.
36. Niesenbaum RA (1988) The ecology of sporulation by the macroalga *Ulva lactuca* L. (Chlorophyceae). Aquat Bot 32: 155–166.

37. Mantri VA, Singh RP, Bijo AJ, Kumari P, Reddy CRK, et al. (2011) Differential response of varying salinity and temperature on zoospore induction, regeneration and daily growth rate in *Ulva fasciata* (Chlorophyta, Ulvales). J Appl Phycol 23: 243–250.

38. Ruangchuay R, Dahamat S, Chirapat A, Notoya M (2012) Effects of culture conditions on the growth and reproduction of Gut Weed, *Ulva intestinalis* Linnaeus (Ulvales, Chlorophyta). Songklankarin J Sci Technol 34: 501–507.

39. Shin HW (2008) Rapid attachment of spores of the fouling alga *Ulva fasciata* on biofilms. J Environ Biol 29: 613–619.

40. Imchen T (2012) Recruitment potential of a green alga *Ulva flexuosa* Wulfen dark preserved zoospore and its development. PLoS One 7(3): e32651.

41. de Paula Silva PH, McBride S, de Nys R, Paul NA (2008) Integrating filamentous "green tide" algae into tropical pond-based aquaculture. Aquaculture 284: 74–80.

42. Lawton RJ, Mata L, de Nys R, Paul NA (2013) Algal bioremediation of waste waters from land-based aquaculture using *Ulva*: selecting target species and strains. PLoS One 8(10): e77344.

43. Shimada S, Yokoyama N, Arai S, Hiraoka M (2008) Phylogeography of the genus *Ulva*(Ulvophyceae, Chlorophyta), with special reference to the Japanese freshwater and brackish taxa. J Appl Phycol 20: 979–989.

44. Luo MB, Liu F (2011) Salinity-induced oxidative stress and regulation of antioxidant defense system in the marine macroalga *Ulva prolifera*. J Exp Mar Bio Ecol 409: 223–228.

45. Poole LJ, Raven JA (1997) The biology of *Enteromorpha*. In: Round FE, Chapman DJ, editors. Progress in Phycological Research. Bristol: Biopress Ltd. 1–148.

46. Malta E-J, Draisma SGA, Kamermans P (1999) Free-floating *Ulva* in the southwest Netherlands: species or morphotypes? A morphological, molecular and ecological comparison. Eur J Phycol 34: 443–454.

47. Hebert PDN, Ratnasingham S, de WaardJR (2003) Barcoding animal life: cytochrome c oxidase subunit 1 divergences among

closely related species. Proc R Soc London Ser B Biol Sci 270: S96–S99.

48. Scardino AJ, Guenther J, de Nys R (2008) Attachment point theory revisited: the fouling response to a microtextured matrix. Biofouling 24: 45–53.
49. Nilsen G, Nordby Ø (1975) A sporulation-inhibiting substance from vegetative thalli of the green alga *Ulva mutabilis*, Føyn. Planta 125: 127–139.
50. Han T, Choi G-W (2005) A novel marine algal toxicity bioassay based on sporulation inhibition in the green macroalga *Ulva pertusa* (Chlorophyta). Aquat Toxicol 75: 202–212.
51. Thompson SEM, Callow JA, Callow ME, Wheeler GL, Taylor AR, et al. (2007) Membrane recycling and calcium dynamics during settlement and adhesion of zoospores of the green alga *Ulva linza*. Plant, Cell Environ 30: 733–744.
52. Clarke K, Gorley R (2006) PRIMER v6: User manual/tutorial. Plymouth (UK): Primer-E. 190.
53. Forbord S, Skjermo J, Arff J, Handå A, Reitan KI, et al. (2012) Development of *Saccharina latissima* (Phaeophyceae) kelp hatcheries with year-round production of zoospores and juvenile sporophytes on culture ropes for kelp aquaculture. J Appl Phycol 24: 393–399.
54. Lüning K (1988) Photoperiodic control of sorus formation in the brown alga *Laminaria saccharina*. Mar Ecol Prog Ser 45: 137–144.
55. Lersten NR, Voth PD (1960) Experimental control of zoid discharge and rhizoid formation in the green alga *Enteromorpha*. Bot Gaz 122: 33–45.
56. Han T, Han Y-S, Kim K-Y, Kim J-H, Shin H-W, et al. (2003) Influences of light and UV-B on growth and sporulation of the green alga *Ulva pertusa* Kjellman. J Exp Mar Bio Ecol 290: 115–131.
57. Wichard T, Oertel W (2010) Gametogenesis and gamete release of *Ulva mutabilis* and *Ulva lactuca* (Chlorophyta): regulatory effects and chemical characterization of the "swarming inhibitor. ". J Phycol 46: 248–259.
58. Titlyanov EA, Titlyanova TV, Lüning K (1996) Diurnal and

circadian priodicity of mitosis and growth in marine macroalgae. II. The green alga *Ulva pseudocurvata*. Eur J Phycol 31: 181–188.

59. Stratmann J, Paputsoglu G, Oertel W (1996) Differentiation of *Ulva mutabilis*(Chlorophyta) gametangia and gamete release are controlled by extracellular inhibitors. J Phycol 32: 1009–1021.
60. Scardino AJ, Zhang H, Cookson DJ, Lamb RN, de Nys R (2009) The role of nano-roughness in antifouling. Biofouling 25: 757–767.
61. Hiraoka M, Ohno M, Kawaguchi S, Yoshida G (2004) Crossing test among floating *Ulva*thalli forming "green tide" in Japan. Hydrobiologia 512: 239–245.
62. de Nys R, Steinberg PD, Willemsen P, Dworjanyn SA, Gabelish CL, et al. (1995) Broad spectrum effects of secondary metabolites from the red alga *Delisea pulchra* in antifouling assays. Biofouling 8: 259–271.
63. McArthur DM, Moss BL (1979) Gametogenesis and gamete structure of *Enteromorpha intestinalis* (L.) Link. Br Phycol J 14: 43–57.
64. Nordby Ø (1977) Optimal conditions for meiotic spore formation in *Ulva mutabilis* Føyn. Bot Mar 20: 19–28.
65. Kalita TL, Titlyanov EA (2011) The effect of temperature on infradian rhythms of reproduction in *Ulva fenestrata* Postels et Ruprecht, 1840 (Chlorophyta: Ulvales). Russ J Mar Biol 37: 52–61.
66. Hiraoka M, Shimada S, Uenosono M, Masuda M (2003) A new green-tide-forming alga, *Ulva ohnoi* Hiraoka et Shimada sp. nov. (Ulvales, Ulvophyceae) from Japan. Phycol Res 51: 17–29.
67. Hiraoka M, Yoshida G (2010) Temporal variation in isomorphic phase and sex ratios of a natural population of *Ulva pertusa* (Chlorophyta). J Phycol 46: 882–888.
68. Hiraoka M, Ichihara K, Zhu W, Ma J, Shimada S (2011) Culture and hybridization experiments on an *Ulva* clade including the Qingdao strain blooming in the Yellow Sea. PLoS One 6(5): e19371.
69. Callow JA, Callow ME (2006) The *Ulva* spore adhesive system. In: Smith AM, Callow JA, editors. Biological adhesives. Heidelberg: Springer-Verlag. 63–78.

CHAPTER

8

PRODUCTION OF GENETICALLY AND DEVELOPMENTALLY MODIFIED SEAWEEDS: EXPLOITING THE POTENTIAL OF ARTIFICIAL SELECTION TECHNIQUES

Bénédicte Charrier[1], Elodie Rolland[1], Vishal Gupta[2], and C. R. K. Reddy[2]

[1]Centre National de la Recherche Scientifique, Sorbonne Université, UPMC Univ Paris 06, UMR 8227, Integrative Biology of Marine Models, Station Biologique de Roscoff, Roscoff, France

[2]Seaweed Biology and Cultivation Group, Division of Marine Biotechnology and Ecology, CSIR-Central Salt and Marine Chemicals Research Institute, Bhavnagar, India

ABSTRACT

Plant feedstock with specific, modified developmental features has been a quest for centuries. Since the development and spread of agriculture, there has been a desire for plants producing disproportionate—or more abundant and more nutritional—biomass that meet human needs better than their native counterparts. Seaweed aquaculture, targeted for human consumption and the production of various raw materials, is a rapidly expanding field and its stakeholders have increasing vested interest for cost-effective and lucrative seaweed cultivation processes. Thus, scientific research on seaweed development is particularly timely: the potential for expansion of seaweed cultivation depends on the sector's capacity to produce seaweeds with modified morphological features (e.g., thicker blades), higher growth rates or delayed (or even no) fertility. Here, we review the various technical approaches used to modify development in macroalgae, which have attracted little attention from developmental biologists to date. Because seaweed (or marine macroalgae) anatomy is much less complex than that of land plants and because seaweeds belong to three different eukaryotic phyla, the mechanisms controlling their morphogenesis are key to understanding their development. Here, we present efficient sources of developmentally and genetically modified seaweeds—somatic variants, artificial hybrids and mutants—as well as the future potential of these techniques.

Keywords: Seaweed, somatic hybridization, mutagenesis, varietal improvement, genetic manipulation

SOCIETAL IMPORTANCE OF SEAWEEDS

Marine macroalgae (seaweeds) are one of the renewable resources in marine ecosystems and carry out several key ecosystem functions that contribute to the productivity of the oceans. Seaweeds are traditionally consumed as human food in several Asian countries, where they are cultivated on a large scale. A wide range of seaweeds belonging to different genera have been known as a food source since prehistoric times (Mouritsen et al., 2013). In most cases, the entire thallus of the seaweed is consumed (e.g., species of *Porphyra* and *Enteromorpha*), although, in some cases, only certain parts are edible (e.g., in *Caulerpa lentillifera*, only the ramuli are edible). In larger seaweeds, such as *Laminaria digitata* and *Undaria pinnatifida*, the stipe is used for preparing local dishes, such as soup. *Chondrus crispus* and certain species of *Gracilaria* are consumed in salad and in these cases only tender tips are used. Seaweeds are good sources of high-quality digestible proteins with a balanced amino acid composition, polyunsaturated fatty acids including omega-3 and omega-6 and important vitamins and minerals, have a caloric content similar to land crops and are high in fiber (MacArtain et al., 2007).

According to recent statistics published by the FAO (2014), seaweed production has increased from less than 4 million wet metric tons (t) in 1980 to almost 20 million wet t in 2012, with more than 50% of total production is used for direct human consumption. The indirect products derived from seaweeds, polysaccharides in particular, have unique applications in processed as well as functional foods, pet foods, feed, fertilizers, cosmetics, and medicines. Furthermore, recent bioprocessing techniques have also unequivocally demonstrated that seaweeds are potential feedstock for production of biofuels and commodity products (Baghel et al., 2015).

The major seaweed species that are currently farmed include *Kappaphycus alvarezii* and *Eucheuma* spp. both known as "cottonii" (>8 million wet t) followed by *Saccharina japonica* (formerly *Laminaria japonica*)—known as "kombu" (5.6 million wet t), *Gracilaria* spp. (~3 million wet t), *U. pinnatifida*—known as "wakame" (2 million wet t) and *Porphyra* spp.—known as "nori" (1.8 million wet t) (Buchholz et al., 2012). The use of *Kappaphycus* enhances the texture of fish cutlets and pork patties and further increases the nutritional value of foodstuffs by providing minerals (Senthil et al., 2005; Jeon and Choi, 2012). Other species such as *Palmaria*, *Chondrus*, and *Ulva* are produced on a lower scale (FAO, 2014). The global utilization of various seaweed species is described in Supplementary Table 1. Seaweed cultivation is now perceived as an excellent alternative source of revenue for coastal fishermen, particularly in light of its advantages, including a shorter production cycle, low capital outlay and relatively simple farming techniques. There have been some biotechnological advances to improve economically important seaweeds. The present article briefly reviews the various research efforts that have been made to produce morphological variants of economically important seaweeds and study the genetic basis behind the changes in their morphology.

GENETIC MANIPULATION AND IMPROVEMENT IN SEAWEEDS

Somaclonal Variants

Cellular biotechnology in seaweeds was initiated in the 1980s and lags far behind that of terrestrial plants.

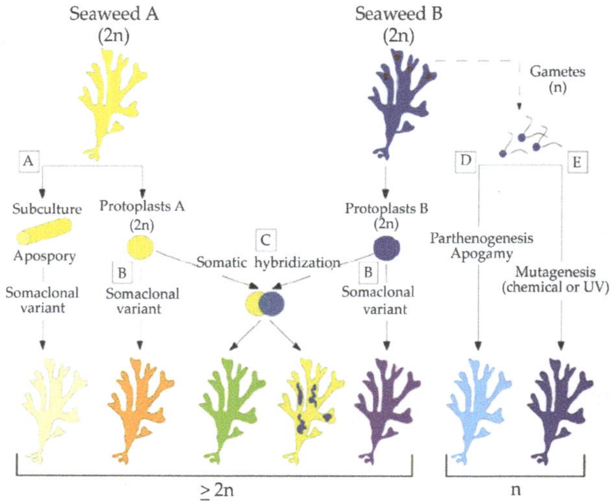

Figure 1. Summary of the different technological approaches for generating morphologically modified macroalgae. Diploid sporophytes are shown as the initial material. Change in color states indicates change in morphology. The more the color is different from the original plant, the more distinct the phenotype is. Dashed lines show cases that involve an alternation of generations (e.g., from sporophyte to gametophyte). (A)Sub-culture of seaweed fragments and apospory; (B) Protoplast preparation. (C) Somatic hybridization producing thalli with either novel characters (green) or chimeric or mixed parental characters (yellow and blue patches). (D) Parthenogenesis giving rise to morphological variants. (E) Chemical or UV-mediated mutagenesis on gametes. Ploidy often increases when (A–C) are employed.

In macroalgae, the development of *in vitro* culture systems facilitates mass proliferation of biomass all year round for the production of valuable compounds under controlled conditions. Culture systems are also a source of novel genetic variants with useful traits arising from somaclonal variation (Figure 1A). The techniques for somatic embryogenesis or dedifferentiation of somatic cells often induce morphological variants in seaweeds. Such morphological and developmental variation has advantages for genetic improvement programs and also can be used as an efficient means of germplasm storage and selection. The new variation can be transient, reversible or permanent. Temporary changes are mainly due to either epigenetic or physiological changes, which can be reversible, even after being heritable for a few generations (Kaeppler et al., 2000). The underlying molecular mechanisms involved in permanent somaclonal variants are rarely investigated and little understood (Larkin and Scowcroft, 1981). Nevertheless, most morphological variation observed during *in vitro* culture of seaweeds is transient and is not passed on to progeny. There are many examples where same genotype can produce different morphological phenotypes. For example: *in vitro* tissue culture of kelp (order Laminariales) sporophytes have a frequent developmental pattern in which outgrowths of aposporous gametophyte-like filaments with differentiated fertile branches can give rise directly to sporophytes (Ar Gall et al., 1996). Similarly, meristem cultures of *Laminaria* regenerate into one of three different body types: (1) uniseriate filaments; (2) thalloid-like structures; (3) dark green, compact calli. Analogously, early development of embryonic germlings from 22 species of Fucaceae can show up to six different developmental types.

Plant protoplasts are amenable to *in vitro* genetic manipulation techniques for developing genetically improved strains of agronomic crops. There have been numerous studies on the isolation and regeneration of protoplasts from a wide variety of seaweed body types ranging from simple leafy thalli to complex, cylindrical, branched thalli (see review Reddy et al., 2008). Unlike higher plants, seaweed protoplasts regenerate and differentiate directly into a complete thallus without any addition of phytohormones in the culture medium. Protoplasts from green seaweeds nevertheless have different types of morphogenetic patterns (Figure 1B) and give rise to several phenotypically variable morphotypes, such as free-living sporangia, microthalli, or saccate (or spherical), tubular (or spindle), irregular, or frondose thalli with various life spans (Reddy et al., 1989; Huang et al., 1996; Chen, 1998; Krishna Kumar et al., 1999; Chen and Shih, 2000; Rusig and Cosson, 2001). In the red alga *Porphyra*, three different types of protoplast regeneration patterns have been described: (1) callus form, (2) filamentous form and (3) conchocelis form (Polne-Fuller et al., 1984; Fujita and Migita, 1985; Waaland et al., 1990; Dipakkore et al., 2005). Callus-like outgrowth (i.e., an unorganized cellular mass) has been reported from protoplast culture of various brown macroalgal species. Similarly, two distinct patterns of development have been observed for protoplast-based development of *Gracilaria gracilis*, giving rise to plants that differ in appearance and life span. Regenerated plants either resemble parental plants, with slender, branched thalli, or remain small with thick, unbranched thalli, many of which die. Although the underlying mechanisms have been poorly studied, this variation in developmental morphological abnormalities has been attributed to a variety of factors, including the type of donor tissue from which protoplasts are prepared and the culture conditions employed for regeneration. For example, protoplasts isolated from the vegetative thalli of *Monostroma latissimum* regenerate into normal thalli, whereas protoplasts isolated from the holdfast develop into filaments (Chen, 1998). Protoplasts from *Ulva fasciata* develop into microthalli when cultured in high density (Chen and Shih, 2000). For brown seaweeds in the order Laminariales, protoplast regeneration into normal sporophytes can occur via different developmental processes, such as direct regeneration into plantlets (in *L. digitata*; Benet et al., 1997, *U. pinnatifida*; Matsumura et al., 2001 and *L. japonica*; Sawabe and Ezura, 1996; Sawabe et al., 1997; Matsumura et al., 2000), or indirect regeneration, depending on water temperature in some species. Indirect regeneration occurs either after de-differentiation of the tissue into a filament (in *U. pinnatifida*; Matsumura et al., 2001, *L. saccharina*; Benet et al., 1997) or after the development of callus-like masses [*U. pinnatifida*; (Matsumura et al., 2001) and *L. japonica*; Matsumura et al., 2000]. Gupta et al. (2012) provide the first report of epigenetic regulation of morphology in protoplast-derived germlings, with DNA methylation acting as an underlying molecular mechanism in protoplast-derived germlings in *Ulva reticulata* (Figure 2). Axenic culture conditions may also cause the development of abnormal thalli. Most studies have shown that seaweed-associated microflora produce certain morphogenetic substances that, in turn, lead to normal thallus structure (Matsuo et al., 2005; Spoerner et al., 2012). However, most commercial seaweed cultivation is currently based on simple vegetative propagation due to economic and farming advantages. The *in vitro* culture techniques currently being developed for seaweeds can create new genetic variants or promote clonal propagation in photobioreactors for high end applications.

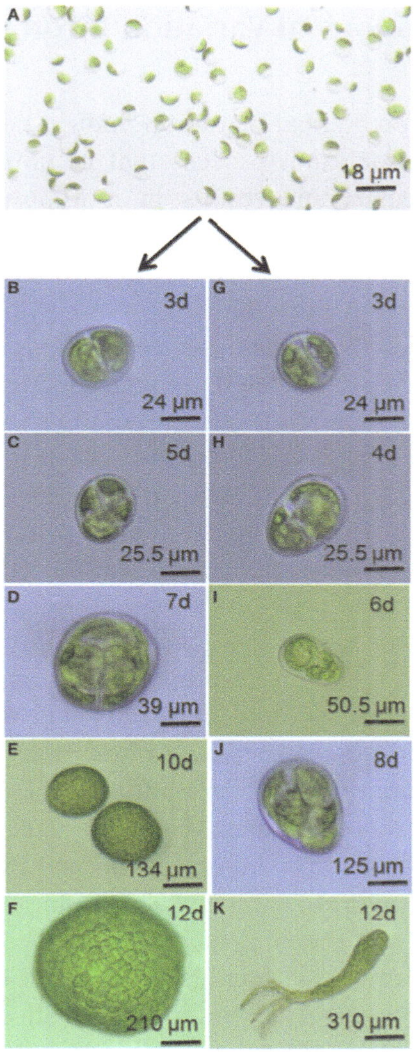

Figure 2. Illustration of developmental variation in marine macroalgae upon protoplast generation. Developmental variants among protoplast-derived germlings of *Ulva reticulata* Forsskål C (Source: Gupta et al., 2012). *U. reticulata* protoplasts (A) regenerate either into a callus (D–F) due partly to symmetrical cell divisions (B,C) or into an organism displaying a developmental pattern similar to that of native organisms (G–K). Note the asymmetrical division taking place in (G), better seen in (H) after cell enlargement.

Inducing Morphological Variation through Somatic Hybridization

In addition to morphological variation induced by *in vitro* culture, hybridization is an important process that combines phylogenetically distinct genetic lineages and results in morphotypes that are either intermediate to the parental species or completely novel (Figure 1C). In seaweeds, species are not always accurately defined and are constantly being delineated through molecular genotyping or sequencing (Maggs et al., 2007). Through the crossing of two genetically distinct species, hybridization can bring together divergent genetic lineages (Hodge et al., 2010). The resulting hybrid either combines parental phenotypes or results in a new phenotype quite distinct from the parental type. In addition to natural hybridization events that lead to speciation, somatic hybridization via protoplast fusion offers great promise for achieving wide crosses between species that are difficult or impossible to hybridize conventionally (Davey et al., 2005). Fusing protoplasts of different origins harnesses natural genetic diversity and provides novel genetic combinations resulting in the potential improvement of various functional traits. However, protoplast fusion leads to either heterokaryon (fusion of nuclei of different species) or homokaryon (fusion of nuclei of same species) recombination. In both cases, the increase in ploidy itself can improve agronomic traits. Most studies use color differences to distinguish heterokaryons from homokaryons and from unfused parental protoplasts. Somatic hybridization has shown unprecedented success in terrestrial plants but few attempts have been made in seaweeds (Reddy et al., 2008). Furthermore, a detailed description of cell division and developmental stages is still lacking in almost all the protoplast fusion studies published to date. Below, we provide a brief overview of the regeneration of fusion products so far reported from different seaweed species.

Intrageneric Hybridization

The first report of protoplast fusion and fusion product regeneration between two color morphs of *Porphyra* was that of Fujita and Migita (1987) using the polyethylene glycol (PEG) method. Protoplast fusion from two different color types (green and reddish purple) in *Porphyra yezoensis* UEDA gave rise to callus-like masses that were composed of greenish and reddish-purple cells of various diameters. The young thalli grown from these cell masses attained a length of about 60 cm and were composed of microscopically chimeral tissues (as observed in the calli), irregularly variegated with greenish and reddish purple cell groups and possibly due to the independent segregation of two chloroplast types from the fusion partners. Further studies on sexual crosses between green mutant and wild-type red *P. yezoensis*demonstrate the production of uniform wild-type color and green mutants along with mutants consisting both of wild-type patches and green mutant patches (Niwa et al., 2002). The mechanisms behind the formation of chimeric thalli from protoplast fusion clearly differ from those involved in natural crosses. Protoplast fusions between other *Porphyra* species such as *P. yezoensis* and *P. pseudolinearis* give rise to plants resembling either one or the other parent or occasionally mosaic chimeric plants with different color contours (Fujita and Saito, 1990). Similar mechanisms of independent segregation of chloroplasts have been reported in various studies on higher plants (Cocking, 1983; Davey et al., 2005). These results reveal that crossing two species by protoplast fusion or normal sexual crosses shows different developmental processes. The possible reason for the development of chimeric thalli may be chromosomal complementation from both fusion partners, whereas in natural crosses, only one set of chromosomes from each fusion partner is found in the resulting genotype.

Protoplast fusion between two *Gracilaria* species, i.e., between *G. tikvahiae* (green pigmented) and *G. chilensis* (red pigmented) result in bicolor, red and green chimeric plants (Cheney, 1990). The hybrids exhibit several unusual morphological features, such as branching morphology unlike either of the parental plants. Limitations in the regeneration of somatic hybrids have led some researchers to fuse newly released spores. Kapraun (1989, 1990) developed parasexual hybrids from the fusion of zoospores from *Enteromorpha* (now *Ulva*) and *Ulvaria*. Most hybrid germlings showed morphology similar to *Enteromorpha*, although a few showed multinucleate thalli with "giga" characteristics, including larger cells with greatly enlarged vacuoles (nuclei in "gigas" vegetative cells are 8–10 μm diameter compared with 5–6 μm diameters in control plants). Mizukami et al. (1995) demonstrated protoplast fusion between *P. suborbiculata* and *P. yezoensis*. The hybrids initially showed repeated cell divisions and formed a multicellular body, which, upon subsequent culturing, produced rhizoids. The germinated hybrids showed thalli morphologically different from each other as well as from the fusion partners. The thallus of *P. suborbiculata* is brownish and roundish, and possesses spike-like serrations on the margin, whereas the thallus of *P. yezoensis* is greenish and slender and has smooth margins. Hybrid thalli were usually brownish and had a roundish, long and slender shape—thus combining features of the parental thalli—and wrinkled margins.

Intergeneric Hybridization

Subsequent studies on intergeneric protoplast fusion have been carried out between morphologically different strains of *Ulva pertusa* and *Enteromorpha prolifera* (Reddy et al., 1992). Presumptive heteroplasmic fusion products were identified based on their larger size and the presence of twin chloroplasts. Subsequent analysis on regeneration patterns of fused protoplasts showed that they were similar to normal (unfused) protoplast development. Most of the regenerated plants from fusion products had a thallus similar to either *U. pertusa* or *E. prolifera*. However, the thalli of some plants had a characteristic irregular and dentate margin, which has never been observed in the parental type.

Transdivisional Hybridization

Kito et al. (1998) published the first report of successful transdivisional protoplast fusion between *Monostroma* and *Porphyra*. Protoplast fusion products of these two species showed different regeneration processes and characteristics. Although the fusion partners displayed distinct monostromatic and distromatic thalli, the regenerated hybrids were green with a distromatic structure. Initial heterokaryons were identified based on clearly distinguishable chloroplast colors, but were indistinguishable after 5 days of culture. Of the hybrids generated, one of the heterofusant plants grew into a multicellular body followed by the development of rhizoid-like and bud-like organs. Finally, the plant grew into a long stringy plant (1.5 m in length and 1.5 cm in width) and was named "cyojo." Another hybrid grew a multicellular body, which, upon subsequent culturing, separated into individual cells, each one growing into a long stringy plant; this mutant was named "kattsunbo." A third fusion product grew into a bud. This bud-like plant became a thallus (7.5 cm in length and 6.5 cm in width) and was named "nigo."

These few examples illustrate the range of morphological alterations that can be generated by somatic hybridization in seaweeds. Furthermore, procedures have been developed to isolate seaweed protoplasts and bring them to the full thallus regeneration stage on a wide range of seaweeds including the most anatomically complex taxa such as *Laminaria*, *Undaria*, *Gracilaria*, and *Kappaphycus*—which are some of the most commercially important seaweed genera. Nevertheless, somatic hybrids have not been widely developed and have not yet produced "cultivars" for field cultivation. However, the progress made thus far with the development of homo- or hetero-karyons via somatic hybridization provides useful groundwork for continued research in the development of macroalgae with improved functional traits.

Inducing Morphological Variation through Parthenogenesis

Parthenogenesis can occur either through apospory (Figure 1A) or apogamy (Figure 1D) without any ploidy changes. In the former, diploid gametophytes are produced directly by sporophytic cells whereas, in the latter, haploid sporophytes produced directly from gametophytic cells. In most cases, parthenogenetic plants are identical to wild types but occasionally give rise to phenotypic variants. Parthenogenetic proliferation (asexual reproduction) has been extensively reported for several taxa of brown and green seaweeds cultured *in vitro*, but are rarely reported in red algae. Tatarenkov et al. (2005) observed dwarf morphs of *Fucus vesiculosus* along with common morphs in wild populations itself in the brackish water Baltic Sea. The dwarf morphs may have evolved from common morphs in response to the prevailing low salinity habitat. Shan et al. (2013) reported induction of larger parthenogenetic sporophytes from gametophytic clones of *U. pinnatifida*. Furthermore, the blades of parthenogenetic sporophytes were uniformly smooth and without wrinkles on either side of the mid-rib. The genetic features of parthenogenetic phenotypes have also been investigated for *Laminaria* and *Undaria*. Parthenogenetic plants with morphological variation can lead to new insights into the genetics of the relationship between ploidy level and morphology. These parthenogenetic plants can also be used as a potential resource for breeding studies for the genetic improvement of macroalgal strains of aquaculture importance.

Mutagenesis-Mediated Morphological Variations

Macroalgal mutants have been little studied. In 1958, Ralph Lewin emphasized the importance of macroalgal genetics (Lewin, 1958); now, more than 50 years later, still very few mutants have been isolated and analyzed. In addition to the difficulties of growing macroalgae in laboratory conditions and their complicated and relatively long life cycles, there is a surprising lack of interest for this field of study. As a result—and despite the amenability of haploid organisms for genetic and molecular analyses of underlying mechanisms and pathways—the identification of genes involved in the control of marine macroalgal growth and development lags far behind that of other studied multicellular organisms (Maluszynski et al., 1995; Howell, 1998).

The most advanced genetic characterization of developmental mutants in a multicellular alga has been carried out in the Chlorophyceae taxon *Volvox carteri* (sub-division Chlorophyta; Leliaert et al., 2012). This microscopic freshwater alga is composed of about 2000 bi-flagellated cells stuck together within a thick extracellular gelatinous matrix, forming a moving spherical body. While most cells remain flagellated and vegetative, 16 cells differentiate into non-flagellated, larger asexual reproductive cells (gonidia) from asymmetric cell divisions (Starr, 1969). Subsequent developmental steps lead to embryo inversion, externalizing the flagellated somatic cells and internalizing the gonidia.

Chemical mutagenesis and transposon-tagging in *V. carteri* produced several morphologically impaired mutants (Sessoms and Huskey, 1973). The *glsA* gene, coding for a chaperone protein involved in both protein translation and transcriptional regulation (Miller and Kirk, 1999; Pappas and Miller, 2009), controls the asymmetric cell division giving rise to the gonidia cells. Additional mutants led to the identification of transcription factors controlling genes specific to either the gonidial cells (*lag* gene) or the somatic cells (*regA* gene) (Kirk, 2003). In addition, the *invB* mutant, impaired in the inversion process, has been shown to code for a nucleotide-sugar transporter (Ueki and Nishii, 2009) necessary for the expansion of the glycoprotein-rich gonidia vesicle, which tightly surrounds the multicellular sphere before inversion.

The use of temperature-sensitive transposons to generate both tagged mutants and revertants, and the use of genetic transformation to complement the mutants and further analyze them at the molecular level are important assets of *Volvox* genetics because they allow rapid molecular identification of the causal genes (Ueki and Nishii, 2008). A similar approach can be used in macroscopic and marine macroalgae. To date, only a very few morphological mutants have been characterized—primarily at the phenotypical level. In compensation for their complex life cycles and sometimes challenging culture conditions, these mutants often display similar or increased reproductive capabilities in lab conditions compared to the wild types (*Ulva mutabilis*, Fjeld and Løvlie, 1976). Mutants were obtained either from sampling natural stocks, or among offspring of mutagenized populations.

Naturally-Induced Mutations in Marine Macroalgae

Some macroalgal lines are particularly prone to spontaneous mutations. The resulting phenotypes often revert to the wild type, indicating that the mutations are unstable, with most being attributed to a particularly unstable locus. However, there have been some stable mutants that have lent themselves to genetic analyses (Föyn, 1961). One laboratory strain of the green marine macroalga *U. mutabilis* has been shown to be particularly inclined to spontaneous morphological changes with a rate 10 times higher than other strains (Fjeld and Løvlie, 1976). Some generated mutants displaying growth and morphological features different from the wild type parent have been reported (Föyn, 1959). The *slender* mutant (Figure 3) displays a higher growth rate and impairment in cell differentiation and cell enlargement, because only small and undifferentiated cells are observed (Løvlie, 1969). Similarly, cells show altered polarized division and cell differentiation in the *lumpy* mutant, which develops as a loose aggregate of undifferentiated cells with no specific organs (Bryhni, 1973, Figure 3). In this mutant, the composition of the cell wall is modified, with an increased proportion of water-extractable polysaccharides, resulting in higher plasticity of the cell wall (Bryhni, 1978). It is not clear whether the inhibition of cell differentiation is a result of increased cell wall plasticity. In contrast, the *globose* mutant is small and seems to have a more active differentiation process, developing many more rhizoids and a darker thallus than the wild type (Föyn, 1961, not shown). There are other distinct phenotypes, less well described (e.g., non-sporulating mutant, Figure 3). Løvlie (1978) also showed developmental variations in *U. mutabilis* temperature-sensitive mutants. The mutants develop normal phenotypes, forming a filament consisting of a row of cells at 22°C, whereas at 15°C they produce abnormal phenotypes.

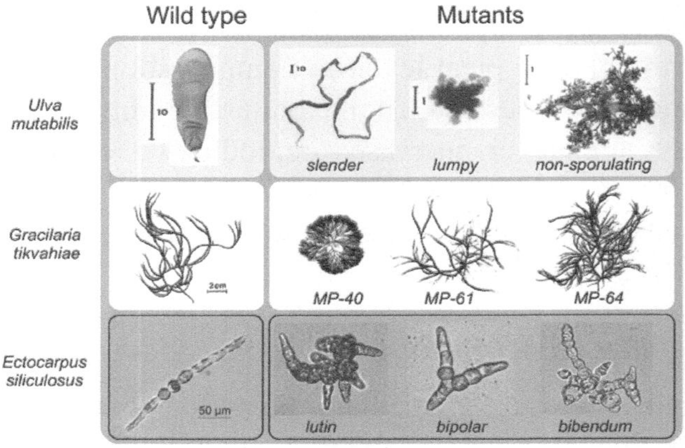

Figure 3. Morphology of some marine macroalgal mutants. Examples of some morphological mutants of the green macroalga *Ulva mutabilis*, the red macroalga *Gracilaria tikvahiae* at mature stages, and the brown macroalga *Ectocarpus siliculosus* at early stages. Unless otherwise specified, the scale bar is the same as for the wild-type individual. Permission to reproduce the images of *Ulva mutabilis* (Fjeld and Løvlie, 1976) and *Gracilaria tikvahiae*(Patwary and van der Meer, 1982) was granted by John Wiley & Sons, Inc. and the journal *Botany*, respectively. *Ectocarpus*: personal photos (B. Charrier).

In the brown alga *Ectocarpus siliculosus*, the natural *immediate upright* mutant shows a flaw in the early morphogenesis of the filamentous sporophyte, in which the prostrate basal body is reduced, thereby resembling a gametophyte with mainly upright filaments (Peters et al., 2008). This phenotype is due to a single-locus mutation, emerging in the offspring population (gametophytes) from a wild-type sporophyte, probably by spontaneous random mutation.

Chemically or UV-Light Induced Mutations in Marine Macroalgae

Various mutagenic agents have been used in marine macroalgae (Figure 1E). Chemical agents, such as N-methyl-N'-nitro-N-nitrosoguanidine and ethyl-methane sulfonate (EMS) are efficient. EMS has been used to produce red macroalga *G. tikvahiae* mutants (van der Meer, 1979), which display altered branching patterns and thallus thickness, potentially due to the modification of the size of the medullary and sub-cortical cells of the algal thallus (Patwary and van der Meer, 1982). These morphological alterations are accompanied by a modification in the agar composition: agar strength is higher in the MP-40 mutant, but lower in the MP-61 mutant (Patwary and van der Meer, 1983a) (Figure 3). Interesting biological issues therefore lie in the potential functional link between the biophysical characteristics of agar and the morphological alterations observed in these mutants. The demonstration of this functional link requires adequate segregation analyses of these characters in the progeny of the mutants, which, unfortunately, were not carried out at that time. Nevertheless, and of interest for aquaculture purposes, the MP-40 mutant displays a faster growth rate than the wild-type even at high densities. It also fixes ammonium better from seawater and is more resistant to epiphytic colonization (Patwary and van der Meer, 1983b). In contrast to most generated mutations that were single locus and recessive, the MP-40 mutant displayed a partly dominant mutation (Patwary and van der Meer, 1982). Unavailable in the 1980s, modern technologies can now use this genetic feature to identify the mutated gene.

The use of chemical mutagens in liquid media represents a major drawback due to the risk of contamination during waste elimination. UV irradiation is safer and as efficient, and has been used on the green macroalga *U. mutabilis* and the brown macroalga *E. siliculosus*. Half-life curves related to the mutagen dose have been established (Le Bail and Charrier, 2013). Exposition of gametes to UV-irradiation doses corresponding to 50% survival produce only single-locus mutations in *Ectocarpus* (*étoile* mutant, Le Bail et al., 2011, *orobourous* mutant, Coelho et al., 2011; *mut* mutant, Billoud et al., 2015), making them amenable to relatively simple genetic analyses. A library of 60 morphological mutants has been constructed for *E. siliculosus*, which all show a stable phenotype after five rounds of sub-culturing or five parthenogenetic generations (Le Bail and Charrier, 2013). The observed morphological defaults cover all the developmental steps in this alga, ranging from alterations in cell differentiation and filament growth polarity, to positional and stage modification in the branching process (Figure 3). The transition to the reproductive phase is altered in some mutants. Interestingly, most mutants are affected in several of these developmental features, indicating that the function of the wild-type gene is not constrained to a single, precise developmental stage, but spans different developmental phases. This is consistent with the low level of architectural complexity of the *Ectocarpus* body, which is mainly composed of filaments (Le Bail et al., 2008).

Because these mutations were generated randomly and were not tagged, their identification requires forward genetic approaches and molecular tools, unavailable at the time. Today however, the advent of cost-effective and improved molecular and sequencing techniques can help tackle the genetic characterization of these mutants, and even identify the causal mutation, and the gene directly responsible for the morphological alteration. The recently described next-generation sequencing-based mapping approach developed for *E. siliculosus* (Billoud et al., 2015) should pave the way to the other macroalgal mutants generated from seaweeds with preferably small genome sequences.

CONCLUSION AND FUTURE PROSPECTS

Evidence that intensive cultivation of inbred seaweed lines results in both an impoverishment of wild biodiversity (possibly by gene swamping) and an outbreak of infections and epiphytism in culture plants (Loureiro et al., 2015), has raised awareness in aquaculture stakeholders and policy-makers that the inventory, the maintenance and the exploitation of seaweed genetic resources should be fully deployed (Pullin, 2006; Pullin and White, 2011; FAO, 2013). Hence, initiation and development of long-term seaweed genetic improvement programs are recommended, which include continuous selection processes targeting the maintenance of genetic diversity throughout the production of improved lines (Robinson et al., 2013). Macroalgal stocks with altered morphological traits can either be continuously collected from the wild or generated on demand in culture facilities using a combination of techniques likely to modify the developmental traits in a stable way. Although somatic hybridization frequently gives unpredictable results, generating hybrids between macroalgal partners displaying interesting and stable morphological features—obtained for example by induced mutagenesis—can combine both traits in a single organism. Using this approach, Mizukami et al. (1995) generated hybrids between a wild type and a mutant strain of *P. yezoensis*.

In addition to the interest of these techniques for aquaculture, production of morphologically altered macroalgae is a very valuable tool for studying the genetics of basic developmental mechanisms. In comparison with the wild-type organism, a mutant can provide information about cell types, specific developmental steps and gene expression patterns that play a role in the impaired developmental process, and may lead to the identification of the causal gene itself. In this regard, the development of genetic studies should be undertaken to address some biological issues that can only otherwise just be touched upon. Genetic transformation, of little use in the aquaculture sector (Robinson et al., 2013), can target specific cellular mechanisms by selectively modifying the expression of key genes, as it has been performed for decades in land plants. Genetic transformation of macroalgae is now possible in red and green seaweeds (Mikami, 2014; Wichard et al., 2015) and should be further developed, especially in brown algae, so that genetics and transgenesis can be used to their full potential in the field of macroalgal developmental biology.

Altogether, the current stage of technological knowledge now allows access to new and more powerful ways of generating morphologically altered macroalgae, using both traditional methods, such as the production of somaclonal variants (whose phenotype cannot be predicted and controlled), and newly developed gene-targeted methods, such as creating macroalgal variants with specific, precise and predictable phenotypes. The combined use of these methods will help bring the field of macroalgal development up to speed, to match that of land plants.

Conflict of Interest Statement

The authors declare that the research was conducted in the absence of any commercial or financial relationships that could be construed as a potential conflict of interest.

ACKNOWLEDGMENTS

CSIR, New Delhi, Department of Biotechnology, New Delhi and Department of Science and Technology, New Delhi are gratefully acknowledged for their long term funding support that led to advancement of Indian seaweed research.

REFERENCES

1. Ar Gall, E., Asensi, A., Marie, D., and Kloareg, B. (1996). Parthenogenesis and apospory in the *Laminariales*: a flow cytometry analysis. *Eur. J. Phycol.* 31, 369–380.
2. Baghel, R. S., Trivedi, N., Vishal, G., Neori, A., Reddy, C. R. K., Lali, A., et al. (2015). Biorefining of marine macroalgal biomass for production of biofuel and commodity chemicals. *Green Chem.* doi: 10.1039/C4GC02532F
3. Benet, H., Gall, A. E., Asensi, A., and Kloareg, B. (1997). Protoplast regeneration from gametophytes and sporophytes of some species in the order *Laminariales* (Phaeophyceae). *Protoplasma* 199, 39–48. doi: 10.1007/BF02539804
4. Billoud, B., Jouanno, É., Nehr, Z., Carton, B., Rolland, É., Chenivesse, S., et al. (2015). Localisation of causal locus in the genome of the brown macroalga Ectocarpus: NGS-based mapping and positional cloning approaches. *Front. Plant Sci.* 6:68. doi: 10.3389/fpls.2015.00068
5. Bryhni, E. (1973). Differences in cell-walls of *Ulva mutabilis* wild-type and its disorganized mutant *lumpy*. *J. Ultrastruct. Res.* 42, 412.
6. Bryhni, E. (1978). Quantitative differences between polysaccharide compositions in normal differentiated *Ulva mutabilis* and the undifferentiated mutant *lumpy*. *Phycologia* 17, 119–124. doi: 10.2216/i0031-8884-17-2-119.1
7. Buchholz, C. M., Krause, G., and Buck, B. H. (2012). “Seaweed and man,” in *Seaweed Biology: Novel Insights into Ecophysiology, Ecology, and Utilization. Ecological Studies*, Vol. 219, eds C. Wiencke and K. Bischof (Berlin; Heidelberg: Springer-Verlag), 471–493.
8. Chen, Y. C. (1998). Development of protoplasts from holdfasts and vegetative thalli of *Monostroma latissimum* (Chlorophyta, Monostromatacae) for algal seed stock. *J. Phycol.* 34, 1075–1081. doi: 10.1046/j.1529-8817.1998.341075.x
9. Chen, Y. C., and Shih, H. C. (2000). Development of protoplasts of *Ulva fasciata* (Chlorophyta) for algal seed stock. *J. Phycol.* 36, 608–615. doi: 10.1046/j.1529-8817.2000.99128.x

10. Cheney, D. (1990). “Genetic improvement of seaweeds through protoplasts fusion,” in *Economically Important Marine Plants of the Atlantic: their Biology and Cultivation*, eds C. Yarish, C. Penniman, and P. van Patten (Storrs, CT: Univ Conn Sea Grant Program), 15–25.
11. Cocking, E. C. (1983). The enzymatic isolation of plant protoplasts: historical perspective. *Int. Rev. Cytol.* 16, 1–3.
12. Coelho, S. M., Godfroy, O., Arun, A., Le Corguillé, G., Peters, A. F., and Cock, J. M. (2011). OUROBOROS is a master regulator of the gametophyte to sporophyte life cycle transition in the brown alga *Ectocarpus*. *Proc. Natl. Acad. Sci. U.S.A.* 108, 11518–11523. doi: 10.1073/pnas.1102274108
13. Davey, M. R., Anthony, P., Power, J. B., and Lowe, K. C. (2005). Plant protoplasts: status and biotechnological perspectives. *Biotechnol. Adv.* 23, 131–171. doi: 10.1016/j.biotechadv.2004.09.008
14. Dipakkore, S., Reddy, C. R. K., and Jha, B. (2005). Production and seeding of protoplast of *Porphyra okhaensis* (Bangials, Rhodophyta) in laboratory culture. *J. Appl. Phycol.* 17, 331–337. doi: 10.1007/s10811-005-7291-8
15. FAO. (2013). *CGRFA-14/13/Report- Report of the Fourteenth Regular Session of the Commission on Genetic Resources for Food and Agriculture Rome*, Italy, 15–19.
16. FAO. (2014). *The State of World Fisheries and Aquaculture: opportunities and challenges*. Rome.
17. Fjeld, A., and Løvlie, A. (1976). “Genetics of multicellular marine algae,” in *The Genetics of Algae Botanical Monographs*, eds R. A. Lewin (Oxford; London; Edinburgh; Melbourne, VIC: Blackwell Scientific Publication), 219–235.
18. Föyn, B. (1959). Geschlechtskontrollierte Vererbung bei der marinen Grünalga *Ulva mutabilis*. *Arch. Protistenkd.* 104, 236–253.
19. Föyn, B. (1961). Globose, a recessive mutant in *Ulva mutabilis*. *Bot. Mar.* 3, 60–64.
20. Fujita, Y., and Migita, S. (1985). Isolation and culture of protoplasts from seaweeds. *Bull. Fac. Fish. Nagasaki Univ.* 57, 39–45.
21. Fujita, Y., and Migita, S. (1987). Fusion of protoplasts from thalli

of two different color types in *Porphyra yezoensis* Ueda and development of fusion products. *Jpn. J. Phycol.* 35, 201–208.

22. Fujita, Y., and Saito, M. (1990). Protoplasts isolation and fusion in *Porphyra* (Bangiales, Rhodophyta). *Hydrobiologia* 204/205, 161–166. doi: 10.1007/BF00040228
23. Gupta, V., Bijo, A. J., Kumar, M., Reddy, C. R. K., and Jha, B. (2012). Detection of epigenetic variations in the protoplast-derived germlings of *Ulva reticulata* using methylation sensitive amplification polymorphism (MSAP). *Mar. Biotechnol.* 14, 692–700. doi: 10.1007/s10126-012-9434-7
24. Hodge, F. J., Buchanan, J., and Zuccarello, G. C. (2010). Hybridization between the endemic brown algae *Carpophyllum maschalocarpum* and *Carpophyllum angustifolium* (Fucales): genetic and morphological evidence. *Phycol. Res.* 58, 239–247. doi: 10.1111/j.1440-1835.2010.00583.x
25. Howell, S. H. (1998). *Molecular Genetics of Plant Development*. Cambridge, UK: Cambridge University Press.
26. Huang, X., Weber, J. C., Hinson, T. K., Mathieson, A. C., and Minocha, S. C. (1996). Transient expression of the GUS reporter gene in the protoplasts and partially digested cells of *Ulva lactuca L*. (Chlorophyta). *Bot. Mar.* 39, 467–474.
27. Jeon, M. R., and Choi, S. H. (2012). Quality characteristics of pork patties added with seaweed powder. *Korean J. Food Sci. Anim. Resour.* 32, 77–83. doi: 10.5851/kosfa.2012.32.1.71
28. Kaeppler, S. M., Kaeppler, H. F., and Rhee, Y. (2000). Epigenetic aspects of somaclonal variation in plants. *Plant Mol. Biol.* 43, 179–188. doi: 10.1023/A:1006423110134
29. Kapraun, D. F. (1989). Recovery and development of parasexual fusion products in *Enteromorpha linza* (L.) J Ag (Ulvales, Chlorophyta). *J. App. Phycol.* 1, 239–246. doi: 10.1007/BF00003649
30. Kapraun, D. F. (1990). Parasexual fusion products in green algae: *Enteromorpha* and *Ulvaria* (Ulvales, Chlorophyta). *Hydrobiologia*204/205, 151–159. doi: 10.1007/BF00040227
31. Kirk, D. L. (2003). Seeking the ultimate and proximate causes of volvox multicellularity and cellular differentiation. *Integr. Comp. Biol.* 43, 247–253. doi: 10.1093/icb/43.2.247

32. Kito, H., Kunimoto, M., Kamanishi, Y., and Mizukami, Y. (1998). Protoplasts fusion between *Monostroma nitidum* and *Porphyra yezoensis* and subsequent growth of hybrid plants. *J. Appl. Phycol.* 10, 15–21. doi: 10.1023/A:1008063415548

33. Krishna Kumar, G. R., Addepalli, M. K., and Reddy, C. R. K. (1999). Regeneration of the thallus of *Monostroma oxyspermum*(Chlorophyta) from protoplasts in axenic culture. *Phycologia* 38, 503–507. doi: 10.2216/i0031-8884-38-6-503.1

34. Larkin, P. J., and Scowcroft, W. R. (1981). Somaclonal variation—a novel source of variability from cell cultures for plant improvement. *Theor. Appl. Genet.* 60, 197–214.

35. Le Bail, A., and Charrier, B. (2013). "Culture methods and mutant generation in the filamentous brown algae *Ectocarpus siliculosus*," in *Plant Organogenesis Methods in Molecular Biology*, ed I. De Smet (Humana Press), 323–332. Available at: http://dx.doi.org/10.1007/978-1-62703-221-6_22.

36. Le Bail, A., Billoud, B., Le Panse, S., Chenivesse, S., and Charrier, B. (2011). ETOILE regulates developmental patterning in the filamentous brown alga *Ectocarpus siliculosus*. *Plant Cell* 23, 1666–1678. doi: 10.1105/tpc.110.081919

37. Le Bail, A., Billoud, B., Maisonneuve, C., Peters, A., Cock, J. M., and Charrier, B. (2008). Initial pattern of development of the brown alga *Ectocarpus siliculosus* (Ectocarpales, Phaeophyceae) sporophyte. *J. Phycol.* 44, 1269–1281. doi: 10.1111/j.1529-8817.2008.00582.x

38. Leliaert, F., Smith, D. R., Moreau, H., Herron, M. D., Verbruggen, H., Delwiche, C. F., et al. (2012). Phylogeny and molecular evolution of the green algae. *Crit. Rev. Plant Sci.* 31, 1–46. doi: 10.1080/07352689.2011.615705

39. Lewin, R. A. (1958). "Genetics and marine algae," in *Perspectives in Marine Biology*, eds L. A. Berkeley and A. A. Buzzati-Traverso (Berkeley, CA: University of California Press), 547–557.

40. Loureiro, R., Gachon, C. M. M., and Rebours, C. (2015). Seaweed cultivation: potential and challenges of crop domestication at an unprecedented pace. *New Phytol.* doi: 10.1111/nph.13278. [Epub ahead of print].

41. Løvlie, A. (1969). Cell size, nucleic acids and synthetic efficiency in the wild type and a growth mutant of the multicellular alga *Ulva mutabilis* Föyn. *Dev. Biol.* 20, 349–367. doi: 10.1016/0012-1606(69)90019-0
42. Løvlie, A. (1978). On the genetic control of cell cycles during morphogenesis in *Ulva mutabilis*. *Dev. Biol.* 64, 164–177. doi: 10.1016/0012-1606(78)90068-4
43. MacArtain, P., Gill, C. I., Brooks, M., Campbell, R., and Rowland, I. R. (2007). Nutritional value of edible seaweeds. *Nutr. Rev.* 65, 535–543. doi: 10.1111/j.1753-4887.2007.tb00278.x
44. Maggs, C. A., Verbruggen, H., and De Clerck, O. (2007). "Molecular systematics of red algae: building future structures on firm foundations," in *Unravalling the Algae: The Past, Present and Future of Algal Systematics*, eds J. Brodie and J. M. Lewis (Boca Raton, FL; London, UK; New York, NY: Taylor and Francis), 103–121.
45. Maluszynski, M., Ahloowalia, B. S., and Sigurbjörnsson, B. (1995). "Application of *in vivo* and *in vitro* mutation techniques for crop improvement," in *The Methodology of Plant Genetic Manipulation: Criteria for Decision Making*. Developments in Plant Breeding, eds A. C. Cassells and P. W. Jones (Dordrecht: Springer), 303–315.
46. Matsumura, W., Yasui, H., and Yamamoto, H. (2000). Mariculture of *Laminaria japonica* (Laminariales, Phaeophyceae) using protoplast regeneration. *Phycol. Res.* 48, 169–176. doi: 10.1111/j.1440-1835.2000.tb00213.x
47. Matsumura, W., Yasuj, H., and Yamamoto, H. (2001). Successful sporophyte regeneration from protoplasts of *Undaria pinnatifida*(laminariales, phaeophyceae). *Phycologia* 40, 10–20. doi: 10.2216/i0031-8884-40-1-10.1
48. Matsuo, Y., Imagawa, H., Nishizawa, M., and Shizuri, Y. (2005). Isolation of an algal morphogenesis inducer from a marine bacterium. *Science* 307, 1598. doi: 10.1126/science.1105486
49. Mikami, K. (2014). A technical breakthrough close at hand: feasible approaches toward establishing a gene-targeting genetic transformation system in seaweeds. *Front. Plant Sci.* 5:498. doi: 10.3389/fpls.2014.00498

50. Miller, S. M., and Kirk, D. L. (1999). *glsA*, a *Volvox* gene required for asymmetric division and germ cell specification, encodes a chaperone-like protein. *Dev. Camb. Engl.* 126, 649–658.
51. Mizukami, Y., Okauchi, M., Kito, H., Ishimoto, S., Ishida, T., and Fuseya, M. (1995). Culture and development of electrically fused protoplasts from red marine algae, *Porphyra yezoensis* and *P. suborbiculata*. *Aquaculture* 132, 361–367. doi: 10.1016/0044-8486(95)93662-T
52. Mouritsen, O. G., Dawczynski, C., Duelund, L., Jahreis, G., Vetter, W., and Schröder, M. (2013). On the human consumption of the red seaweed dulse (*Palmaria palmate* (L.) Weber & Mohr). *J. Appl. Phycol.* 25, 1777–1791. doi: 10.1007/s10811-013-0014-7
53. Niwa, K., Mizuta, A., and Aruga, Y. (2002). Genetic characterization of a spontaneous green-type pigmentation mutant of *Porphyra yezoensis* and the significance of using heterozygous conchocelis in nori farming. *Fish. Sci.* 68, 729–735. doi: 10.1046/j.1444-2906.2002.00486.x
54. Pappas, V., and Miller, S. M. (2009). Functional analysis of the *Volvox carteri* asymmetric division protein GlsA. *Mech. Dev.* 126, 842–851. doi: 10.1016/j.mod.2009.07.007
55. Patwary, M. U., and van der Meer, J. P. (1982). Genetics of *Gracilaria tikvahiae* (Rhodophyceae). VIII: phenotypic and genetic characterization of some selected morphological mutants. *Can. J. Bot.* 60, 2556–2564. doi: 10.1139/b82-310
56. Patwary, M. U., and van der Meer, J. P. (1983a). Genetics of *Gracilaria tikvahiae* (Rhodophyceae). IX: some properties of agars extracted from morphological mutants. *Bot. Mar.* 26, 295–299.
57. Patwary, M. U., and van der Meer, J. P. (1983b). Growth experiments on morphological mutants of *Gracilaria tikvahiae* (Rhodophyceae). *Can. J. Bot.* 61, 1654–1659.
58. Peters, A. F., Scornet, D., Ratin, M., Charrier, B., Monnier, A., Merrien, Y., et al. (2008). Life-cycle-generation-specific developmental processes are modified in the immediate upright mutant of the brown alga *Ectocarpus siliculosus*. *Dev. Camb. Engl.* 135, 1503–1512.
59. Polne-Fuller, M., Biniaminov, M., and Gibor, A. (1984). Vegitative

propagation of *Porphyra perforata*. *Proc. Int. Seaweed Symp.* 11, 308–313. doi: 10.1007/978-94-009-6560-7_60

60. Pullin, R. S. V. (2006). "Genetic resources for aquaculture: status and trends," in *Workshop on Status and Trends in Aquatic Genetic Resources. A Basis for International Policy*, eds D. M. Bartley, B. J. Harvey, and R. S. V. Pullin (FAO Fisheries Proceedings), 109–143.
61. Pullin, R., and White, P. (2011). *Climate Change and Aquatic Genetic Resources for Food and Agriculture: State of Knowledge, Risks and Opportunities. Background Study Paper No. 55*. Rome: FAO Commission on Genetic Resources for Food and Agriculture.
62. Reddy, C. R. K., Gupta, M. K., Mantri, V. A., and Jha, B. (2008). Seaweed protoplasts: status, biotechnological perspectives and needs. *J. Appl. Phycol.* 20, 619–632. doi: 10.1007/s10811-007-9237-9
63. Reddy, C. R. K., Iima, M., and Fujita, Y. (1992). Induction of fast growing and morphologically different strains through intergeneric protoplasts fusion of *Ulva* and *Enteromorpha* (Ulvales, chlorophyta). *J. Appl. Phycol.* 4, 57–65. doi: 10.1007/BF00003961
64. Reddy, C. R. K., Migita, S., and Fujita, Y. (1989). Protoplasts isolation and regeneration of three species of *Ulva* in axenic culture. *Bot. Mar.* 32, 483–490. doi: 10.1515/botm.1989.32.5.483
65. Robinson, N., Winberg, P., and Kirkendale, L. (2013). Genetic improvement of macroalgae: status to date and needs for the future. *J. Appl. Phycol.* 25, 703–716. doi: 10.1007/s10811-012-9950-x
66. Rusig, A., and Cosson, J. (2001). Plant regeneration from protoplasts of *Enteromorpha intestinalis* (Chlorophyta, Ulvophyceae) as seed stock for macroalgal culture. *J. Appl. Phycol.* 13, 103–108. doi: 10.1023/A:1011136712938
67. Sawabe, T., and Ezura, Y. (1996). Regeneration from *Laminaria japonica* Areschoug (Laminariales, Phaeophyceae) protoplasts isolated with bacterial alginase. *Plant Cell Rep.* 15, 892–895.
68. Sawabe, T., Ezura, Y., and Yamamoto, H. (1997). Plant regeneration from protoplasts of *Laminaria japonica* Areschoug (Laminariales, Phaeophyceae) in a continuous-flow culture system. *Plant Cell Rep.* 17, 109–112.

69. Senthil, A., Mamatha, B. S., and Mahadevaswamy, M. (2005). Effect of using seaweed (*Euchema*) powder on the quality of fish cutlet. *J. Food Sci. Nutr.* 56, 327–335. doi: 10.1080/09637480500224205

70. Sessoms, A. H., and Huskey, R. J. (1973). Genetic control of development in *Volvox*: isolation and characterization of morphogenetic mutants. *Proc. Natl. Acad. Sci. U.S.A.* 70, 1335–1338. doi: 10.1073/pnas.70.5.1335

71. Shan, T. S., Pang, S. J., and Gao, S. Q. (2013). Novel means for variety breeding and sporeling production in the brown seaweed *Undaria pinnatifida* (Phaeophyceae): crossing female gametophytes from parthenosporophytes with male gametophyte clones. *Phycol. Res.* 61, 154–161. doi: 10.1111/pre.12014

72. Spoerner, M., Wichard, T., Bachhuber, T., Stratmann, J., and Kendal, W. S. (2012). *Ulva mutabilis* (Chlorophyta) depends on a combination of two bacterial species excreting regulatory factors. *J. Phycol.* 48, 1433–1447. doi: 10.1111/j.1529-8817.2012.01231.x

73. Starr, R. C. (1969). Structure, reproduction and differentiation in *Volvox carteri* f. nagariensis Iyengar, HK9 and 10. *Arch. Protistenkd.* 11, 204–222.

74. Tatarenkov, A., Bergstrom, L., Jonsson, R. B., Serrao, E. A., Kautsky, L., and Johannesson, K. (2005). Intriguing asexual life in marginal populations of the brown seaweed *Fucus vesiculosus*. *Mol. Ecol.* 14, 647–651. doi: 10.1111/j.1365-294X.2005.02425.x

75. Ueki, N., and Nishii, I. (2008). Idaten is a new cold-inducible transposon of *Volvox carteri* that can be used for tagging developmentally important genes. *Genetics* 180, 1343–1353. doi: 10.1534/genetics.108.094672

76. Ueki, N., and Nishii, I. (2009). Controlled enlargement of the glycoprotein vesicle surrounding a *Volvox* embryo requires the InvB nucleotide-sugar transporter and is required for normal morphogenesis. *Plant Cell* 21, 1166–1181. doi: 10.1105/tpc.109.066159

77. van der Meer, J. P. (1979). Genetics of *Gracilaria* sp. *(Rhodophyceae, Gigartinales). V. Isolation and characterization of mutant strains. Phycologia* 18, 47–54. doi: 10.2216/i0031-8884-18-1-47.1

78. Waaland, J. R., Dickson, L. G., and Watson, B. A. (1990). Protoplasts isolation and regeneration in the marine red alga *Porphyra nereocystis*. *Planta* 181, 522–528. doi: 10.1007/BF00193005

79. Wichard, T., Charrier, B., Mineur, F., Bothwell, J. H., De Clerck, O., and Coates, J. C. (2015). The green seaweed Ulva: a model system to study morphogenesis. *Front. Plant Sci.* 6:72. doi: 10.3389/fpls.2015.00072

CHAPTER

9

TRANSCRIPTOMIC ANALYSIS OF THE RED SEAWEED LAURENCIA DENDROIDEA (FLORIDEOPHYCEAE, RHODOPHYTA) AND ITS MICROBIOME

Louisi Souza de Oliveira[1], Gustavo Bueno Gregoracci[1], Genivaldo Gueiros Zacarias Silva[2], Leonardo Tavares Salgado[3], Gilberto Amado Filho[3], Marcio Alves-Ferreira[4], Renato Crespo Pereira[5], and Fabiano L Thompson[1]

[1]Departamento de Biologia Marinha, Instituto de Biologia, Universidade Federal do Rio de Janeiro (UFRJ) Av. Carlos Chagas Filho, 373-CCS - IB - BLOCO A (ANEXO) A3- 202, Rio de Janeiro 21941-599, Brazil.

[2]Laboratório de Bioinformática e Biologia Evolutiva, Universidade Federal de Pernambuco. Av. Prof. Moraes Rego 1235, Cidade Universitária, Recife 50670-901PE, Brazil.

[3]Instituto de Pesquisa Jardim Botânico do Rio de Janeiro, Rua Pacheco Leão, 915. Jardim Botânico, Rio de Janeiro 22460-030RJ, Brazil.

4Departamento de Genética. Instituto de Biologia. Av. Prof. Rodolpho Paulo Rocco, s/n, CCS, Sala A2-93, Universidade Federal do Rio de Janeiro (UFRJ), Rio de Janeiro 21941-599RJ, Brazil.

[5]Departamento de Biologia Marinha, Universidade Federal Fluminense (UFF). Morro do Valonguinho, s/n. Centro, Niteroi 24001-970RJ, Brazil.

ABSTRACT

Background

Seaweeds of the *Laurencia* genus have a broad geographic distribution and are largely recognized as important sources of secondary metabolites, mainly halogenated compounds exhibiting diverse potential pharmacological activities and relevant ecological role as anti-epibiosis. Host-microbe interaction is a driving force for co-evolution in the marine environment, but molecular studies of seaweed-associated microbial communities are still rare. Despite the large amount of research describing the chemical compositions of *Laurencia* species, the genetic knowledge regarding this genus is currently restricted to taxonomic markers and general genome features. In this work we analyze the transcriptomic profile of *L. dendroidea* J. Agardh, unveil the genes involved on the biosynthesis of terpenoid compounds in this seaweed and explore the interactions between this host and its associated microbiome.

Results

A total of 6 transcriptomes were obtained from specimens of *L. dendroidea* sampled in three different coastal locations of the Rio de Janeiro state. Functional annotations revealed predominantly basic cellular metabolic pathways. Bacteria was the dominant active group in the microbiome of *L. dendroidea*, standing out nitrogen fixing Cyanobacteria and aerobic heterotrophic Proteobacteria. The analysis of the relative contribution of each domain highlighted bacterial features related to glycolysis, lipid and polysaccharide breakdown, and also recognition of seaweed surface and establishment of biofilm. Eukaryotic transcripts, on the other hand, were associated with photosynthesis, synthesis of carbohydrate reserves, and defense mechanisms, including the biosynthesis of terpenoids through the mevalonate-independent pathway.

Conclusions

This work describes the first transcriptomic profile of the red seaweed *L. dendroidea*, increasing the knowledge about ESTs from the Florideophyceae algal class. Our data suggest an important role for *L. dendroidea* in the primary production of the holobiont and the role of

Bacteria as consumers of organic matter and possibly also as nitrogen source. Furthermore, this seaweed expressed sequences related to terpene biosynthesis, including the complete mevalonate-independent pathway, which offers new possibilities for biotechnological applications using secondary metabolites from *L. dendroidea*.

Keywords: Red seaweed, Terpene, Bacteria, Holobiont, Metabolic pathway, EST

BACKGROUND

Laurencia dendroidea is a red seaweed species widespread in the Atlantic Ocean, whose type locality is in Brazil. It is found from the intertidal to the subtidal zone at 3m depth. The thalli are erect, forming dense tufts 4–20 cm high, brown-purple or violet-greenish in color [1]. The genus *Laurencia* [2] was recognized, since the first studies on natural products in the 1960s, as an important source of secondary metabolites, mainly halogenated compounds [3, 4]. The secondary metabolites of *Laurencia* play a relevant ecological role as chemical defenses against bacterial colonization and infection [5, 6, 7].

Seaweeds are especially susceptible to microbial colonization due to the biosynthesis and release of large amounts of organic compounds, which may serve as chemo-attractants and nutrient source for microbes [8]. In this context, secondary metabolites and exudates may act together selecting the microbial community associated with the surfaces and tissues of seaweeds [9, 10]. Host-microbe interaction is widely recognized as one of the main driving forces for co-evolution in the marine environment, leading to the establishment of beneficial microbiomes. For instance, microbes associated with seaweed tissues may possess the ability to fix nitrogen, mineralize the organic substrates and also supply the seaweeds with carbon dioxide and growth factors [11, 12, 13, 14]. The microbiome on seaweeds tends to be species-specific and different from the surrounding seawater [15]. However, the characterization of the microbial community living at the surface of macroalgae is still limited and the molecular studies of these communities are rare [15, 16, 17].

The untapped diversity of the secondary metabolites of *Laurencia*, particularly terpenes, has attracted considerable attention of different research groups worldwide. The pharmacological potential of these compounds comprises the strong antibiotic [18, 19], antiviral [20], antimalarial [21], antitrypanosomal [22], antileishmanial [23], anti-inflammatory [24] and anti-carcinoma [25, 26, 27] activities. A major secondary metabolite of *L. dendroidea* is the sesquiterpene (C15) (-)-elatol, a substance that has a high biocidal and anti-epibiosis activity and could be used for the preparation of antifouling paints, or for the development of antimicrobials [28, 29, 30]. A first attempt for the commercial application of (-)-elatol resulted in the filing of the patent in Brazil to use this compound as an antifouling agent. However, technological developments are still needed to ensure its commercial viability [31]. This obstacle stems from the low yield of the extraction process, the complexity of the organic total synthesis of (-)-elatol in laboratory [32], and the failure of the large-scale cultivation of this species. A possible alternative to circumvent this problem is the synthesis of (-)-elatol in the laboratory using genetically modified organisms [31]. The cellular location and the environmental factors that induce the production of this compound by *L. dendroidea* are known [33, 34], but the genes involved in the biosynthesis of this compound were not yet determined, representing a new research frontier in the technological use of (-)-elatol. Recent studies have determined some of the genes responsible for the biosynthesis of terpenes (i.e. cyclases or synthases) in bacteria [35], fungi [36], and plants [37]. The sequence homology observed among at least some classes of terpene synthases from these organisms [38] may facilitate the search for homolog genes in *L. dendroidea*.

Despite the large number of studies based on the chemical composition of *Laurencia* species, the genetic knowledge regarding this genus is currently restricted to taxonomic markers [39, 40]. The genome size of *L. dendroidea* is estimated to be about 833 Mbp, based on a study of another species of the same genus [41], but gene sequences from this species have not previously been described. In this work we analyze the transcriptomic profile of *L. dendroidea*at different geographic locations, unveil the genes involved on the biosynthesis of terpenoid compounds in this seaweed and also explore the interactions between the alga and the associated microbiome.

METHODS

Specimens Collection

Specimens of *L. dendroidea* were randomly collected in the intertidal zone during high tide at Azedinha (22°44’28.76”S, 41°52’55.70”W) and Forno beaches (22°45’42.72”S, 41°52’29.81”W), both in Búzios, and at Ibicuí beach (22°57’45.02”S, 41°01’29.05”W) located in Mangaratiba, all these places on the coast of the Rio de Janeiro state, Brazil (Figure 1). Seaweeds were collected from nearly the same depth in two subsequent days, at approximately the same hour, with the same climatic characteristics to minimize the variation in abiotic factors. The collected thalli were rapidly cleaned of macroscopic epiphytes using tweezers, without damage to the host seaweed, and the samples were immediately frozen in liquid nitrogen, to better preserve the holobiont.

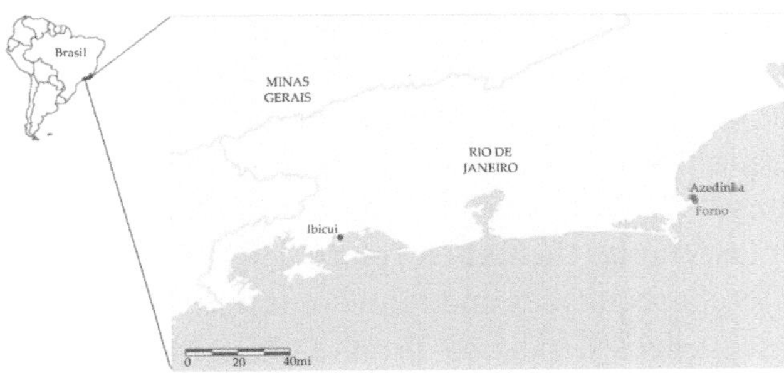

Figure 1. Collection sites of specimens of *L.dendroidea*in Búzios and Mangaratiba, on the coast of the Rio de Janeiro state, Brazil. Scale bar presented in miles (mi).

RNA Extraction, Reverse-Transcription and Pyrosequencing

Two specimens of. *L. dendroidea* from each location were separately ground in liquid nitrogen using a mortar and pestle to obtain a fine powder. Then, 100 mg of powder from each sample was suspended in 1 mL of extraction buffer (6.5 M guanidinium hydrochloride, 100 mM Tris-HCl pH 8.0, 0.1 M sodium acetate pH 5.5, 0.1 M β-mercaptoethanol, 0.2 M KOAc). Total RNA was extracted following the method previously proposed for another red seaweed [42], but we performed an extra centrifugation step and transferred the supernatant phase before adding the chloroform, which improved the RNA quality. In order to eliminate DNA residues, all the samples were treated with DNAse (RNAse free, PROMEGA, Madison, USA). The double-stranded cDNAs (ds cDNAs) were synthesized and amplified using the SMARTer cDNA synthesis kit and the Advantage2 polymerase (Clontech, Califórnia, USA) starting from 1 μg of total RNA. The optimal number of amplification cycles was determined to be 23. This amplification did not exclude the prokaryotic portion of the holobiont, allowing the study of the microbiome along with the host. The PCR amplification products were purified using the NucleoSpin® Extract II kit (Macherey-Nagel, Düren, Alemanha). Finally the ds cDNAs were eluted in TE buffer (10 mM Tris-HCl pH 7.6; 1 mM EDTA) and sequenced using 454 pyrosequencing technology [43].

Transcriptome Analysis

The sequences from each sample were preprocessed using the software Prinseq [44] to trim poly-A/T tails at least 20 bp long and to remove reads shorter than 75 bp, and then assembled into contigs using the Roche's algorithm Newbler (minimum overlap length=40 bp, minimum overlap identity=95%). In our analysis we annotated both contigs and singlets after assembly (hereafter referred as transcripts), since they contained different sequences and relevant information. We downloaded all the EST sequences deposited for the class Florideophyceae in the NCBI (comprising 11 species) and assembled the reads using the TGICL software from TIGR [45]. Afterwards, the assembly of all sequences derived from *L. dendroidea* was aligned against the Florideophyceae EST NCBI database using the Promer alignment tool (MUMmer 3.0) using the 'maxmatch' parameter [46]. The results were parsed using the show-coords script with - k and - r parameters and only reciprocal matches were considered for calculations. Sequences annotated as Bacteria were treated separately in this analysis, but eventual micro-eukaryotic sequences could not be removed, since the database is not complete regarding eukaryotic marine life and no *Laurencia* sequences aside from taxonomic markers are available.

Taxonomic and functional analysis were performed on assembled sequences from all samples, using the Newbler software, and automatically annotated, using the MG-RAST server, through BLAST, against the GenBank, COG, KEGG and Subsystems databases with maximum e-value cutoff of 10^{-5}[47]. The sequences obtained in this project are publicly available in the MG-RAST database and were organized in a file for each sample, named according to the site of origin, and a file containing the assembler of all reads (http://metagenomics.anl.gov/linkin.cgi?project=1274). To characterize the major phenotypic features of the microbial community associated with *L. dendroidea*, features of bacterial genera identified against Genbank (through MG-Rast) were manually annotated using the Bergey's manuals of Systematic Bacteriology (2nd ed.). Additionally, we explored the relative contributions of Bacteria and Eukarya to the functional profile. Sequences annotated against the Genbank corresponding to these domains were extracted using the Workbench tool from MG-RAST server, and re-annotated against functional hierarchies (COG, Subsystems). The functional profiles of the domains were compared using the Statistical Analysis of Metagenomic Profiles (STAMP) bioinformatics software v2.0 [48]. Statistical significance ($p < 0.05$) was calculated pairwise using two-sided G-test (with Yates' correction) and Fisher's exact test, and the confidence intervals for each proportion were calculated using the asymptotic method with a continuity correction considering the threshold of 95%. Furthermore, a specific search for two profiles using hidden markov models was performed, through the HMMER 3.0 software [49]. The first HMM profile was based on the alignment of all vanadium bromoperoxidases deposited in the protein database of NCBI, and the second one, based on the universal metal-binding domain of terpene synthases (PF03936), was obtained from PFAM as previously described [35].

RESULTS

A total of 6 transcriptomes (235,572 reads, 52 Mbp) were obtained for specimens of the seaweed *L. dendroidea* originated from three different locations in the Rio de Janeiro coast. The assembly of the sequences from each replicate resulted on 500–1,000 contigs and 10,000–16,000 singlets (see Table 1 for detailed information). The COG functional annotation and the GenBank taxonomic annotation indicated that the transcriptomic profile of *L. dendroidea* was highly similar among the samples (Additional files 1 and 2). Since no significant differences were observed, all the reads of the 6 transcriptomes were assembled in order to represent a transcriptomic profile for this species, resulting on 3,887 contigs and 38,010 singlets. A total of 30,585 tentative unigenes (73% of the transcripts) were identified as genes coding for proteins with unknown function, indicating the need for further molecular studies in order to unravel the function of a large portion of the transcriptome of this seaweed. The closest red algal genus with sequences deposited in the database is *Griffithsia*, classified in the order Ceramiales, for which we found only 1,277 ESTs, most of them (99.76%) derived from *Griffithsia okiensis* [50]. Searching at a higher taxonomic level, the total number of ESTs from the class Florideophyceae deposited in NCBI was 37,198, comprising 21,475 unigenes, from which only 5.95% matched with 3.34% unigenes from this study (Figure 2). These numbers include the sequences of Bacteria associated with the *Laurencia* holobiont (1.94%), from which 0.3% matched with 1.39% of the sequences in the Florideophyceae database, indicating that the reference database itself contains bacterial sequences. Excluding those bacterial sequences from our analysis, 3.04% of the remaining sequences are left matching 4.56% of sequences from the Florideophyceae database (Figure 2). Therefore, 95.02% of the sequences provided by this work could potentially enrich our current knowledge regarding Florideophyceae as they represent unknown genes.

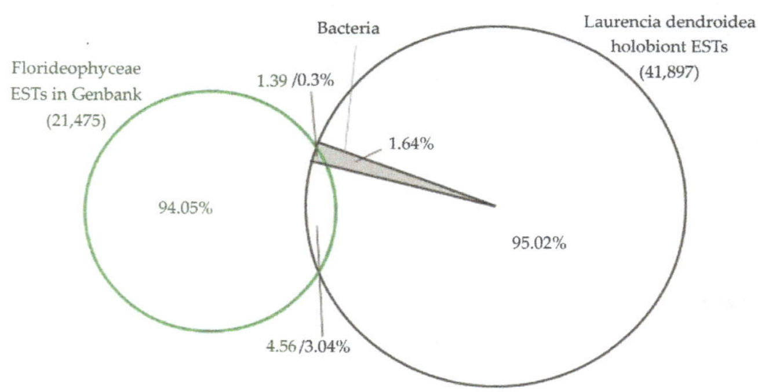

Figure 2. MUMMER-based identification of shared sequences between this study and the dbEST for the class Florideophyceae. The shaded area corresponds to sequences annotated as bacteria in this study.

Table 1. Characteristics of the sequencing and assembly of the cDNA libraries from the*Laurencia dendroidea*holobiont

Location	**Azedinha1**	**Azedinha2**	**Forno1**	**Forno2**	**Ibicuí1**	**Ibicuí2**
Total Nucleotides (basepairs)	11,635,249	9,384,269	11,049,671	7,101,334	5,550,607	8,011,563
N. of Sequences	51,592	42,577	49,001	31,434	24,423	36,545
N. of Contigs	1,079	926	985	556	586	683
Avg. Size of Contigs	492.24 ± 190.19	489.62 ± 195.59	481.88 ± 195.80	466.06 ± 182.92	465.71 ± 164.32	487.58 ± 193.41
N. of Singlets	15,755	14,480	14,830	10,935	10,522	11,719
Avg. Size of Singlets	202.17 ± 78.19	198.52 ± 74.80	198.01 ± 77.76	197.09 ± 76.42	212.72 ± 80.09	195.50 ± 75.30

Major Groups of Transcripts of *L. Dendroidea*

The functional classification of the ESTs revealed that most of the transcripts were related to the basal metabolism of the *Laurencia* holobiont (Figures 3 and 4). The most represented COG categories were associated to Translation, Ribosomal Structure and Biogenesis (18.65%), Posttranslational Modification, Protein Turnover and Chaperones (14.90%), and Amino acid Transport and Metabolism (7.57%). Additionally, functions associated with Energy Production and Conversion were relatively common (7.37%). Moreover, the sequences related to Replication, Recombination and Repair (7.37%), and the ESTs involved in Carbohydrate Transport and Metabolism (5.42%) were among the most represented categories in the transcriptome of *L. dendroidea* (Figure 3). The Subsystems annotation corroborated further the general expression profile of *Laurencia*. The main recognized features are Protein Metabolism (19.20%) and Carbohydrates (13.11%). Transcripts related to Cofactors, Vitamins, Prosthetic Groups, Pigments (8.88%), Amino Acids and Derivatives (8.77%) and RNA Metabolism (8.71%) were also numerous (Figure 4).

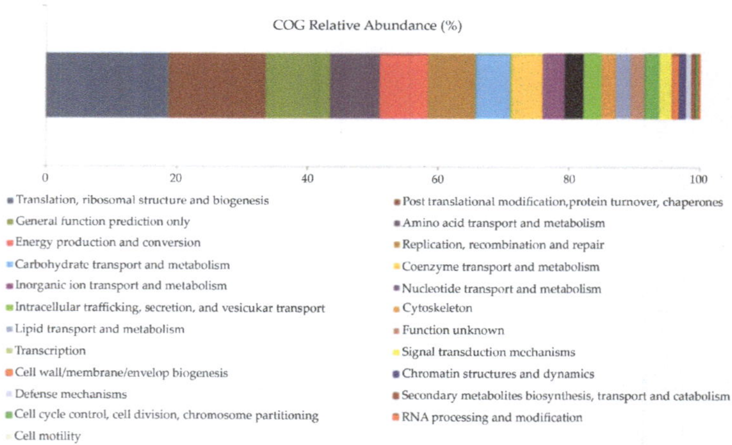

Figure 3. COG functional profile overview of the transcriptome of*L.dendroidea.*

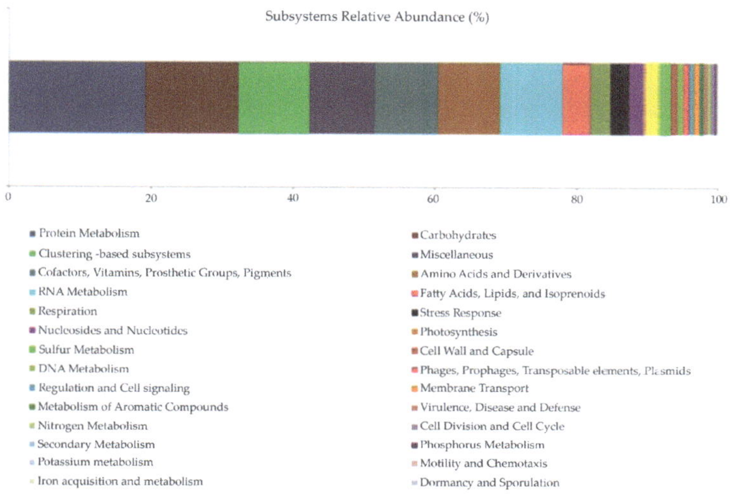

Figure 4. Subsystems functional profile overview of the transcriptome of*L.dendroidea.*

Transcriptome of *L. Dendroidea*-Associated Microbiome

The functional analysis of the transcriptome revealed bacterial genes that are important for surface colonization, such as the transcripts related to flagellum (0.11% of the total), CheY-like receiver domain (0.04% of the total), and S-adenosylmethionine synthetase (0.03% of the total). Indeed, we detected fewer sequences involved in Motility and Chemotaxis (0.11% of the total) in comparison with the ones related to Capsular and extracellular polysaccharides (0.53% of the total).

A total of 6,154 reads (14.69% of the total) were assigned to taxonomic categories using the GenBank database. Among them, 17.26% were classified in the domain Bacteria (Figure 5a). The most abundant bacterial transcripts were assigned to the phylum Cyanobacteria (35.97%), mainly to the orders Chroococcales, Oscillatoriales and Nostocales. The second most represented phylum is Proteobacteria (32.86%) with Gammaproteobacteria and Alphaproteobacteria as the dominant classes (Figure 5b).

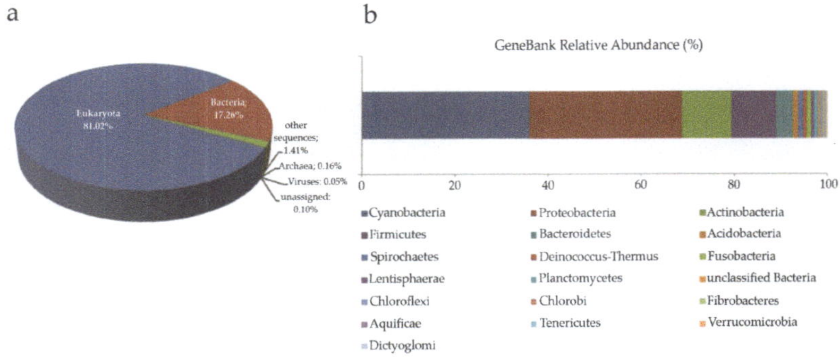

Figure 5. Taxonomic classification for the transcriptome of*L.dendroidea.*(a) Taxonomy overview. (b) Relative abundance of bacterial phyla.

Manual annotation revealed the majority of the bacterial transcripts (to which a description of respiratory metabolism could be found in Bergey's manuals) as ascribed to aerobic (62.30%) or aerotolerant groups (14.00%). We also verified a higher abundance of transcripts related to respiration (2.96%) in comparison with the ones involved in the fermentative metabolism (0.64%). Furthermore, Bacteria expressed genes, such as Superoxide dismutase (0.51%), Glutaredoxins (0.42%), Alkyl hydroperoxide reductase (0.21%), and the chaperones GroEL (3.17%), DnaJ (1.37%) and DnaK (0.84%), related to protection from reactive oxygen species produced during aerobic metabolism (Additional file 3).

Genes involved in Photosynthesis (3.18%) and in the biosynthesis of starch (0.66%) were more abundant in eukaryotes, while ESTs related to Carbohydrate (5.63%) and Lipid Transport and Metabolism (3.58%), and to Energy Production and Conversion (11.38%) were more represented in Bacteria. Transcripts associated to Amino acid metabolism (11.50%) were also more represented in Bacteria, except for the glutamate biosynthesis that was preferentially expressed by Eukarya (0.58%, Additional file 3).

Additionally, several transcripts were attributed to bacterial genera known to be heterotrophs (> 51.9%) or motile (> 28.4%). Furthermore, 25.4% of the heterotroph-associated transcripts belong to genera recognized as pathogens or closely associated to eukaryotes. Along this context, the Hmmer search for vanadium-dependent bromoperoxidases, which could be involved in response to infection, resulted on 10 hits, and their functional classification was confirmed by Blastx.

Terpenoid Biosynthesis in the Holobiont

Within the functional annotations, 34 transcripts associated to the terpenoid backbone biosynthesis in *L. dendroidea* were found, representing all the required enzymes involved in the mevalonate-independent pathway (Table 2, Figure 6). The identified genes participate in important steps for the synthesis of dimethylallyl diphosphate (EC 2.2.1.7; EC: 1.1.1.267; EC: 2.7.7.60; EC: 2.7.1.148; EC 4.6.1.12; EC: 1.17.7.1; EC: 1.17.1.2), its isomerization to isopentenyl diphosphate (EC: 5.3.3.2), and the condensation of these two C5-units, through the action of prenyltransferases, generating geranyl diphosphate (GDP, EC: 2.5.1.1), farnesyl diphosphate (FDP, EC: 2.5.1.10), and geranylgeranyl diphosphate (GGDP, EC: 2.5.1.29). We also found genes involved in the subsequent steps to the synthesis of chlorophylls (EC: 1.3.1.83), plastoquinone, phylloquinone, ubiquinone (EC: 2.5.1.84, EC: 2.5.1.85, EC: 2.5.1.91) and N-glycans, (EC: 2.5.1.87). The Hmmer search for the metal binding conserved domain (PF03936) in the transcriptome of *L. dendroidea* resulted on 3 hits, and the subsequent manual annotation confirmed their classification as terpene synthases.

Table 2. Description of the enzymes involved on terpenoid backbone biosynthesis

Enzyme codes	Enzyme names	Databases
EC 2.2.1.7	1-deoxy-D-xylulose-5-phosphate synthase.	SEED
EC: 1.1.1.267	1-deoxy-D-xylulose-5-phosphate reductoisomerase.	KEGG/SEED
EC: 2.7.7.60	2-C-methyl-D-erythritol 4-phosphate cytidylyltransferase	KEGG
EC: 2.7.1.148	4-(cytidine 5'-diphospho)-2-C-methyl-D-erythritol kinase.	KEGG/SEED
EC 4.6.1.12	2-C-methyl-D-erythritol 2,4-cyclodiphosphate synthase	SEED
EC: 1.17.7.1	(E)-4-hydroxy-3-methylbut-2-enyl-diphosphate synthase	KEGG/SEED
EC: 1.17.1.2	4-hydroxy-3-methylbut-2-enyl diphosphate reductase	KEGG/SEED
EC: 5.3.3.2	Isopentenyl-diphosphate Delta-isomerase.	KEGG/SEED
EC: 2.5.1.1	Dimethylallyltranstransferase.	KEGG/SEED
EC: 2.5.1.10	(2E,6E)-farnesyl diphosphate synthase.	KEGG/SEED

EC: 2.5.1.29	Geranylgeranyl diphosphate synthase	KEGG/SEED
EC: 2.5.1.87	Ditrans,polycis-polyprenyl diphosphate synthase ((2E,6E)-farnesyl diphosphate specific)	KEGG
EC: 1.3.1.83	Geranylgeranyl diphosphate reductase.	KEGG/SEED
EC: 2.5.1.85	All-trans-nonaprenyl-diphosphate synthase (geranylgeranyl-diphosphate specific)	KEGG
EC: 2.5.1.84	All-trans-nonaprenyl-diphosphate synthase (geranyl-diphosphate specific)	KEGG
EC: 2.5.1.91	All-trans-decaprenyl-diphosphate synthase.	KEGG

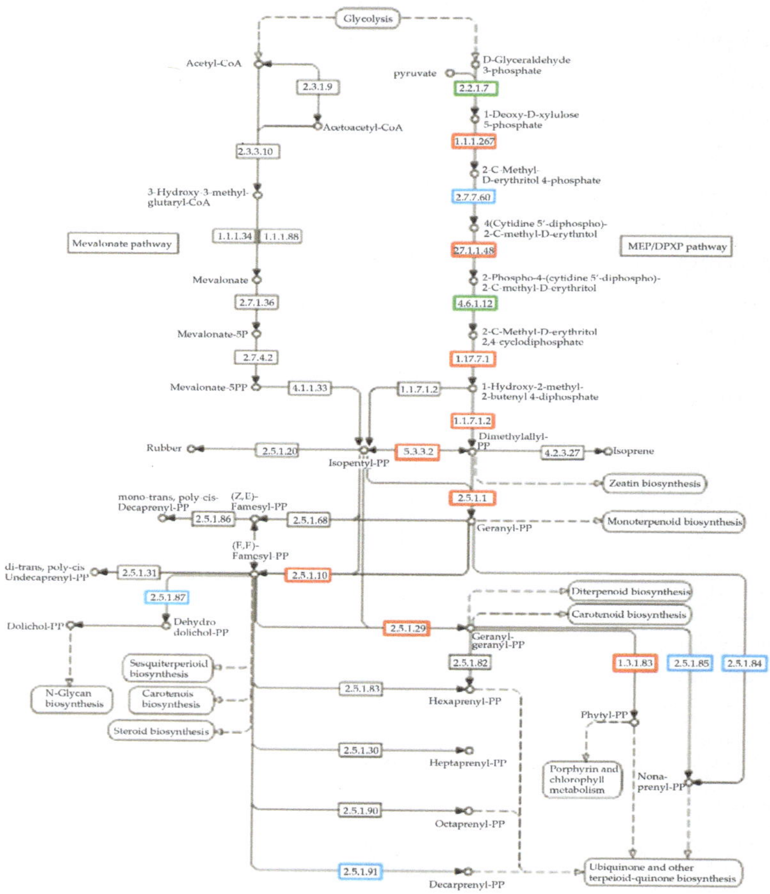

Figure 6. Terpenoid backbone biosynthetic pathway. Blue squares represent the genes identified through the KEGG database, green squares points the genes identified using the SEED database and red squares highlight the genes identified using both databases.

DISCUSSION

The present study provides the largest transcriptome dataset for the class Florideophyceae and represents the first transcriptomic characterization of the seaweed *Laurencia dendroidea*. The presented numbers could be an overestimate of the contribution of *L. dendroidea* to the Florideophyceae database, since we worked with complex samples. Nevertheless, at least some of the sequencing projects in the Florideophyceae dbEST are also based on non-axenic field samples [51, 52], hampering the achievement of a more accurate estimate. Indeed, it is notable the presence of sequences deposited in this database that matched our bacterial sequences.

Recent advances in the field of algal genomics included only the complete sequencing of the nuclear genome of the microalgae *Cyanidioschyzon merolae* [53], *Ostreococcus tauri* [54], *Chlamydomonas reinhardtii* [55], and *Cyanophora paradoxa* [56] and the brown macroalga *Ectocarpus siliculosus* [57]. Moreover, EST projects have provided valuable information in the transcriptomic profile of some species of Rhodophyta [50, 51, 58, 59, 60, 61, 62] in the phylogenetic relationships among photosynthetic eukaryotes [63, 64] and have also unveiled genes involved in stress response [52, 65, 66] and in life phase differentiation [67, 68, 69, 70].

The transcriptomic profile of *L. dendroidea* and its corresponding associated microbiome was closely similar among all the samples, regardless of their place of origin. Likewise, a previous study verified a higher similarity between bacterial populations from seaweeds of the same species sampled at different sites than between those from different species growing at the same habitat, emphasizing the specificity of this association [71]. Our data reinforces these findings as we observed a high similarity in the taxonomic composition of the active microbiome associated with *L. dendroidea* in different sample sites.

Major Groups of Transcripts of *L. Dendroidea*

The functional annotation of the transcripts revealed predominantly basic cellular metabolic pathways. In general, functions related to translation and protein synthesis, from amino acid precursors to post-translational modifications are the most abundantly expressed in the transcriptome of *L. dendroidea*. Besides, complete pathways for energy production were well represented, mainly related to the pyruvate dehydrogenase complex, electron transfer, thioredoxins, citric acid cycle and NADH dehydrogenase. The ESTs involved in carbohydrate transport and metabolism (mainly glycolysis, starch and sucrose metabolism, and pentose phosphate pathway), Cofactors, Vitamins, Prosthetic Groups, Pigments (including Folate and Pterines, Tetrapyrroles and Pyridoxine), RNA Metabolism (mainly RNA Processing and Modification) were among the most represented categories in the transcriptome of *L. dendroidea*. Other relevant features in this transcriptome are related to DNA replication, recombination and repair, which are important to the survival and growth of the seaweed, especially in the rocky-shore coastal environment where the organisms are subject to high UVB levels that causes serious damages to DNA [72]. The ability to resist to UV-exposure influences the vertical distribution of seaweeds [73], and *L. dendroidea* typically grows in the lower midlittoral zone where UV-damage repair may be necessary. The same set of expressed sequences relevant in the transcriptome of *L. dendroidea* are among the most represented in the EST databases of *Gracilaria gracilis*[62], *G. changii*[51], *G. tenuistipitata*[60], *Porphyra yezoensis*[61, 67], *P. haitanensis*[59], *Eucheuma denticulatum*[74], *Furcellaria lumbricalis*[52], and *Kappaphycus alvarezii*[66], possibly indicating a general pattern of expression in red seaweeds.

Transcriptome of *L. Dendroidea*-Associated Microbiome

Seaweeds are especially susceptible to epibiosis because they inhabit environments with strong competition for space [75], and release large amounts of organic compounds that induce the microbial colonization [76], but the interaction between seaweeds and their microbiomes is little known to the molecular level.

The functional analysis of the holobiont transcriptome revealed the expression of bacterial genes involved on cell motility and chemotaxis, for example the ESTs related to flagellum and CheY-like receiver domain which are important, respectively, for the recognition of the surface of the seaweed and the establishment of the biofilm [77, 78]. However, the relatively low abundance of these transcripts in comparison with the ones involved in extracellular polysaccharide synthesis suggests a mature biofilm with some level of detachment, possibly of dispersal cells [79]. Transcripts coding for the enzyme S-adenosylmethionine synthetase, which participates in the synthesis of quorum sensing autoinducers, were also detected [80]. Quorum sensing (QS) is a bacterial cell to cell communication mechanism based on the release and perception of signaling molecules such as oligopeptides, N-acyl homoserine lactones (AHL) and autoinducers that allow bacteria to monitor their own population density and to coordinate swarming, biofilm formation, stress resistance, and biosynthesis of toxins and secondary metabolites [81], and it exhibits an important role in the interactions between bacteria and their eukaryotic hosts. Several red seaweeds are able to inhibit bacterial QS signaling, such as *Delisea pulchra*[82] and *Ahnfeltiopsis flabelliformis*[83], and a small inhibitory activity against QS signaling was previously detected in the ethyl acetate extract from a *Laurencia* sp. [84].

The taxonomic analysis of the transcriptome showed Bacteria as the dominant active group in the microbiome of *L. dendroidea*, with Cyanobacteria and Proteobacteria as the most represented bacterial phyla. These groups were also verified as predominant in the evaluation of the microbial diversity associated with four functional groups of seaweeds through metagenomics [17].

Among the cyanobacterial transcripts associated with the thalli of *L. dendroidea*, the Chroococcales, Oscillatoriales and Nostocales were the dominant orders, all of them comprising nitrogen fixing species. In a previous study, Phlips and Zeman [85] reported the occurrence and the nitrogen fixing activity of epiphytic forms of *Oscillatoria* associated to *Sargassum*thalli. Nitrogen can be the limiting nutrient in coastal ecosystems [86] and under this situation, nitrogen fixing cyanobacteria may be favored and gain in growth and reproductive success. In fact, Hoffman [87] pointed that despite their important contribution to benthic primary production, the main role of Cyanobacteria in the tropical marine ecosystems appears to be as nitrogen fixers. However, no sequences related to nitrogen fixation were observed in our data. This is expected since our data clearly indicates an oxygenic environment, and the nitrogenase expression is inhibited by oxygen [88]. Our samples, collected near the peak of photosynthetic activity (right before midday) should have a very low expression of this nitrogenase [89]. In fact, the most abundant cyanobacteria genus were *Synechococcus* and *Cyanothece*, which together with *Lyngbya* and *Synechocystis* were previously reported to rely on temporal separation between photosynthesis and nitrogen fixation, the last occurring mainly at night [90, 91]. Further studies on the diel variation of the transcriptome profile could verify this hypothesis.

Analyzing the functional relative contribution of specific domains, we noticed a higher involvement of Bacteria in the Amino acid metabolism, except for the biosynthesis of glutamate, more represented in eukaryotes. Such situation was reported for *Rhizobium*nodules, where plants provide glutamate and a carbon source and in turn the nitrogen fixing Bacteria provide ammonium and amino acids such as alanine and aspartate for asparagine biosynthesis in the plant cytosol [92]. Although specialized mechanisms like nodules are not known in red algae, our data suggests a similar interaction between the seaweed and the associated microbiome, involving the exchange of nitrogen compounds.

Proteobacteria was the second largest active group with assigned sequences mostly to the classes Gammaproteobacteria and Alphaproteobacteria. The higher abundance of these classes was previously reported for the surface microbiome of the macroalgae *Ulva australis*[93] and *Laminaria hyperborean*[94], through denaturing gradient gel electrophoresis (DGGE) analysis. Predominantly heterotrophs, these groups would be opportunists, exploring an oxic productive environment [95]. The high prevalence of aerobic and aerotolerant groups reflects a photosynthesizing environment, also noted by Barott *et al.*[17]. The predominance of respiration over fermentative metabolism in the holobiont transcriptomic profile reinforces these findings. The aerobic metabolism generates reactive oxygen species (ROS) [96] that can damage DNA, lipids, and proteins [97]. In order to cope with oxygen toxicity and grow in aerobic conditions, Bacteria expressed genes correlated to oxidative stress, such as Superoxide dismutase, Glutaredoxins and Alkyl hydroperoxide reductase [98], and also stress related chaperones such as GroEL, DnaJ and DnaK [99, 100].

Transcripts associated to photosynthesis and to the biosynthesis of carbohydrate reserves, such as starch, were more represented in eukaryotes, which indicate an important role of *L. dendroidea* in the primary production of the holobiont, generating carbon in excess to its immediate demand. The typical starch from Rhodophyta is called floridean starch and it shows structural similarities with starch granules from higher plants except for the lack of amylose in most of the species

[101]. On the other hand the Bacteria contributed more to Carbohydrate and Lipid Transport and Metabolism, and to Energy Production and Conversion, standing out genes related to glycolysis and also to lipid and polysaccharide breakdown, reinforcing the role of Bacteria as consumers of organic matter in this holobiont [102].

Despite the beneficial or neutral interaction processes depicted here between *L. dendroidea*and its microbiome, some bacteria may also offer threats to the health and survival of seaweeds in their natural environment [103]. As such, defense mechanisms, such as the aforementioned secondary compounds of *L. dendroidea* [18], may have been evolutionarily selected. The expression of vanadium-dependent bromoperoxidases, involved on the halogenation and cyclization of terpenes in Rhodophyta [104], was detected in the transcriptomic profile of *L. dendroidea*. Additionally the previously reported increase on the bromination activity of red algae in response to infection signals, such as agar oligosaccharide [105], indicates an important role of this enzyme in the chemical defense of Rhodophyta.

Terpenoid Biosynthesis in the Holobiont

The biosynthesis of terpenoid backbones provides precursors for the biosynthesis of diverse compounds that display relevant roles in plant and algal physiology [106]. The identified genes are involved in important steps for the biosynthesis of the building blocks dimethylallyl diphosphate, isopentenyl diphosphate and the higher-order building blocks geranyl diphosphate, farnesyl diphosphate and geranylgeranyl diphosphate, which are the precursors of monoterpenoids (C10), sesquiterpenoids (C15), and diterpenoids (C20), respectively [107]. The subsequent addition of isoprene units leads to the biosynthesis of sterols (isoprenoids with a C30 backbone) which are components of cell membranes; carotenoids (C40) and chlorophylls (with a C20 isoprenoid side-chain) that act as photosynthetic pigments; and plastoquinone, phylloquinone and ubiquinone (with long isoprenoid side-chains) that participate in electron transport systems for respiration or photosynthesis [106]. Terpenoid backbones are also required for the biosynthesis of N-glycans, important components for the proper folding of proteins in eukaryotic cells [108]. The biosynthesis of isopentenyl pyrophosphate (IPP) and dimethylallyl pyrophosphate (DMAPP), the central intermediates in the biosynthesis of isoprenoids, occur through two different pathways in plants, one dependent (MVA) and other independent of mevalonate (DOXP/MEP). The mevalonate (MVA) pathway, located in the cytosol, is responsible for the production of sterols, triterpenes and some sesquiterpenes [109]. The MVA-independent pathway operates in plastids and provides the precursors to monoterpenes, diterpenes, certain sesquiterpenes, carotenoids and the side chains of chlorophyll and plastoquinone [110]. This division between isoprenoids derived from plastids and cytoplasm was also observed in red algae [111, 112]. Despite the occurrence of both biosynthetic routes in Rhodophyta, this study found only transcripts associated with the mevalonate-independent pathway. Furthermore, three transcripts were identified containing the terpene synthase family metal-binding domain [35], representing new possible targets for further functional clarification. Phylogenetic reconstruction based on genes of terpene synthases was attempted, using the fragments (50–310 amino acids) we obtained from our whole transcriptome strategy (data not shown). However, it is difficult to infer a phylogenetic relationship among taxonomic groups using the gene fragments of this pathway because, in nearly all cases, the bootstrap support for the branches is low when homologous sequences were available for analysis. Nevertheless, it is notable that in most cases, the sequences from *L. dendroidea* holobiont and other red algae cluster together with a relatively high bootstrap support.

These findings associated to the reconstruction of a complete pathway for the biosynthesis of terpenoid backbones in *L. dendroidea* are important steps to enable the heterologous biosynthesis of terpenes of interest, such as (-)-elatol, in genetically modified organisms. The molecular engineering of *Escherichia coli* and *Saccharomyces cerevisiae* has recently allowed the use of these microorganisms as cell factories to synthesize plant terpenes such as the antimalarial drug artemisinin [113, 114], opening up new avenues for the scalable biosynthesis of terpenoid compounds. Our research provides a comparative basis for prospecting more specific terpene synthases genes for (-)-elatol and other commercially relevant terpenes, which could be explored in cell factories. This could be accomplished through the use of high producing strains of *L. dendroidea* under favorable conditions.

CONCLUSIONS

Our work describes the first transcriptomic profile of the red seaweed *L. dendroidea*, increasing the knowledge of ESTs from the Florideophyceae class. Basic cellular metabolic functions were the most represented in this profile, as observed in other seaweeds. The associated microbial transcriptome was independent of the location of collect, and the holobiont transcriptome indicated interesting interactions such as biofilm formation, the possible exchange of nitrogen compounds between bacteria and eukaryotes, the role of *L. dendroidea* in photosynthesis and of bacteria as consumers of excess carbon, and the bacterial molecular strategies to cope with the oxidative stress generated during aerobic metabolism. In addition, seaweeds defense mechanisms were also suggested with the disclosure of a complete mevalonate-independent pathway. The present study is a first contribution to the transcriptomic analysis of *L. dendroidea*, and opens up new avenues for biotechnological applications using this seaweed.

ACKNOWLEDGMENTS

We thank the financial support of CAPES, CNPq, and FAPERJ.

AUTHORS' CONTRIBUTIONS

LSO carried out the samples collection, and RNA extraction, participated in the bioinformatic analysis and drafted the manuscript. GBG participated in the bioinformatic analysis and in the discussions and draft of the manuscript. GGZS carried out the bioinformatic analysis and participated in the discussion of the results. LTS participated in the sample collection, the discussion of the results, and the acquisition of funding. GAF participated in the acquisition of funding, the work planning and the discussion of the results, MAF participated in RNA extraction, EST library construction and discussion of the results. RCP participated in the acquisition of funding, work planning, discussion of the results, and draft of the manuscript. FLT participated in the acquisition of funding, work planning, data interpretation and draft of the manuscript. All authors read and approved the final manuscript.

REFERENCES

1. Agger S, Lopez-Gallego F, Schmidt-Dannert C: Diversity of sesquiterpene synthases in the basidiomycete *Coprinus cinereus*. Mol Microbiol. 2009, 72 (5): 1181-1195. 10.1111/j.1365-2958.2009.06717.x.
2. Asamizu E, Nakajima M, Kitade Y, Saga N, Nakamura Y, Tabata S: Comparison of RNA expression profiles between the two generations of Porphyra yezoensis (Rhodophyta), based on expressed sequence tag frequency analysis. J Phycol. 2003, 39 (5): 923-930. 10.1046/j.1529-8817.2003.03003.x.
3. Aspilla PS, Antonio A, Zuccarello GC, Rojas NRL: A partial expressed sequence tag (EST) library of the economically important red alga Eucheuma denticulatum (NL Burham) FC Collins and Hervey. Phil Sci Lett. 2010, 3: 109-120.
4. Barott KL, Rodriguez Brito B, Janouškovec J, Marhaver K, Smith JE, Keeling P, Rohwer FL: Microbial diversity associated with four functional groups of benthic reef algae and the reef building coral Montastraea annularis. Environ Microbiol. 2011, 13 (5): 1192-1204. 10.1111/j.1462-2920.2010.02419.x.
5. Bengtsson MM, Sjøtun K, Øvreås L: Seasonal dynamics of bacterial biofilms on the kelp Laminaria hyperborea. Aq Microb Ecol. 2010, 60 (1): 71-83. 10.3354/ame01409.
6. Berman-Frank I, Chen Y-B, Gerchman Y, Dismukes G, Falkowski P: Inhibition of nitrogenase by oxygen in marine cyanobacteria controls the global nitrogen and oxygen cycles. Biogeoscs Discuss. 2005, 2: 261-273. 10.5194/bgd-2-261-2005.
7. Berman-Frank I, Lundgren P, Falkowski P: Nitrogen fixation and photosynthetic oxygen evolution in cyanobacteria. Res Microbiol. 2003, 154 (3): 157-164. 10.1016/S0923-2508(03)00029-9.
8. Bischof K, Gomez I, Molis M, Hanelt D, Karsten U, Lüder U, Roleda MY, Zacher K, Wiencke C: Ultraviolet radiation shapes seaweed communities. Life in Extreme Environments. 2007, 5: 187-212.
9. Burke C, Steinberg P, Rusch D, Kjelleberg S, Thomas T: Bacterial community assembly based on functional genes rather than species. Proc Natl Acad Sci. 2011, 108 (34): 14288-14293. 10.1073/pnas.1101591108.

10. Cane DE, Ikeda H: Exploration and mining of the bacterial terpenome. Accounts Chem Res. 2011, 45 (3): 463-472.

11. Carter-Franklin JN, Parrish JD, Tschirret-Guth RA, Little RD, Butler A: Vanadium haloperoxidase-catalyzed bromination and cyclization of terpenes. J Am Chem Soc. 2003, 125 (13): 3688-3689. 10.1021/ja029271v.

12. Cassano V, Metti Y, Millar AJK, Gil-Rodríguez MC, Sentíes A, Díaz-Larrea J, Oliveira MC, Fujii MT: Redefining the taxonomic status of Laurencia dendroidea (Ceramiales, Rhodophyta) from Brazil and the Canary Islands. Eur J Phycol. 2012, 47 (1): 67-81. 10.1080/09670262.2011.647334.

13. Chan CX, Yang EC, Banerjee T, Yoon HS, Martone PT, Estevez JM, Bhattacharya D: Red and green algal monophyly and extensive gene sharing found in a rich repertoire of red algal genes. Curr Biol. 2011, 21 (4): 328-333. 10.1016/j.cub.2011.01.037.

14. Chan CX, Zauner S, Wheeler G, Grossman AR, Prochnik SE, Blouin NA, Zhuang Y, Benning C, Berg GM, Yarish C, et al: Analysis of Porphyra membrane transporters demonstrates gene transfer among photosynthetic eukaryotes and numerous sodium-coupled transport systems. Plant Physiol. 2012, 158 (4): 2001-2012. 10.1104/pp.112.193896.

15. Chatter R, Othman RB, Rabhi S, Kladi M, Tarhouni S, Vagias C, Roussis V, Guizani-Tabbane L, Kharrat R: In vivo and in vitro anti-inflammatory activity of neorogioltriol, a new diterpene extracted from the red algae Laurencia glandulifera. Mar Drugs. 2011, 9 (7): 1293-1306. 10.3390/md9071293.

16. Clean coastal waters: understanding and reducing the effects of nutrient pollution. Edited by: Howarth RW, Anderson DM, Church TM, Greening H, Hopkinson CS, Huber WC, Marcus N, Naiman RJ, Segerson K, Sharpley AN. 2000, National Academy Press, Washington, DC

17. Cock JM, Sterck L, Rouzé P, Scornet D, Allen AE, Amoutzias G, Anthouard V, Artiguenave F, Aury JM, Badger JH: The Ectocarpus genome and the independent evolution of multicellularity in brown algae. Nature. 2010, 465 (7298): 617-621. 10.1038/nature09016.

18. Cole JJ: Interactions between bacteria and algae in aquatic

ecosystems. Annu Rev Ecol Syst. 1982, 13: 291-314. 10.1146/annurev.es.13.110182.001451.

19. Collén J, Roeder V, Rousvoal S, Collin O, Kloareg B, Boyen C: An expressed sequence tag analysis of thallus and regenerating protoplasts of Chondrus crispus (Gigartinales, Rhodophyceae). J Phycol. 2006, 42 (1): 104-112. 10.1111/j.1529-8817.2006.00171.x.
20. Collén PN, Collén J, da Silva Reis M, Pedersén M, Setubal JC, Varani AM, Colepicolo P, Oliveira MC: Analysis of expressed sequence tags from the agarophyte Gracilaria tenuistipitata (Rhodophyta). J Appl Phycol. 2012, 24 (4): 641-647. 10.1007/s10811-011-9681-4.
21. Cottrell MT, Kirchman DL: Natural assemblages of marine proteobacteria and members of the Cytophaga-Flavobacter cluster consuming low- and high-molecular-weight dissolved organic matter. Appl Environ Microbiol. 2000, 66 (4): 1692-1697. 10.1128/AEM.66.4.1692-1697.2000.
22. Croft MT, Lawrence AD, Raux-Deery E, Warren MJ, Smith AG: Algae acquire vitamin B12 through a symbiotic relationship with bacteria. Nature. 2005, 438 (7064): 90-93. 10.1038/nature04056.
23. Da Gama BAP, Pereira RC, Carvalho AGV, Coutinho R: The effects of seaweed secondary metabolites on biofouling. Biofouling. 2002, 18 (1): 13-20. 10.1080/08927010290017680.
24. Da Gama BAP, Pereira RC, Soares AR, Teixeira VL, Yoneshigue-Valentin Y: Is the mussel test a good indicator of antifouling activity? a comparison between laboratory and field assays. Biofouling. 2003, 19 (SUPPL): 161-169.
25. Degenhardt J, Kollner TG, Gershenzon J: Monoterpene and sesquiterpene synthases and the origin of terpene skeletal diversity in plants. Phytochemistry. 2009, 70: 1621-1637. 10.1016/j.phytochem.2009.07.030.
26. Derelle E, Ferraz C, Rombauts S, Rouzé P, Worden AZ, Robbens S, Partensky F, Degroeve S, Echeynié S, Cooke R: Genome analysis of the smallest free-living eukaryote Ostreococcus tauri unveils many unique features. Proc Natl Acad Sci. 2006, 103 (31): 11647-11652. 10.1073/pnas.0604795103.
27. Dobretsov S, Teplitski M, Paul V: Mini-review: quorum sensing

in the marine environment and its relationship to biofouling. Biofouling. 2009, 25 (5): 413-427. 10.1080/08927010902853516.

28. Falcão VR, Tonon AP, Oliveira MC, Colepicolo P: RNA Isolation method for polysaccharide rich algae: agar producing Gracilaria tenuistipitata (Rhodophyta). J Appl Phycol. 2008, 20 (1): 9-12. 10.1007/s10811-007-9174-7.

29. Farr SB, Kogoma T: Oxidative stress responses in Escherichia coli and Salmonella typhimurium. Microbiol Rev. 1991, 55 (4): 561-585.

30. Fredriksson A, Ballesteros M, Dukan S, Nystrom T: Defense against protein carbonylation by DnaK/DnaJ and proteases of the heat shock regulon. J Bacteriol. 2005, 187 (12): 4207-4213. 10.1128/JB.187.12.4207-4213.2005.

31. Gil-Rodríguez M, Sentíes A, Díaz-Larrea J, Cassano V, Fujii MT: Laurencia marilzae sp. nov. (Ceramiales, Rhodophyta) from the Canary Islands, Spain, based on morphological and molecular evidence. J Phycol. 2009, 45 (1): 264-271. 10.1111/j.1529-8817.2008.00624.x.

32. Goecke F, Labes A, Wiese J, Imhoff JF: Chemical interactions between marine macroalgae and bacteria. Mar Ecol Prog Ser. 2010, 409: 267-300.

33. Guiry MD, Guiry GM: AlgaeBase. 2010, National University of Ireland, Galway,http://wwwalgaebaseorg,

34. Haas AF, Nelson CE, Kelly LW, Carlson CA, Rohwer F, Leichter JJ, Wyatt A, Smith JE: Effects of coral reef benthic primary producers on dissolved organic carbon and microbial activity. PLoS One. 2011, 6 (11): e27973-10.1371/journal.pone.0027973.

35. Hanzelka BL, Greenberg E: Quorum sensing in Vibrio fischeri: evidence that S-adenosylmethionine is the amino acid substrate for autoinducer synthesis. J Bacteriol. 1996, 178 (17): 5291-

36. HMMER Biosequence analysis using profile hidden Markov models.http://hmmer.org/,

37. Hoffman L: Marine cyanobacteria in tropical regions: diversity and ecology. Eur J Phycol. 1999, 34 (4): 371-379.

38. Huang TC, Lin RF, Chu MK, Chen HM: Organization and expression

of nitrogen-fixation genes in the aerobic nitrogen-fixing unicellular cyanobacterium Synechococcus sp. strain RF-1. Microbiology. 1999, 145 (Pt 3): 743-753.

39. Huovinen P, Gómez I, Lovengreen C: A five year study of solar ultraviolet radiation in Southern Chile (39°S): Potential impact on physiology of coastal marine algae?. Photochem Photobiol. 2006, 82 (2): 515-522. 10.1562/2005-07-05-RA-601.

40. Ignea C, Cvetkovic I, Loupassaki S, Kefalas P, Johnson CB, Kampranis SC, Makris AM: Improving yeast strains using recyclable integration cassettes, for the production of plant terpenoids. Microb Cell Fact. 2011, 10: 4-10.1186/1475-2859-10-4.

41. Irie T, Suzuki M, Kurosawa E, Masamune T: Laurinterol, debromolaurinterol and isolaurinterol, constituents of Laurencia intermedia Yamada. Tetrahedron. 1970, 26 (13): 3271-3277. 10.1016/S0040-4020(01)92906-0.

42. Irie T, Suzuki M, Masamune T: Laurencin, a constituent from Laurencia species. Tetrahedron Lett. 1965, 6: 1091-1099. 10.1016/S0040-4039(00)90038-8.

43. Kamiya M, Kawai H, Moon D, Goff LJ: Isolation and characterization of phase-specific cDNAs from carposporophytes of Gracilariopsis andersonii (Gracilariales, Rhodophyta). Eur J Phycol. 2011, 46 (1): 27-35. 10.1080/09670262.2010.548101.

44. Kapraun DF: Nuclear DNA content estimates in multicellular green, red and brown algae: phylogenetic considerations. Ann Bot. 2005, 95 (1): 7-44. 10.1093/aob/mci002.

45. Kim J, Kim Y, Seo Y, Park S: Quorum sensing inhibitors from the red alga, Ahnfeltiopsis flabelliformis. Biotechnol Bioproc E. 2007, 12 (3): 308-311. 10.1007/BF02931109.

46. Kim MM, Mendis E, Kim SK: Laurencia okamurai extract containing laurinterol induces apoptosis in melanoma cells. J Med Food. 2008, 11 (2): 260-266. 10.1089/jmf.2007.575.

47. Kirkland JB: Lipid peroxidation, protein thiol oxidation and DNA damage in hydrogen peroxide-induced injury to endothelial cells: role of activation of poly(ADP-ribose)polymerase. Biochim Biophys Acta. 1991, 1092 (3): 319-325. 10.1016/S0167-4889(97)90007-0.

48. Kladi M, Xenaki H, Vagias C, Papazafiri P, Roussis V: New

cytotoxic sesquiterpenes from the red algae Laurencia obtusa and Laurencia microcladia. Tetrahedron. 2006, 62 (1): 182-189. 10.1016/j.tet.2005.09.113.

49. Kostamo K, Olsson S, Korpelainen H: Search for stress-responsive genes in the red alga Furcellaria lumbricalis (Rhodophyta) by expressed sequence tag analysis. J Exp Mar Biol Ecol. 2011, 404 (1–2): 21-25.

50. Kurtz S, Phillippy A, Delcher AL, Smoot M, Shumway M, Antonescu C, Salzberg SL: Versatile and open software for comparing large genomes. Genome Biol. 2004, 5 (2): R12-10.1186/gb-2004-5-2-r12.

51. Kuzuyama T: Mevalonate and nonmevalonate pathways for the biosynthesis of isoprene units. Biosci Biotechnol Biochem. 2002, 66 (8): 1619-1627. 10.1271/bbb.66.1619.

52. Lachnit T, Blumel M, Imhoff JF, Wahl M: Specific epibacterial communities on macroalgae: phylogeny matters more than habitat. Aquat Biol. 2009, 5 (2): 181-186.

53. Largo DB, Fukami K, Nishijima T: Occasional pathogenic bacteria promoting ice-ice disease in the carrageenan-producing red algae Kappaphycus alvarezii and Eucheuma denticulatum (Solieriaceae, Gigartinales, Rhodophyta). J Appl Phycol. 1995, 7 (6): 545-554. 10.1007/BF00003941.

54. Laule O, Fürholz A, Chang HS, Zhu T, Wang X, Heifetz PB, Gruissem W, Lange M: Crosstalk between cytosolic and plastidial pathways of isoprenoid biosynthesis in Arabidopsis thaliana. Proc Natl Acad Sci USA. 2003, 100 (11): 6866-10.1073/pnas.1031755100.

55. Lee H, Lee HK, An G, Lee YK: Analysis of expressed sequence tags from the red alga Griffithsia okiensis. J Microbiol (Seoul). 2007, 45 (6): 541-

56. Lhullier C, Falkenberg M, Ioannou E, Quesada A, Papazafiri P, Antunes Horta P, Paulo Schenkel E, Vagias C, Roussis V: Cytotoxic halogenated metabolites from the Brazilian red alga Laurencia catarinensis. J Nat Prod. 2010, 73 (1): 27-32. 10.1021/np900627r.

57. Liu C, Wang X, Huang X, Liu J: Identification of hypo-osmotically induced genes in Kappaphycus alvarezii (Solieriaceae, Rhodophyta) through expressed sequence tag analysis. Bot Mar. 2011, 54 (6): 557-562.

58. Lluisma AO, Ragan MA: Expressed sequence tags (ESTs) from the marine red alga Gracilaria gracilis. J Appl Phycol. 1997, 9 (3): 287-293. 10.1023/A:1007920607900.
59. Lodwig EM, Hosie AH, Bourdes A, Findlay K, Allaway D, Karunakaran R, Downie JA, Poole PS: Amino-acid cycling drives nitrogen fixation in the legume-Rhizobium symbiosis. Nature. 2003, 422 (6933): 722-726. 10.1038/nature01527.
60. Manefield M, de Nys R, Naresh K, Roger R, Givskov M, Peter S, Kjelleberg S: Evidence that halogenated furanones from Delisea pulchra inhibit acylated homoserine lactone (AHL)-mediated gene expression by displacing the AHL signal from its receptor protein. Microbiology. 1999, 145 (2): 283-291. 10.1099/13500872-145-2-283.
61. Margulies M, Egholm M, Altman WE, Attiya S, Bader JS, Bemben LA, Berka J, Braverman MS, Chen YJ, Chen Z, et al: Genome sequencing in microfabricated high-density picolitre reactors. Nature. 2005, 437 (7057): 376-380.
62. Martin VJ, Pitera DJ, Withers ST, Newman JD, Keasling JD: Engineering a mevalonate pathway in Escherichia coli for production of terpenoids. Nat Biotechnol. 2003, 21 (7): 796-802. 10.1038/nbt833.
63. Martin-Lescanne J, Rousseau F, De Reviers B, Payri C, Couloux A, Cruaud C, Le Gall L: Phylogenetic analyses of the Laurencia complex (Rhodomelaceae, Ceramiales) support recognition of five genera: Chondrophycus, Laurencia, Osmundea Palisada and Yuzurua stat. nov. Eur J Phycol. 2010, 45 (1): 51-61. 10.1080/09670260903314292.
64. Matsuzaki M, Misumi O, Shin-i T, Maruyama S, Takahara M, Miyagishima S, Mori T, Nishida K, Yagisawa F, Nishida K: Genome sequence of the ultrasmall unicellular red alga Cyanidioschyzon merolae 10D. Nature. 2004, 428 (6983): 653-657. 10.1038/nature02398.
65. McCracken D, Cain J: Amylose in floridean starch. New Phytol. 1981, 88 (1): 67-71.
66. McDougald D, Rice SA, Barraud N, Steinberg PD, Kjelleberg S: Should we stay or should we go: mechanisms and ecological

consequences for biofilm dispersal. Nat Rev Microbiol. 2011, 10 (1): 39-50.

67. McGarvey DJ, Croteau R: Terpenoid metabolism. Plant Cell. 1995, 7 (7): 1015-

68. Merchant SS, Prochnik SE, Vallon O, Harris EH, Karpowicz SJ, Witman GB, Terry A, Salamov A, Fritz-Laylin LK, Maréchal-Drouard L: The Chlamydomonas genome reveals the evolution of key animal and plant functions. Science. 2007, 318 (5848): 245-251. 10.1126/science.1143609.

69. Messner KR, Imlay JA: The identification of primary sites of superoxide and hydrogen peroxide formation in the aerobic respiratory chain and sulfite reductase complex of Escherichia coli. J Biol Chem. 1999, 274 (15): 10119-10128. 10.1074/jbc.274.15.10119.

70. Meyer F, Paarmann D, D›Souza M, Olson R, Glass EM, Kubal M, Paczian T, Rodriguez A, Stevens R, Wilke A, et al: The metagenomics RAST server - a public resource for the automatic phylogenetic and functional analysis of metagenomes. BMC Bioinformatics. 2008, 9: 386-10.1186/1471-2105-9-386.

71. Miller TR, Hnilicka K, Dziedzic A, Desplats P, Belas R: Chemotaxis of Silicibacter sp. strain TM1040 toward dinoflagellate products. Appl Environ Microbiol. 2004, 70 (8): 4692-10.1128/AEM.70.8.4692-4701.2004.

72. Morrone D, Chambers J, Lowry L, Kim G, Anterola A, Bender K, Peters R: Gibberellin biosynthesis in bacteria: separate ent-copalyl diphosphate and ent-kaurene synthases in *Bradyrhizobium japonicum*. FEBS Lett. 2009, 583 (2): 475-480. 10.1016/j.febslet.2008.12.052.

73. Nikaido I, Asamizu E, Nakajima M, Nakamura Y, Saga N, Tabata S: Generation of 10,154 expressed sequence tags from a leafy gametophyte of a marine red alga, Porphyra yezoensis. DNA Res. 2000, 7 (3): 223-227. 10.1093/dnares/7.3.223.

74. O'Toole GA, Kolter R: Flagellar and twitching motility are necessary for Pseudomonas aeruginosa biofilm development. Mol Microbiol. 1998, 30 (2): 295-304. 10.1046/j.1365-2958.1998.01062.x.

75. Parks DH, Beiko RG: Identifying biologically relevant differences

between metagenomic communities. Bioinformatics. 2010, 26 (6): 715-721. 10.1093/bioinformatics/btq041.

76. Pereira RC, Costa-Lotufo LV: Bioprospecting of bioactives from seaweeds: potential, obstacles and alternatives. Braz J Pharmacog. 2012, 22 (4): 894-905.
77. Pereira RC, Da Gama BA, Teixeira VL, Yoneshigue-Valentin Y: Ecological roles of natural products of the Brazilian red seaweed Laurencia obtusa. Braz J Biol. 2003, 63 (4): 665-672.
78. Persson F, Svensson R, Nylund GM, Fredriksson NJ, Pavia H, Hermansson M: Ecological role of a seaweed secondary metabolite for a colonizing bacterial community. Biofouling. 2011, 27 (6): 579-588. 10.1080/08927014.2011.589001.
79. Pertea G, Huang X, Liang F, Antonescu V, Sultana R, Karamycheva S, Lee Y, White J, Cheung F, Parvizi B: TIGR Gene indices clustering tools (TGICL): a software system for fast clustering of large EST datasets. Bioinformatics. 2003, 19 (5): 651-652. 10.1093/bioinformatics/btg034.
80. Phlips E, Zeman C: Photosynthesis, growth and nitrogen fixation by epiphytic forms of filamentous cyanobacteria from pelagic Sargassum. B Mar Sci. 1990, 47 (3): 613-621.
81. Price DC, Chan CX, Yoon HS, Yang EC, Qiu H, Weber AP, Schwacke R, Gross J, Blouin NA, Lane C, et al: Cyanophora paradoxa genome elucidates origin of photosynthesis in algae and plants. Science. 2012, 335 (6070): 843-847. 10.1126/science.1213561.
82. Ren X, Zhang X, Sui Z: Identification of phase relative genes in tetrasporophytes and female gametophytes of Gracilaria/Gracilariopsis lemaneiformis (Gracilariales, Rhodophyta). Electron J Biotechn. 2006, 9 (2): 127-132. 10.2225/vol9-issue2-fulltext-11.
83. Rohmer M: The discovery of a mevalonate-independent pathway for isoprenoid biosynthesis in bacteria, algae and higher plants. 1999, 16: 565-574.
84. Rosenberg G, Paerl H: Nitrogen fixation by blue-green algae associated with the siphonous green seaweed Codium decorticatum: effects on ammonium uptake. Mar Biol. 1981, 61 (2): 151-158. 10.1007/BF00386654.

85. Sakemi S, Higa T, Jefford CW: Venustatriol. A new, anti-viral, triterpene tetracyclic ether from Laurencia venusta. Tetrahedron Lett. 1986, 27 (36): 4287-4290. 10.1016/S0040-4039(00)94254-0.
86. Salgado LT, Viana NB, Andrade LR, Leal RN, Da Gama BAP, Attias M, Pereira RC, Amado Filho GM: Intra-cellular storage, transport and exocytosis of halogenated compounds in marine red alga Laurencia obtusa. J Struct Biol. 2008, 162 (2): 345-355. 10.1016/j.jsb.2008.01.015.
87. Sandin SA, McNamara DE: Spatial dynamics of benthic competition on coral reefs. Oecologia. 2011, 168 (4): 1079-1090.
88. Santos AO, Veiga-Santos P, Ueda-Nakamura T, Dias-Filho BP, Sudatti DB, Bianco EM, Pereira RC, Nakamura CV: Effect of elatol, isolated from red seaweed Laurencia dendroidea, on Leishmania amazonensis. Mar Drugs. 2010, 8 (11): 2733-2743. 10.3390/md8112733.
89. Schmieder R, Edwards R: Quality control and preprocessing of metagenomic datasets. Bioinformatics. 2011, 27 (6): 863-864. 10.1093/bioinformatics/btr026.
90. Schwarz F, Aebi M: Mechanisms and principles of N-linked protein glycosylation. Curr Opin Struct Biol. 2011, 21 (5): 576-582. 10.1016/j.sbi.2011.08.005.
91. Schwender J, Zeidler J, Groner R, Muller C, Focke M, Braun S, Lichtenthaler FW, Lichtenthaler HK: Incorporation of 1-deoxy-D-xylulose into isoprene and phytol by higher plants and algae. FEBS Lett. 1997, 414: 129-134. 10.1016/S0014-5793(97)01002-8.
92. Shen S, Zhang G, Li Y, Wang L, Xu P, Yi L: Comparison of RNA expression profiles on generations of Porphyra yezoensis (Rhodophyta), based on suppression subtractive hybridization (SSH). BMC Res Not. 2011, 4 (1): 428-10.1186/1756-0500-4-428.
93. Skindersoe ME, Ettinger-Epstein P, Rasmussen TB, Bjarnsholt T, De Nys R, Givskov M: Quorum sensing antagonism from marine organisms. Marine Biotechnol. 2008, 10 (1): 56-63. 10.1007/s10126-007-9036-y.
94. Staufenberger T, Thiel V, Wiese J, Imhoff JF: Phylogenetic analysis of bacteria associated with Laminaria saccharina. FEMS Microbiol Ecol. 2008, 64 (1): 65-77. 10.1111/j.1574-6941.2008.00445.x.

95. Steinberg PD, de Nys R: Chemical mediation of colonization of seaweed surfaces. J Phycol. 2002, 38 (4): 621-629. 10.1046/j.1529-8817.2002.02042.x.

96. Stevens H, Stubner M, Simon M, Brinkhoff T: Phylogeny of Proteobacteria and Bacteroidetes from oxic habitats of a tidal flat ecosystem. FEMS Microbiol Ecol. 2005, 54 (3): 351-365. 10.1016/j.femsec.2005.04.008.

97. Sudatti DB, Fujii MT, Rodrigues SV, Turra A, Pereira RC: Effects of abiotic factors on growth and chemical defenses in cultivated clones of Laurencia dendroidea J. Agardh (Ceramiales, Rhodophyta). Mar Biol. 2011, 158 (7): 1439-1446. 10.1007/s00227-011-1660-4.

98. Tanaka Y, Ogawa H, Miyajima T: Bacterial decomposition of coral mucus as evaluated by long-term and quantitative observation. Coral Reefs. 2011, 30 (2): 443-449. 10.1007/s00338-011-0729-3.

99. Teo SS, Ho CH, Teoh S, Lee WW, Tee JM, Rahim RA, Phang SM: Analyses of expressed sequence tags from an agarophyte, Gracilaria changii (Gracilariales, Rhodophyta). Eur J Phycol. 2007, 42 (1): 41-46. 10.1080/09670260601012461.

100. Toepel J, Welsh E, Summerfield TC, Pakrasi HB, Sherman LA: Differential transcriptional analysis of the cyanobacterium Cyanothece sp. strain ATCC 51142 during light-dark and continuous-light growth. J Bacteriol. 2008, 190 (11): 3904-3913. 10.1128/JB.00206-08.

101. Topcu G, Aydogmus Z, Imre S, Gören AC, Pezzuto JM, Clement JA, Kingston DGI: Brominated sesquiterpenes from the red alga Laurencia obtusa. J Nat Prod. 2003, 66 (11): 1505-1508. 10.1021/np030176p.

102. Tujula NA, Crocetti GR, Burke C, Thomas T, Holmström C, Kjelleberg S: Variability and abundance of the epiphytic bacterial community associated with a green marine Ulvacean alga. ISME J. 2009, 4 (2): 301-311.

103. Vairappan CS, Anangdan SP, Tan KL, Matsunaga S: Role of secondary metabolites as defense chemicals against ice-ice disease bacteria in biofouler at carrageenophyte farms. J Appl Phycol. 2010, 22 (3): 305-311. 10.1007/s10811-009-9460-7.

104. Vairappan CS, Daitoh M, Suzuki M, Abe T, Masuda M: Antibacterial

halogenated metabolites from the Malaysian Laurencia species. Phytochemistry. 2001, 58 (2): 291-297. 10.1016/S0031-9422(01)00243-6.

105. Vairappan CS, Kawamoto T, Miwa H, Suzuki M: Potent antibacterial activity of halogenated compounds against antibiotic-resistant bacteria. Planta Med. 2004, 70 (11): 1087-1090. 10.1055/s-2004-832653.

106. Vairappan CS, Suzuki M, Ishii T, Okino T, Abe T, Masuda M: Antibacterial activity of halogenated sesquiterpenes from Malaysian Laurencia spp. Phytochemistry. 2008, 69 (13): 2490-2494. 10.1016/j.phytochem.2008.06.015.

107. Vairappan CS: Potent antibacterial activity of halogenated metabolites from Malaysian red algae, Laurencia majuscula (Rhodomelaceae, Ceramiales). Biomol Eng. 2003, 20 (4–6): 255-259.

108. Veiga-Santos P, Pelizzaro-Rocha KJ, Santos AO, Ueda-Nakamura T, Dias Filho BP, Silva SO, Sudatti DB, Bianco EM, Pereira RC, Nakamura CV: In vitro anti-trypanosomal activity of elatol isolated from red seaweed Laurencia dendroidea. Parasitology. 2010, 137 (11): 1661-1670. 10.1017/S003118201000034X.

109. Wanke M, Skorupinska-Tudek K, Swiezewska E: Isoprenoid biosynthesis via 1-deoxy-D-xylulose 5-phosphate/2-C-methyl-D-erythritol 4-phosphate (DOXP/MEP) pathway. Acta Biochim Polon. 2001, 48: 663-672.

110. Weinberger F, Coquempot B, Forner S, Morin P, Kloareg B, Potin P: Different regulation of haloperoxidation during agar oligosaccharide-activated defence mechanisms in two related red algae, Gracilaria sp. and Gracilaria chilensis. J Exp Bot. 2007, 58 (15–16): 4365-4372.

111. White DE, Stewart IC, Grubbs RH, Stoltz BM: The catalytic asymmetric total synthesis of elatol. J Am Chem Soc. 2008, 130 (3): 810-811. 10.1021/ja710294k.

112. Xiaolei F, Yongjun F, Songnian H, Guangce W: Generation and analysis of 5318 expressed sequence tags from the filamentous sporophyte of Porphyra haitanensis (Rhodophyta). J Phycol. 2007, 43 (6): 1287-1294. 10.1111/j.1529-8817.2007.00415.x.

113. Yang H, Mao YX, Kong FN, Yang GP, Ma F, Wang L: Profiling of the transcriptome of Porphyra yezoensis with Solexa sequencing technology. Chinese Sci Bull. 2011, 56 (20): 2119-2130. 10.1007/s11434-011-4546-4.

114. Zheng M, Storz G: Redox sensing by prokaryotic transcription factors. Biochem Pharmacol. 2000, 59 (1): 1-6. 10.1016/S0006-2952(99)00289-0.

SECTION IV

COMPOSITION, CONSTITUENTS AND USES

CHAPTER

10

FATTY ACID COMPOSITION AND PROSTAGLANDIN CONTENT OF THE RED SEAWEED GRACILARIA SP. FROM INDONESIA

Muhammad Ikbal Illijas[1], Arifuddin[1], Luqman Saleh[1], and Yutaka Itabashi[2]

[1]Department of Aquaculture, Pangkep State Polytechnic of Agriculture, Indonesia

[2]Graduate School of Fisheries Sciences, Hokkaido University, Japan

ABSTRACT

High content of polyunsaturated fatty acids (PUFAs) such as arachidonic and eicosapentaenoic acids are typical for the red alga. Analysis of fatty acid composition and prostaglandin content was conducted in the red alga *Gracilaria* sp. from Indonesia. Total lipid of the alga was extracted with $CHCl_3$-MeOH (2:1, v/v). Analysis of the fatty acids composition was performed on gas chromatography (GC) equipped with omega wax column (30 m x 0,32 mm i.d., Supelco, PA, USA) and analysis of prostaglandins were carried out by HPLC on ODS column (Mightysil RP-18 GP, 250 mm x 4.6 mm, 5 μm). The content of fatty acids high for were palmitic acid (50%) and arachidonic acid (26.9%), whereas prostaglandin E2 was identified and found lower concentration (44.2 μg/gram total lipid).

Keywords: Fatty acids; prostaglandins; Gracilaria; seaweed

INTRODUCTION

Red algae are well recognized as sources of polyunsaturated fatty acids (PUFAs) with 20 carbon atoms, mainly arachidonic acid (20:4n- 6) and eicosapentaenoic acid (20:5n-3). These PUFAs have unique biological activities and are the precursors in biosynthesis of prostaglandins, leucotrienes and other eicosaenoids, which are important to the maintenance of normal mammalian physiology (Gerwick et al., 1993; Gerwick & Bernet, 1993). The majority of the red algal species are rich in 20:5n-3 or contain about equal amounts of 20:5n-3 and 20:4n- 6. However, *G. verrucosa* (*G. vermiculophyla*) contains higher of 20:4n-6 than 20:5n-3 (Pohl et al., 1968; Takagi et al., 1985; Araki et al., 1986; Khotimchenko et al., 1991; Dawes et al., 1993), Khotimchenko & Levchenko, 1997).

The red alga G. verrucosa, which is intensively cultured in coastal areas of shrimp and fish shrimp ponds in Indonesia, has luck of information in detail regarding its lipid bioactive contents. Although there were some species of *Gracilaria* have been published in detail about their lipid bioactive contents those collected from Japanese Waters, such as *G. vermiculophylla* (Itabashi et al., 2006), *G. gigas* (Hsu et al., 2007), and *G. asiatica* (Sajiki, 1996). Another species is *G. chilensis* (Lion et al., 2006) collected from Chile, which contained different lipids bioactive from other *Gracilaria* species.

The aim of study was to evaluate fatty acids and eicosanoid composition of Indonesian red alga *G. verrucosa*.

MATERIALS AND METHODS

Seaweed

The red seaweed *Gracilaria* sp. was collected from shrimp ponds located in Bone District, South Sulawesi. A part of the alga was extracted at

Department of Aquaculture, Pangkep State Polytechnic of Agriculture. Another part of the alga was freezed for further extraction at Bioanalytical Chemistry Laboratory, Hokkaido University, Japan

Lipid Extraction

Each Gracilaria sp. sample was cut into pieces (3-5 mm) and homogenized for 5 min. at room temperature. Lipids was extracted by soaking the homogenate overnight in $CHCl_3$/MeOH/H_2O (2:1:0.8 by vol.). After filtration, the solvent will removed at 25°C under reduced pressure using a rotary evaporator, and then the residual lipids was made up to a known concentration with $CHCl_3$/MeOH (2:1, v/v) and stored at -30°C until use.

Analysis of FFA Composition

FFAs were converted to methyl esters by heating at 95°C for 1 h in 5% HCl/MeOH (Christie, 2003). Analysis of the fatty acid methyl esters was carried out using a Shimadzu GC-14A gas chromatograph (Shimadzu) equipped with an Omegawax 320 column (30 m x 0.32 mm i.d., Supelco, PA, USA). Helium was used as the carrier gas at a constant flow rate of 1 mL/min. The split ratio was 1:50. The column temperature was maintained at 160°C for 17 min, then elevated to 230°C at a ramp rate of 5°C/min. The final temperature was kept for 30 min. The injector and flame-ionization detector (FID) temperatures were set at 240°C. Peaks were monitored on a Chromatopac CR6A (Shimadzu) and identified by comparing retention data of the known fatty acids from some marine organisms including seaweeds (Takagi et al., 1985; Takagi et al., 1986).

Identification of Prostaglandins

For identification of eicosanoid compounds, LC/MS was used. Total lipids was subjected to LC/MS equipped with a Mightysil column, RP-18 GP (250 mm x 4.6 mm, 5 μm). Identification of eicosanoid compounds was conducted by comparing their mass spectra with those of authentic standard. For complete identification of the eicosanoid compounds, co-chromatography using authentic compounds was employed.

Determination of Prostaglandin Contents

The alga was finely sliced and placed in 100 mL-bottle. To which 30 mL of ethyl acetate and 150 μL of 1 M HCl in methanol was added. The mixtures were shaken for 5 min. and then centrifuged at 2,500 rpm for 5 min. at 20°C. The supernatants were pipetted and evaporated under reduced pressure using a rotary evaporator at 25°C. This procedure was repeated once more. The residue was dissolved in 200 μL of methanol as test solution. For determination of PGs, the test solution (5 μL) was subjected to HPLC (Hitachi Ltd, Tokyo, Japan) on a Mightysil column, RP-18 GP (250 mm x 4.6 mm, 5 μm). Determination of PGs was carried out by comparing the peak of PGs extracted from samples and authentic standards of PG. The contents of PGs in the samples was calculated from standard curve of the PG standards.

HPLC and LC Conditions

HPLC was performed at 40°C using a gradient elution from acetonitrile/ water (40:60, v/v) containing 0.02% acetic acid (solvent A) to 100% acetonitrile (solvent B). The mobile phase system is as follows: 0-20 min. (solvent A), 20-60 min. (gradient of solvent A and B) and 60-80 min. (solvent B). The flow rate is 0.5 mL/ min. The PG peaks were monitored by a diode array detector (Model L-7455 LaChrom, Hitachi Ltd, Tokyo, Japan) set at 196 nm.

RESULTS AND DISCUSSION

Fatty Acid Composition

Analysis of fatty acid compositions of the red alga by GC was shown in Table 1. The result showed that the dominant fatty acids were palmitic acid (C16:0, 50%) and arachidonic acid (C20:4n-6, 29.6%). There were significant amounts of the fatty acids, namely stearic acid (C14:0) dan oleic acid (C18:1n-9).

Table 1. Fatty acid composition of the red alga *Gracilaria* sp.

Fatty acid	Composition
14:0	3.0±0.3
iso 15:0	0.2±0.0
anteiso 15:0	0.1±0.1
15:0	05±0.2
15:1	nd
iso 16:0	0.1±0.0
anteiso 16:0	0.1±0.0
16:0	50.0±4.4
iso 17:0	0.4±0.0
17:0	0.2±0.1
18:0	1.1±0.3
20:0	0.1±0.0
22:0	0.2±0.0
24:0	nd
Saturates	**56.0**
14:1n-9	nd
16:1n-7.9	1.8±0.2
16:1n-5	nd
18:1n-9	3.3±0.4
18:1n-7	1.3±0.1
18:1n-5	nd
24:1n-9	nd
Monoenes	**6.4**
16:2n-4	nd
16:3n-4	0.2±0.2
18:2n-6	0.7±0.0
18:3n-6	0.4±0.1
18:3n-3	0.1±0.0
18:4n-3	0.2±0.3
20:2n-6	0.2±0.1
20:3n-6	2.4±0.1
20:4n-6	29.6±2.6
20:3n-3	nd
20:4n-3	nd
20:5n-3	0.1±0.0
22:6n-3	nd
Polyenes	**34.0**
Others	3.6±0.9

nd : not detected
tr : trace (< 0.1)

C16:0 is saturated fatty acid, which is abundantly found in many kind of seaweed (Stefanov et al., 1988; Aknin et al., 1990; Khotimchenko, 1998). Whereas, the polyunsaturated acid, C20:4n-6 is a typical fatty acid for the red algae (Khotimchenko et al., 1990; Araki et al., 1990; Illijas et al., 2009). This fatty acid is synthesized from hydrolysis of lipid membrane, glycoglycerolipids, such as monogalactosyldiasylglyserol (MGDG), digalactosyldiasylglyserol (DGDG) and sulfoquinovosyldiasylglyserol (SQDG) catalized by glycerolip acyl-hydrolase (Figure 1, Illijas et al., 2008). Function of the arachidonic acid is a precursor for biosynthesis of prostaglandins and other eicosanoic compounds in the red algae (Hsu et al., 2007; Illijas, 2008).

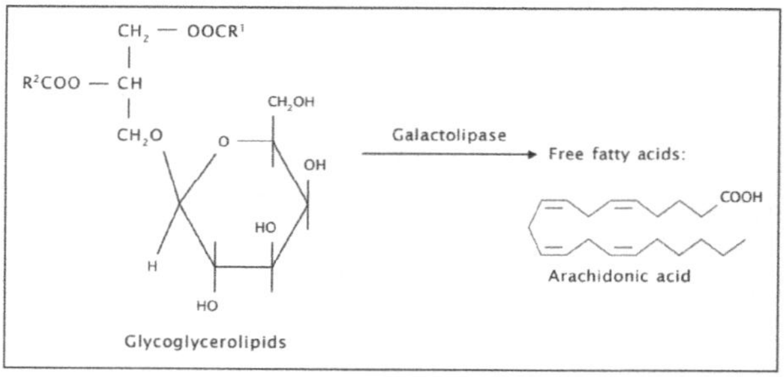

Figure 1. Biosynthesis pathway of arachidonic acid in the red alga G. vermiculophylla (Illijas et al., 2008).

Prostaglandin

In the HPLC chromatograms (Figure 2), one of peaks was identified as prostaglandin E_2 (PGE_2). The identification was conducted by comparing the chromatograms with HPLC chromatogram of prostaglandin standards (Figure 3).

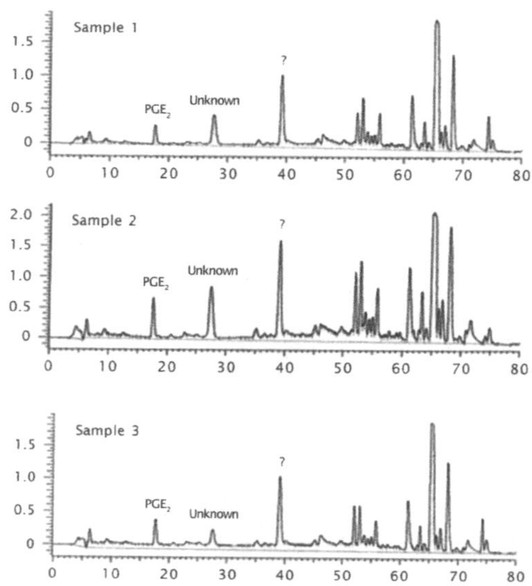

Figure 2. HPLC chromatograms of the red alga Gracilaria sp. extracts ($CHCl_3$-MeOH extracts)

This prostaglandin is also found abundantly in many species of the red algae, such as G. gigas (Hsu et al., 2007), G. asiatica (Sajiki et al., 1998), G. vermiculophylla (Illijas, 2008).

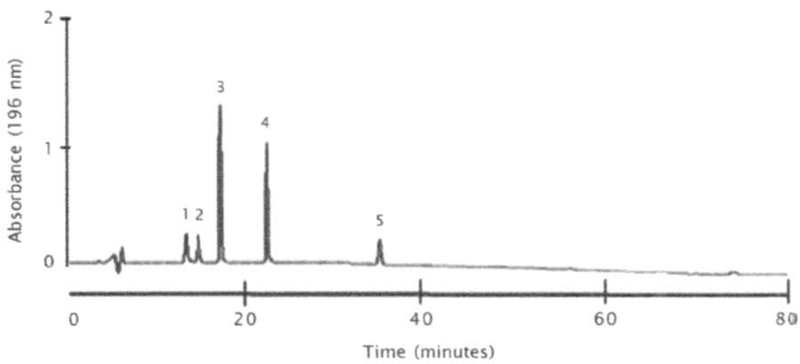

Figure 3. HPLC chromatogram of prostaglandin standards. Peak 1: PGE_3; Peak 2: PGF_{2a}; Peak 3: PGE_2; Peak 4: 15-keto-PGE_2; Peak 5: PGA_2.

Biosynthesis pathway of prostaglandin in seaweed is still unclear. However, it was found that prostaglandin is formed from oxidation of arachidonic acid, which is likely involve cyclooxygenase as catalyst, because prostaglandin formation decreased as addition of aspirin, an anti cyclooxygenase compound, was conducted to reaction mixtures of G. vermiculophylla extract and free arachidonic acid (Figure 4) (Illijas, 2008).

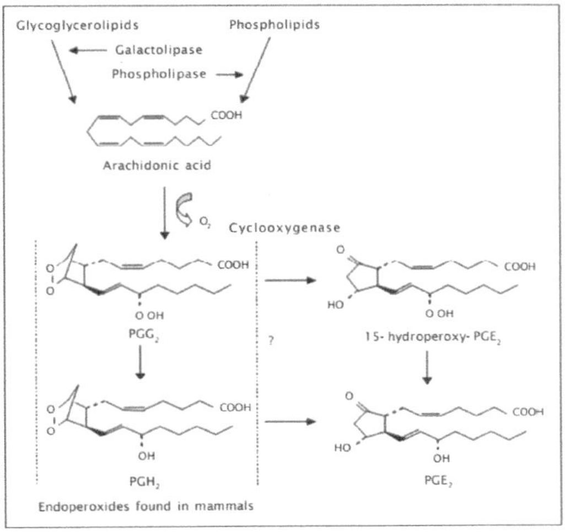

Figure 4. Proposed biosynthesis pathway of prostaglandin in the red alga G. vermiculophylla (Illijas, 2008).

Prostaglandin function is also still unclear in the seaweed. However, several result of researches showed that formation of prostaglandins occurred when the seaweed was physically treated (Nakajima et al., 1998) so that the prostaglandins also known as secondary metabolites. The prostaglandins have also been found to be produced along with other eicosanoid compounds when the red alga Chondrus crispus was incubated with pathogen extract (Bouarab et al., 2004; Gaquerel et al., 2007). In mammals and human, prostaglandins play an important role as hormone, which control several kinds of metabolism (Samuelsson, 1975).

CONCLUSION

Arachidonic acid, the dominant polyunsaturated fatty acid found in the seaweed, is precursor of synthesis of prostaglandin E_2, the only eicosanoic compound could be identified in this study.

ACKNOWLEDGMENTS

This study was part funded by Indonesian Management of Higher Education for relevance and efficiency (I-MHERE) Project B.I Batch III Directorate General of Higher Educatioan and World Bank, awarded to Pangkep State Polytechnic of Agriculture, South Sulawesi, Indonesia.

REFERENCES

1. Aknin, M., Miralles, J., & Kornprobst, J-M. 1990. Sterol and fatty acid distribution in red algae from the Senegalase coast. Comp. Biochem. Physiol., 96B: 559-563.
2. Araki, S., Sakurai, T., Oohusa, T., Kayama, M., & Nisizawa, K. 1990. Content of arachidonic and eicosapentaenoic acids in polar lipids from Gracilaria (Gracilariales, Rhodophyta). Hydrobilogia, 204/205: 513-519.
3. Bouarab, K., Adas, F., Gaquerel, E., Kloareg, B., Salaun, J-P., & Potin, P. 2004. The innate immunity of a marine red alga involves oxylipin from both the eicosanoid and octadecanoid pathways. Plant Physiology, 135: 1,838-1,848.
4. Christie, W.W. 2003. Preparation of derivatives of fatty acids. In

Lipid analysis, third edition; The Oily Press: Bridgwater, England, p. 205-224.

5. Gaquerel, E., Herve, C., Labriere, C., Boyen, C., Potin, P., & Salaun, J-P. 2007. Evidence for oxylipin synthesis and induction of a new polyunsaturated fatty acid hydroxylase activity in Chondrus crispus in response to methyljasmonate. Biochemica et Biophysica Acta, 1771: 565-575.
6. Hsu, B-Y., Tsao, C-Y., Chiou, T-C., Hwang, P-A., & Hwang, D-F. 2007. HPLC determination for prostaglandins from seaweed Gracilaria gigas. Food Control, 18: 639-645.
7. Illijas, M.I., Nakamura, & Terasaki, M. 2008. Purification and characterization of glicerolipid acyl-hydrolase from the red alga Gracilaria vermiculophylla. Fisheries Science, 74: 670- 676.
8. Illijas, M.I., Rimber, R.I., Yasui, T., & Itabashi, Y. 2009. Lipid class dan fatty acid composition of a little-known and rarely collected alga Exophyllum wentii Weber-van Bosse from Bali Island, Indonesia. Olei Science, 58(3): 103-110.
9. Illijas, M.I. 2008. Lipid and prostaglandin in the red alga, Gracilaria vermiculophylla. Thesis, Graduate School of Fisheries Sciences, Hokkaido University, Japan.
10. Khotimchenko, S.V. 1998. Fatty acids of brown algae from Russian far east. Phytochemistry, 49: 2,363-2,369.
11. Lion, U., Wiesemier, T., Weinberger, F., Beltran, J., Flores, V., Faugeron, S., Correa, J., & Pohnert, G. 2006. Phospholipases and galactolipases trigger oxylipin-mediated wound-activated defence in the red alga Gracilaria chilensis against epiphytes. Chem. Bio. Chem., 7: 457-462.
12. Sajiki, J. & Kakimi, H. 1998. Identification of eicosanoids in the red algae, Gracilaria asiatica, using high-performance liquid chromatography and electrospray ionization mass spectrometry. Journal of Chromatography A, 795: 227-237.
13. Samuelsson, B., Granstr6m, E., Green, K., Hamberg, M., & Hammarstrom, S. 1975. Prostaglandins. Annual Reviews.

14. Stefanov, K., Konaklieva, M., Brechany, E.Y., & Christie, W.W. 1988. Fatty acid composition of some algae from the black sea. Phytochemistry, 27: 3,495-3,497.
15. Takagi, T., Asahi, M., & Itabashi, Y. 1985. Fatty acid composition of twelve algae from Japanese waters. Yukagaku, 34: 1,008-1,012.

CHAPTER

12

BIOACTIVITY AND CONSTITUENTS OF SEVERAL COMMON SEAWEEDS

Zhang Yi[1,2], Han Jinyuan[1], Mu Jun[1], Feng Yan[1], Gu Xiaojie[1], And Ji Yuanxi[1]

[1]School of Environmental and Chemical Engineering, Dalian Jiaotong University, Dalian 116028, China

[2]School of Life Science and Biotechnology, Dalian University of Technology, Dalian 116024, China

ABSTRACT

Bioactivity and constituents of 8 common seaweeds from Dalian intertidal zone of northern Yellow Sea were investigated. In the anti-methicillin-resistant *Staphylococcus aureus* (MRSA) test, *Symphyocladia latiuscula* and *Enteromorpha intestinalis* showed obvious activities with MICs much lower than 1.0 mg mL^{-1}. In the DNA damage repair test (DDRT), *Chondrus ocellatu* showed selective inhibitory activity against the DNA repair-defective *E. coli* strain vs. the wild-type *E. coli* strain; while *Sym. latiuscula*, *Enteromorpha intestinalis* and *Sar. kjellmanianum* showed significant anti-*E. coli* activity with MICs of 64–128 μg mL^{-1}. In the anti-*Pyricularia oryzae* test, *Sym. latiuscula* and *Rh. confervoides* strongly inhibited the germination of the spores of *P. oryzae* on agar plate. In the brine shrimp larvicidal test, *Sym. latiuscula, Rh. confervoides* and *Sar. kjellmanianum* exhibited potent toxicity against brine shrimp larvae, with LC_{50} much lower than 1 mg mL^{-1}. The HPTLC analysis revealed their diversified secondary metabolites. The HPLC-DAD-MS analysis of the strongest species *Sym. latiuscula* and database searching showed that it can produce quite diversified metabolites, including halogenated ones, some of which may be new natural products. The results demonstrated the potentials of these seaweeds in the development of new antibiotics, antitumor drugs, agricultural fungicides and pesticides.

Keywords: Saweeds, biomass, antibacterial, antifungal, larvicidal, HPTLC, HPLC-DAD-MS

Macroalgae are important primary producers of marine and coastal ecosystem and are also useful medicine and pesticide resources due to their potentials in accumulating large quantities of bioactive metabolites [1]. As is known to all, algal diversity and biomass depend on the environmental factors such as temperature, light, and nutrition, which can be influenced by human activities (e.g. introducing species and pollutions in the form of heat and eutrophication). Therefore, it is important to investigate and utilize the algal resources in coastal zones.

Dalian, an important harbor and industrial city, is located on the tip of the Liaodong Peninsula, between China's inner Bohai Sea and the north Yellow Sea. Its marine ecological environment is influenced by strong sea currents and rivers, which provides diversified habitats for seaweeds and also brings about human influences. In this article, we evaluated the biomass and their biomedical and agrichemical potentials as new antibiotics, antitumor drugs, agricultural fungicides, and pesticides in some common seaweeds from Dalian coastline by determining several bioactivities of them, including anti-methicillin-resistant Staphylococcus aureus (MRSA), anti-Escherichia coli DNA repair-defective/wildtype strains, anti-Pyricularia oryzae, and anti-brine shrimp (Artemia parthenogenetica L.) larvae.

MATERIALS AND METHODS

Materials and Biomass Calculation

Eight species of seaweeds were collected in the intertidal zone of Fujiazhuang Beach (121°36′13.82″E, 38°48′ 36.66″N) of Dalian in July of 2009, as shown in Table 1. These species were selected for their remarkably higher occurrence frequency. Their biomasses were calculated by taking the average weight of air-dried samples collected randomly from five quadrates in their habitats for unit area of 1 m^2 [2]. They were identified by Professor Hongwei Wang, an expert on algal taxonomy in School of Life Sciences, Liaoning Normal University.

Table 1. Biomass and habitats of seaweeds under investigation

Species	Biomass ($g\ m^{-2}$)	Habitats
Chlorophyta		
Ulva pertusa	44.9	Middle intertidal zone
Cladophora fascicularis	21.1	Higher intertidal pools
Enteromorpha intestinalis	39.1	Higher intertidal zone
Rhodophyta		
Chondrus ocellatus	37.2	Lower intertidal zone
Rhodomela confervoides	25.5	Middle intertidal pools
Symphyocladia latiuscula	8.1	Lower intertidal zone
Phaeophyta		
Undaria pinnatifida	26.5	Lower intertidal zone
Sargassum kjellmanianum	41.8	Middle intertidal pools

Indicator Organisms for Bioassay

DDRT strains (E. coli strains AB1157 as wild type and AB3027 with defect in DNA damage repair) were provided by E. coli Genetic Stock Center, Yale University, Methicillin-resistant Staphylococcus aureus (MRSA) strain came from Library of Strains of Dalian Friendship Hospital. Pyricularia oryzae was from the Library of Strains of Chinese Academy of Agricultural Sciences. Brine shrimp eggs (Artemia parthenogenetica L.) were collected at Fuzhou Bay Saltworks of Bohai Sea.

Extraction of Seaweed

The seaweed samples were air dried and extracted with 100 mL of methanol per 20 g of sample for 24 h at room temperature. Then the extracts were filtrated, rotary evaporated in vacuum at 45°C, and finally dissolved in 3 mL of methanol.

Anti-MRSA Activity Test

The preliminary screening was performed using paper disk diffusion method with the dose of 30 mL extract/disk [3]. The re-screening was performed using MIC method [4]. The test concentration covered the range from 0.0005 to 1.024 mg mL^{-1}.

Brine Shrimp Lethality Test

The brine shrimp lethality test can reflect samples' toxicity against brine shrimp larvae, and the results have a high correlation with antitumor tests. It also can be used as a preliminary screening model for agricultural pesticides. The experimental method used was similar to that in our previous study [5]. In the test, the concentration gradients were set to be 1.024, 0.512, 0.256, 0.128, 0.064, 0.032, 0.016, 0.008, 0.004, 0.002, 0.001, and 0.0005 mg mL^{-1}. Then, the semi-lethal concentrations (LC_{50}) were calculated from the lethal curve fit by polynomial regressions [5].

Growth Inhibition Test of *Pyricularia Oryzae*

Pyricularia oryzae is the pathogen of rice, which can be used to screen agricultural fungicides. Furthermore, Kobayashi et al. [6] found that P. oryzae's spore germination or hyphal dysmorphism was associated with anti-tumor activity, and is often used to screen antitumor drugs.

Plate confrontation method and spore spreading method were adopted in present study.

In plate confrontation method, P. oryzae was inoculated at the center of each Petri dish containing Sabouraud's medium and cultivated at 30°C till the colony reached 1 cm in diameter. Then air dried paper disks loaded with samples at the dose of 30 mL extract/disk were placed 1 cm from the edge of the colonies. The plates were incubated at 30°C for 24–48 h and the inhibition of mycelium growth was observed.

In spore spreading method, P. oryzae was first cultured on Petri dished for 10 d at 30°C and then the spore suspension was harvested by scraping mycelia with a sterile blade over the surface of the plate containing 15 mL of sterile water. Afterwards, the crude suspension was filtrated with three layers of sterile gauze to give pure spore suspension. The 100 mL of this suspension was spread onto the surface of each Sabouraud's culture plate. The paper disks with samples at dose of 30 mL extract per disk were air dried and placed onto the pre-coated plates. The results were expressed as the diameter of inhibition zones observed and measured after incubation at 30°C for 24–48 h.

DNA Damage Repair Test

DNA damage repair test (DDRT) is mainly used to detect active substances with DNA damaging ability. This test, using two E. coli strains, AB1157 for wild type and AB3027 with defect in DNA damage repair [7], consisted of preliminary screening and re-screening.

The preliminary screening used AB3027 strain and was performed by paper disk diffusion method [3]. On each piece of filter paper, the dose was 30 mL extract/disk.

The re-screening used both two E. coli strains and was performed by MIC method [4], in which the test concentration covered the range from 0.0005 to 1.024 mg mL^{-1}.

The selectivity of inhibition was calculated by ratio between MIC against AB1157 and MIC against AB3027 and was used to assess if the sample had potential DNA damaging ability.

HPTLC Analysis

For these samples, 20 mL of each (20 mg mL^{-1}) was applied on high-performance thin-layer chromatographic plates (HPTLC silica gel 60 F254, Germany) with a capillary. Then the HPTLC plate was developed by a mixture of $CHCl_3$/MeOH (10:1, v/v). Afterwards, the plate was colorized by H_2SO_4-anisaldehyde reagent at 105°C for 2 min and by acidic $FeCl_3$ ethanol solution at room temperature for 5 min, respectively. The images were scanned as results.

HPLC-DAD-MS Analysis

Symphyocladia latiuscula, as the species with strong bioactivities in all the four assays, was analyzed by HPLC-DADMS. The experiment was performed on an Agilent 1200 HPLC-DAD-MS apparatus with an Ultra Trap System XCT 6330 LC/MS detector. Stationary phase: Nucleosil 100 C18, 3 μm, 100×2 mm ID; equipped with a guard column, 10×2 mm ID, same stationary phase. Mobile phases: A=0.1% formic acid, B=acetonitrile with 0.06% formic acid; gradient was from 10% B to 100% B in 15 min and 100% B to the end; flow rate: 400 mL min^{-1}; column temperature: 40°C. MS condition: mode: ESI, positive and negative, alternating; capillary voltage: 3.5 kV; temperature: 350°C. The injection volume was 10 mL (5 mg mL^{-1}).

RESULTS

Biomass and Distribution

The sampling and calculation showed that in July of 2009 (Table 1), the eight common seaweeds growing at rocky intertidal zone of Fujiazhuang Beach all produced biomass over 20 g m^{-2}, except Symphyocladia latiuscula. In the higher intertidal zone, Enteromorpha intestinalis and Cladophora fascicularis were dominant species. In the middle intertidal zone, Ulva pertusa, Sargassum kjellmanianum, and Rhodomela confervoides contributed the most to biomass. For the lower intertidal zone, Chondrus ocellatus, Undaria pinnatifida, and Sym. latiuscula were the dominant species. The results reflected the biomass production and biodiversity of common seaweeds in this coastal area. According to Shao and Li's report [8], some species, including Ch. ocellatus, En. intestinalis, and Ul. pertusa, have also become the dominant species in most sea areas, probably due to their adaptability to environmental changes caused by human activities (e.g. global warming and eutrophication).

In the development of new marine drugs, sustainable supply of drugs resources is always a bottleneck. Compared to some invertebrates such as sponges and corals, seaweeds, especially those with large biomass, may be a more realistic pharmaceutical resource. Thus, the biomass of seaweeds, naturally occurring or increased due to human activities, should be considered for the development of new medicines or agrichemicals.

The potential of seaweeds as new drugs is shown by the following bioassays.

Anti-MRSA Activity Test

Methicillin-resistant Staphylococcus aureus (MRSA) is an important cause of bacteremia with an increasing incidence worldwide and also a much studied target organism in new drug discovery [9].

In the present investigation, some seaweed samples showed remarkable anti-MRSA potentials (Table 2). In the preliminary screening, Sym. latiuscula showed the strongest activity with the inhibition zone up to 32 mm in diameter. Rh. confervoides, Ch. Ocellatus and Sar. kjellmanianum also showed powerful inhibitory activity; while Cl. fasciculari, En. Intestinalis and Un. pinnatifida only showed moderate activity. In re-screening, Sym. latiuscula and En. intestinalis both exhibited strong anti-MRSA activity with MICs much lower than 1 mg mL^{-1}. The improved activity of En. intestinalis in the re-screening might be due to its possible active constituents with low polarity, which was difficult to diffuse and take effect in paper disk method but easier to show activity in broth microdilution method. In addition, the remaining five seaweeds also showed moderate activities in re-screening with MICs of about 1 mg mL^{-1}, which are comparable to those in other reports on antimicrobial activities of seaweed extracts [10]. Therefore, most of these seaweeds may be used as sources of new antibiotics against drug-resistant bacteria.

Growth Inhibition Test of *Pyricularia Oryzae*

In the screening of these seaweeds for their inhibitory activity against P. oryzae, two methods were applied (Table 2). In plate confrontation method, none of these seaweeds showed inhibitory activity, indicating that these samples can not inhibit the growth of P. oryzae mycelium. However, in spore spreading method, Sym. latiuscula and Rh. confervoides showed inhibition zones with diameter of 35 and 20 mm, respectively. The results indicated their strong inhibitory activity against spore germination of P. oryzae. The rest of the seaweeds did not exhibit remarkable activity.

Table 2. The bioactivity against MRSA and Pyricularia oryzae of seaweeds

Species	Inhibitory activity against MRSA		Inhibition against *P. oryzae*	
	Inhibition zone in preliminary screening (Φ, mm)	MIC in re-screening (mg mL^{-1})	Diameter of Inhibition zone in Plate confrontation method (mm)	Diameter of Inhibition zone in Spore spreading method (mm)
Ulva pertusa	7.5±0.7	8.192	–	–
Cladophora fascicularis	8.0±0.0	1.024	–	–
Enteromorpha intestinalis	8.0±0.0	0.512	–	–
Chondrus ocellatus	10.5±0.7	1.024	–	–
Rhodomela confervoides	12.0±0.0	1.024	–	20.0±1.0
Symphyocladia latiuscula	32.0±0.0	0.256	–	35.0±2.0
Undaria pinnatifida	9.0±0.0	1.024	–	–
Sargassum kjellmanianum	10.5±0.7	1.024	–	–

Brine Shrimp Lethality Test

As a simple, fast and cheap model, brine shrimp lethality test has been widely used in the screening for antitumor drugs and pesticides [11,12]. This test was also employed in the current study, and the results showed that three species, Rh. confervoides, Sar. Kjellmanianum and Sym. latiuscula, exhibited strong toxicity with lethality ratio of 100% in the preliminary screening using 40 mL of extract and LC_{50} of 287, 354, and 430 μg mL^{-1}, respectively, in the quantitative re-screening (Figure 1). Therefore, these seaweeds may be used to develop new antitumor agents and pesticides since many previous reports have displayed good correlation between the results of brine shrimp lethality test and antitumor or anti-pest assays [13].

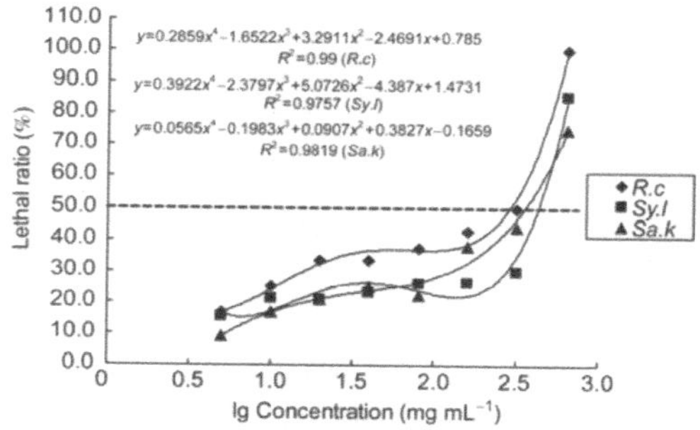

Figure 1. Lethal curves of three active seaweed samples in brine shrimp lethality test (under concentrations higher than 640 μg mL^{-1}, i.e. at logarithm of concentration higher than 2.8, the lethal ratios all reached 100%).

E. coli DDRT

Application of DDRT model in antitumor pre-screening is based on the mechanism that many anticancer drugs act by damaging DNA. Two E. coli strains, AB1157 as wild type and AB3027 with defect in DNA damage repair, were used in this test. Therefore, if a sample showed anti-tumor activity via the mechanism of DNA damage, stronger inhibition would be observed against AB3027 than against AB1157.

In preliminary screening (Table 3), Sym. latiuscula, En. intestinalis, Ch. ocellatus and Sar. kjellmanianum showed inhibitory activity against AB3027(-) strain. In the rescreening, the selectivity value of Ch. Ocellatus reached 2, indicating that its inhibitory activity against the defect-type (-) E. coli was greater than against the wild-type (+). Therefore, it may have anti-tumor potential via the mechanism of DNA damage and serve as a possible source of anti-tumor drugs. The MICs of Rh.confervoides, Sar. kjellmanianum, and Enteromorpha intestinalis against two E. coli strains were 128 or 64 μg mL^{-1}, respectively, which indicated their strong antimicrobial activity and potentials as sources of antibiotic drugs. However, their selectivity values did not all exceed 1, showing that they may not act by damaging DNA.

Table 3. The bioactivity of seaweeds in DNA damage repair test

Species	Inhibition zone against AB3027 (–) (Φ, mm)	MIC against AB1157 (+) (μg mL^{-1})	MIC against AB3027 (–) (μg mL^{-1})	Selectivity
Ulva pertusa	–			
Cladophora fascicularis	–			
Enteromorpha intestinalis	11.2±1.3	64	128	0.5
Chondrus ocellatus	13.3±2.1	512	256	2
Rhodomela confervoides	–			
Symphyocladia latiuscula	18.0±1.0	128	128	1
Undaria pinnatifida	–			
Sargassum kjellmanianum	12.2±1.4	64	64	1

HPTLC Analysis

The colorization by H_2SO_4-anisaldehyde reagent showed that these seaweeds contained quite diverse metabolites. The grey spots with Rf value in the range 0.55–0.73 should usually be steroids or terpenoids, while the main tailing spots with Rf value 0.51 may be some fatty acid derivatives (marked with arrows in Figure 2(a)).

The acid $FeCl_3$ reagent further displayed typical blue spots with Rf value 0.23 and 0.38 in two red algae Rh. confervoides (R.C.) and Sym. latiuscula (S.L.) extracts, showing the presence of phenolic compounds (marked with arrows in Figure 2(b)).

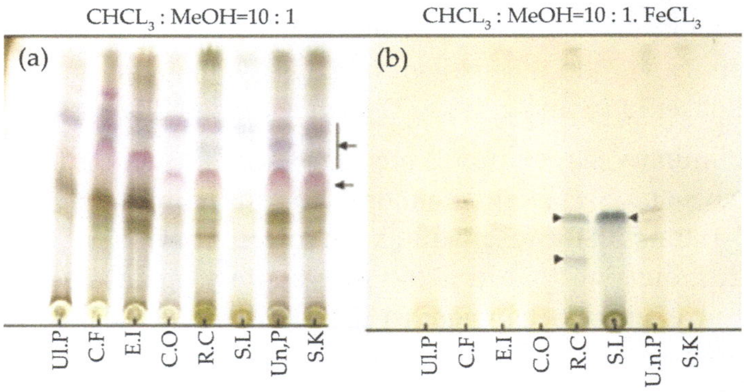

Figure 2. The HPTLC of seaweeds extracts. (a) Dyed by H_2SO_4-anisaldehyde; (b) dyed by acid $FeCl_3$ reagent.

HPLC-DAD-MS Analysis

Since Symphyocladia latiuscul showed strong bioactivities in all the four assays, its extract was further analyzed by HPLC-DAD-MS. The molecular weight of each main peak was deduced by quasi-molecular peaks in both positive and negative mode mass spectra (Figure 3). Then their molecular weights were used to search the plant metabolites in the Dictionary of Natural Products on DVD (DNP2011) [14]. During the searching, factors like wavelength of maximum UV absorbance or conjugated systems for those hits without UV data, biological resource, polarity's reasonability, and elemental constituents for some obvious halogenated compounds, were taken into account to rule out unreasonable hits. The analysis results are summarized in Table 4.

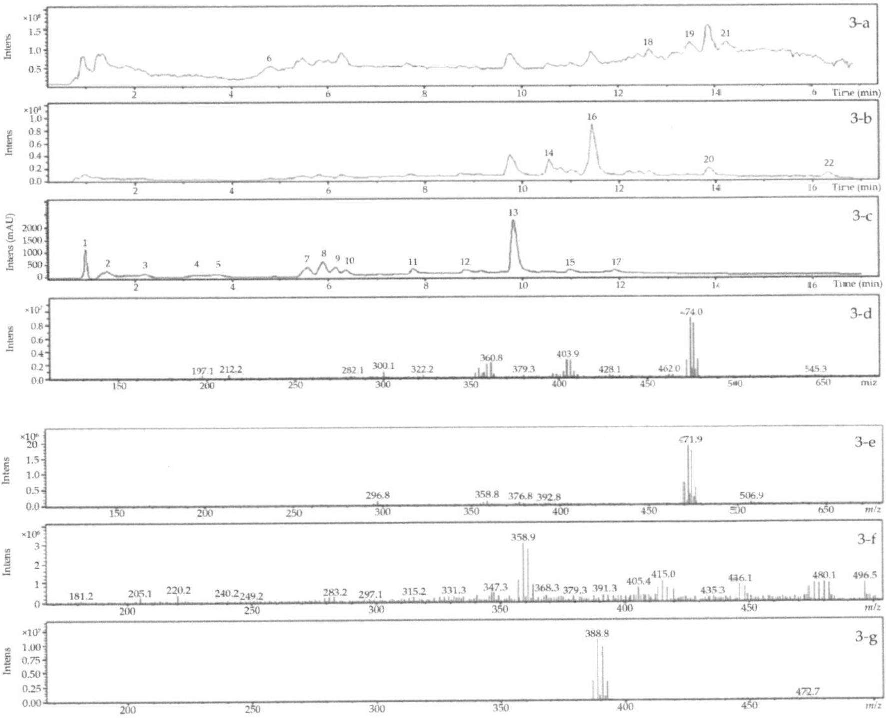

Figure 3. HPLC-DAD-MS of Symphyocladia latiuscula extract. (a) TIC(−); (b) TIC(-); (c) chromatographic spectrum under 224–235 nm; (d)–(e) MS(+) and MS(-) of peak 10; (f)–(g) MS(+) and MS(-) of peak 13.

Among the 22 main peaks, 7 showed obvious mass characteristics of brominated compounds, i.e., isotopic peak clusters with successive mass difference of two Daltons and typical intensity ratios, including 1:2:1 clusters for dibrominated, 1:3:3:1 for tribrominated, and 1:4:6:4:1 for tetrabrominated compounds. Among them, main component of peak 8 and peak 13 may be possibly known metabolites respectively from red algae Rhodomela confervoides and Symphyocladia latiuscul, both of which belong to the family of Rhodomelaceae, a group known for their ability in producing halogenated compounds [15]. The other five peaks may be new brominated natural products.

Table 4. LC-DAD-MS analysis and database searching of main peaks in Symphyocladia latiuscula extract

Peak Num.	Rt (min)	UV maxima (nm)	Molecular weight	Number of brominated positions	Key mass peaks (*m/z*) (for brominated peaks: also giving ratio of abundance)[a]	Possible hits from database DNP2011 (Structure type; molecular formula; CRC number; biological resource; bioactivity)
1	1.0	325	346		$[M+H]^{+}$: 347; $[M+K]^{+}$: 385; ESI^{-}: 267	Alkaloid $C_{14}H_{22}N_2O_8$ (HJY35) from red alga *Porphyra tenera*
2	1.4	330	360		$[M+H]^{+}$: 361; $[M-H]^{-}$: 359	No hit
3	2.2	<230	376		$[M+Na]^{+}$: 399; $[M-H]^{-}$: 375	Diterpenoids $C_{22}H_{32}O_5$ (DOS43 & DOS46) from red alga *Laurencia saitoi*
4	3.2	<230	392		$[M+Na]^{+}$: 415; $[M+K]^{+}$: 431; $[M-H+F.A.]^{-}$:437	No hit
5	3.7	<230	586		$[M+H]^{+}$: 587; $[M-H]^{-}$: 585	Phycoerythrins' Chromophore $C_{33}H_{38}N_4O_6$ (CRR89) from blue-green algae and most red alga
6	4.8	285	339	di-	$[M+H]^{+}$: 338/340/342=1:2:1; $[M-H]^{-}$:336/338/340=1:2:1	No hit
7	5.6	<230	434	tri-	$[M+H]^{+}$: 432/434/436/438=1:3:3:1; $[M-H]^{-}$:430/432/434/436=1:3:3:1	No hit
8	5.9	<230	426		$[M+H]^{+}$: 427; $[M-H]^{-}$: 425	No hit
			409	di-	$[M+H]^{+}$: 408/410/412=1:2:1; $[M-H]^{-}$: 406/408/410=1:2:1	Bromophenol $C_{12}H_{11}Br_2NO_5$ (OML64) from red alga *Rhodomela confervoides*
9	6.1	<230	356		$[M+H]^{+}$: 357; $[M-H]^{-}$: 355	No hit
10	6.3	295	474	tri-	$[M+H]^{+}$: 472/474/476/478=1:3:3:1; $[M-H]^{-}$: 470/472/474/476=1:3:3:1	No hit
11	7.8	293	476	tri-	$[M+H]^{+}$: 474/476/478/480=1:3:3:1, $[M+H-CH_2]^{+}$:460/462/464/468=1:3:3:1,	No hit
12	8.8	290	445	tetra-	$[M+H]^{+}$: 442/444/446/448/450=1:4:6:4:1, $[M+H-CO_2]^{+}$: 398/400/402/404/406=1:4:6:4:1; ESI^{-}: 309/311/313=1:2:1	No hit
13	9.8	295	391	tri-	$[M+H-CH_4O]^{+}$: 357/359/361/363=1:3:3:1; $[M-H]^{-}$: 387/389/391/393=1:3:3:1	Bromophenol $C_8H_7Br_3O_3$ (KBY88) from *Symphyocladia latiuscula*
14	10.6	280	528		$[M+Na]^{+}$: 551; $[M-H]^{-}$: 527	No hit
15	11.0	291	500		$[M+Na]^{+}$: 523; $[M-H]^{-}$: 499	Triterpenoid $C_{32}H_{52}O_4$ (OQD71) from red alga *Galaxaura* sp.
16	11.4	290	556		$[M+Na]^{+}$: 579; $[M-H]^{-}$: 555	Sulfoglycolipid $C_{25}H_{48}O_{11}S$ (OHX57) from red alga *Caulacanthus ustulatus*, brown algae *Sargassum thunbergii* and *Sar. wightii* Sulfoglycolipid $C_{25}H_{48}O_{11}S$ (PTF38) from red alga *Chondria armata*
17	11.9	292	436		$[M+K]^{+}$: 475; $[M-H+F.A.]^{-}$: 481	No hit
18	12.6	<230	654		$[M+Na]^{+}$: 677; $[M-H]^{-}$:653; $[M-H+HCl]^{-}$: 689	No hit
19	13.5	<230	504		$[M+Na]^{+}$: 527; $[M-H+F.A.]^{-}$: 549	No hit
20	13.8	<230	492		$[M+Na]^{+}$: 515; $[M-H]^{-}$: 491	Triterpenoid $C_{30}H_{52}O_5$ (KHT44) from red alga *Laurencia obtuse*; cytotoxic
21	14.2	<230	543		$[M+H]^{+}$: 544; $[M-H+F.A.]^{-}$: 588	No hit
22	16.3	<230	766		$[M+Na]^{+}$: 789; $[M-H]^{-}$: 765	No hit

a) F.A.: formic acid.

Besides, six non-brominated peaks' main component showed identical molecular weights, and similar UV and polarity characteristics with some algal metabolites, including alkaloid, diterpenoids, triterpenoids, photosynthetic segment and sulfoglycolipids. One triterpenoid was reported to exhibit cytotoxicity. Additionally, other 10 peaks did not give proper matches in DNP2011 and may be some new natural products.

Of course, if standard substances had been available, thrse known compounds could have been better confirmed by comparison of retention times and UV spectra.

DISCUSSION

As exhibited by the results, the seaweeds under investigation produced large biomass and also showed high biological activity. Sym. latiuscula in Rhodophyta showed strong anti-MRSA and anti-brine shrimp activities, and inhibitory activity against the spore germination of P. oryzae; besides, it exhibited remarkable inhibitory activity against the two E. coli DDRT strains. Rh. confervoides showed remarkable anti-brine shrimp activity, moderate anti-MRSA activity, and inhibitory activity against the spore germination of P. oryzae. Ch. ocellatus showed moderate anti-MRSA activity and strongest selective inhibitory activity against E. coli strain AB3027 with defect in DNA damage repair. Sar. kjellmanianum in Phaeophyta exhibited powerful anti-brine shrimp activity, moderate anti-MRSA activity and non-selective activity against (+/–) E. coli strains. However, Un. pinnatifida only showed moderate anti-MRSA activity. As to Chlorophyta, En. intestinalis exhibited anti-E. coli and anti-MRSA activities. Cl. fascicularis and Ul. pertusa only inhibited MRSA moderately. In general, the bioactivities of seaweeds from Rhodophyta and Phaeophyta were relatively stronger than those from Chlorophyta.

The seaweeds with anti-MRSA activity or non-selective activity against (+/–) E. coli strains can be used for the development of new drugs to treat infection caused by drugresistant G^+ bacteria or G^- bacteria, respectively. As for the seaweeds possessing both of the two bioactivities, they are potential resources of broad-spectrum antibiotics. To species exhibiting selective inhibitory activity in DDRT, activity against spore germination of P. oryzae or anti-brine shrimp larvae activity, they may have potential anti-tumor activity and are expected to serve as sources of new anticancer drugs. Besides, the algae that can inhibit P. oryzae or kill brine shrimp larvae may be developed as potential antifungal agents or pesticides for medical and agricultural purposes.

Fan's group has reported the diversified bioactivities of seaweeds in China [16,17]. In their reports, the ethyl acetate partitions of alcohol extracts of Sar. kjellmanianum, Rh. confervoides, and Ul. pertusa collected at Shandong coastline also showed strong anti-Staphylococcus aureus (common test strain) activity; the aqueous partition of Sym. Latiuscula's alcohol extract also inhibited E. coli; the alcohol extracts of Sym. Latiuscula, Rh. confervoides, and Ul. pertusa also exhibited cytotoxicity to cancer cell lines. Likewise, our samples from Dalian coastline also showed similar or related activities except that Ul. Pertusa from Dalian did not show significant activity as the same Shandong species did. However, in our test, these seaweeds also showed anti-drug resistant Staphylococcus aureus, anti-fungal, and pesticidal ability.

In our preliminary investigation on their chemical constituents, the HPTLC results showed that these seaweeds can produce quite diverse metabolites and red algae Rh. confervoides and Sym. latiuscula were also found to contain phenolic compounds. HPLC-DAD-MS analysis suggested that Sym. Latiuscula, with the broadest strong activity in present assays, may contain different types of known or new metabolites, including rich brominated compounds such as two known brominated phenols. There are lots of reports on bioactivity of marine halogenated compounds [15]. So, it is possible that the strong bioactivity of Sym. latiuscula has some relationship with its rich brominated metabolites.

In the past decades, there have been more and more reports on bioactivity of seaweeds. *For example*, Sar. kjellmanianum displayed antioxidant effect [18]; Rh. confervoides and Sym. latiuscula showed anti-tumor, antibacterial and antioxidant activities [16]. However, their advantages have not been fully utilized till now. Un. pinnatifida is widely cultivated but is only used to produce food up to now. Sar. kjellmanianum also has large biomass in the Yellow Sea region, but is only used as the raw material of algin industry. If their biomedical and agrichemical applications can be fully exploited, the high value utilization will be greatly promoted. En. intestinalis is widely distributed along coastlines. Enteromorpha species often multiply at a surprising rate when the condition is appropriate. Since 2008, green tides caused by overgrowing Enteromorpha species occurred nearly every summer in Qingdao, China, and researchers have begun to develop ways to turn wastes into wealth. In our study, we found that En. intestinalis may be used as source of antibiotics.

CONCLUSION

In the current study, the common seaweeds from Dalian coastline of the North Yellow Sea were found to yield considerable biomass and some natural constituents with remarkable antibacterial, antifungal, and larvicidal activities as well as possible antitumor potentials. By natural collection or artificial cultivation, their sustainable supply can be ensured, which lays the basis for further chemical research on and sustainable exploitation of their bioactive constituents for medicinal and agrichemical purposes. Furthermore, some waste or harmful algae can also produce wealth. Thus, their higher value utilization is expected in the future.

REFERENCES

1. Li P, Gao P, Wang J X. The research of bioactive substances from marine microorganisms (in Chinese). Hebei Fisheries, 2009, (2): 43–47
2. McMinn A, Martin A, Ryan K. Phytoplankton and sea ice algal biomass and physiology during the transition between winter and spring (McMurdo Sound, Antarctica). Polar Biol, 2010, 33: 1547–1556
3. Shan W R, Li J X, Liu H F. Study on the inhibitory effects of different active materials screened by the filter paper on cotton verticillium wilt (in Chinese). Chin Agricult Sci Bull, 2010, 26: 285–289
4. National Committee for Clinical Laboratory Standards. Methods for dilution antimicrobial susceptibility tests for bacteria that grow aerobically. Approved Standard. 6th Ed. NCCLS document M7-A6. Wayne, Pennsylvania: NCCLS, 2003
5. Zhang Y, Mu J, Han J Y, et al. An improved brine shrimp larvae lethality microwell test method. Toxicol Mechan Methods, 2012, 22: 23–30
6. Kobayashi H, Namikoshi M, Yoshimoto T, et al. A screening method for an antimitotic and antifungal substances using conidia of *Pyriculavia oryzae*, modification and application to tropicalsarine fungi. J Antibiot, 1996, 49: 873–879
7. Alam M Z, Ahmad S, Malik A. Genotoxic and mutagenic potential of agricultural soil irrigated with tannery effluents at Jajmau (Kanpur), India. Arch Environ Contam Toxicol, 2009, 57: 463–476
8. Shao K S, Li X Y. Seasonal variation of benthic seaweed community in the intertide in Dalian (in Chinese). J Dalian Fisheries University, 2000, 15: 29–34

9. Wang L, Li W P, Liu B, et al. Research progress methicillin-resistant *Staphylococcus aureus* (in Chinese). Chin Nurs Res, 2011, 25: 1695–1697

10. Manilal A, Sujith S, Kiran G S, et al. Biopotentials of seaweeds collected from southwest coast of India. J Marine Sci Technol, 2009, 17: 67–73

11. Meyer B N, Ferrigni N R, Putnam G E, et al. Brine shrimp: A convenient general bioassay for active plant constituents. Plant Med, 1982, 45: 31–34

12. Hu Z Y, Liu S Z, Huang H, et al. A rapid screening assay for insecticidal antibiotic produced by marine actinomycetes (in Chinese). Mar Sci Bull, 2000, 19: 36–41

13. Yang C Y. A contrastive study of the screening of endophytic fungi for anticancer activity with the methods of the saline shrimp and A549 cancer cell (in Chinese). J Yunnan Univ National (Nat Sci), 2009, 18: 250–254

14. Buckingham J. Dictionary of Natural Products on DVD. London, UK: Chapman and Hall/CRC, 2011

15. Cabrita M T, Vale C, Rauter A P. Halogenated compounds from marine algae. Mar Drugs, 2010, 8: 2301–2317

16. Xu X L, Fan X, Song F H. Study on bioactivity of Chinese marine alga extracts (in Chinese). Oceanol Limnol Sin, 2004, 35: 55–63

17. Xu N J, Fan X, Han L J, et al. Study on the antibacterial activities of the ethanol extracts of seaweed (in Chinese). Oceanol Limnol Sin, 2002, 33: 265–270

18. Tian Y Y, Yu J. Progress in research on brown seaweed in medicine and food (in Chinese). Mar Sci Bull, 1996, 15: 81–86

CHAPTER

12

USE OF SEAWEED (SARGASSUM TENERRIMUM) EXTRACT AS GEL ENHANCER FOR LESSER SARDINE (SARDINELLA BRACHIOSOMA) SURIMI

Snehal Subhash Shitole[1], Amjad Khansaheb Balange[1], and Shardul Sham Gangan[2]

[1]Central Institute of Fisheries Education, Mumbai, India

[2]Khar Land Research Station, Panvel, India

ABSTRACT

Surimi is a Japanese word for washed fish mince which can be used as a raw material for the preparation of various analogue products like shrimp, lobster, crab analogue, etc. which have a very good demand in the international market. Most of the lean fishes in tropical environment have good gel-forming ability, compared with fatty fishes but due to overexploitation of lean fishes its stock got depleted. However, fatty fish like lesser sardine have low gel-forming ability due to high fat content. Recently, plant phenolic compounds have been used successfully as protein cross-linkers. Therefore, an attempt has been made in the present investigation to extract phenolic compounds from seaweed which is abundantly available at the west coast of India and to use it as protein cross-linker in fatty fish, i.e. lesser sardine surimi. Water seaweed (*Sargassum tenerrimum*) extract (WSE) contained 16.24 mg tannin/g of dry seaweed powder. Effects of WSE at different levels (0.5–2.5 % of Surimi) on the properties of gels from lesser sardine (*Sardinella brachiosoma*) surimi were investigated in comparison with surimi gel without seaweed extract. Gels added with 2.0 % WSE had the increases in gel strength by 76.27 %, compared with the control (without addition of extracts). The lowered expressible moisture content was observed in surimi gels incorporated with 2 % WSE. Slight decreases in whiteness were observed with increasing seaweed extract concentration. Protein solubility % also indicates that, sample prepared with 2 % WSE have low solubility in 0.6 M KCL. There was no significant difference in the pH values of treated surimi gel and control surimi gel samples. Therefore, it may be concluded that the extract of seaweed can be used as gel enhancer in lesser sardine surimi with coincidental increase in texture likeness and had no negative effect on colour and odour likeness.

KEYWORDS Cross-linking Phenolic compounds Gelation Seaweed Lesser sardine Surimi

INTRODUCTION

In recent times, importance of total utilization of fish landings, including under-utilized fish species as human food, has been realized because of the diminishing marine catch and increasing consumer interest in fish products. While several methods are available for value addition of fish, surimi technology has been recognized as one of the most successful techniques for low-cost fish utilization. Surimi gel is a three-dimensional myofibrillar protein network. The textural properties developed during gelation are normally expressed in terms of gel strength, which is the basic parameter for determining the quality and price of surimi (Benjakul et al. 2004a). To increase the gel strength of surimi, various food grade ingredients have been used. However, the addition of these ingredients poses adverse effect on the surimi gel, particularly on off-flavor or off-color development (Rawdkuen and Benjakul 2008). In general, the lean fish have been used for the surimi production but overexploitation of the lean fish has resulted in the insufficiencies of those species as raw material. The use of under-utilized small pelagic fish species, such as sardine and mackerel could be a better alternative for the lean fish but their use for surimi production is limited mainly due to the large quantity of lipids and myoglobin in the muscle tissue (Chaijan et al. 2004). Furthermore, pelagic fish has been found to possess the high proteolytic activity, which is associated with gel softening. To alleviate the problem, protein additives have been widely used to enhance the gel strength of the surimi via inhibition of proteolysis caused by an endogenous proteinase (Benjakul et al. 2004b). Recently, the interactions between phenolic compounds and proteins have been paid more attention in the processing of certain food products. There have been a few studies describing the cross-linking ability of phenolic compounds with proteins (Rawel et al. 2002; Strauss and Gibson 2004). Balange and Benjakul (2009a) reported a significant increase in the gel strength of Bigeye snapper surimi when commercial phenolic compounds in oxidised forms were added. Among all oxidised phenolic compounds used, oxidised tannic acid exhibited the highest gel strengthening effect, compared with oxidised ferulic acid, oxidised catechin and oxidised caffeic acid (Balange and Benjakul 2009a).

Brown seaweed (*Sargassum tenerrimum*) is very common in the west coast of India. Seaweed contains phenol level up to 20 % of their dry weight (Connan and Stengel 2007). Tannin substances with phenolic character occur in marine algae in the physodes of Phaeophyta, such as *Sargassum* species (Vimalabai et al. 2004). Seaweeds have been identified as a rich source of bioactive compounds and their extracts also exhibit antimicrobial potential against the pathogenic microbes. Pharmacologist, physiologists and chemist have been paying increasing attention to the marine organisms particularly on seaweeds for screening bioactive substances (Arunkumar et al. 2010). The bioactive molecules—phenolic compounds, i.e. the secondary metabolites, mainly as a phlorotannins are found at high level in marine brown algae (Ragan and Glombitza 1986). The preparation of seaweed extract containing phenolic compounds could increase the value of those seaweeds and the novel natural additives can be applied in food industry, especially surimi industry.

However, there is a little information on the utilisation of seaweed extract as the cross-linking agents in food proteins, particularly myofibrillar proteins. Therefore, the objectives of this research were to extract and quantify total phenolic compounds in the seaweed (*S. tenerrimum*) and to use the extracts as gel enhancer in surimi from lesser sardine (*Sardinella brachiosoma*).

METHODS

Chemicals

Tannic acid Powder Pure (Molychem, Product code: 19280, Mumbai, India), sodium carbonate (Thomas Dekker, Mumbai, India), Folin–Ciocalteu Phenol reagent (Sisco Research Laboratory, Mumbai, India) were obtained.

Preparation of Seaweed Extracts

Brown seaweed (*S. tenerrimum*) species available along Ratnagiri coast, Maharashtra, India were collected by hand-picking method. Collected samples were packed in polyethylene bags and transported to the Lab. Collected seaweeds were washed with fresh water and were subjected to sun drying for 14 days.

Standardization of Extraction of Phenolic Compounds from Dry Seaweed

The sundried seaweeds were ground using a portable grinding machine (Kenstar, Senator, Japan). Then seaweed powder was sieved with Test sieve (Jayant scientific IND, India) using a screen with a diameter of 0.07 mm. Seaweed extract was then prepared according to the method described by Zahra et al. (2007) with slight modification. The different weights of seaweed powder, i.e. 0.5, 1, 1.5 g were taken and dissolved in 100 ml distilled water and labelled as A, B and C, respectively. Then the mixtures were autoclaved at 121 °C for 15 min in Autoclave of 'Equitron' brand (medical instruments mfg. Co, Mumbai). The autoclaved treatment followed by centrifugation (2,000×*g* for 10 min), with centrifuge (Hettich Zentrifugen, D-78532; Tuttlingen, Germany) and then the supernatant was evaporated directly on flame, to get a final volume 10 ml and extract was collected in glass bottles.

Quantification of Total Phenolic Content

Quantification of total phenolic compounds in different seaweed extracts was carried out according to the method of Kuda et al. (2005). Briefly, 0.4 ml aliquot of the each seaweed extract from different dilutions of each sample was transferred into a test tube containing 0.8 ml of the 10 % Folin–Ciocalteu phenol reagent. After 3 min, 1.6 ml of the 10 % sodium carbonate solution was added. The contents were mixed routinely, using glass rod and left to stand at room temperature for 1 h in dark. Absorbance was recorded at 750 nm using a spectrophotometer (Thermo Scientific Genesys 10 UV, USA). Tannic acid was used as standard for quantification of phenolic content in seaweed. Absorbance for tannic acid solution was recorded at different concentrations and used as the standard. Estimation of the phenolic contents was carried out in triplicates and the results were expressed as mg of Tannic acid equivalent (TA)/g dry sample. Blank for each extract was prepared in the same manner, except that distilled water was used instead of Folin–Ciocalteu reagent.

Surimi Gel Preparation

Fresh surimi from lesser sardine (*S. brachiosoma*) (without prior freezing and cryoprotectants) was purchased from M. D. Naik, Pethkilla, Ratnagiri, Maharashtra, India. Surimi gel was prepared according to the method described by Balange and Benjakul (2009c) with slight modification. Chopped surimi was mixed with seaweed extract (Phenolic compounds) containing 16.24 mg tannin/g of dry seaweed powder, at different concentrations, i.e. 0.0, 0.5, 1.0, 1.5, 2.0 and 2.5 % (% based on weight of surimi) and was then chopped for 5 min at low temperature to obtain a homogeneous surimi along with seaweed extract. Prepared surimi samples (with seaweed extract) and surimi with cryoprotectants (without seaweed extract) were then packed and stored overnight into cold storage maintained at −20 °C.

On next day, frozen surimi was tempered by keeping at room temperature for 1 h, until the core temperature reached 0–2 °C. The surimi was then cut into small pieces with an approximate thickness of 1 cm. Chopped surimi was placed in a mincer and mixed with 3 % salt (73 % surimi moisture content). The final moisture content for all surimi gel samples were adjusted to 80 %. The mixture was chopped for 20 min at low temperature to obtain a homogeneous sol. The sol was then stuffed into krehalon casing (UK) with a diameter of 2.5 cm and both ends of the casing were sealed tightly. Sols were incubated at 40 °C for 30 min, followed by heating at 90 °C for 20 min in water bath (Bio-Technics India, Mumbai). The control gels were prepared in similar way but without seaweed extract addition. After heating, prepared surimi gels were chilled in ice for 20 min and stored overnight in an incubator (Metalab, Mumbai, India) at 4 °C prior to analysis.

Measurement of Gel Properties

Textural Analysis

Textural analysis of gels was performed using a texture analyzer: RHEO TEX (Type: SD-700, Japan). Gels were tested at room temperature. Prepared surimi gels were cut into five cylindrical shaped pieces of 2.5 cm in length. The breaking force (gel strength) and deformation (elasticity/deformability) were measured for each sample by keeping the pieces of each sample into the texture analyzer equipped with a spherical plunger (5 mm diameter; 60 mm/min plunger speed). The probe was pressed into the cut surface of a gel specimen perpendicularly at a constant speed, until puncture occurred. The force in gram (g) required to puncture into the gel (breaking force) and the distance (in mm) at which the ball probe punctured into the gel (deformation) were recorded. Gel strength for each surimi gel was measured from respective breaking force and deformation.

Determination of Whiteness

Whiteness was measured using whiteness meter (MINIOLTA, Chroma meter CR-400; Japan), used in surimi industry for whiteness measurement.

Determination of Expressible Moisture Content

Expressible moisture content was measured according to the method of Balange and Benjakul (2009b) with a slight modification. Gel sample with a thickness of 0.5 cm was weighed (X) and placed between three sheets of Whatman paper No. 4 at the bottom and two sheets on the top of the sample. The standard weight (5 kg) was placed at the top and held for 2 min. The sample was then removed from the papers and weighed again

(Y). Expressible moisture content was calculated using the following equation:

Expressible moisture content(%) = 100 [(X − Y)/X]

Determination of pH

pH was measured using pH meter (Sentex, California, USA). Surimi gel (5 g) was weighed and homogenized with 45 ml of distilled water at a speed of 800 rpm for 1 min and filtered using a filter paper (Whatman paper No. 1). The pH of filtrate was recorded using a pH meter.

Sensory Evaluation

Lesser sardine gels without and with seaweed extract were equilibrated at room temperature and evaluated for appearance, color, taste, texture, odor and overall liking by 30 untrained panelists, in comparison with the control gel (without phenolic compound). A nine-point hedonic scale, in which a score of 1 = not like very much, 5 = neither like nor dislike and 9 = like extremely, was used for evaluation. High score indicated good quality and vice versa (Meilgaard et al. 1990).

Solubility Determination

Solubility of protein in surimi gel was determined as described by Benjakul et al. (2001). Finely chopped gel sample (1 g) was solubilised with 0.6 M KCl. The mixture was homogenized for 1 min, boiled for 2 min and stirred for 4 h at room temperature (28–30 °C) using a magnetic stirrer (Analisis, Mumbai, India). The mixture was centrifuged at 10,000×g for 30 min. Two ml of 50 % (w/v) cold trichloroacetic acid (TCA) were added to 10 ml of supernatant. The mixture was kept at 4 °C for 18 h prior to centrifugation at 10,000×g for 20 min. The precipitate was washed with 10 % (w/v) TCA, followed by solubilizing in 0.5 M NaOH. Protein concentration was determined by the Biuret method (Robinson and Hodgen 1940). Solubility of protein in surimi samples was expressed as the percentage of total protein in surimi gels solubilized directly in 0.5 M NaOH.

SDS–Polyacrylamide Gel Electrophoresis

Protein patterns of surimi gels added with or without WSE at various concentrations were analyzed by SDS–PAGE according to the method of Laemmli (1970). To prepare the protein sample, 27 ml of 5 % (w/v) SDS solution heated to 85 °C were added to the sample (3 g). The mixture was then homogenized using a homogeniser (Polytron, Kinematica, Switzerland) at a speed 11,000 rpm for 2 min and incubated at 85 °C for 1 h to dissolve total proteins. The samples were centrifuged at 3,500×*g* for 20 min to remove undissolved debris. Protein concentration of the supernatant was determined by the Biuret method (Robinson and Hodgen 1940) using bovine serum albumin as a standard. The sample was then mixed with sample buffer (4 ml of 10 % SDS, 2 ml of glycerol, 1 ml of β-Mercaptoethanol, 2.5 ml of 0.5 M Tris–HCl (pH 6.8), and 0.03 g Bromophenol blue) at 1:1 ratio (v/v). The samples (20 μg protein) were loaded onto the polyacrylamide gel made of 10 % running gel and 4 % stacking gel and subjected to electrophoresis at a constant current of 15 mA per gel using a Hoefer Mini Electrophoresis (San Francisco, USA). After separation, the proteins were stained with 0.02 % (w/v) Coomassie Brilliant Blue R-250 in 50 % (v/v) methanol and 7.5 % (v/v) acetic acid and destained with 50 % methanol (v/v) and 7.5 % (v/v) acetic acid, followed by 5 % methanol (v/v) and 7.5 % (v/v) acetic acid.

Scanning Electron Microscopy (SEM)

Microstructure of surimi gels without WSE, i.e. control (Sample A) and with 2 % WSE (Sample B) was determined using scanning electron microscope. The surimi gel samples were cut into small pieces and dried with pressure treatment inside the scanning electron microscope. The specimens were observed with a scanning electron microscope (FEI QUANTA 200 ESEM AND EDAX SYSTEM, Oregon, USA) at an accelerated voltage of 20 kV.

Statistical Analysis

Data were subjected to analysis of variance (ANOVA). Comparison of means was carried out by Duncan's multiple-range tests. Analysis was performed using a SPSS package (SPSS 10.0 for Windows, SPSS Inc, Chicago, IL, USA).

RESULTS AND DISCUSSION

Quantification of Total Phenolic Content

The total phenolic content extracted with water for three different sizes of seaweed powder is depicted in Table 1. It was observed that the extract prepared with 0.5 g dry seaweed powder contained 12.70 mg tannin/g of dry seaweed powder. Extract prepared with 1.0 g dry seaweed powder contained 16.24 mg tannin/g of dry seaweed powder. The higher amount of phenolic content was found in the extract prepared with 1.5 g seaweed powder was 21.85 mg tannin/g of dry seaweed powder. From results it was concluded that, with increasing quantity of seaweed powder, phenolic content in extract was also increased. This was in agreement with earlier reports about total phenolic content in the seaweed extract prepared with 1.0 g seaweed powder (Zahra et al. 2007). However, with 1.5 g seaweed powder, the color of the extract became dark as compared with 0.5 and 1.0 g seaweed powder extract. So, it was decided to use 1.0 g seaweed powder as standard sample size for extraction of phenolic compound and for further use in determining quality of surimi.

Table 1. Total phenolic content in seaweed extract at different concentrations

Sample	mg tannin/g of dry seaweed powder
A	$12.70 \pm 0.76^{a,*}$
B	16.24 ± 0.32^{b}
C	21.85 ± 0.67^{c}

±SE of mean ($n = 3$)

* Different superscripts in the same column indicates the significant difference ($P < 0.05$)

Hwang et al. (2010) reported that, freeze dried hot water extract from *Sargassum hemiphyllum* contained 0.240 mg phenolic compound/ml of extract. Zubia et al. (2008) observed the phenolic content in *Sargassum mangarevense* was 2.45 %, whereas Zahra et al. (2007) found total phenolic compounds in *Sargassum boveanum* water extract was about 17 ± 0.492 mg catechin equivalent/g of dry sample and also recorded the total phenolic content in ethanol extract for same species was about 0.9 mg catechin equivalent/g of dry sample.

Effect of WSE on Gel Strength of Lesser Sardine Surimi Gel

Lesser sardine surimi gels were prepared from 0, 0.5, 1.0, 1.5, 2.0 and 2.5 % seaweed extract concentration containing with total phenolic compound of 16.24 mg tannin/g of dry seaweed powder (Fig. 1). It was observed that there was increase in gel strength with the increase in concentration of the WSE up to a particular level. An experiment was conducted to see the effect of overnight frozen storage on the surimi samples kept without and with cryoprotectants at −20 °C. All the surimi gels prepared with WSE and without addition of any cryoprotectants and stored at −20 °C have good gel formation compared with gel prepared with cryoprotectants and without any WSE (data not shown). The overnight storage at −20 °C without any cryoprotectants might not have affected the gel-formation ability of myofibrillar proteins. Surimi gel with 2 % WSE had the significant ($P < 0.05$) increase in gel strength up to 104 g/cm, compared with that of the control gel, i.e. 51 g/cm. For gels added with 0.5 % of WSE, the minor increase in gel strength up to 97 g/cm was observed compared with that of the control gel, i.e. 51 g/cm. Progressive increase in gel strength was observed when WSE concentrations were used from 0.5 to 2.0 %. With 2.5 % WSE concentration, gel strength decreased, but the gel strength was higher than that of control. Use of WSE at 2 % concentration, increased the lesser sardine surimi gel strength by 76.27 %; compared to that of control surimi gel.

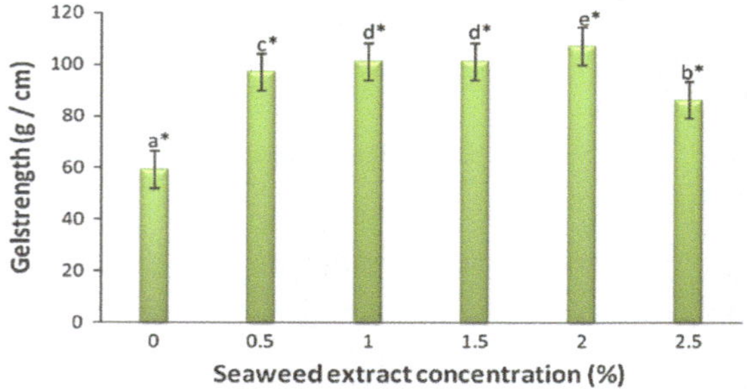

Figure 1. Effect of WSE on lesser sardine surimi gel strength. **Different letters* indicate the significant difference ($P < 0.05$). ± Standard error of mean ($n = 3$).

Lesser sardine surimi gel prepared with addition of 2 % WSE showed significant increase in gel strength by 76.27 % as compared with control. This increase in gel strength might be attributed to the cross-linking activity of phenolic compounds present in the seaweed extract which could induce the formation of both covalent and non-covalent bonds of gel matrix (Prigent et al. 2003). The result is in accordance with that of Balange and Benjakul (2009b) who reported the increases in breaking force and deformation of mackerel surimi with the addition of kiam wood extract containing phenolic compound. The decrease in gel strength of surimi gel with increasing concentration of WSE above 2.0 % in the present investigation may be associated with self-aggregation of phenolic compounds, leading to the loss in capability of protein cross-linking. De Freitas and Mateus (2001) found that the high concentration of phenolic compounds lowers the efficiency in interacting with protein.

Effect of WSE on Expressible Moisture

The lowest expressible moisture content of lesser sardine surimi was found when WSE at optimum level (2 %) was added (Fig. 2). Increase in expressible moisture content was noted in surimi gel prepared without WSE. Moreover, expressible moisture content was high in surimi prepared with WSE above optimum level. Among the WSE, the extract added at 2 % in surimi yielded the gel with the lowest expressible moisture content ($P < 0.05$).

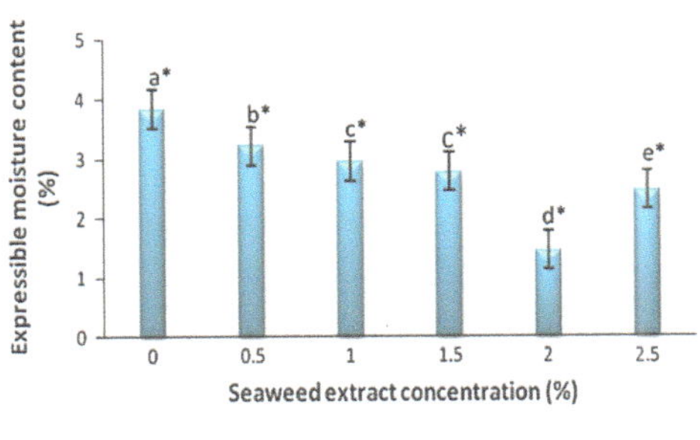

Figure 2. Effect of WSE on expressible moisture content of lesser sardine surimi gel. **Different letters* indicate the significant difference ($P < 0.05$). ± Standard error of mean ($n = 3$).

At the optimal level, the cross-linking of proteins in the lesser sardine surimi gels could be enhanced. This resulted in the formation of stronger network with greater water-holding capacity. Among the extracts WSE at a level of 2.0 % yielded the gel with the lowest expressible moisture content. This reconfirmed that WSE addition resulted in gel strengthening. As a result, gel network with capability of imbibing water could be obtained. Balange and Benjakul (2009b) also reported lowest expressible moisture content of mackerel surimi gel with water kiam wood extract added at optimum level. Also, increase in expressible moisture content was found in surimi gels added with water kiam wood extract above optimum level. Balange and Benjakul (2009c) also reported lowest expressible moisture content in gels with addition of 0.50 % oxidised tannic acid.

Effect of WSE on Whiteness of Lesser Sardine Surimi Gel

Decrease in whiteness of lesser sardine surimi gels was observed as the levels of the WSE were increased (Fig. 3). From the observations, it was clear that with increasing concentration of WSE the whiteness decreased. It was 44.60° when compared with that of control of 47.81°. It was observed that, there was slight difference among treatment effect on whiteness of surimi gel ($P < 0.05$).

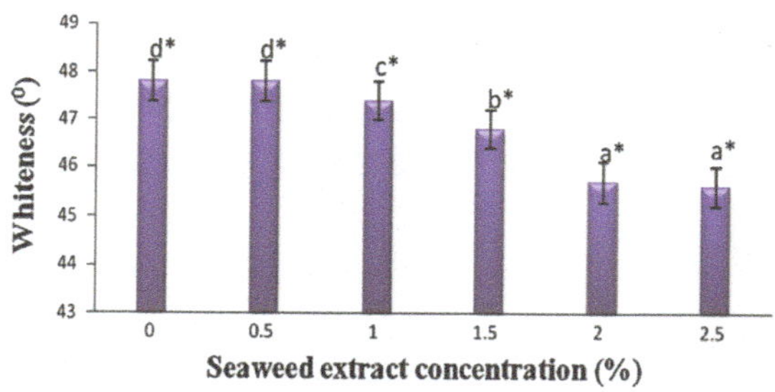

Figure 3. Effect of WSE on lesser sardine surimi whiteness. **Different letters* indicate the significant difference ($P < 0.05$). ± Standard error of mean (n = 3).

Evaporation of water extract at high temperature for a long time enhanced the darkening of water extract. Pansera et al. (2004) used the process of hydrosolubilisation at 100 °C for the extraction of tannin and found that the extraction process at high temperature motivates a hydrocracking of sugar and other organic compounds with darkening of the final product. From the result, surimi gel added with WSE had a slight decrease in whiteness.

Effect of WSE on the pH of Lesser Sardine Surimi Gel

In the present study, addition of WSE-containing phenolic compounds showed no significant difference in pH ($P > 0.05$) with increasing concentration of WSE in surimi gel (Fig. 4). The WSE added in lesser sardine surimi gel contained higher amount of phenolic compounds, i.e. polyphenols, possessing one or more aromatic rings bearing hydroxyl substituent (Parr and Bolwell 2000), which might have resulted in slight increase in pH of surimi gel in the present investigation which was not significant compared with control surimi gel sample.

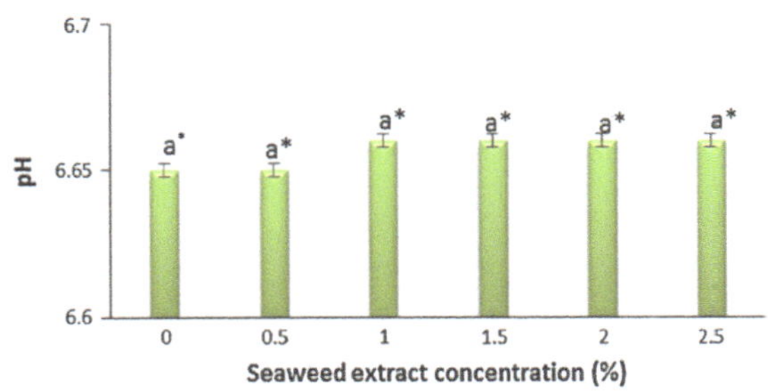

Figure 4. Effect of dry WSE on lesser sardine surimi pH. Not significant difference ($P > 0.05$). ± Standard error of mean ($n = 3$).

Effect of WSE on the Protein Solubility of Lesser Sardine Surimi Gel

Solubility of surimi gels added with different concentrations of WSE is shown in Fig. 5. Solubility was found to be low in surimi gel prepared with 2 % WSE. However, all the gel samples prepared with WSE have low solubility as compared with control gel. In the gels with and without WSE, the decrease in solubility suggests the formation of protein aggregates during setting and heating (Balange and Benjakul 2009a). Also hydrophobic interactions may occur between phenolic compounds and hydrophobic amino acids such as alanine, valine, isoleucine, leucine, methionine, phenylalanine, tyrosine, tryptophan, cysteine and glycine residues (Prigent et al. 2003). The lower solubility obtained in the present investigation is very well correlating with higher gel strength and lower expressible moisture when added with 2 % WSE in surimi gel.

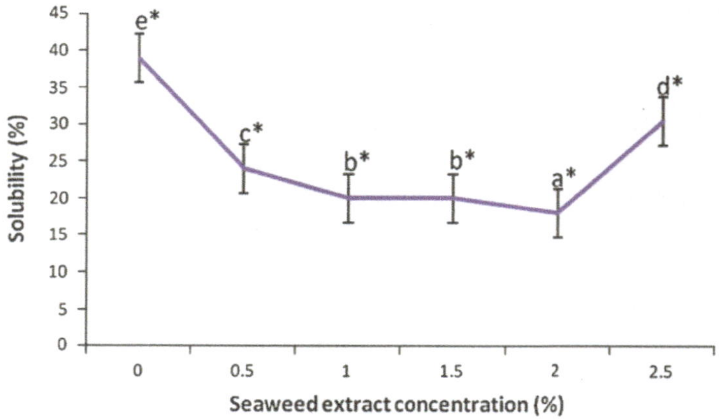

Figure 5. Effect of WSE on lesser sardine surimi gel solubility. **Different letters* indicate the significant difference ($P < 0.05$). ± Standard error of mean (n = 3).

Effect of WSE on the Sensory Properties of Lesser Sardine Surimi Gel

Likeness scores of the control gels (without extract) and those added with the WSE are depicted in Fig. 6. The results revealed that lesser sardine surimi gel prepared with WSE at the 2 % concentration had highest texture and overall acceptability score. The trend observed for textural scores are well reflected by the similar trends for the increasing gel strength and decreasing expressible water values. The explanation given elsewhere for gel strength and expressible water holds good here. Organoleptically, panellist could not distinguish product for their colour but more preference was found for appearance, odour, taste and texture of surimi gels prepared with seaweed extract. Balange and Benjakul (2009b) also reported the highest overall acceptability for surimi gel prepared with 0.15 % ethanolic kiam wood extract, which exhibited highest surimi gel strength.

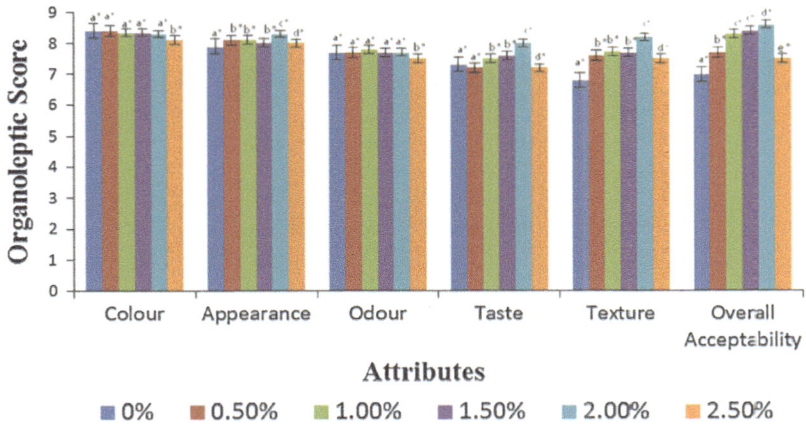

Figure 6. Sensory score of lesser sardine surimi gels added with and without WSE. **Different letters* indicate the significant difference ($P < 0.05$). ± Standard error of mean ($n = 3$).

Effect of WSE on the Protein Pattern of Lesser Sardine Surimi Gel

The protein patterns of surimi gels without and with the addition of WSE are depicted in Fig. 7. For gels with addition of 2.0 % WSE, the MHC band disappeared almost completely. The result suggested that MHC was cross-linked by phenolic compounds to a high extent via non-disulphide covalent bonds. MHC was most susceptible to cross-linking during setting (Benjakul et al. 2004a). No marked changes in actin band intensity were observed between the control gel and those with addition of different concentrations of WSE. Additionally, dark flesh fish were reported to possess high autolytic activity (Benjakul et al. 2004b), which is associated with the poor gel properties. It was postulated that phenolic compounds might partially lower the proteolysis caused by endogenous proteinases. Cross-linked proteins were more likely less susceptible to proteolysis. This might be associated with gel strengthening in addition to enhanced protein cross-linking.

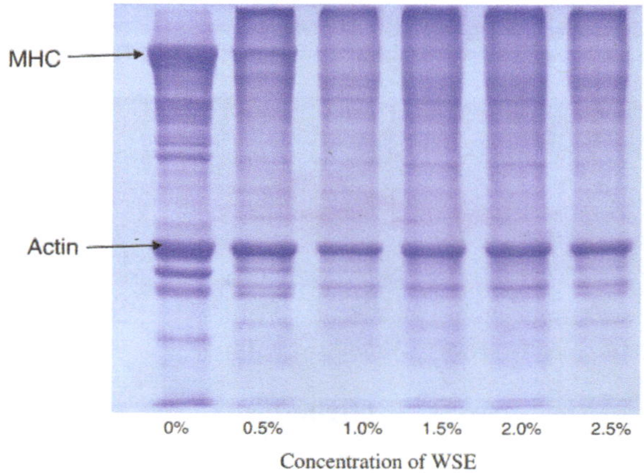

Figure 7. Effect of WSE on protein pattern of lesser sardine surimi gel.

Effect of WSE on the Microstructure of Lesser Sardine Surimi Gel

Microstructures of control gel without WSE (A) and with 2 % WSE (B) are illustrated in Fig. 8. Surimi gel containing 2 % WSE had a finer and continuous matrix than the control. This suggested that phenolic compounds might induce the cross-linking of protein, in which the protein filaments could be formed. Those filaments further underwent polymerization effectively, leading to a gel network with a fibrillar structure and finer strands with high capacity of imbibing water. This might be attributed to the greater number of binding sites in phenolic compounds, which in turn caused a higher aggregation. The finer and more ordered structure of the gel with 2 % WSE added correlated with the highest gel strength (Fig. 1) as well as the lowest expressible moisture content (Fig. 2).

Sample A (Control i.e. without WSE)

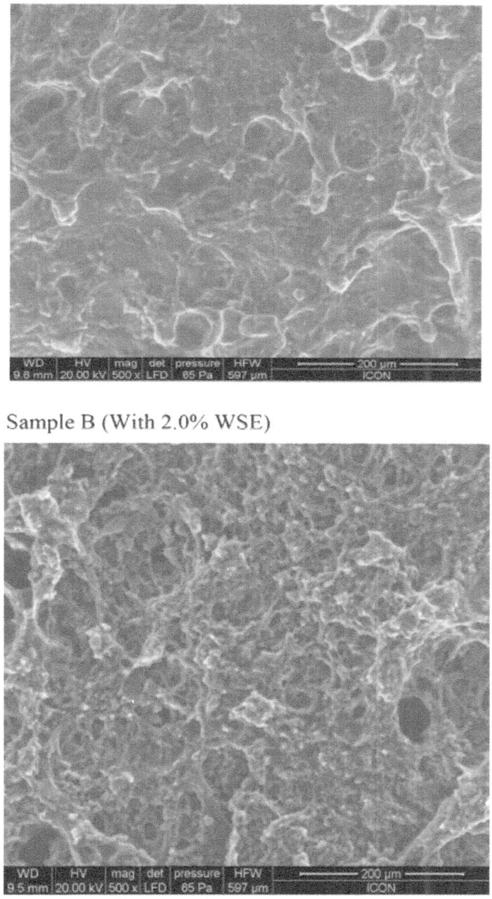

Figure 8. Effect of WSE on microstructure of lesser sardine surimi gel.

CONCLUSIONS

Seaweed extract had a potential in strengthening the gel of lesser sardine surimi when the optimum level (2.0 %) was introduced. Addition of seaweed extract had no detrimental effect on sensory properties of surimi gel. Thus, the extract from seaweed can be used as a natural gel enhancer for lesser sardine surimi industry.

ACKNOWLEDGMENTS

The authors would like to express their sincere thanks to The Director, CIFE, Mumbai and Associate Dean, Collage of Fisheries, Ratnagiri. This research was supported by the International Foundation of Science (IFS), Stockholm, Sweden through a grant to Dr Amjad K Balange.

REFERENCES

1. Arunkumar K, Sivakumar S, Rengasamy R (2010) Review of bioactive potential in seaweeds (marine macroalgae): a special emphasis on bioactivity of seaweeds against plant pathogens. Asian J Plant Sci 9:227–240
2. Balange A, Benjakul S (2009a) Enhancement of gel strength of bigeye snapper (*Priacanthus tayenus*) surimi using oxidised phenolic compounds. Food Chem 113:61–70
3. Balange AK, Benjakul S (2009b) Use of kiam wood extract as gel enhancer for mackerel (*Rastrelliger kanagurta*) surimi. Int J Food Sci Technol 44:1661–1669
4. Balange AK, Benjakul S (2009c) Effect of oxidized phenolic compounds on the gel property of mackerel (*Rastrelliger kanagurta*) surimi. LWT Food Sci Technol 42:1059–1064
5. Benjakul S, Visessanguan W, Srivilai C (2001) Gel properties of bigeye snapper (*Priacanthus tayenus*) surimi as affected by setting and porcine plasma protein. J Food Qual 24:453–471
6. Benjakul S, Visessanguan W, Chantarasuwan C (2004a) Effect of high temperature setting on gelling characteristics of surimi from some tropical fish. Int J Food Sci Technol 39:671–680
7. Benjakul S, Visessanguan W, Tueksuban J, Tanaka M (2004b) Effect of some protein additives on proteolysis and gel-forming ability of lizardfish (*Saurida tumbil*). Food Hydrocoll 18:395–401
8. Chaijan M, Benjakul S, Visessanguan W, Faustman C (2004) Characteristics and gel properties of muscles from sardine (*Sardinella gibbosa*) and mackerel (*Rastrelliger kanagurta*) caught in Thailand. Food Res Int 37:1021–1030
9. Connan S, Stengel DB (2007) Environmental control of brown algal phenol production and assessment of their metal binding properties.

1st environmental change institute research day, NUI Galway

10. De Freitas V, Mateus N (2001) Structural features of procyanidin interactions with salivary proteins. J Agric Food Chem 49:940–945
11. Hwang P, Wu C, Gau S, Chien S, Hwang D (2010) Antioxidant and immune-stimulating activities of hot water extract from seaweed *Sargassum hemmiphyllum*. J Mar Sci Technol 18:41–46
12. Kuda T, Tsunekawa M, Hishi T, Araki Y (2005) Antioxidant properties of dried 'kayamo-nori', a brown alga Scytosiphon lomentaria (*Scytosiphonales, Phaeophyceae*). Food Chem 89:617–622
13. Laemmli UK (1970) Cleavage of structural proteins during assembly of head of bacteriophage T4. Nature 227:680–685
14. Meilgaard M, Civille GV, Carr BT (1990) Sensory Evaluation Techniques. CRC Press Inc., Florida
15. Pansera MR, Iob GA, Atti-Santos AC, Rossato M, Atti-Serafini L, Cassel E (2004) Extraction of tannin by *Acacia mearnsii* with supercritical fluids. Braz Arch Biol Technol 47:1–7
16. Parr AJ, Bolwell GP (2000) Phenols in the plant and in man. The potential for possible nutritional enhancement of the diet by modifying the phenol content or profile. J Sci Food Agric 80:985–1012
17. Prigent SVE, Gruppen H, Visser AJ, Van Koningsveld GA, de Jong GAH, Voragen AGJ (2003) Effects of non-covalent interactions with 5-*O* (ortho)-caffeoylquinic acid (chlorogenic acid) on the heat denaturation and solubility of globular proteins. J Agric Food Chem 51:5088–5095
18. Ragan M, Glombitza K (1986) Phlorotannins, brown algal polyphenols. In: Round FE, Chapman DJ (eds) Progress in Phycological Research, Biopress, Bristol 4:129–241
19. Rawdkuen S, Benjakul S (2008) Whey protein concentrate: autolysis inhibition and effects on the gel properties of surimi prepared from tropical fish. Food Chem 106:1077–1084
20. Rawel HM, Czajka D, Rohn S, Kroll J (2002) Interactions of different phenolic acids and flavonoids with soy proteins. Int J Biol Macromol 30:137–150

21. Robinson HW, Hodgen CG (1940) The biuret reaction in the determination of serum protein. I. A study of condition necessary for the production of the stable colour which bears a quantitative relationship to the protein concentration. J Biol Chem 135:707–725
22. Strauss G, Gibson SM (2004) Plant phenolics as cross-linkers of gelatin gels and gelatin-based coacervates for use as food ingredients. Food Hydrocoll 18:81–89
23. Vimalabai C, Prathiba RA, Sumithra P (2004) Phenolic compounds in brown seaweeds from Tuticorin, southeast coast of India. Seaweed Res Util 26(1–2):93–98
24. Zahra R, Mehrnaz M, Farzaneh V, Kohzad S (2007) Antioxidant activity of extract from a brown alga, *Sargassum bovenum*. Afr J Biotechnol 6(24):2740–2745
25. Zubia M, Payri C, Deslandes E (2008) Alginate, mannitol, phenolic compounds and biological activities of two range extending brown algae, *Sargassum mangarevense* and *Turbinaria ornata* (Phaeophyta: Fucales), from Tahiti (French Polynesia). J Appl Phycol 20:1033–1043

CHAPTER

13

THERAPEUTIC IMPORTANCE OF SULFATED POLYSACCHARIDES FROM SEAWEEDS: UPDATING THE RECENT FINDINGS

Seema Patel

Department of Biotechnology, Lovely Professional University, Jalandhar, India

ABSTRACT

Seaweeds, being prolific sources of bioactive components have garnered unprecedented interest in recent times. The complex polysaccharides from the brown, red and green seaweeds possess broad spectrum therapeutic properties. Especially, the sulfated polysaccharides, viz. fucans, carrageenans and ulvans have exhibited strong antioxidant, antitumor, immunostimulatory, anti-inflammatory, pulmonary fibrosis anticoagulant/antithrombotic, lipid lowering, antiviral, antibacterial, antiprotozoan, hyperplasia prevention, gastrointestinal, regenerative and nano medicine applications. Considering the immense biomedical prospects of sulfated polysaccharides, the profound and emerging functional properties published in recent times will be discussed here with experimental evidences. The limitations of the seaweed-derived sulfated polysaccharides in healthcare will be summarized. Strategies to maximize extraction and bioavailability will be pondered.

KEYWORDS Sulfated polysaccharides, Antioxidant, Antitumor, Anti-coagulant, Antiviral

INTRODUCTION

In recent years, much attention has been focused on polysaccharides isolated from natural sources. During the last decade, numerous bioactive polysaccharides with interesting functional properties have been discovered from seaweeds (Fig. 1). Several algal species belonging to phaeophyta, rhodophyta and chlorophyta divisions have been recognized as crucial sources of sulfated polysaccharides (SP). These SP constitute an important ingredient of cell walls and get harvested by suitable extraction or precipitation method, followed by purification, characterization and biological studies (Fig. 2). The biological features of the SP reported till now are antioxidant, antitumor, immunomodulatory, inflammation, anticoagulant, antiviral, antiprotozoan, antibacterial, antilipemic. Currently, the regenerative medicine and tissue engineering application of the SP has become a hot research area. Jiménez-Escrig et al. (2011) have reviewed the vital role of SP from seaweeds in human health.

Figure 1. Seaweeds growing on the California Coast of the Pacific Ocean.

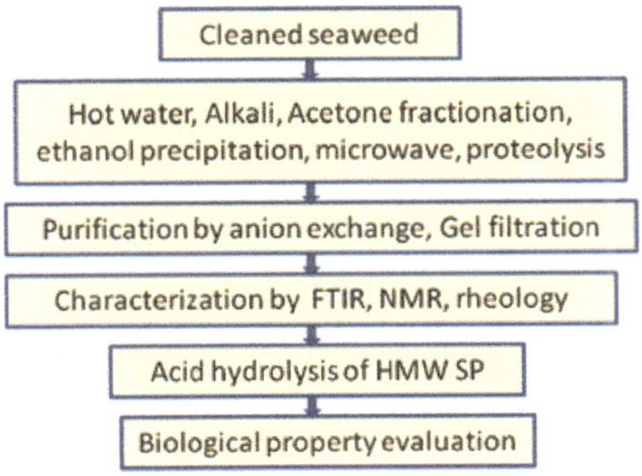

Figure 2. A flowchart depicting the sequential steps for sulfated polysaccharide preparation and biological activity evaluation.

Bioactive SP extracted from seaweeds can be classified into three types. The major fucan yielding brown seaweeds genera are *Fucus*, *Sargassum*, *Laminaria*, *Undaria*, *Lessonia*, *Dictyota*, *Dictyopteris*,*Ascophyllum*, *Eclonia*, *Canistrocarpus*, *Lobophota*, *Turbinaria*, *Padina*, *Adenocystis*, *Sphacelaria*, *Cystoseira*, etc. Fucan represents a family of water soluble, SP rich in sulfated l-fucose, extracted from extracellular matrix of these weeds (Li et al. 2008; Costa et al. 2011a). Fucoidan, the sulfated alpha-l-fucan (term often interchangeably used with fucans) has demonstrated a wide range of pharmacological activities. Carrageenans are a family of linear SP, extracted from red seaweeds, viz. *Gracialaria*, *Gigartina*, *Gelidium*, *Lomentaria*, *Corallina*, *Champia*, *Solieria*, *Gyrodinium*, *Nemalion*,*Sphaerococcus*, *Boergeseniella*, *Sebdenia*, *Scinaia*, etc. This group of polysaccharides has a backbone of alternating 3-linked β-d-galactose and 4-linked α-d-galactose residues (Tuvikene et al. 2006). Three categories of carrageenans, kappa (κ), iota (ι), and lambda (λ) have been identified till now based on their sulfation degree, solubility and gelling properties (Leibbrandt et al. 2010). Ulvan is the major water soluble, sulfated polysaccharide, extracted from the cell wall of green algae, viz. *Ulva*, *Enteromorpha*, *Monostroma*, *Caulerpa*, *Codium*, *Gayralia*. Ulvans are composed of disaccharide repetition moieties made up of sulfated rhamnose linked to either glucuronic acid, iduronic acid, or xylose and represent about 8–29 % of the algal dry weight (Lahaye and Robic 2007). The above-described SP have been illustrated in Fig. 3.

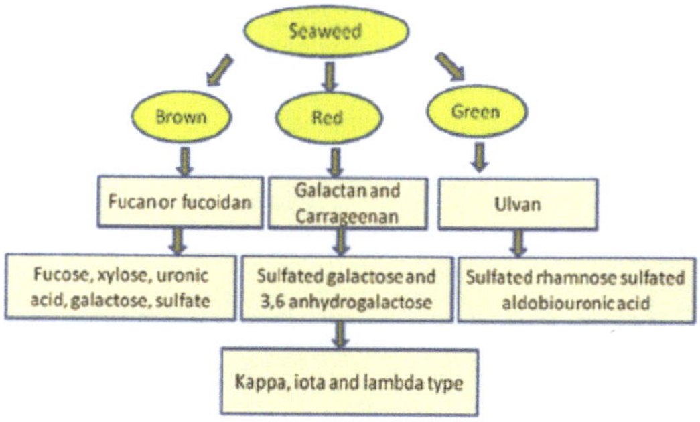

Figure 3. Classification of the bioactive sulfated polysaccharides.

The therapeutic mechanisms of these SP vary, hence it is yet to be studied precisely. For anticoagulation potency, the formation of the SP/protease protein complex and the associated non-specific polar interaction between the negatively and positively charged groups in the polysaccharide and protein is responsible for anticoagulant activity. The anticoagulant activity is mainly attributed to thrombin inhibition mediated by heparin cofactor II, with different effectiveness depending on the compound. Similarly, selectin blockade, inhibition of enzyme and complement cascade seem to be the triggers leading to anti-inflammation. Combating viral infection has been shown by adsorption and internalization steps (Kim et al. 2011, 2012).

Ion exchange, gel filtration, FTIR, NMR analyses are employed to elucidate the composition and structure of SP. Cutting edge technologies, viz. MTT assay, flow cytometry, western blot analysis, BCA protein assay, SDS-PAGE and gelatin zymography has been employed for analysis of their functional properties (Jiang and Guan 2009). Although the use of the seaweed-derived polysaccharides in food industry as thickening, gelling agents, and stable excipients for control release tablets are well established, the clinical use is still to gain ground. Manifold increase in the published findings on this aspect in recent time is evidence enough for the craze over this highly promising domain. Recently, Senni et al. (2011) have reviewed the advancement in therapeutic potential of marine polysaccharides. However, this report was not confined to seaweeds and dealt only with the tissue engineering applications. Also, Wijesekara et al. (2011) have published an overview of clinically crucial SP extracted from marine algae. Keeping with the hot trend and in an attempt to present a new perspective, the present review summarizes the up-to-date literature data and discusses the pharmaceutical potential of different SP extracted from brown, red and green seaweeds.

THERAPEUTIC POTENTIAL OF SULFATED POLYSACCHARIDES

Researchers across the globe are waking up to the discovery that seaweed-derived bioactive products are a storehouse of healthy attributes. Recent times have seen a surge in interest to tap these unexploited marine sources to develop novel therapeutics. The SP of algal origin have exhibited miraculous biological properties. The common seaweeds, their SP and observed bioactivity spectra have been presented in Table 1.

Table 1. The studied seaweeds, their bioactive sulfated polysaccharides and therapeutic properties

Biological properties	Seaweed	Sulfated polysaccharide	References
Antioxidant	*Gracilaria birdiae*(red)	Fucoidan	Souza et al. (2012)
	Fucus vesiculosus(brown)	Galactan	Veena et al. (2007)
	Gigartina skottsbergii(red)	Carrageenan	Barahona et al. (2011)
	Schizymenia binderi(red)	Rhamnan	Magalhaes et al. (2011)
	Lessonia vadosa(brown)		Costa et al. (2011a, b)
	Dictyopteris delicatula (brown)		Wang et al. (2008)
	Sargassum filipendula (brown)		Devaki et al. (2009)
	Laminaria japonica(brown)		Camara et al. (2011)
	Ulva lactuca (green)		Hu et al. (2010)
	Canistrocarpus cervicornis (brown)		Yang et al. (201[illegible])
	Undaria pinnitafida(brown)		Costa et al. (2010)
	Corallina officinalis(red)		
	Corallinasertularioide(red)		
	Dictyotacervicornis(brown)		
	Sargassumfilipendula(brown)		
	Dictyopterisdelicatula(brown)		

Antitumor	*Saccharina japonica*(brown)	Galactofucan	Vishchuk et al. (2011)
	Undaria pinnatifida(brown)	Mannogluc-uronofucan	Costa et al. (2010)
	Sargassumfilipendula(brown)		Charles et al. (2007)
	Dictyopterisdelicatula(brown)		Ye et al. (2008)
	Caulerpaprolifera(green)		Croci et al. (2011)
	Dictyotamenstrualis(brown)		Ermakova et al. (2011)
	Monostroma nitidum(green)		Costa et al. (2011a, b)
	Sargassum pallidum(brown)		Magalhaes et al. (2011)
	Laminaria saccharina(brown)		Jin et al. (2010)
	Ecklonia cava(brown)		Lins et al. (2009)
	Sargassum hornery(brown)		Foley et al. (2011)
	Costaria costata(brown)		Haneji et al. (2005)
	Sargassum filipendula (brown)		
	Dictyopteris delicatula (brown)		
	Champia feldmannii(red)		
	Ascophyllum nodosum (brown)		
	Cladosiphon *okamuranus* Tokida		
Immunostimulatory	*Enteromorpha prolifera* (green)	Fucoidan	Kim et al. (2011, 2012)
	Champia feldmannii(red)	κ-carrageenan	Lins et al. (2009)
	Fucus vesiculosus(brown)	Oligosaccharides	Kawashima et al. (2011)
	Kappaphycus striatum (red)		Kima and Joo (2008)
Antiinflammation and antinociceptive	*Solieria filiformis*(red)	Galactan	de Araújo et al. (2011)
	Gelidium crinale (red)	Mannoglucuronofucans	Farias et al. (2011)

	Sargassum hemiphyllum (brown)	κ-carrageenan	de Sousa et al. (2011a)
	Gracilaria cornea(red)	Oligosaccharides	Hwang et al. (2011)
	Gracilaria birdiae(red)		Coura et al. (2011)
	Laminaria saccharina(brown)		Croci et al. (2011)
	Lobophora variegate(brown)		Medeiros et al. (2008)
	Turbinaria ornata(brown)		Ananthi et al. (2009)
	Padina gymnospora(brown)		Marques et al. (2012)
			Jiang and Guan (2009)
Anticoagulation and antithrombosis)	*Ecklonia cava*(brown)	Arabinogalactans	Wijesinghe et al. (2011)
	Dictyota cervicornis(brown)	Rhamnan	Costa et al. (2010)
	Caulerpa cupresoides(green)	Galactan	Ciancia et al. (2007)
	Codium fragile(green)		Li et al. (2011)
	Codium vermilara(green)		Mao et al. (2008)
	Monostroma latissimum (green)		Camara et al. (2011)
	Monostroma nitidum(green)		Albuquerque et al. (2004)
	Canistrocarpus cervicornis (brown)		Pushpamali et al. (2008)
	Dictyota menstrualis(brown)		Croci et al. (2011)
	Lomentaria catenata(red)		
	Laminaria saccharina(brown)		
Lipid lowering	*Ulva lactuca* (green)	Fucoidan	Kim et al. (2010)
	Sargassum polycystum (brown)		Sathivel et al. (2008)
	Sargassum wightii (brown)		Raghavendran et al. (2005)
	Laminaria japonica(brown)		Huang et al. (2010)

Antiviral (Influenza, herpes, HIV)	*Gyrodinium impudium* (red)	Galactan	Ghosh et al. (2009)
	Nemalion helminthoides (red)	Mannans	Kim et al. (2011, 2012)
	Gayralia oxysperma(green)	Heterorhamnan	Recalde et al. (2009)
	Sphaerococcus coronopifolius (red)	Xylomannan sulfate	Cassolato et al. (2008)
	Boergeseniella thuyoides (red)	Xylogalactofucan	Bouhlal et al. (2011)
	Sebdenia polydactyla(red)	Xylomannan	Bandyopadhyay et al. (2011)
	Sphacelaria indica(brown)		Mandal et al. (2007)
	Cystoseira indica(brown)		
	Grateloupia indica(red)		Chattopadhyay et al. (2007)
	Laminaria angustata(brown)		Trinchero et al. (2009)
	Adenocystis utricularis (brown)		Mandal et al. (2008)
	Scinaia hatei (red)		
Antibacterial (ampicillin resistant *E. coli)* Antiprotozoan (cryptosporidiosis, malaria)	*Kappaphycus alvarezii* (red)	Fucoidan	Kumaran et al. (2010)
	Padina boergessenii(brown)		Maruyama et al. (2007)
	Undaria pinnatifida(brown)		Chen et al. (2009)
Prevent hyperplasia	Brown seaweeds	Fucoidan	Hlawaty et al. (2011)
			Freguin-Bouilland et al. (2007)
Cause gastrointestinal contraction	*Halymenia floresia*(red)	Galactan	Graça et al. (2011)
	Cladosiphon okamuranus Tokida (brown)	Fucoidan	Matsumoto et al. (2004)
Regenerative and nano medicine	Brown seaweeds	Fucoidan	Sezer et al. (2008)
	Ulva rigida (green)	Ulvan	Murakami et al. (2010)
			Nakamura et al. (2008)
			Fukuta and Nakamura (2008)
			Toskas et al. 2011)

Antioxidant

Souza et al. (2012) isolated a SP by aqueous extraction from the red seaweed *Gracilaria birdiae*and observed that the slimy substance exhibits moderate antioxidant properties as measured by DPPH free-radical scavenging effect. Veena et al. (2007) evaluated the efficacy of fucoidan from edible seaweed *Fucus vesiculosus* in Wistar rats (5 mg/kg body wt.). Advocation of the SP enhanced the antioxidant status, thereby preventing membrane injury and averting stone formation. Barahona et al. (2011) evaluated the antioxidant capacity of sulfated galactans from red seaweed *Gigartina skottsbergii* and *Schizymenia binderi*, commercial carrageenans, and fucoidan from brown seaweed *Lessonia vadosa* by the oxygen radical absorbance capacity (ORAC) method. Fucoidan from *L. vadosa* and the sulfated galactan from *S. binderi* exhibited the highest antioxidant capacity. The antioxidant capacity was also evaluated by ABTS and hydroxyl radical scavenging assays. *Corallinasertularioide*, *Dictyotacervicornis*, *Sargassumfilipendula* and *Dictyopterisdelicatula* were studied and found to have SP having immense antioxidant potential in the form of total antioxidant, reducing power and ferrous ion chelating activities (Costa et al. 2010). Two SP fractions rich in galactose and xylose from *Corallina officinalis* demonstrated considerable antioxidant properties (Yang et al. 2011). Hu et al. (2010) isolated two sulfated rhamnose-rich polysaccharide fractions from *Undaria pinnatifida* and evaluated their antioxidant abilities in vitro. It was revealed that the SP possessed strong antioxidant properties. Ye et al. (2008) evaluated the antioxidant activities of SP from *Sargassum pallidum* by DPPH (2,2-diphenyl-1-picrylhydrazyl)-free-radical scavenging assay and reported activity, though low at the tested concentration. Camara et al. (2011) extracted heterofucans from *Canistrocarpus cervicornis* by proteolytic digestion followed by sequential acetone precipitation. The SP exhibited total antioxidant capacity, low hydroxyl radical scavenging activity, good superoxide radical scavenging efficiency and excellent ferrous chelating ability. Devaki et al. (2009) studied the liver mitochondrial and microsomal fraction from rats to evaluate the antioxidative effect of oral gavaging with *Ulva lactuca* polysaccharide extract (200 mg/kg body weight, daily for 21 days). Electron microscopy of rat liver tissue intoxicated with d-galactosamine revealed the swelling and loss of mitochondrial cristae. However, the rats pre-treated with the SP overcame the d-galactosamine challenge without significant abnormality of TCA, microsomal enzymes and mitochondria structural aberrations. These results suggested that the SP play crucial role in stabilizing the functional status of mitochondrial and microsomal membrane by prevention of the oxidative stress induced by d-galactosamine. Fucoidan was extracted from *Laminaria japonica* through anion-exchange column chromatography and their antioxidant activities were investigated. Superoxide and hydroxyl radical scavenging activity, chelating ability and reducing power analysis showed that all fractions possessed considerable antioxidant activity (Wang et al. 2008). Gao et al. (2011) investigated the effects of fucoidan on improving learning and memory impairment in rats induced by infusion of beta-amyloid peptide, Aβ (1–40) and its possible mechanisms. The results indicated that fucoidan could ameliorate Aβ-induced cognitive disorders in neural maladies like Alzheimer's. The mechanisms appeared to regulate the cholinergic system (increasing the activity of choline acetyl transferase), reduce the oxidative stress (reduced malondialdehyde content in hippocampal tissue of brain) and inhibit the cell apoptosis (increase of Bcl-2/Bax ratio and a decrease of caspase-3 activity). Hong et al. (2011) investigated the protective effect of fucoidan on dimethylnitrosamine-induced liver fibrogenesis in rats. When administered (100 mg/kg, 3 times per week), fucoidan improved liver fibrosis by inhibiting the expression of transforming growth factor beta 1 [TGF-β (1)]/Smad3 and the tissue inhibitor of metalloproteinase 1 (TIMP-1), and increasing the expression of metalloproteinase-9 (MMP-9). Fucoidan also significantly decreased the accumulation of the extracellular matrix and collagen, confirming its anti-fibrotic effect. Costa et al. (2011b) obtained five sulfated heterofucans from *S. filipendula* by proteolytic digestion followed by sequential acetone precipitation, which displayed considerable antioxidant potential. Magalhaes et al. (2011) obtained six families of SP from seaweed *D. delicatula* employing above-mentioned protocols, followed by molecular sieving on Sephadex G-100. Some fractions of the heterofucans showed high ferrous ion chelating activity and some fractions showed reasonable reducing power, about 53.2 % of the activity of vitamin C. These results clearly indicate the beneficial effects of SP from seaweeds in antioxidant status of consumers.

Antitumor

Vishchuk et al. (2011) isolated fucoidans from brown seaweeds *Saccharina japonica* and *U. pinnatifida* and tested their antitumor activity against human breast cancer T-47D and melanoma SK-MEL-28 cell lines. The highly branched partially acetylated sulfated galactofucan, built up of (1 → 3)-α-l-fucose residues from *S. japonica* and *U. pinnatifida*distinctly inhibited proliferation and colony formation in both breast cancer and melanoma cell lines in a dose-dependent manner. These results indicated that the fucoidan from the studied seaweeds may be a potential approach toward cancer treatment. After 72-h incubation of HeLa cell with SP (0.01–2 mg/ml), the proliferation was inhibited between 33.0 and 67.5 % by *S. filipendula*; 31.4 and 65.7 % by *D.delicatula*; 36.3 and 58.4 % by *Caulerpaprolifera*, and 40.2 and 61.0 % by *Dictyotamenstrualis*. Costa et al. (2010) inferred that the antiproliferative efficacy of SP positively correlated with the sulfate content. In Sprague–Dawley rats fed with *Monostroma nitidum* diet, significant increase in UGT1A1 and UGT1A6 mRNA levels was found, indicating potential application in chemoprevention medicine (Charles et al. 2007). Ye et al. (2008) evaluated the antitumor activities of SP from *S. pallidum* by MTT [3-(4,5-dimethylthiazol-2-yl)-2,5-diphenyltetrazolium bromide] assay, which showed a significantly high antitumor activity against the human hepatocellular carcinoma (HepG2), human lung adenocarcinoma epithelial (A549) and human gastric carcinoma (MGC-803) cells. Croci et al. (2011) explored the possible antitumor activities of SP from the brown seaweed *Laminaria saccharina.* The incorporation of the parent SP and the sulfated fucans into Matrigel plugs containing melanoma cells induced a significant reduction in hemoglobin content as well as the frequency of tumor-associated blood vessels. Also, these two SP administrations resulted in a significant reduction of tumor growth when inoculated into mice. The sulfated fucan fraction markedly inhibited breast cancer cell adhesion to human platelet-coated surfaces. Ermakova et al. (2011) showed that fucoidans from brown algae *Eclonia cava, Sargassum hornery* and *Costaria costata* play an inhibitory role in colony formation in human melanoma and colon cancer cells. Costa et al. (2011b) observed antiproliferative activity of fucan from *S. filipendula*against HeLa cells by MTT test. The heterofucan was extracted from the brown seaweed by proteolytic digestion followed by sequential acetone precipitation. This SP showed antiproliferative activity on Hela cells and induced apoptosis by mitochondrial release of apoptosis-inducing factor (AIF) into cytosol. In addition, it decreased the expression of anti-apoptotic protein Bcl-2 and increased expression of apoptogenic protein Bax. Magalhaes et al. (2011) obtained six families of SP from seaweed *D. delicatula* by proteolytic digestion, followed by acetone fractionation and molecular sieving on Sephadex G-100. A fraction of the heterofucan showed high antiproliferative activity inhibiting almost 100 % of HeLa cell proliferation. Jin et al. (2010) investigated the effects of fucoidan on the apoptosis of human promyeloid leukemic cells and fucoidan-mediated signaling pathways. Fucoidan induced apoptosis of human promyelocytic leukemia (HL-60), human promyelocytic (NB4) and THP-1 (human acute monocytic leukemia) cell line. Fucoidan treatment of HL-60 cells induced activation of caspases 8, 9, and 3, the cleavage of Bid, and altered mitochondrial membrane permeability. Buthionine-[*R*,*S*]-sulfoximine rendered HL-60 cells more sensitive to fucoidan. It was concluded that the activation of MEKK1, MEK1, ERK1/2 and JNK, depletion of glutathione and production of NO are important mediators in fucoidan-induced apoptosis of human leukemic cells. Lins et al. (2009) investigated the in vitro and in vivo antitumor properties of a SP isolated from the seaweed *C. feldmannii*. The SP did not show any significant in vitro cytotoxicity at the experimental dose, but showed in vivo antitumor effect. The inhibition rates of sarcoma 180 tumor development were 48.62 and 48.16 % at the doses of 10 and 25 mg/kg, respectively. It also increased the response elicited by anti-cancer drug, 5-fluorouracil (5-FU) from 48.66 to 68.32 %. Though liver and kidney were moderately affected, the enzymatic activity of alanine aminotransferase or urea/creatinine levels was not disturbed. Leucopenia associated with 5-fluorouracil treatment was prevented when the chemotherapeutic was administered along with SP. An unfractionated fucoidan was extracted from the brown alga *Ascophyllum nodosum* and its effect on the apoptosis of human HCT116 colon carcinoma cells was studied and the signaling pathways involved were investigated. Fucoidan decreased cell viability and induced apoptosis of the carcinoma cells, through activation of caspases 9 and 3 and the cleavage of PARP (Foley et al. 2011). Haneji et al. (2005) examined the effect of fucoidan from the brown seaweed *Cladosiphon okamuranus* Tokida against an incurable form of cancer, the adult T-cell leukemia (ATL). It was observed that fucoidan inhibited the growth of peripheral blood mononuclear cells of ATL patients and caused apoptosis of HTLV-1-infected T-cell lines through a cascade of down regulations. In vivo treatment of the cancer transplanted in mice also showed partial inhibition of the tumors. Now that, cancer has assumed an epidemic proportion and the treatment scenario is still bleak, the SP from the marine weeds hold the promise for novel anticancer formulae.

Immunostimulatory

Water-soluble SP extracted from *Enteromorpha prolifera* and fractionated using ion-exchange chromatography was investigated to determine their in vitro and in vivo immunomodulatory activities. Some fractions stimulated a macrophage cell line Raw 264.7 inducing considerable nitric oxide (NO) and various cytokine production via up-regulated mRNA expression. The in vivo experiment results showed increase in IFN-γ and IL-2 secretions, suggesting that the SP is a strong immunostimulator. It is implied that the SP can activate T cells by up-regulating Th-1 response (Kim et al. 2011). Lins et al. (2009) demonstrated that SP extracted from *C. feldmannii* is an immunomodulatory agent, evident from the increase in the production of specific antibodies. Kawashima et al. (2011) demonstrated that fucoidan enhances the probiotic effects of lactic acid bacteria on immune functions. In vitro test results showed that fucoidan amplified interferon (IFN)-γ production mediated by IL-12 production from Peyer's patch and spleen cells in response to a strain of LAB, *Tetragenococcus halophilus* KK221. In vivo study showed that Th1/Th2 immunobalance was significantly improved by oral administration of both fucoidan and KK221 to ovalbumin-immunized mice. Kima and Joo (2008) observed that fucoidan from *F. vesiculosus* shows immunostimulating and maturing effects on dendritic cells (DCs) via a pathway involving nuclear factor-κB (NF-κB). κ-Carrageenan oligosaccharides from red algae *Kappaphycus striatum* have immunomodulation effects on S180 tumor-bearing mice. The sulfated derivative (200 μg/g/day) showed an increase in natural killer cells (NK cells) up to 76.1 %. It suggested that chemical modification (especially sulfation) of carrageenan oligosaccharides can enhance their antitumor effect and boost their antitumor immunity. Yuan et al. (2011) reported not only the capacity of SP to elicit cellular immunity but also the importance of chemical modification of the parent polysaccharide.

Anti-Inflammation/Antinociception/Inhibition of Pulmonary Fibrosis

de Araújo et al. (2011) studied the antiinflammatory and antinociception (less sensitivity to painful stimulus) properties of seaweed *Solieria filiformis* in vivo. Male Swiss mice pre-treated with the SP, on receiving an injection of 0.8 % acetic acid, 1 % formalin or 30 min prior to a thermal stimulus, showed significantly reduced number of writhes. It showed antinociceptive action through a peripheral mechanism; however, did not show any significant anti-inflammatory effect. The SP from the brown seaweed *Spatoglossum schroederi* was assayed for the antinociceptive effect on Swiss mice. The SP purified by anion-exchange chromatography inhibited both phases of the formalin test. In the first phase the maximum 45 % reduction in paw licking was observed. This inhibitory effect suggested a mixed mechanism similar to morphine, which was not confirmed in the hot-plate test. It was concluded that the pronounced antinociceptive effect of SP could be developed as a new source of analgesic drugs (Farias et al. 2011). The SP galactan extracted from the red marine alga *Gelidium crinale* was purified by ion-exchange chromatography and tested by intravenous route in rodent experimental models of inflammation and nociception. The anti-inflammatory activity was evaluated in the model of rat paw edema induced by different inflammatory stimuli. Antinociceptive effect was assessed in models of nociception/hyperalgesia elicited by chemical (formalin test), thermal (hot plate), and mechanical (von Frey) stimuli in mice. It was observed that SP inhibited the time course of dextran-induced paw edema and showed a maximal effect at 1 mg/kg (42 %). At the highest dose, the SP also inhibited the paw edema induced by histamine (49 %) and phospholipase A(2) (44 %). The galactan inhibited both neurogenic and inflammatory phases of the formalin test and the treatment was well tolerated by the test animals (de Sousa et al. 2011a). Hwang et al. (2011) explored SP from brown seaweed *Sargassum hemiphyllum* for possible anti-inflammatory effect. The SP was administered against the mouse macrophage cell line (RAW 264.7) activated by lipopolysaccharide (LPS). The secretion profiles of pro-inflammatory cytokines, including IL-1β, IL-6, TNF-α, and NO, were found significantly to be reduced in 1–5 mg/ml dose ranges of SP treatments. RT-PCR analysis suggested that the SP inhibits the LPS-triggered mRNA expressions of IL-β, iNOS and COX-2 in a dose-dependent manner. It was concluded that the anti-inflammatory properties of SP may be attributed to the down-regulation of NF-κB in nucleus. Coura et al. (2011) evaluated the effects of SP from the red seaweed *Gracilaria cornea* in nociceptive and inflammatory mice models. At all tested doses, the SP significantly reduced nociceptive responses, as measured by the number of writhes. In a formalin test, the SP significantly reduced licking time in both phases of the test at a dose of 27 mg/kg. In a hot-plate test, the antinociceptive effect was observed only in animals treated with 27 mg/kg of SP, suggesting that the analgesic effect occurs through a central action mechanism at the highest dose. The lower doses of SP (3 and

9 mg/kg) caused only a slight reduction in neutrophil migration in the rat peritoneal cavity but significantly inhibited paw edema induced by carrageenan, especially at 3 h after treatment. Reduction in edema was confirmed by myeloperoxidase activity in the affected paw tissue. After 14 consecutive days of intraperitoneal administration of the SP (9 mg/kg), the biochemical, hematological and histopathological evaluations of the internal organs are performed and no systemic damage was found. de Sousa et al. (2011b) investigated the involvement of the hemoxygenase-1 (HO-1) pathway in the anti-inflammatory action of a SP from the red seaweed *G. birdiae.* The SP was administered at various concentrations to Wistar rats and observed that at 10 mg/kg concentration, it exerted an anti-inflammatory effect. A remarkable decrease in leukocytes in the peritoneal cavity was also observed. The SP also reduced the paw edema induced by carrageenan and inhibited the paw edema induced by dextran in the first half-hour. The *O*-sulfated mannoglucuronofucans and sulfated fucan fractions from the brown seaweed *L. saccharina* were evaluated for possible treatment of inflammation in vivo. Both types of SP exhibited inhibition of leukocyte rush into the sites of inflammation in the murine models (Croci et al. 2011). Medeiros et al. (2008) extracted a sulfated heterofucan from the brown seaweed *Lobophora variegata* by proteolytic digestion, followed by acetone fractionation, molecular sieving, and ion-exchange chromatography. The fucoidan revealed that it inhibits leukocyte migration to the inflammation site. Ear swelling caused by croton oil was also inhibited when sulfated polysaccharides from *F. vesiculosus* and *L. variegata* were used. Ananthi et al. (2009) investigated the anti-inflammatory effect of crude SP from brown alga *Turbinaria ornata* against carrageenan-induced paw edema in rats and vascular permeability in mice. Oral administration of SP reduced the paw edema and showed inhibitory effect on vascular permeability considerably, in a dose-dependent manner. SP extracted from brown algae *Padina gymnospora* showed efficacy in reducing leukocyte influx into the peritoneal cavity in mice at 10 mg/kg body weight, causing a decrease of 60 %, without any cytotoxicity (Marques et al. 2012). Idiopathic pulmonary fibrosis is a pathological condition characterized by accumulation of excess fibroblasts, deposition of collagen and inflammation in lungs. The pro-fibrogenic cytokine transforming growth factor-beta 1 (TGF-beta1) has attracted much attention for its potential role in the etiology of this serious lung injury. MS80, a new kind of sulfated

oligosaccharide extracted from seaweed, inhibits TGF-beta1-induced pulmonary fibrosis in vitro and bleomycin-induced pulmonary fibrosis in vivo. The oligosaccharide competitively inhibited heparin/HS-TGF-beta1 interaction through its high binding affinity for TGF-beta1, also arrested human embryo pulmonary fibroblast (HEPF) cell proliferation and collagen deposition. MS80 proved to be a potent suppressor of bleomycin-induced rat pulmonary fibrosis in vivo (Jiang and Guan 2009). Du et al. (2010) reported that efficacy of MS80 lies in targeting the CD40 signal pathway by blocking RIP2. The precise mechanism of functionality is not clear; nevertheless, the sulfated polysaccharides studied above promise therapeutic potential in inflammatory disorders.

Anticoagulation

Batteries of assays for assessment of anticoagulation properties of SP from seaweeds have been conducted in recent times. Tests ranging from activated partial thromboplastin time (APTT), thrombin time (TT), prothrombin time (PT), antithrombin to anticoagulation factor Xa activities have been performed and compared with heparin. Wijesinghe et al. (2011) purified a SP from brown seaweed *Ecklonia cava* and investigated its anticoagulant activity in vitro and in vivo. It extended the coagulation time in Wistar rats in a dose- and time-dependent manner. Costa et al. (2010) evaluated in vitro anticoagulant activities of marine algae SP by APTT test. *D. cervicornis* SP prolonged the coagulation time, only 1.4-fold lesser than Clexane®, a low molecular weight commercial heparin. In the prothrombin time (PT) test, which evaluates the extrinsic coagulation pathway, *Caulerpa cupresoides* showed aggression. *Codium fragile* and *Codium vermilara* water-soluble sulfated arabinogalactans prevented coagulation, but they induced platelet aggregation. It was observed that anticoagulant activity was higher in SP samples with higher sulfate content. In this regard, *C. vermilara* proved to be superior with a higher degree of sulfation and arabinose content (Ciancia et al. 2007). The hot water extract of green alga *Monostroma latissimum* gives a sulfated rhamnan polysaccharide with an anticoagulant activity. The anticoagulant activity as evaluated by assays of the APTT and thrombin time promises that it can be a potential source of anticoagulant (Li et al. 2011). Mao et al. (2008) isolated two sulfated, rhamnose-containing polysaccharides from marine green algae *M. nitidum* and evaluated their anticoagulant activities. The results showed that both the SP possess high anticoagulant activities, and were potent thrombin inhibitors mediated by heparin cofactor II. They also hastened thrombin and coagulation factor Xa inhibition by potentiating antithrombin III. Camara et al. (2011) extracted sulfated heterofucans from *C. cervicornis* which prolonged APTT. Four sulfated polysaccharides doubled APTT with only 0.1 mg/ml of plasma, only 1.25-fold less than Clexane®. Albuquerque et al. (2004) extracted heterofucans from the brown seaweed *D. menstrualis* by proteolytic digestion, followed by sequential acetone precipitation. The anticoagulant activities of these heterofucans were determined by APTT test. A fucan fraction (20 g/ml) demonstrated significant anticoagulant activity, about 4.88-fold lesser than Clexane® (4.1 g/ml). Pushpamali et al. (2008) isolated a highly sulfated (21.76 %), 100–500 kDa molecular weight galactan anticoagulant from microbial-fermented freeze-dried red algae *Lomentaria catenata*. It demonstrated that the anticoagulant compound showed better efficacy than heparin and prolonged activity toward APTT and PT assays. Croci et al. (2011) studied that the SP from the brown seaweed *L. saccharina* shows promising activity on thrombosis. Fernández et al. (2012) studied the anticoagulation efficacy of sulfated β-d-mannan extracted from green seaweed *C. vermilara*and reported that higher sulfate content leads to more pronounced effect. Fucoidan has been proposed as a potential substitute of the anticoagulant heparin, with added merits. Unlike mammalian mucosa-derived heparin, fucoidan is extracted from plants, so less likely to contain infectious agents, such as viruses or prions (Boisson-Vidal et al. 1995). The current findings promise a host of possible candidates for natural anticoagulant preparation.

Lipid Lowering

Fucoidan has been reported to affect the development of adipocytes. To elucidate the role of fucoidan in adipogenesis, its inhibitory effect on adipocyte differentiation via mitogen-activated protein kinase (MAPK) signaling pathway in 3T3-L1 preadipocytes was studied. Fucoidan treatment inhibited the adipocyte differentiation, evidenced by decreased lipid accumulation and down-regulation of adipocyte markers. Also, it inhibited the expression of adipogenic transcription factors, α (C/EBPα), γ (PPARγ) and AP2, crucial for adipocyte development (Kim et al.

2010). Sathivel et al. (2008) evaluated the anti-peroxidative and anti-hyperlipidemic property of *U. lactuca* polysaccharide extract against d-galactosamine (500 mg/kg body weight)-induced anomaly in rat. d-Galactosamine-intoxicated rats showed significant liver damage with acute aberration in serum lipid profile, hepatic protein thiols, deposits of lipid droplets and abnormal appearance of mitochondria. Rats pretreated with ulvan (30 mg/kg body weight/day/for 21 days) showed a significant inhibition against abnormality induced by d-galactosamine. The effect of *Sargassum polycystum* crude SP extract on lipid metabolism was examined against acetaminophen-induced hyperlipidemia in experimental rats. The prior oral administration of *S. polycystum* (200 mg/kg body wt./day for a period of 15 days) crude SP extract showed considerable prevention in the severe disturbances of lipid profile and metabolizing enzymes (serum lecithin cholesterol acyl transferase and hepatic triglyceride lipase) triggered by acetaminophen. Liver histology also supported their protective nature against fatty changes induced during acetaminophen intoxication (Raghavendran et al. 2005). Josephine et al. (2007) studied the possible capacity of SP in normalizing hyperlipidemia induced by the immunosuppressant drug cyclosporine A (25 mg/kg body weight, orally for 21 days) in Wistar rat kidney. As a side effect of the drug, lipid profile showed fluctuation resulting in nephrotoxicity manifested by the enhanced urinary excretion of urea, uric acid and creatinine. The SP-treated groups (5 mg/kg body weight, subcutaneously) showed a normalized lipid profile and lipid metabolizing enzymes. Moreover, this group of rats showed a normal concentration of urinary constituents. Huang et al. (2010) investigated the effect of fucoidan from *L. japonica* on hyperlipidemic rats. The SP reduced the concentration of serum total cholesterol, triglyceride and low-density lipoprotein cholesterol and increased the concentration of high-density lipoprotein cholesterol of the studied rats. The activities of lipoprotein lipase, hepatic lipoprotein and lecithin cholesterol acyltransferase were also enhanced. Above findings corroborate that the SP from seaweeds are ideal option for effective abatement of the lipid abnormalities.

Antiviral

Many viruses display affinity for cell surface heparan sulfate proteoglycans playing crucial role in virus entry. This raises the possibility of the application of SP in antiviral therapy (Ghosh et al. 2009). Kim et al. (2012) purified a SP, p-KG03, from the red marine microalga, *Gyrodinium impudium*. The galactan conjugated to uronic acid and sulfated groups had showed inhibition of encephalomyocarditis virus. The inhibitory activity of the SP against influenza virus was examined. The results of a cytopathic effect reduction assay using MDCK cells demonstrated that p-KG03 exhibited the 50 % effective concentration (EC50) values of 0.19–0.48 μg/ml against influenza type A virus infection. The antiviral activity of p-KG03 was deduced to be directly associated with its interaction with viral particles, interfering with its adsorption and internalization into host cell. It was expected to be a candidate for antiviral drug development. The soluble fractions of a sulfated, (1 → 3)-linked α-d-mannans obtained by hot water extraction from *Nemalion helminthoides* showed appreciable antiherpetic activity (Recalde et al. 2009). A homogeneous branched sulfated heterorhamnan was obtained by aqueous extraction, followed by ultrafiltration from the green seaweed *Gayralia oxysperma* which exerted high specific activity against herpes simplex virus (HSV-1) (Cassolato et al. 2008). Treatment of human immunodeficiency virus type 1 (HIV-1), the dreaded etiological agent of AIDS poses tough challenges. The limitations encountered in therapeutic strategy are toxicity, resistance and high costs. Water-soluble sulfated galactans isolated from two red algae *Sphaerococcus coronopifolius* (Gigartinales, Sphaerococcaceae) and *Boergeseniella thuyoides*(Ceramiales, Rhodomelaceae) inhibited in vitro replication of the human immunodeficiency virus (HIV) at 12.5 μg/ml. In addition, the studied polysaccharides were capable of inhibiting the in vitro replication of HSV-1 on Vero cells. The adsorption step of HSV-1 to the host cell seemed to be the specific target for the SP action. While for HIV-1, these results suggest a direct inhibitory effect on HIV-1 replication by controlling the appearance of the new generations of virus and potential virucidal effect (Bouhlal et al. 2011). Ghosh et al. (2009) studied that xylomannan sulfate and its sulfated derivatives purified from *Sebdenia polydactyla* showed strong activity against HSV-1. The IC50 values were in the range 0.35–2.8 μg/ml and they did not exert cytotoxicity at concentrations up to 1,000 μg/ml. Many xylogalactofucan- and alginic acid-containing fractions from marine alga *Sphacelaria indica* showed antiherpetic activity. The IC50 values of their chemically sulfated derivatives against HSV-1 were in the range of 0.6–10 μg/ml and they lacked cytotoxicity at concentrations up to 200 μg/ml (Bandyopadhyay et al. 2011). Sulfated fucan-containing fractions isolated from the brown seaweed *Cystoseira indica*showed potent antiviral activity against HSV-1 and 2 HSV-2 without cytotoxicity for Vero cell cultures. Chemical, chromatographic and spectroscopic methods showed that the anti-herpetic activity of the SP is by inhibition of the virus adsorption (Mandal et al. 2007). Chattopadhyay et al. (2007) analyzed the SP fractions isolated from crude water extract of *Grateloupia indica* and showed their potent anti-HSV activity. The SP, xylogalactofucan fractions extracted from *Laminaria angustata*, after addition of sulfate groups showed enhanced capability to inhibit HSV-1. The IC50 values of these fractions against HSV-1 were in the range of 0.2–25 μg/ml and they lacked cytotoxicity at concentrations up to 1,000 μg/ml (Saha et al. 2012). SP fractions from brown seaweed *Adenocystis utricularis* were analyzed for their in vitro anti-HIV-1 activity. Two of the five studied fractions showed potent anti-HIV-1 activity both against wild type and drug-resistant HIV-1 strains, mediated by blockade of early events of viral replication (Trinchero et al. 2009). The antiviral activity was dependent on the sulfate contents of the polysaccharides. Kazłowski et al. (2012) conducted both in vitro and in vivo studies on Japanese encephalitis virus prevention property of novel SP from *Gracilaria* sp. and *M. nitidum*. During in vitro studies performed by MTT or plaque assays, low-degree-polymerization SP showed a remarkably high positive effect on survivability in JEV-infected C3H/HeN mice. The in vivo antiviral activity was assumed to be a resultant of better absorption of low-DP SP than undigested PS. The results support the feasibility of antiviral drug development from various SP and their derivatives.

Antibacterial and Antiprotozoan

Kumaran et al. (2010) studied that SP extracted from red alga *Kappaphycus alvarezii* and brown alga *Padina boergessenii* exert promising inhibitory response against antimicrobial-resistant *Escherichia coli* strains and, in particular, the inhibitory response of ampicillin-resistant *E. coli*, isolated from local fish markets and seafood processing plants. Maruyama et al. (2007) investigated the effects of fucoidan isolated from the sporophyll of *U. pinnatifida* on the *Cryptosporidium parvum* adhesion to the cultured human intestinal cells and its infection in neonatal mice. The *C. parvum* adhesion to human intestinal 407 cells was significantly suppressed by a low dose (1 mg/ml) of fucoidan (1 μg/ml). The results of the in vivo experiments revealed that *C. parvum* oocysts in the fucoidan-treated mice was reduced to nearly one-fifth of the oocysts number treated with phosphate buffered saline. It was concluded that fucoidan might inhibit cryptosporidiosis through the direct binding of fucoidan to the *C. parvum*-derived functional mediators in the intestinal epithelial cells in neonatal mice. Chen et al. (2009) investigated the inhibitory effects of fucoidan from the edible brown seaweed *U. pinnatifida*, on the growth of Plasmodium parasites. The antimalarial activity of fucoidan was assessed against the cultured *Plasmodium falciparum* parasites in vitro and on *Plasmodium berghei*-infected mice in vivo. Fucoidan significantly inhibited the invasion of erythrocytes by *P. falciparum* merozoites. Its 50 % inhibition concentration was similar to those for the chloroquine-sensitive *P. falciparum* 3D7 strain and the chloroquine-resistant K1 strain. Four-day suppressive testing in *P. berghei*-infected mice with fucoidan resulted in a 37 % suppressive effect versus the control group and a delay in death associated with anemia.

Prevent Hyperplasia

Hlawaty et al. (2011) investigated the therapeutic potential of low molecular weight fucoidan on vascular smooth muscle cell and human vascular endothelial cell proliferation and migration in vitro and in vivo. Sprague–Dawley rats with induced thoracic aorta injury were treated with SP (5 mg/kg/day) for 14 days. Results showed that SP prevented intimal hyperplasia in rat thoracic aorta. In situ zymography showed that the activity of matrix metalloproteinase (MMP)-2 in the neo-intima is significantly reduced. Fucoidans have been shown to mobilize bone marrow-derived progenitor cells via stimulation of stromal-derived factor (SDF)-1 release. Mobilized progenitor cells have been suggested to repair intimal lesions after immune-mediated endothelial injury and thus prevent intimal proliferation. Freguin-Bouilland et al. (2007) evaluated the therapeutic effect of these SP, in Brown Norway and Lewis rat aortic allograft model of transplant arteriosclerosis. The recipient rats were treated with SP (5 mg/kg/day) for 30 days. In contrast to untreated aortic allografts, the SP-treated allografts showed significantly less intimal proliferation. The SP treatment stimulated allograft reendothelialization, as evidenced by strong intimal endothelial nitric oxide synthase antibody and CD31 signals.

Gastrointestinal Functions

Graça et al. (2011) showed that a sulfated galactan isolated from red algae *Halymenia floresia*has promising effects on gastrointestinal (GI) motor functions mediated by voltage-gated Ca^{2+}channels. So, it is suggested that the SP can be useful when gastrointestinal contraction is necessary during motility-related disorders. Inflammatory bowel disease caused by enteric pathogens is a severe form of gastric disease characterized by excess production of proinflammatory cytokine IL-6. Fucoidan derived from brown algae *C. okamuranus* Tokida imparts LPS tolerance and prevents the expression of IL-6 mRNA as evidenced by in vitro and in vivo tests (Matsumoto et al. 2004).

In Regenerative and Nano Medicine

Sezer et al. (2008) prepared a fucoidan–chitosan hydrogel by swelling the polymers in acidic solution and investigated its dermal burn treatment efficiency. Dermal burns were inflicted on male New Zealand white rabbits and the prepared hydrogel was applied on the wounds. Histopathological evaluation of the biopsy samples was done at intervals. No edema was seen in tested groups after 3-day treatment and fibroplasia and scar were fixed after 7-day treatment. The best regeneration on dermal papillary formation and the fastest closure of the wounds were observed in fucoidan–chitosan hydrogels after 14-day treatment. Murakami et al. (2010) developed a hydrogel sheet by blending alginate, chitosan and fucoidan, for rapid wound healing. The hydrogel absorbed Dulbecco's minimal essential medium (DMEM) and fluid absorption became constant within 18 h. On application, this hydrogel is expected to act as tissue adhesive and heal the wound in a moist milieu. Histological examination showed the advanced granulation tissue and capillary formation in the healing-impaired wounds treated with the hydrogel on day 7. Nakamura et al. (2008) reported that a chitosan/fucoidan complex-hydrogel enhanced the half life of fibroblast growth factor (FGF-2) by shielding it against denaturants as heat and proteolysis. Subcutaneous injection of the FGF-2-containing complex-hydrogel into the back of mice showed controlled release of bioactive protein. Slow diffusion of the growth factor induced neovascularization and fibrous tissue formation near the site of injection after 1 week. The complex-hydrogel was biodegraded after 4 weeks after supplying adequate amount of the angiogenic agents for protection of the ischemic heart. Fukuta and Nakamura (2008) reported that fucoidan and its oligosaccharides have the ability to stimulate production of hepatocyte growth factor (HGF) by induction during translation. So, it is believed that fucoidan may protect tissues and organs by mechanisms involving HGF.

Toskas et al. (2011) evaluated the nanofiber ability of an ulvan-rich extract from the alga *Ulva rigida*. Ulvan-based uniform, crystalline nanofibers of diameter 84 nm were produced by blending them with poly(vinyl alcohol) (PVA). The interesting biological and physicochemical properties of the nanofibers can lead to new biomedical applications such as drug release systems. Taken together, these findings indicate that the SP can revolutionize regenerative and nanomedicine, if exploited properly.

BOTTLENECKS ENCOUNTERED

Extraction yield differs with respect to species, period and season of seaweed harvest (Robic et al. 2009). The SP are extracted from the seaweed biomass by many methods which influence their amount and chemical composition. The fucans of brown algae are highly complex and heterogeneous in structure, rendering their study difficult. Fonseca et al. (2008) compared the galactans from two species of red algae having same structure and size but slight variation in sulfation. Due to the variation in sulfate content, the two SP differed in their anticoagulant and venous antithrombotic activities. From the results it was concluded that slight differences in the proportions of sulfated residues in the galactan chain may be critical for the interaction between proteases, inhibitors and activators of the coagulation system. Also, the variations pose challenges in developing therapeutics. Furthermore, the high molecular weights of SPs pose issue in bio-availability (Jiao et al. 2011).

STRUCTURE–FUNCTION CORRELATION OF SP

It is important to understand the biochemical and molecular mechanism of therapeutic actions of SP, in order to develop effective drugs. The monomeric constituents, molecular size, sulfation site, specific structural motif, degree of branching determination are vital for reproducibility of result. Pomin (2009) has reported that the anticoagulant action of SP lies in its ability to inhibit plasma proteases via allosteric changes. The stereospecificities of the carbohydrate–protein complexes hinge on the number of residues in the repeating units, sulfation pattern, anomeric configuration, glycosidic linkage position and molecular mass. Also, the heterogeneities, such as acetylation, methylation and pyruvilation contribute in eliciting variations in functionality (Bilan et al. 2007). A single structural change has been traced to result considerable qualitative difference in results. Pomin and Mourao (2008) reported that preparation of oligosaccharides with well-defined chemical structures from sulfated fucan helps in the studies of carbohydrate–protein interaction. Fonseca et al. (2008) reported that algal sulfated galactans have a procoagulant effect along with the serpin-dependent anticoagulant activity. The procoagulant effect depends on the sulfation pattern of the SP. Slight differences in the proportions and/or distribution of sulfated residues along the galactan chain is critical for the interaction between proteases, inhibitors, and activators of the coagulation system, resulting in a distinct pattern in anti- and procoagulant activities. Identification of structural attributes of SP vital for their biological activities has been limited by their heterogeneous structures. Alasalvar et al. (2010) reported the strong correlation between structure of SP and their antioxidant potency. The monomeric constitution, degree of sulfation and their position, type of glycosidic linkage were held chief determining factors for variation in activity. High sulfate content and low molecular size were studied to exert stronger radical scavenging activities. Frenette and Weiss (2000) determined that sulfation is critical for efficacy of fucoidan in hematopoietic progenitor activity. The desulfated fucoidan failed to promote angiogenesis in vitro or to induce immature CD34+ cell mobilization in vivo. Fucoidan inhibits the human complement system mediated through interactions with certain proteins belonging to the classical pathway, particularly the protein C4. NMR spectra showed that the branched fucoidan oligosaccharides display a better anticomplementary activity compared to linear structures. Spectroscopy and molecular modeling of fucoidan oligosaccharides indicated that the presence of side chains reduces the flexibility of the backbone, mimicking a conformation recognized by the protein C4 (Clement et al. 2010). Leiro et al. (2007) observed that immunostimulatory activity of ulvan-like SP extracted from *U. rigida* was decreased significantly after desulfation of the SP, suggesting the importance of the functional group in eliciting immune response. To tackle the problem of heterogeneity of algal SP, a new approach has been established. The information obtained from studies of invertebrate SP that have a regular structure can be used to deduce the functionally of algal SP (Jiao et al. 2011).

MAXIMIZATION OF THE EXTRACTION AND IMPROVEMENT IN BIOAVAILABILITY

Aqueous (Ghosh et al. 2009) and acetone extraction (Marques et al. 2012) are the most prevalent techniques in SP production from seaweeds. Due to the variations in active growth parameters and extraction conditions, every new SP purified is a unique compound with signature structural features, promising a potential new drug. Rodriguez-Jasso et al. (2011) extracted fucoidan from brown seaweed *F. vesiculosus* by microwave-assisted extraction. Extraction at 120 psi, 1 min, using 1 g/25 ml water proved optimum condition for maximum fucoidan recovery. It was concluded that pressure, extraction time and alga/water ratio affected the SP yield (Rodriguez-Jasso et al. 2011). Supercritical CO_2 extraction, ultrasonic-aid extraction and membrane separation technology may be applied to harvest SP from the seaweeds. Short extraction times, and non-corrosive solvents, cost effective an environmentally benign technique are required for maximum yield. Acid hydrolysis of high molecular weight fucans into low molecular weight compounds facilitates their structural investigation. Further, the low molecular weight fucoidans can be obtained by fucoidanase (E.C.3.2.1.44) treatment. This enzyme sourced from hepatopancreas of invertebrates, marine bacteria and fungi has an added advantage of hydrolyzing the SP without messing with its side substitute groups (Qianqian et al. 2011). Endolytic enzymes, such as ulvan lyases isolated from the flavobacteria *Persicivirga ulvanivorans* cleave the glycosidic bond between the sulfated rhamnose and a glucuronic or iduronic acid in the ulvans (Collen et al. 2011). Alkali modifications of carrageenans are suggested for improved application potential (Campo et al. 2009). Success of commercial reproducibility of highly diverse fucoidan lies in proper characterization with the help of powerful analytical tools (Fitton 2011).

CONCLUSION

The research on SP from seaweeds and their wide biological spectrum have skyrocketed in recent years. Their clinical evaluation for possible noble therapeutics development is catching momentum like never before. For above goals to materialize, the underlying molecular mechanisms need to be understood precisely and elucidated clearly. The relation between structure and function should be unraveled by intensive studies. This up-to-date review on this emerging technique is expected to contribute significantly in supplementing background knowledge, kindling interest for future explorations. Further purification steps and investigation on structural features as well as in vivo experiments are needed to test the viability of their use as therapeutic agents. The SP with appreciably few side effects and myriad benefits could potentially be exploited for complementary medicine use and disease management.

REFERENCES

1. Alasalvar C, Shahidi F, Miyashita K, Wanasundara U (2010) Seafood quality, safety, and health applications. In: Alasalvar C, Shahidi F, Miyashita K, Wanasundara U (eds) An overview, in handbook of seafood quality, safety and health applications. Wiley-Blackwell, Oxford. doi:10.1002/9781444325546.ch1
2. Albuquerque IR, Queiroz KC, Alves LG, Santos EA, Leite EL, Rocha HA (2004) Heterofucans from *Dictyota menstrualis* have anticoagulant activity. Braz J Med Biol Res 37:167–171
3. Ananthi S, Raghavendran HR, Sunil AG, Gayathri V, Ramakrishnan G, Vasanthi HR (2009) In vitro antioxidant and in vivo anti-inflammatory potential of crude polysaccharide from *Turbinaria ornata* (Marine Brown Alga). Food Chem Toxicol 48:187–192
4. Bandyopadhyay SS, Navid MH, Ghosh T, Schnitzler P, Ray B (2011) Structural features and in vitro antiviral activities of sulfated polysaccharides from *Sphacelaria indica*. Phytochemistry 72:276–283
5. Barahona T, Chandía NP, Encinas MV, Matsuhiro B, Zúñiga EA (2011) Antioxidant capacity of sulfated polysaccharides from seaweeds. A kinetic approach. Food Hydrocol 25:529–535
6. Bilan MI, Vinogradova EV, Shashkov AS, Usov AI (2007) Structure of a highly pyruvylated galactan sulfate from the Pacific green alga *Codium yezoense* (Bryopsidales, Chlorophyta). Carbohydr Res 342:586–596
7. Boisson-Vidal C, Haroun F, Ellouali M, Blondin C, Fischer AM,

De Agostini A, Jozefonvicz J (1995) Biological activities of polysaccharide from marine algae. Drugs Fut 20:1237–1249

8. Bouhlal R, Haslin C, Chermann JC, Colliec-Jouault S, Sinquin C, Somin G, Cerantola S, Riadi H, Bourgougnon N (2011) Antiviral activities of sulfated polysaccharides isolated from *Sphaerococcus coronopifolius* (Rhodophytha, Gigartinales) and *Boergeseniella thuyoides*(Rhodophyta, Ceramiales). Mar Drugs 9:1187–1209
9. Camara RB, Costa LS, Fidelis GP, Nobre LT, Dantas-Santos N, Cordiro SL, Costa MS, Alves LG, Rocha HA (2011) Heterofucans from the brown seaweed *Canistrocarpus cervicornis* with anticoagulant and antioxidant activities. Mar Drugs 24:124–138
10. Campo VL, Kawano DF, Silva DBD Jr, Carvalho I (2009) Carrageenans: biological properties, chemical modifications and structural analysis—a review. Carbohydr Polym 77:167–180
11. Cassolato JEF, Noseda MD, Pujol CA, Pellizzari FM, Damonte EB, Duarte MER (2008) Chemical structure and antiviral activity of the sulfated heterorhamnan isolated from the green seaweed *Gayralia oxysperma*. Carbohydr Res 343:3085–3095
12. Charles AL, Chang C-K, Wu M-L, Huang T-C (2007) Studies on the expression of liver detoxifying enzymes in rats fed seaweed (*Monostroma nitidum*). Food Chem Toxicol 45:2390–2396
13. Chattopadhyay K, Mateu CG, Mandal P, Pujol CA, Damonte EB, Ray B (2007) Galactan sulfate of *Grateloupia indica*: isolation, structural features and antiviral activity. Phytochemistry 86:1428–1435
14. Chen JH, Lim JD, Sohn EH, Choi YS, Han ET (2009) Growth-inhibitory effect of a fucoidan from brown seaweed *Undaria pinnatifida* on *Plasmodium* parasites. Parasitol Res 104:245–250
15. Ciancia M, Quintana I, Vizcargüénaga MI, Kasulin L, de Dios A, Estevez JM, Cerezo AS (2007) Polysaccharides from the green seaweeds *Codium fragile* and *C. vermilara* with controversial effects on hemostasis. Int J Biol Macromol 41:641–649
16. Clement MJ, Tissot B, Chevolot L, Adjadj E, Du Y, Curmi PA, Daniel R (2010) NMR characterization and molecular modeling of fucoidan showing the importance of oligosaccharide branching in its anticomplementary activity. Glycobiol 20:883–894

17. Collen PN, Sassi J-F, Rogniaux H, Marfaing H, Helbert W (2011) Ulvan lyases isolated from the flavobacteria *Persicivirga ulvanivorans* are the first members of a new polysaccharide lyase family. J Biol Chem 286:42063–42071

18. Costa LS, Fidelis GP, Cordeiro SL, Oliveira RM, Sabry DA, Câmara RBG, Nobre LTDB, Costa MSSP, Almeida-Lima J, Farias EHC, Leite EL, Rocha HAO (2010) Biological activities of sulfated polysaccharides from tropical seaweeds. Biomed Pharmacother 64:21–28

19. Costa LS, Telles CB, Oliveira RM, Nobre LT, Dantas-Santos N, Camara RB, Costa MS, Almeida-Lima J, Melo-Silveira RF, Albuquerque IR, Leite EL, Rocha HA (2011a) Heterofucan from *Sargassum filipendula* induces apoptosis in HeLa cells. Mar Drugs 9:603–614

20. Costa LS, Fidelis GP, Telles CB, Dantas-Santos N, Camara RB, Cordiro SL, Costa MS, Almeida-Lima J, Melo-Silveira RF, Oliveira RM, Albuquerque IR, Andrade GP, Rocha HA (2011b) Antioxidant and antiproliferative activities of heterofucans from the seaweed *Sargassum filipendula*. Mar Drugs 9:952–966

21. Coura CO, de Araújo IW, Vanderlei ES, Rodrigues JA, Quinderé AL, Fontes BP, de Queiroz IN, de Menezes DB, Bezerra MM, E Silva AA, Chaves HV, Jorge RJ, Evangelista JS, Benevides NM (2011) Antinociceptive and anti-inflammatory activities of sulphated polysaccharides from the red seaweed *Gracilaria cornea*. Basic Clin Pharmacol Toxicol. doi:10.1111/j.1742-7843.2011.00811.x

22. Croci DO, Cumashi A, Ushakova NA, Preobrazhenskaya ME, Piccoli A, Totani L, Ustyuzhanina Ne, Bilan MI, USov AI, Grachev AA, Morozevich GE, Berman AE, Sanderson CJ, Kelly M, DiGregorio P, Rossi C, Tinari N, Iacobelli S, Rabinovich GA, Nifantiev NE (2011) Fucans, but not fucomannoglucuronans, determine the biological activities of sulfated polysaccharides from *Laminaria saccharina* brown seaweed. PLoS One 28:e17283. doi:10.1371/journal.pone.0017283

23. de Araújo IWF, Vanderlei ESO, Rodrigues JAG, Coura CO, Quinderé ALG, Fontes BP, de Queiroz INL, Jorge RJB, Bezerra MM, Silva AAR, Chaves HV, Monteiro HSA, de Paula RCM, Benevides

NMB (2011) Effects of a sulfated polysaccharide isolated from the red seaweed *Solieria filiformis* on models of nociception and inflammation. Carbohydr Polym 86:1207–1215

24. de Sousa AA, Benevides NM, de Freitas Pires A, Fiúza FP, Queiroz MG, Morais TM, Pereira MG, Assreuy AM (2011a) A report of a galactan from marine alga *Gelidium crinale* with in vivo anti-inflammatory and antinociceptive effects. Fundam Clin Pharmacol. doi:10.1111/j.1472-8206.2011.01001.x
25. de Sousa OVE, de Araújo IW, Quinderé AL, Fontes BP, Eloy YR, Rodrigues JA, Silva AA, Chaves HV, Jorge RJ, de Menezes DB, Evangelista JS, Bezerra MM, Benevides NM (2011b) The involvement of the HO-1 pathway in the anti-inflammatory action of a sulfated polysaccharide isolated from the red seaweed *Gracilaria birdiae*. Inflamm Res 60:1121–1130
26. Devaki T, Sathivel A, BalajiRaghavendran HR (2009) Stabilization of mitochondrial and microsomal function by polysaccharide of *Ulva lactuca* on d-Galactosamine induced hepatitis in rats. Chem Biol Interact 177:83–88
27. Du X, Jiang S, Xin X, Li J, Geng M, Jiang H (2010) MS80, a novel sulfated polysaccharide, inhibits CD40-NF-kappaB pathway via targeting RIP2. Mol Cell Biochem 337:277–285
28. Ermakova S, Sokolova R, Kim SM, Um BH, Isakov V, Zvyagintseva T (2011) Fucoidans from brown seaweeds *Sargassum hornery*, *Eclonia cava*, *Costaria costata*: structural characteristics and anticancer activity. Appl Biochem Biotechnol 164:841–850
29. Farias WR, Lima PC, Rodrigues NV, Siqueira RC, Amorim RM, Pereira MG, Assreuy AM (2011) A novel antinociceptive sulphated polysaccharide of the brown marine alga *Spatoglossum schroederi*. Nat Prod Commun 6:863–866
30. Fernández PV, Estevez JM, Cerezo As, Ciancia M (2012) Sulfated β-d-mannan from green seaweed *Codium vermilara*. Carbohydr Polym 87:916–919
31. Fitton JH (2011) Therapies from fucoidan; multifunctional marine polymers. Mar drugs 9:1731–1760
32. Foley SA, Mulloy B, Tuohy MG (2011) An unfractionated fucoidan from *Ascophyllum nodosum*: extraction, characterization, and

apoptotic effects in vitro. J Nat Prod 74:1851–1861

33. Fonseca RJ, Oliveira SN, Melo FR, Pereira MG, Benevides NM, Mourao PA (2008) Slight differences in sulfation of algal galactans account for differences in their anticoagulant and venous antithrombotic activities. Thromb Haemost 99:539–545
34. Freguin-Bouilland C, Alkhatib B, David N, Lallemand F, Henry JP, Godin M, Thuillez C, Plissonnier D (2007) Low molecular weight fucoidan prevents neointimal hyperplasia after aortic allografting. Transplantation 15:1234–1241
35. Frenette PS, Weiss L (2000) Sulfated glycans induce rapid hematopoietic progenitor cell mobilization: evidence for selectin-dependent and independent mechanisms. Blood 96:2460–2468
36. Fukuta K, Nakamura T (2008) Induction of hepatocyte growth factor by fucoidan and fucoidan-derived oligosaccharides. J Pharm Pharmacol 60:499–503
37. Gao Y, Li C, Yin J, Shen J, Wang H, Wu Y, Jin H (2011) Fucoidan, a sulfated polysaccharide from brown algae, improves cognitive impairment induced by infusion of Aβ peptide in rats. Environ Toxicol Pharmacol. doi:10.1016/j.etap.2011.12.022
38. Ghosh T, Pujol CA, Damonte EB, Sinha S, Ray B (2009) Sulfated xylomannans from the red seaweed *Sebdenia polydactyla*: structural features, chemical modification and antiviral activity. Antivir Chem Chemother 19:235–242
39. Graça JRV, Bezerra MM, Lima V, Ariévilo J, Rodrigues G, Monteiro DLS, Quinderé ALG, Amorim RCDN, de Paula RCM, Benevides NMB (2011) Effect of a crude sulfated polysaccharide from *Halymenia floresia* (Rhodophyta) on gastrointestinal smooth muscle contractility. Braz Arch Biol Technol 54:907–916
40. Haneji K, Matsuda T, Tomita M, Kawakami H, Ohshiro K, Uchihara JN, Masuda M, Takasu N, Tanaka Y, Ohta T, Mori N (2005) Fucoidan extracted from *Cladosiphon okamuranus* Tokida induces apoptosis of human T-cell leukemia virus type 1-infected T-cell lines and primary adult T-cell leukemia cells. Nutr Cancer 52:189–201
41. Hlawaty H, Suffee N, Sutton A, Oudar O, Haddad O, Olliver V, Laguillier-Morizot C, Gattegno L, Letourneur D, Charnaux N (2011) Low molecular weight fucoidan prevents intimal hyperplasia

in rat injured thoracic aorta through the modulation of matrix metalloproteinase-2 expression. Biochem Pharmacol 15:233–243

42. Hong SW, Jung KH, Lee HS, Zheng HM, Choi MJ, Lee C, Hong SS (2011) Suppression by fucoidan of liver fibrogenesis via the TGF-β/Smad pathway in protecting against oxidative stress. Biosci Biotechnol Biochem 75:833–840
43. Hu T, Liu D, Chen Y, Wu J, Wang S (2010) Antioxidant activity of sulfated polysaccharide fractions extracted from *Undaria pinnitafida* in vitro. Int J Biol Macromol 46:193–198
44. Huang L, Wen K, Gao X, Liu Y (2010) Hypolipidemic effect of fucoidan from *Laminaria japonica* in hyperlipidemic rats. Pharm Biol 48:422–426
45. Hwang PA, Chien SY, Chan YL, Lu MK, Wu CH, Kong ZL, Wu CJ (2011) Inhibition of Lipopolysaccharide (LPS)-induced inflammatory responses by *Sargassum hemiphyllum*sulfated polysaccharide extract in RAW 264.7 macrophage cells. J Agric Food Chem 59:2062–2068
46. Jiang HD, Guan HS (2009) MS80, a novel sulfated oligosaccharide, inhibits pulmonary fibrosis by targeting TGF-beta1 both in vitro and in vivo. Acta Pharmacol Sin 30:973–979
47. Jiao G, Yu G, Zhang J, Ewart HS (2011) Chemical structures and bioactivities of sulfated polysaccharides from marine algae. Mar Drugs 9:196–223
48. Jiménez-Escrig A, Gómez-Ordóñez E, Rupérez P (2011) Seaweed as a source of novel nutraceuticals: sulfated polysaccharides and peptides. Adv Food Nutr Res 64:325–337
49. Jin JO, Song MG, Kim YN, Park JI, Kwak JY (2010) The mechanism of fucoidan-induced apoptosis in leukemic cells: involvement of ERK1/2, JNK, glutathione, and nitric oxide. Mol Carcinog 49:771–782
50. Josephine A, Veena CK, Amudha G, Preetha SP, Varalakshmi P (2007) Protective role of sulphated polysaccharides in abating the hyperlipidemic nephropathy provoked by cyclosporine A. Arch Toxicol 81:371–379
51. Kawashima T, Murakami K, Nishimura I, Nakano T, Obata A (2011) A sulfated polysaccharide, fucoidan, enhances the

immunomodulatory effects of lactic acid bacteria. Int J Mol Med. doi:10.3892/ijmm.2011.854

52. Kazłowski B, Chiu Y-H, Kazłowska K, Pan C-L, Wu C-J (2012) Prevention of Japanese encpalitis virus infections by low-degree-polymerisation sulfated saccharides from *Garcilaria* sp. and *Monostroma nitidum*. Food Chem. doi:10.1016/j.foodchem.2012.01.106
53. Kim K-J, Lee O-H, Lee B-Y (2010) Fucoidan, a sulfated polysaccharide, inhibits adipogenesis through the mitogen-activated protein kinase pathway in 3T3-L1 preadipocytes. Life Sci 86:21–22
54. Kim J-K, Cho ML, Karnjanapratum S, Shin I-S, You SG (2011) In vitro and in vivo immunomodulatory activity of sulfated polysaccharides from *Enteromorpha prolifera*. Int J Biol Macromol 49:1051–1058
55. Kim M, Yim JH, Kim S-Y, Kim HS, Lee WG, Kim SJ,Kang P-S, Lee C-K (2012) In vitro inhibition of influenza A virus infection by marine microalga-derived sulfated polysaccharide p-KG03. Antiviral Res 93:253–259
56. Kima MH, Joo HG (2008) Immunostimulatory effects of fucoidan on bone marrow-derived dendritic cells. Immunol Lett 115:138–143
57. Kumaran S, Deivasigamani B, Alagappan K, Sakthivel M, Karthikeyan R (2010) Antibiotic resistant *Escherichia coli* strains from seafood and its susceptibility to seaweed extracts. Asian Pac J Trop Med 3:977–981
58. Lahaye M, Robic A (2007) Structure and functional properties of ulvan, a polysaccharide from green seaweeds. Biomacromolecules 8:1765–1774
59. Leibbrandt A, Meier C, König-Schuster M, Weinmüllner R, Kalthoff D, Pflugfelder B, Graf P, Frank-Gehrke B, Beer M, Fazekas T, Unger H, Prieschl-Grassauer E, Grassauer A (2010) Iota-carrageenan is a potent inhibitor of influenza A virus infection. PLoS One 5(12):e14320. doi:10.1371/journal.pone.0014320
60. Leiro JM, Castro R, Arranz JA, Lamas J (2007) Immunomodulating activities of acidic sulphated polysaccharides obtained from the seaweed *Ulva rigida* C. Agardh. Int Immunopharmacol 7:879–888
61. Li B, Lu F, Wei X, Zhao R (2008) Fucoidan: structure and bioactivity.

Molecules 13:1671–1695

62. Li H, Mao W, Zhang X, Qi X, Chen Y, Chen Y, Xu J, Zhao C, Hou Y, Yang Y, Li N, Wang C (2011) Structural characterization of an anticoagulant-active sulfated polysaccharide isolated from green alga *Monostroma latissimum*. Carbohydr Polym 85:394–400
63. Lins KO, Bezerra DP, Alves AP, Alencar NM, Lima MW, Torres VM, Farias WR, Pessoa C, de Moraes MO, Costa-Lotufo LV (2009) Antitumor properties of a sulfated polysaccharide from the red seaweed *Champia feldmannii* (Diaz-Pifferer). J Appl Toxicol 29:20–26
64. Magalhaes KD, Costa LS, Fidelis GP, Oliveira RM, Nobre LT, Dantas-Santos N, Camara RB, Albuquerque IR, Cordeiro SL, Sabry DA, Costa MS, Alves LG, Rocha HA (2011) Anticoagulant, antioxidant and antitumor activities of heterofucans from the seaweed *Dictyopteris delicatula*. Int J Mol Sci 12:3352–3365
65. Mandal P, Mateu CG, Chattopadhyay K, Pujol CA, Damonte EB, Ray B (2007) Structural features and antiviral activity of sulphated fucans from the brown seaweed *Cystoseira indica*. Antivir Chem Chemother 18:153–162
66. Mandal P, Pujol CA, Carlucci MJ, Chattopadhyaya K, Damonte EB, Ray B (2008) Anti-herpetic activity of a sulfated xylomannan from *Scinaia hatei*. Phytochemistry 69:2193–2199
67. Mao W-J, Fang F, Li H-Y, Qi X-H, Sun H-H, Chen Y, Guo S-D (2008) Heparinoid-active two sulfated polysaccharides isolated from marine green algae *Monostroma nitidum*. Carbohydr Polym 74:834–839
68. Marques CT, de Azevedo TCG, Nascimento MS, Medeiros VP, Alves LG, Benevides NMB, Rocha HAO, Leite EL (2012) Sulfated fucans extracted from algae *Padina gymnospora* have anti-inflammatory effect. Rev Bras Farmacogn 22. doi:10.1590/S0102-695X2011005000206
69. Maruyama H, Tanaka M, Hashimoto M, Inoue M, Sasahara T (2007) The suppressive effect of Mekabu fucoidan on an attachment of *Cryptosporidium parvum* oocysts to the intestinal epithelial cells in neonatal mice. Life Sci 80:775–781
70. Matsumoto S, Nagaoka M, Hara T, Kimura-Takagi I, Mistuyama K,

Ueyama S (2004) Fucoidan derived from *Cladosiphon okamuranus* Tokida ameliorates murine chronic colitis through the down-regulation of interleukin-6 production on colonic epithelial cells. Clin Exp Immunol 136:432–439

71. Medeiros VP, Queiroz KC, Cardoso ML, Monteiro GR, Oliveira FW, Chavante SF, Guimaraes LA, Rocha HA, Leite EL (2008) Sulfated galactofucan from *Lobophora variegata*: anticoagulant and anti-inflammatory properties. Biochemistry (Mosc.) 73:1018–1024
72. Murakami K, Aoki H, Nakamura S, Nakamura S, Takikawa M, Hanzawa M, Kishimoto S, Hattori H, Tanaka T, Sato Y, Ishihara M (2010) Hydrogel blends of chitin/chitosan, fucoidan and alginate as healing-impaired wound dressings. Biomaterials 31:83–90
73. Nakamura S, Nambu M, Ishizuka T, Hattori H, Kanatani Y, TAkase B, Kishimoto S, Amano Y, Aoki H, Kiyosawa T, Ishihara M, Maehara T (2008) Effect of controlled release of fibroblast growth factor-2 from chitosan/fucoidan micro complex-hydrogel on in vitro and in vivo vascularization. J Biomed Mater Res A 85:619–627
74. Pomin VH (2009) An overview about the structure-function relationship of marine sulphated polysaccharides with regular chemical structures. Biopolymers 91(8):601–609
75. Pomin VH, Mourao PAS (2008) Structure, biology, evolution, and medical importance of sulfated fucans and galactans. Glycobiology 18:1016–1027
76. Pushpamali WA, Nikapitiya C, De Zoysa M, Whang I, Kim SJ, Lee J (2008) Isolation and purification of an anticoagulant from fermented red seaweed *Lomentaria catenata*. Carbohydr Polym 73:274–279
77. Qianqian W, Shuang M, Hourong X, Min Z, Jingmin C (2011) Purification and the secondary structure of fucoidanase from *Fusarium sp.* LD8. Evid Based Complement Altern Med. doi:10.1155/2011/196190
78. Raghavendran HR, Sathivel A, Devaki T (2005) Effect of *Sargassum polycystum*(Phaeophyceae)-sulphated polysaccharide extract against acetaminophen-induced hyperlipidemia during toxic hepatitis in experimental rats. Mol Cell Biochem 276:89–96
79. Recalde MP, Noseda MD, Pujol CA, Carlucci MJ, Matulewicz MC (2009) Sulfated mannans from the red seaweed *Nemalion*

helminthoides of the South Atlantic. Phytochemistry 70:1062–1068

80. Robic A, Sassi J-F, Dion P, Lerat Y, Lahaye M (2009) Seasonal variability of physicochemical and rheological properties of ulvan in two *Ulva* species (Chlorophyta) from the Brittany coast. J Phycol 45:962–973
81. Rodriguez-Jasso RM, Mussatto SI, Pastrana L, Aguilar CN, Teixeira JA (2011) Microwave-assisted extraction of sulfated polysaccharides (fucoidan) from brown seaweed. Carbohydr Polym 86:1137–1144
82. Saha S, Navid MH, Bandyopadhyay SS, Schnitzler P, Ray B (2012) Sulfated polysaccharides from *Laminaria angustata*: structural features and in vitro antiviral activities. Carbohydr Polym 87:123–130
83. Sathivel A, Raghavendran HR, Srinivasan P, Devaki T (2008) Anti-peroxidative and anti-hyperlipidemic nature of *Ulva lactuca* crude polysaccharide on d-galactosamine induced hepatitis in rats. Food Chem Toxicol 46:3262–3267
84. Senni A, Pereira J, Gueniche F, Delbarre-Ladrat C, Sinquin C, RAtiskol J, Godeau G, Fischer A-M, Helley D, Colliec-Jouault S (2011) Marine polysaccharides: a source of bioactive molecules for cell therapy and tissue engineering. Mar Drugs 9:1664–1681
85. Sezer AD, Cevher E, Hatipoglu F, Ogurtan Z, Bas AL, Akbuga J (2008) Preparation of fucoidan-chitosan hydrogel and its application as burn healing accelerator on rabbits. Biol Pharm Bull 31:2326–2333
86. Souza BWS, Cerqueira MA, Bourbon AI, Pinheiro AC, Martins JT, Teixeira JA, Coimbra MA, Vicente AA (2012) Chemical characterization and antioxidant activity of sulfated polysaccharide from the red seaweed *Gracilaria birdiae*. Food Hydrocol 27:287–292
87. Toskas G, Hund R-D, Laourine E, Cherif C, Smyrniotopoulos V, Roussis V (2011) Nanofibers based on polysaccharides from the green seaweed *Ulva rigida*. Carbohydr Polym 84:1093–1102
88. Trinchero J, Ponce NM, Córdoba OL, Flores ML, Pampuro S, Stortz CA, Salomón H, Turk G (2009) Antiretroviral activity of fucoidans extracted from the brown seaweed *Adenocystis utricularis*. Phytother Res 23:707–712

89. Tuvikene R, Truus K, Vaher M, Kailas T, Martin G, Kersen P (2006) Extraction and quantification of hybrid carrageenans from the biomass of the red algae *Furcellaria lumbricalis*and *Coccotylus truncatus*. Proc Estonian Acad Sci Chem 55:40–53

90. Veena CK, Josephine A, Preetha SP, Varalakshmi P (2007) Beneficial role of sulfated polysaccharides from edible seaweed *Fucus vesiculosus* in experimental hyperoxaluria. Food Chem 100:1552–1559

91. Vishchuk OS, Ermakova SP, Zvyagintseva TN (2011) Sulfated polysaccharides from brown seaweeds *Saccharina japonica* and *Undaria pinnatifida*: isolation, structural characteristics, and antitumor activity. Carbohydr Res 346:2769–2776

92. Wang J, Zhang Q, Zhang Z, Li Z (2008) Antioxidant activity of sulfated polysaccharide fractions extracted from *Laminaria japonica*. Int J Biol Macromol 42:127–132

93. Wijesekara I, Pangestuti R, Kim SK (2011) Biological activities and potential health benefits of sulphated polysaccharides derived from marine algae. Carbohydr Polym 84:14–21

94. Wijesinghe WAJP, Athukorala Y, Jeon Y-J (2011) Effect of anticoagulative sulfated polysaccharide purified from enzyme-assistant extract of a brown seaweed *Ecklonia cava* on Wistar rats. Carbohydr Polym 86:917–921

95. Yang Y, Liu D, Wu J, Chen Y, Wang S (2011) In vitro antioxidant activities of sulfated polysaccharide fractions extracted from *Corallina officinalis*. Int J Biol Macromol 49:1031–1037

96. Ye H, Wang K, Zhou C, Liu J, Zeng X (2008) Purification, antitumor and antioxidant activities in vitro of polysaccharides from the brown seaweed *Sargassum pallidum*. Food Chem 111:428–432

97. Yuan H, Song J, Li X, Ning L, Song L (2011) Enhanced immunostimulatory and antitumor activity of different derivatives of κ-carrageenan oligosaccharides from *Kappaphycus striatum*. J Appl Phycol 23:59–65

CHAPTER

14

IOTA-CARRAGEENAN IS A POTENT INHIBITOR OF RHINOVIRUS INFECTION

Andreas Grassauer[1,2], Regina Weinmuellner[1], Christiane Meier[1,2], Alexander Pretsch[1,2], Eva Prieschl-Grassauer[1,2], and Hermann Unger[2]

[1]Marinomed Biotechnologie GmbH, Veterinaerplatz 1/HA, A-1210 Vienna, Austria

[2]Laboratory of Tropical Veterinary Medicine, Veterinary University Vienna, Veterinaerplatz 1, A-1210 Vienna, Austria

ABSTRACT

Background

Human rhinoviruses (HRVs) are the predominant cause of common cold. In addition, HRVs are implicated in the worsening of COPD and asthma, as well as the loss of lung transplants. Despite significant efforts, no antiviral agent is approved for the prevention or treatment of HRV-infection.

Results

In this study we demonstrate that Iota-Carrageenan, a sulphated polysaccharide derived from red seaweed, is a potent anti-rhinoviral substance in-vitro. Iota-Carrageenan reduces HRV growth and inhibits the virus induced cythopathic effect of infected HeLa cells. In addition, Iota-Carrageenan effectively prevents the replication of HRV1A, HRV2, HRV8, HRV14, HRV16, HRV83 and HRV84 in primary human nasal epithelial cells in culture. The data suggest that Iota-Carrageenan acts primarily by preventing the binding or the entry of virions into the cells.

Conclusion

Since HRV infections predominately occur in the nasal cavity and the upper respiratory tract, a targeted treatment with a product containing Iota-Carrageenan is conceivable. Clinical trials are needed to determine whether Iota-Carrageenan-based products are effective in the treatment or prophylaxis of HRV infections.

BACKGROUND

The family Picornaviridae comprises some notable members, including human rhinovirus (HRV), which infects humans more frequently than any other virus. Infections with HRV lead to the common cold with symptoms such as sore throat, rhinitis, nasal congestion, and cough [1]. The National Institutes of Health (NIH) estimates that there are more than a billion cases of common colds in the USA each year. Besides the self-limiting infection, HRV is implicated as a cause or predisposing agent for otitis media, sinusitis and exacerbations of asthma, as well as other lower respiratory tract disorders [1, 2, 3, 4].

Despite significant efforts no anti-viral agent is approved for the prevention or treatment of HRV-infection. A number of anti-viral compounds have been evaluated for the management of HRV induced colds, including the capsid binders pirodavir and Pleconaril [3, 5, 6, 7]. Studies with biologicals such as intranasal Tremacamra a soluble intercellular adhesion molecule 1 (ICAM-1) and alpha interferon have shown that targeting HRV is possible especially when the drugs are applied prophylactically or the intervention is early. [8, 9, 10].

Another approach targets the HRV proteases 2A and 3C with small molecules. Protease 3C is an enzyme necessary for the posttranslational cleavage of viral precursor polyproteins. Studies with experimental

HRV infection showed promising results for Ruprintrivir a compound developed by (Agouron/Pfizer) [6]. Development of Tremacamra and Ruprintrivir has not advanced to phase III clinical trials until today.

To effectively inhibit the HRV induced inflammatory cascade of the common cold the treatment needs to be initiated rapidly after the first symptoms or even before. Since the HRV infection is self-limiting and not life threatening in most cases a potential therapy has to be safe and effective with an almost unrecognizable level of side effects.

Polymers from various sources are substances that might bear these desired safety properties. In particular sulphated polysaccharides including Carrageenan, a sulphated polysaccharide extracted from red seaweed has an excellent safety profile and has shown anti-viral efficacy against several viruses. The anti-HIV-1 activity of Lambda-, Kappa- and Iota-Carrageenan and other sulphated polymers has been described previously [11, 12]. In a review, Gonzalez M.E. et al. [11] report an anti-viral efficacy of different sulphated polysaccharides including Iota-Carrageenan against several animal viruses. Iota-Carrageenan showed anti-viral activity against the enveloped viruses Herpes simplex virus type 1 and type 2, Semliki Forest virus (SFV), vaccinia virus, African swine fever virus (ASF), and against encephalomyocarditis (EMC) virus. Iota-Carrageenan had no effect on vesicular stomatitis virus (VSV), measles virus, polio virus type 1 (member of the picornaviridae) and adenovirus type 5. Carlucci et al. [11, 13] demonstrated a protective effect of Lambda-Carrageenan on genital herpes simplex virus infection in mice. Pujol et al. [14] showed the anti-viral activity of a Carrageenan isolated from Gigartina skottsbergii against intraperitoneal murine herpes simplex virus infection.

Carrageenan has been generally recognized as safe by the FDA. In addition, Carrageenan has been extensively used in the food, cosmetic and pharmaceutical industry as a thickener and gelling agent. In this report we show that Iota-Carrageenan inhibits the replication of HRV in tissue culture. Therefore Iota-Carrageenan might be a promising candidate for the evaluation of efficacy against HRV in clinical trials in humans.

RESULTS

Carrageenan Promotes Cell Survival after HRV2 Infection

Carrageenan has been shown to bear anti-viral activity against herpes simplex virus (HSV), cytomegalovirus (CMV), dengue virus, papilloma virus, and human immunodeficiency virus (HIV) [11, 12, 15, 16]. To study the effect of different Carrageenan-subtypes (Lambda-, Kappa-, and Iota-Carrageenan) on HRV, a comparative experiment was performed. HRV infection of cells induces morphological changes and cell death, commonly known as cythopathic effect (CPE). To quantify the virus induced cell death, a proliferation assay was employed. As indicator for cell survival the tetrazolium substrate conversion into a formazan dye was measured (XTT-Test). The resulting optical density (OD) values reflected the metabolic activity of cells. HeLa cells were infected with HRV2 at an amount of input virus of 0,1 $TCID_{50}$/cell. Metabolic activity was measured 48 hours post infection (p.i.), when a CPE of more than 90% was observed by microscopy. OD values of HRV2 infected, untreated HeLa cells were set to 0%. Survival of mock infected cells was set to 100%. Polymers Lambda-, Kappa-, and Iota-Carrageenan were applied at a concentration of 200 μg/ml. The protection against virus-induced cell death was 55% for Lambda-Carrageenan and 62% for Kappa-Carrageenan (Figure. 1A). However, 200 μg/ml Iota-Carrageenan completely blocked the virus induced cell death. All tested Carrageenans did not show cytotoxic effects on uninfected cells up to concentrations of 1000 μg/ml after 48 h (data not shown).

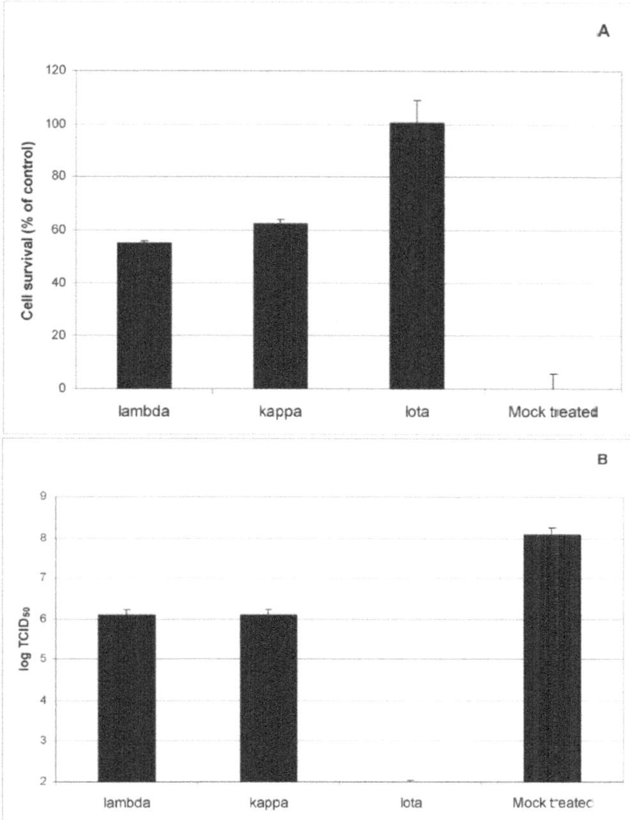

Figure 1. Carrageenans promote cell viability of HRV2 infected HeLa cells and inhibit HRV2 replication in vitro. A. HeLa cells grown in 96-well plates were infected with HRV2 (0,1 $TCID_{50}$/cell) in the presence of Carrageenans (different types are indicated at the x-axis) at a concentration of 200 µg/ml. Plates were incubated at 37°C until cells in the control (no polymer added) showed >90% damage. Cell proliferation was determined with an XTT-assay. OD values (492 nm) obtained from mock infected cells (compare x-axis) were set to 100%, and the viability of cells infected in the absence of polymer was set to 0% (y-axis). The bars represent the mean of a quadruplicate experiment, the standard deviation is indicated. B. HeLa cells in 24-well plates were infected with HRV2 (0,1 $TCID_{50}$/cell) in the presence of Carrageenans (different types are indicated at the x-axis) at a concentration of 200 µg/ml. Viral infectivity in the supernatants

was determined by $TCID_{50}$ assay on HeLa cells (y-axis). Values represent the mean of six parallel titrations, standard deviation is indicated.

Iota-Carrageenan Reduces Production of HRV Particles

HeLa cells were seeded in 24-well plates (2 * 10^4 cells per well) and infected with HRV2 (0,1 $TCID_{50}$/cell) in the presence of Iota-Carrageenan at a concentration of 200 µg/ml. When cell lysis was observed in the untreated control, supernatants were harvested. Viral titers were determined by $TCID_{50}$ assays on HeLa cells. HRV2 replication in untreated control cells resulted in the generation of 10^8 $TCID_{50}$/ml after 48 h (Figure 1B). Lambda- and Kappa-Carrageenan reduced HRV2 titers in cell supernatants by two orders of magnitude. Iota-Carrageenan exceeded the activity of Lambda- and Kappa-Carrageenan and prevented viral titer production for at least 6 orders of magnitude when compared with the untreated control (Figure 1B). Since, the detection limit was 10^2 $TCID_{50}$ in this test an even higher effect cannot be excluded.

The Anti-viral Effect of Iota-Carrageenan is Dependent on the Amount of Input Virus

To test whether the amount of input virus has an effect on the anti-viral properties of Iota-Carrageenan, HeLa cells were infected with HRV2 with three different amounts of input virus. The survival of infected HeLa cells after 72 h was determined with an XTT assay as described above. In one arm of the experiment the infection was performed in the presence of Iota-Carrageenan (prophyalxis experiment). In the other arm of the experiment the infection was performed in the absence of polymer (treatment experiment). 30 minutes after infection the inocula were removed in both groups, cells were washed and polymer was added at concentrations ranging from 0,7 µg/ml to 400 µg/ml. A clear dependency on the amount of input virus was observed in both tested schemes (Figure 2). At an amount of input virus of 0,01 $TCID_{50}$/cell HeLa cells were completely protected from virus induced CPE at a concentration as low as 4 µg/ml in both the prophylaxis and the treatment experiment. At 0,1 $TCID_{50}$/cell the protective concentration of Iota-Carrageenan increased to 10 µg/ml in the prophylaxis experiment and no protection was observed in the treatment experiment. With input virus of 1 $TCID_{50}$/

cell in the prophylaxis experiment only the highest concentration of Iota-Carrageenan (400 µg/ml) resulted in a partial protection and again no effect was observed in the treatment experiment. These data suggest that the anti-viral effect of Iota-Carrageenan is dependent on the amount of input virus. In addition, the polymer is more efficacious in a prophylactic setting compared to a treatment setting.

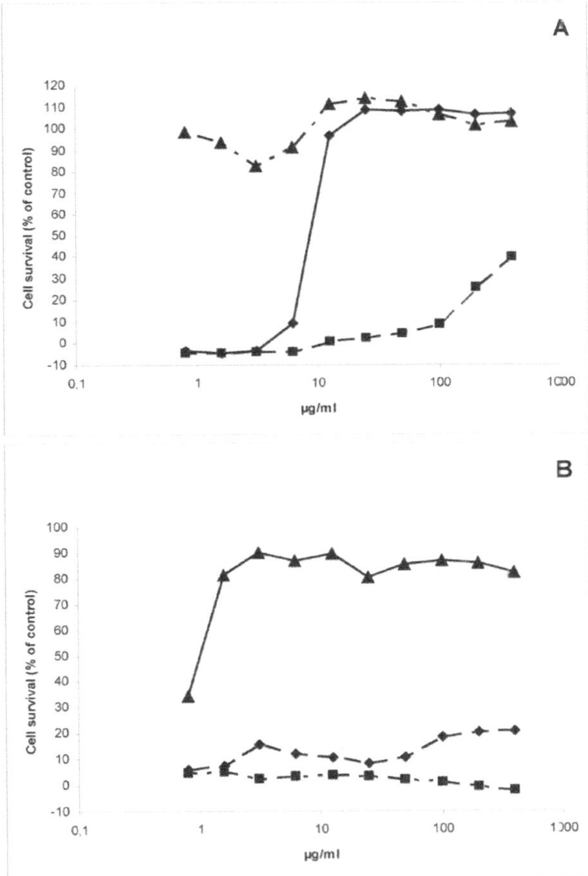

Figure 2. Iota-Carrageenan induced inhibition of HRV2 infected cells is dependent on the amount of virus. A. Preincubation of virus with polymer. HeLa cells grown in 96-well plates were infected with HRV2 in the presence of Iota-Carrageenan at concentrations as indicated at the x-axis. B. Treatment with polymer after infection. HeLa cells grown in 96-well plates were infected with HRV2. 30 minutes after infection medium containing Iota-Carrageenan at the

concentrations indicated at the x-axis was added. Plates were incubated at 37°C until cells in the control (no polymer added) showed >90% damage. Cell proliferation was determined with an XTT-assay. OD values (492 nm) obtained from mock infected cells (compare x-axis) were set to 100%, and the viability of cells infected in the absence of polymer was set to 0% (y-axis). Black triangles indicate an amount of input virus of 0,01 $TCID_{50}$/cell, black diamonds indicate 0,1 $TCID_{50}$/cell and black squares indicate 1 $TCID_{50}$/cell. A representative experiment is shown.

In an analogous experiment the viral titers in supernatant were determined. Cells were infected with HRV2 (0,1 $TCID_{50}$/cell) in the presence or absence of different concentrations of Iota-Carrageenan. 48 h after infection supernatants were harvested and titers were determined with a $TCID_{50}$ assay. When the infection was done in presence of Iota-Carrageenan, a dose-dependent reduction of the viral titer was observed showing a reduction of 3 orders of magnitude at 100 µg/ml (Figure 3A). When the polymer was added after the infection a reduction in the viral titer of 2 orders of magnitude was observed for the concentrations 25, 50 and 100 µg/ml. At the concentrations 12,5 and 6,25 µg/ml the reduction in the viral titer was less pronounced (Figure 3B). Similar to the CPE assays, this experiment showed that titers of supernatants of cells infected in the presence of Iota-Carrageenan are reduced compared to titers of supernatants of cells infected in the absence of Iota-Carrageenan. In order to exclude a direct effect of the polymer to the cells we incubated Hela cells with Iota-Carrageenan three hours before infection. Cells were washed twice with PBS prior to infection with HRV. This treatment of the cells did not result in a significant effect on cell survival and replication of HRV (data not shown). The data imply that the anti-viral effect of Iota-Carrageenan against HRV is due to the inhibition of viral binding or entry into the cells.

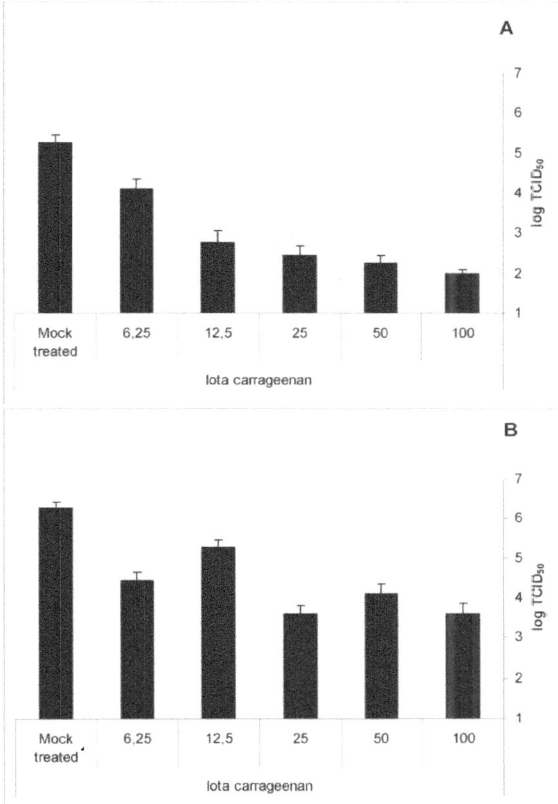

Figure 3. Iota-Carrageenan dose-dependently inhibits HRV2 replication in cell culture. (A) Preincubation of virus with polymer. HeLa cells grown in 12-well plates were infected with HRV2 (0,1 $TCID_{50}$/cell) in the presence of Iota-Carrageenan at the concentration indicated at the x-axis. 30 minutes after infection the inoculum was removed and medium containing Iota-Carrageenan with the concentration indicated was added. Untreated cells were used as control (mock treated). B. Treatment with polymer after infection. HeLa cells grown in 24-well plates were infected with HRV2 (0,1 $TCID_{50}$/cell). 30 minutes after infection the inoculum was removed and medium containing Iota-Carrageenan with the concentration indicated at the x-axis was added. Untreated cells were used as control (mock treated). Viral titers in the supernatants of infected cells were determined after 48 h by $TCID_{50}$ assay on HeLa cells. Values are the means from six parallel titrations, standard deviation is indicated.

Search for Iota-Carrageenan Resistant Variants

We were interested whether Iota-Carrageenan resistant variants occur with high frequency and can be characterized. 6-well plates with HeLa cells ($8 * 10^4$ cells per well) were infected in the presence of polymer with HRV2 (0,1 $TCID_{50}$/cell) and 30 minutes post infection medium containing a given polymer concentration was added. After two days of incubation at 37°C a cytopathic effect was observed in the mock treated control and supernatants were collected. The supernatants of those wells showing partial protection from virus induced cells death (7 μg/ml and 20 μg/ml) were used as inoculum for the next selection round. After ten repetitive infection experiments, the original HRV2 virus was compared with the HRV2 virus from the last passage in a CPE inhibition experiment (for details see material and methods; Figure 4). No significant difference was observed in a CPE inhibition experiment between the original HRV2 viruses that have been obtained after ten passages. This result indicates that escape mutants against Iota-Carrageenan do not occur frequently in HeLa cells.

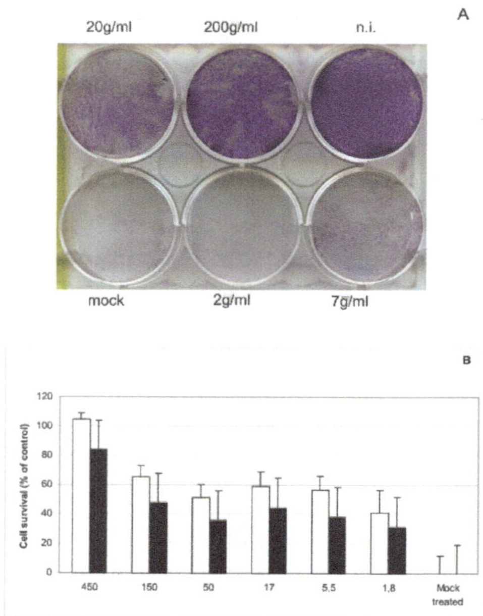

Figure 4. Iota-Carrageenan does not induce HRV2 escape mutants after 10 passages. A. HeLa cells in 6-well plates ($8 * 10^4$ cells per well) were infected

with HRV2 in the presence of Iota-Carrageenan. After infection the cells were washed and medium containing polymer was added at concentrations between 2 μg/ml and 100 μg/ml. Plates were incubated at 37°C until cells in the control (no polymer added) showed >90% damage. Living cells were fixed and stained with crystal violet staining solution. B. Supernatants from infected wells with Carrageenan of 20 μg/ml were used for the next infection round. For the following infection rounds the supernatants of wells with 7 μg/ml or 20 μg/ml were used for the subsequent infection round. After ten repetitive infection experiments the sensitivity of the resulting virus (white bars) to different concentrations of Iota-Carrageenan (x-axis) was compared with that of the original virus (black bars). Cell proliferation was determined with an XTT-assay. Survival of mock infected cells was set to 100%, and that in the absence of polymer was set to 0% (y-axis). The bars represent means of six independent experiments standard deviation is indicated.

Iota-carageenan Blocks Replication of HRV2 in Primary Human Nasal Epithelial Cells (HNep)

In order to study whether the activity against HRV is a tissue culture phenomenon an experiment with human nasal epithelial cells (HNep) was conducted. HNep cells were grown in 24-well plates. The infection with HRV2 (0,1 $TCID_{50}$/cell) was carried out in the presence or absence of Iota-Carrageenan and 30 minutes post infection medium containing polymer in the range of 0,2 μg/ml to 500 μg/ml was added. After 48 hours analysis of viral titers in the supernatants of infected cells revealed that HRV2 replicates to titers of approximately $10^7 TCID_{50}$/ml on HNep cells. The viral titer was below the detection limit of 10^2 $TCID_{50}$/ml when 55 μg/ml of Iota-Carrageenan was already present during the infection (Figure 5A). When the infection was done in the absence of Iota-Carrageenan a concentration of 500 μg/ml was needed to reduce the viral replication below the detection limit (Figure 5B). However, in both cases a significant reduction in the viral titer was observed when the polymer concentration was at least 2 μg/ml. This result shows that Iota-Carrageenan inhibits replication of HRV2 on HNep cells.

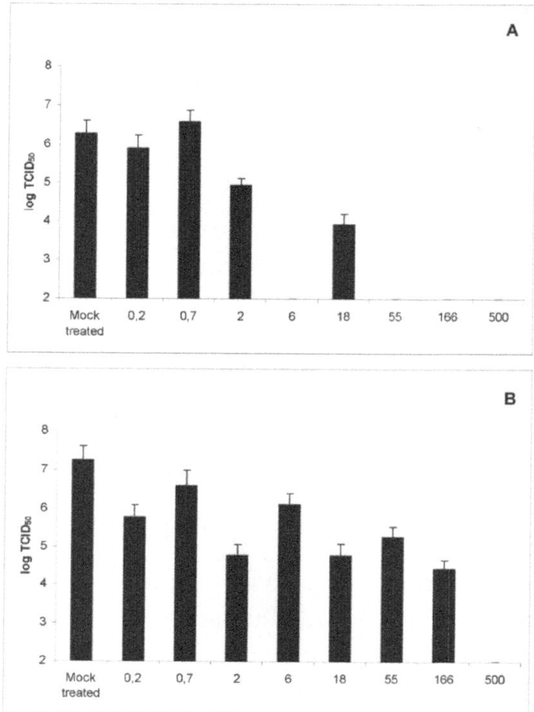

Figure 5. Effect of Iota-carageenan on HRV2 infected human nasal epithelial cells. A. Preincubation of virus with polymer. HNep cells were grown in 24-well plates were infected with HRV2 (0,1 $TCID_{50}$/cell) in the presence of Iota-Carrageenan at the concentration indicated at the x-axis. 30 minutes after infection the inoculum was removed and medium containing Iota-Carrageenan with the concentration indicated was added. B. Treatment with polymer after infection. HNep cells were grown in 24-well plates were infected with HRV2 (0,1 $TCID_{50}$/cell). 30 minutes after infection the inoculum was removed and medium containing Iota-Carrageenan with the concentration indicated at the x-axis was added. Viral titers in the supernatants of infected cells were determined after 48 h by $TCID_{50}$ assay on HeLa cells (y-axis). Bars represent means of four parallel experiments, standard deviation is indicated.

Iota-carrageenan Inhibits Replication of HRV Serotypes 1A, 8, 14, 16, 83, 84 on Primary Human Epithelial Cells

Since more than 100 distinctive HRV serotypes are circulating in humans it was important to reveal whether Iota-Carrageenan is also effective against other strains of HRVs. The work of Ledford et al. shows that the EC_{50} concentration against HRV of the capsid binder Pleconaril has a strain dependent variability between 0,01 µg/ml and >12,5 µg/ml [17]. Based on this work we selected HRV1A, HRV16 and HRV8 for testing. These three viruses belong to the HRV-A virus group and are in contrast to HRV2 relatively insensitive to Pleconaril. From the HRV-B virus group we tested the Pleconaril sensitive strain HRV83, the moderate sensitive strain HRV14, and HRV84, a strain that cannot be inhibited by Pleconaril at a concentration of 12,5 µg/ml. Primary human nasal epithelial cells were seeded in 96-well plates ($4.8 * 10^3$ cells per well) and infected with HRV an amount of input virus of 2 $TCID_{50}$/cell. Supernatants were harvested between 48–72 hours after infection and the viral titers were determined by $TCID_{50}$assays on HeLa cells. While HRV1, HRV14, HRV16 and HRV83 replicated to titers above $10^7 TCID_{50}$/ml, HRV8 reached a titer of $10^{5,1}$ $TCID_{50}$/ml and HRV84 a titer of $10^{4,1}$ $TCID_{50}$/ml (Figure 6). A Iota-Carrageenan concentration of 50 µg/ml was sufficient to reduce the replication on HNep cells of all tested viruses by more than 3 orders of magnitude (99,9%). At a Iota-Carrageenan concentration of 5 µg/ml an inhibition of greater than 99% was observed for HRV1, HRV14, HRV16, HRV83 and HRV84. However, for HRV8 a reduction from $10^{5,2} TCID_{50}$/ml to $10^{3,8}$ $TCID_{50}$/ml was observed at 5 g/ml Iota-Carrageenan. No toxic effects have been observed on HNep cells at the highest tested Iota-Carrageenan concentration of 500 µg/ml. This result demonstrates that Iota-Carrageenan can effectively block the replication of six distinct HRV strain on HNEp cells.

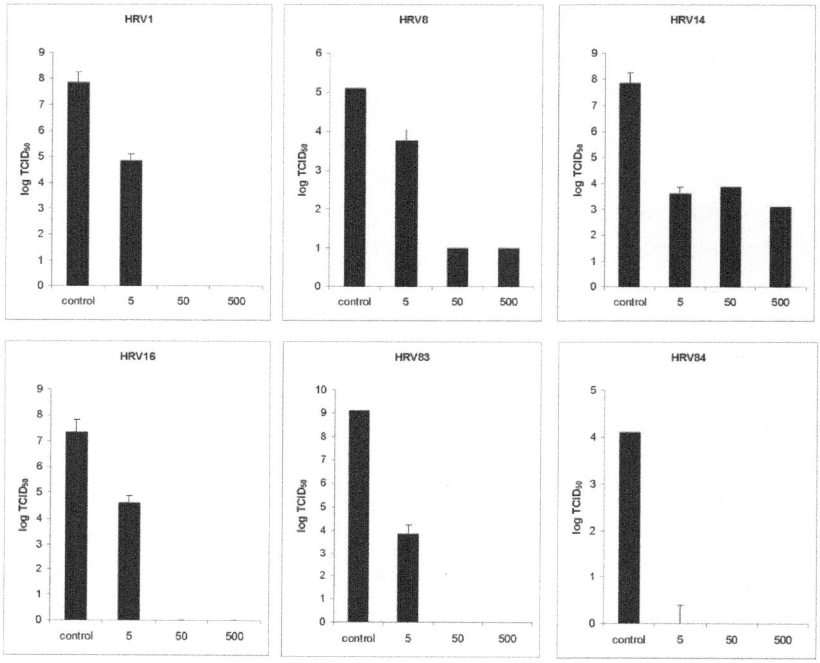

Figure 6. Effect of Iota carageenan on the replication of HRVstrains 1A, 2, 8, 14, 16, 39, 83 and 84 on human nasal epithelial cells. HNep cells were grown in 96-well plates were infected with different HRV strains (indicated at the top of each panel; 0,1 $TCID_{50}$/cell) in the presence of Iota-Carrageenan at the concentrations indicated at the x-axis. 30 minutes after infection the inoculum was removed and medium containing Iota-Carrageenan with the same concentration was added. Viral titers in the supernatants of infected cells were determined after 48 h by $TCID_{50}$ assay on HeLa cells (y-axis). Bars represent means from four parallel experiments, standard deviations are indicated.

DISCUSSION

In this report we demonstrate that Iota-Carrageenan, a commercial thickening agent derived from seaweed, is a potent inhibitor of rhinovirus infectivity in vitro. Two other related polymers Lambda- and Kappa-Carrageenan show moderate effects and did not fully inhibit virus induced cell death in HRV2 infected HeLa cells (Figure 1).

Protection of HeLa cells from virus induced cell death was dependent on the amount of input virus in both cases, when cells were infected in the presence or absence of Iota-Carrageenan (Figure 2). When cells were treated after infection, protection was observed only for the lowest input amount tested (0,01 $TCID_{50}$/cell; Figure 2B). Therefore we conclude that Iota-Carrageenan most likely inhibits binding or entry of the virus into the cells and not a later stage of viral replication. These findings are consistent with previous studies with other viruses that have shown that Carrageenan is active against several viruses in vitro and in vivo [11, 16, 18, 19, 20, 21, 22, 23, 24].

Iota-Carrageenan has been shown to be a potent inhibitor of papillomavirus infection with 50% inhibitory doses in the low ng/ml range [12]. However, when tested against rhinoviruses Iota-Carrageeenan appears to be effective against HRV at concentrations several orders of magnitude higher in the low μg/ml range (Figure 3). This result is comparable with in-vitro data of other viruses such as HIV-1 and Herpes virus [15, 25].

Repeated passage of the HIV virus in the presence of polyanions can lead to resistance mediated by mutations in the envelope glycoprotein gp120, particularly in the V3 loop (K269E, Q278H, N293D), as originally shown for dextran sulphate, and subsequently for Zintevir and negatively charged albumins [26, 27]. While resistant variants emerge relatively fast with HIV-1 we were not able to detect a difference in an in-vitro test between the original virus stock and a HRV2 virus after 10 subsequent passages in the presence of Iota-Carrageenan at concentrations between 7 μg/ml and 20 μg/ml (Figure 4). Although the potential emergence of resistant variants deserves detailed and extensive studies we conclude that Iota-Carrageenan resistant variants do not occur with a high frequency. This result supports the hypothesis that Iota-Carrageenan prevents HRV virions from cell attachment or cell entry in a less specific manner when compared to the results that were obtained by Buck and co-workers for papillomavirus [12]. However, it cannot be excluded that resistant variants of HRV2 may occur at later passages and further studies are needed.

In situ hybridization studies have revealed that the airway epithelial cell is the primary site of HRV infection in vivo [28, 29] and there is growing evidence that virally induced alterations in epithelial cell biology may contribute to disease pathogenesis [30, 31]. Thus we selected HNep cells as target cells for rhinovirus infection studies. Again Iota-Carrageenan was found to be effective against HRV2 on primary human epithelial cells with similar results when compared to the studies on HeLa cells (Figure 5). Our study also shows that viral titers in supernatants of infected HNep cells can vary by several orders of magnitude dependent on the strain (Figure 6). Replication studies with three Type A viruses HRV1A, HRV8, HRV16 and three Type B viruses HRV14, HRV83 and HRV84 revealed that Iota-Carrageenan is effectively inhibiting replication of Type A and Type B rhinoviruses when the polymer is present during infection (Figure 6). Differences between batches of HNep cells resulted in a variation of titers of HRV strains tested. However the anti-viral activity of Iota-Carrageenan was comparable in all tested HNep batches (data not shown).

Our data on primary cells are consistent with the data from infected HeLa cells and thereby support the hypothesis that Iota-Carrageenan interferes with viral replication at a very early stage of viral infection. Most likely, the binding of virions to the cells is hindered. It is not clear whether Carrageenan exerts any additional effects. The inhibitory effect of Iota-Carrageenan might be due to the occlusion of virion surfaces involved in binding to cellular receptors. Alternatively, obligatory conformational changes in the virus may be blocked and in addition, a post-attchment inhibitory effect may exist as described recently for papillomaviruses [12, 32].

Rhinovirus infections yearly account for huge economic losses in terms of lost school and working days. Moreover, recent evidence points towards rhinoviruses as a major cause in exacerbating asthma and COPD. For a review see Drescher et al 2007 [33]. A number of molecules and strategies have been examined in order to combat rhinovirus and the whole family of picornaviruses [5, 12]. Despite this, no therapy has been approved for the treatment of rhinovirus infections yet, and patient care remains symptomatic. Moreover, the efforts in therapeutic development are hampered by the fact that more than 100 distinctive HRV serotypes are circulating in humans. Our studies on primary HNep cells demonstrate that Iota-Carrageenan potently inhibits the replication of the seven distinct rhinovirus strains HRV1, 2, 8, 14, 16, 83 and 84 (Figure 5 and Figure 6). Although we are convinced that the result can be extrapolated for the whole family of human rhinoviruses further experiments are needed proof the efficacy on all strains of HRV.

CONCLUSION

The primary site of infection and replication of HRV in humans is the nasal mucosa. It is tempting to speculate that a targeted treatment of the nasal mucosa with Iota-Carrageenan might create a hostile environment for HRV and thereby block viral entry and replication. Carrageenan is generally recognized as safe for use in food and topical applications. Given the sensitivity and anti-viral effectiveness against several strains of HRV in primary human epithelial cells Iota-Carrageenan deserves consideration as a candidate for clinical trials for the prophylaxis and treatment of rhinovirus induced common cold.

METHODS

Polymers

Lambda carrageen, Kappa carrageen and Iota carrageen were purchased from FMC Biopolymers (Philadelphia, PA). The dry polymer powders were dissolved in cell culture water (PAA, Austria) to a final concentration of 0.4%. This stock solution was sterile filtered through a 0.45 µm filter (Sarstedt, Germany) and stored at 4°C until use.

Viruses, Cell Lines and Media

HRV serotypes (HRV 1A, 2, 8, 14, 16, 83, 84) were obtained from the American Type Culture Collection (Manassas, VA) and grown on HeLa cells. The stocks were frozen at -80°C and virus titers were determined by $TCID_{50}$ assay. The human cervical epithelial carcinoma cell line (HeLa) was obtained from the American Type Culture Collection. The cells were cultivated in Dulbecco's minimal essential medium (PAA) supplemented with 10% fetal bovine serum (PAA) and 1% antibiotic-antimycotic mix (PAA) in a 37°C incubator (Sanyo, Japan; CO_2: 5%, relative humidity: >95%). During virus infection and viral experiments a medium containing 2% fetal bovine serum and 1% antibiotic-antimycotic mix was used.

Human nasal epithelial cells were obtained from PromoCell GesmbH (Heidelberg, Germany) and cultivated in airway epithelial cell growth media (PromoCell).

Inhibition Assays

For determination of anti-viral activity a CPE inhibition assay was performed. HeLa cells were seeded in tissue culture plates 24 hours prior the experiments. At 80% confluence cells were infected with inoculums at indicated amounts of input virus ($TCID_{50}$/cell). In order to test whether the polymers can inhibit viral replication the cells were infected with virus in the presence or absence of polymer. For determination of the CPE plates were washed with PBS (PAA) and stained with 1% crystal violet (Sigma, St. Louis, MO) in 20% ethanol (?) and 3.7% formaldehyde (?). Cell damage was quantified with respect to intensity of the stain retained by living cells in a plate reader (Labsystems, Finland) at 630 nm. Alternatively, the cell metabolism was measured with an XTT reagent kit (Roche, Switzerland). Cell survival in the presence of inhibitors was calculated by setting mock infected cells to 100% survival and cells infected without inhibitor to 0% survival. Virus titers in 50% tissue culture infectious doses ($TCID_{50}$)/ml were determined according to Reed and Muench [34].

Search for Resistant Variants

HeLa cells ($8 * 10^4$ cells per well) were seeded in a 6-well plate and infected with HRV2. The cells were infected with a polymer virus mixture at final concentrations of 200, 20, 7 and 2 µg/ml with an amount of input virus of 0,1 $TCID_{50}$/cell. As control, one well was mock infected with medium and one well was infected with mock treated virus. After an infection time of 30 minutes the plates were washed twice and an overlay with infection medium containing polymer or control medium was added. Samples were taken from each well when a clear cytopathic effect was visible in the untreated control well. For the next selection round the wells showing a clear CPE and difference to the control infection was used.

The infection cycle was repeated 10 times and the resulting virus sample was compared in a CPE inhibition test with the original virus stock.

ACKNOWLEDGEMENTS

We would like to thank Sekina Sadovsky-Sherif for excellent laboratory assistance. This work was supported in part by the Austrian Research Promotion Agency (FFG) grant numbers 813886 and 818252.

COMPETING INTERESTS

The authors AG, CM, RW, EP, are employed by Marinomed. Authors HU and AP are co-founders of Marinomed. AG, CM, AP, and EP are inventors on patent # WO2008067982 held by Marinomed Biotechnologie GmbH that relates to the content of the manuscript. Marinomed Biotechnologie GmbH is financing the processing charge of this manuscript.

AUTHORS' CONTRIBUTIONS

AG, EP and HU participated in design and interpretation of the experiments. RW, CM, AP and AG carried out the viral replication studies on HeLa cells and drafted the manuscript. CM and RW performed the experiments on HNep cells. EP and HU contributed to experimental designs of the study and writing of the manuscript. All authors read and approved of the final manuscript.

REFERENCES

1. Anzueto A, Niederman MS: Diagnosis and treatment of rhinovirus respiratory infections. *Chest* 2003, 123: 1664-1672. 10.1378/chest.123.5.1664
2. Arruda E, Boyle TR, Winther B, Pevear DC, Gwaltney JM Jr, Hayden FG: Localization of human rhinovirus replication in the upper respiratory tract by in situ hybridization. *J Infect Dis* 1995, 171: 1329-1333.
3. Baba M, Snoeck R, Pauwels R, De CE: Sulfated polysaccharides are potent and selective inhibitors of various enveloped viruses, including herpes simplex virus, cytomegalovirus, vesicular stomatitis virus, and human immunodeficiency virus. *Antimicrob Agents Chemother* 1988, 32: 1742-1745.
4. Bardin PG, Johnston SL, Sanderson G, Robinson BS, Pickett MA, Fraenkel DJ, Holgate ST: Detection of rhinovirus infection of the nasal mucosa by oligonucleotide in situ hybridization. *Am J Respir Cell Mol Biol* 1994, 10: 207-213.
5. Brownlee JW, Turner RB: New developments in the epidemiology

and clinical spectrum of rhinovirus infections. *Curr Opin Pediatr* 2008, 20: 67-71. 10.1097/MOP.0b013e3282f41cb6

6. Buck CB, Thompson CD, Roberts JN, Muller M, Lowy DR, Schiller JT: Carrageenan is a potent inhibitor of papillomavirus infection. *PLoS Pathog* 2006, 2: e69. 10.1371/journal.ppat.0020069
7. Cabrera C, Witvrouw M, Gutierrez A, Clotet B, Kuipers ME, Swart PJ, Meijer DK, Desmyter J, De CE, Este JA: Resistance of the human immunodeficiency virus to the inhibitory action of negatively charged albumins on virus binding to CD4.*AIDS Res Hum Retroviruses* 1999, 15: 1535-1543. 10.1089/088922299309829
8. Carlucci MJ, Pujol CA, Ciancia M, Noseda MD, Matulewicz MC, Damonte EB, Cerezo AS: Antiherpetic and anticoagulant properties of carrageenans from the red seaweed Gigartina skottsbergii and their cyclized derivatives: correlation between structure and biological activity. *Int J Biol Macromol* 1997, 20: 97-105. 10.1016/S0141-8130(96)01145-2
9. Carlucci MJ, Scolaro LA, Noseda MD, Cerezo AS, Damonte EB: Protective effect of a natural carrageenan on genital herpes simplex virus infection in mice.*Antiviral Res* 2004, 64: 137-141.
10. Carlucci MJ, Scolaro LA, Noseda MD, Cerezo AS, Damonte EB: Protective effect of a natural carrageenan on genital herpes simplex virus infection in mice.*Antiviral Res* 2004, 64: 137-141.
11. De Palma AM, Vliegen I, De CE, Neyts J: Selective inhibitors of picornavirus replication. *Med Res Rev* 2008.
12. Dreschers S, Dumitru CA, Adams C, Gulbins E: The cold case: are rhinoviruses perfectly adapted pathogens? *Cell Mol Life Sci* 2007, 64: 181-191. 10.1007/s00018-006-6266-5
13. Este JA, Schols D, De VK, Van LK, Vandamme AM, Desmyter J, De CE: Development of resistance of human immunodeficiency virus type 1 to dextran sulfate associated with the emergence of specific mutations in the envelope gp120 glycoprotein. *Mol Pharmacol* 1997, 52: 98-104.
14. Fernandez-Romero JA, Thorn M, Turville SG, Titchen K, Sudol K, Li J, Miller T, Robbiani M, Maguire RA, Buckheit RW Jr, Hartman TL, Phillips DM: Carrageenan/MIV-150 (PC-815), a combination microbicide. *Sex Transm Dis* 2007, 34: 9-14. 10.1097/01.

olq.0000223287.46097.4b

15. Girond S, Crance JM, Van Cuyck-Gandre H, Renaudet J, Deloince R: Antiviral activity of carrageenan on hepatitis A virus replication in cell culture. *Res Virol* 1991, 142: 261-270. 10.1016/0923-2516(91)90011-Q
16. Gonzalez ME, Alarcon B, Carrasco L: Polysaccharides as antiviral agents: antiviral activity of carrageenan. *Antimicrob Agents Chemother* 1987, 31: 1388-1393.
17. Gwaltney JM Jr, Winther B, Patrie JT, Hendley JO: Combined antiviral-antimediator treatment for the common cold. *J Infect Dis* 2002, 186: 147-154. 10.1086/341455
18. Hamasuna R, Eizuru Y, Minamishima Y: Inhibition by iota-carrageenan of the spread of murine cytomegalovirus from the peritoneal cavity to the blood plasma. *J Gen Virol* 1994,75(Pt 1):111-116. 10.1099/0022-1317-75-1-111
19. Hayden FG, Albrecht JK, Kaiser DL, Gwaltney JM Jr: Prevention of natural colds by contact prophylaxis with intranasal alpha 2-interferon. *N Engl J Med* 1986, 314:71-75.
20. Hayden FG, Gwaltney JM Jr: Intranasal interferon-alpha 2 treatment of experimental rhinoviral colds. *J Infect Dis* 1984, 150: 174-180.
21. Hayden FG, Herrington DT, Coats TL, Kim K, Cooper EC, Villano SA, Liu S, Hudson S, Pevear DC, Collett M, McKinlay M: Efficacy and safety of oral pleconaril for treatment of colds due to picornaviruses in adults: results of 2 double-blind, randomized, placebo-controlled trials. *Clin Infect Dis* 2003, 36: 1523-1532. 10.1086/375069
22. Hayden FG, Turner RB, Gwaltney JM, Chi-Burris K, Gersten M, Hsyu P, Patick AK, Smith GJ III, Zalman LS: Phase II, randomized, double-blind, placebo-controlled studies of ruprintrivir nasal spray 2-percent suspension for prevention and treatment of experimentally induced rhinovirus colds in healthy volunteers.*Antimicrob Agents Chemother* 2003, 47: 3907-3916. 10.1128/AAC.47.12.3907-3916.2003
23. Jakiela B, Brockman-Schneider R, Amineva S, Lee WM, Gern JE: Basal cells of differentiated bronchial epithelium are more susceptible to rhinovirus infection. *Am J Respir Cell Mol Biol* 2008,

38: 517-523. 10.1165/rcmb.2007-0050OC

24. Ledford RM, Patel NR, Demenczuk TM, Watanyar A, Herbertz T, Collett MS, Pevear DC: VP1 sequencing of all human rhinovirus serotypes: insights into genus phylogeny and susceptibility to antiviral capsid-binding compounds. *J Virol*2004, 78: 3663-3674. 10.1128/JVI.78.7.3663-3674.2004
25. Neurath AR, Strick N, Li YY: Anti-HIV-1 activity of anionic polymers: a comparative study of candidate microbicides. *BMC Infect Dis* 2002, 2: 27. 10.1186/1471-2334-2-27
26. Pitkaranta A, Hayden FG: Rhinoviruses: important respiratory pathogens.*AnnMed*1998,30:529-537.10.3109/07853899809002600
27. Proud D, Sanders SP, Wiehler S: Human rhinovirus infection induces airway epithelial cell production of human beta-defensin 2 both in vitro and in vivo. *J Immunol* 2004, 172: 4637-4645.
28. Pujol CA, Scolaro LA, Ciancia M, Matulewicz MC, Cerezo AS, Damonte EB: Antiviral activity of a carrageenan from Gigartina skottsbergii against intraperitoneal murine herpes simplex virus infection. *Planta Med* 2006, 72: 121-125. 10.1055/s-2005-373168
29. Reed LJ, Muench H: A simple method of estimating fifty percent endpoints. *Am J Hyg* 1938, 493-497.
30. Spurrell JC, Wiehler S, Zaheer RS, Sanders SP, Proud D: Human airway epithelial cells produce IP-10 (CXCL10) in vitro and in vivo upon rhinovirus infection.*Am J Physiol Lung Cell Mol Physiol* 2005, 289: L85-L95. 10.1152/ajplung.00397.2004
31. Talarico LB, Damonte EB: Interference in dengue virus adsorption and uncoating by carrageenans. *Virology* 2007, 363: 473-485. 10.1016/j.virol.2007.01.043
32. Talarico LB, Pujol CA, Zibetti RG, Faria PC, Noseda MD, Duarte ME, Damonte EB: The antiviral activity of sulfated polysaccharides against dengue virus is dependent on virus serotype and host cell. *Antiviral Res* 2005, 66: 103-110. 10.1016/j.antiviral.2005.02.001
33. Whitton JL, Cornell CT, Feuer R: Host and virus determinants of picornavirus pathogenesis and tropism. *Nat Rev Microbiol* 2005, 3: 765-776. 10.1038/nrmicro1284
34. Witvrouw M, De CE: Sulfated polysaccharides extracted from sea

algae as potential antiviral drugs. *Gen Pharmacol* 1997, 29: 497-511. 10.1016/S0306-3623(96)00563-0

CHAPTER

15

EXTRACTION, ANTIOXIDATIVE, AND ANTIMICROBIAL ACTIVITIES OF BROWN SEAWEED EXTRACTS, TURBINARIA ORNATA AND SARGASSUM POLYCYSTUM, GROWN IN THAILAND

Saowapa Rattaya[1], Soottawat Benjakul[1], and Thummanoon Prodpran[2]

[1]Department of Food Technology, Faculty of Agro-IndustryPrince of Songkla UniversityHat YaiThailand

[2]Department of Material Product Technology, Faculty of Agro-IndustryPrince of Songkla UniversityHat YaiThailand

ABSTRACT

Total phenolic content and antioxidative and antimicrobial activities of methanolic and ethanolic extracts from brown seaweeds, Turbinaria ornata and Sargassum polycystum, were determined. Among all extracts, methanolic extract of T. ornata contained the highest total phenolic content (2.07 mg catechin/g dry seaweed) ($p < 0.05$) and exhibited the highest antioxidative activity as indicated by the greatest ABTS and DPPH radical scavenging activity as well as reducing activity power (RAP), compared with other extracts ($p < 0.05$). When different concentrations of seaweed extracts (100–500 mg/L) were used, antioxidative activities were dose-dependent. Correlations between ABTS and DPPH radical scavenging activity; DPPH radical scavenging activity and RAP; ABTS radical scavenging activity and RAP were observed. Therefore, antioxidants in seaweed extracts possessed the capability of scavenging the radicals together with reducing power. The efficacy in prevention of lipid oxidation of methanolic extract of T. ornata in lecithin-liposome and linoleic oxidation systems was studied. The extract at levels of 100–500 mg/L could retard the oxidation, regardless of chlorophyll removal but its efficacy was lower than that of BHT at levels of 50 and 200 mg/L. For antimicrobial activity, all extracts could not inhibit the growth of Bacillus subtilis, Salmonella enteritidis, and Aspergillus niger, while Staphylococcus aureus was inhibited with the extracts at 500 mg/L.

KEYWORDS Seaweed extract, Antioxidant activity, DPPH, ABTS, RAP, Lipid oxidation, Antimicrobial activity

INTRODUCTION

Lipid oxidation and microbial spoilage are the major deteriorative processes in foods, leading to unacceptability for the consumers and a loss in nutritional value. Additionally, oxidation leads to health disorders such as altherosclerosis and cancerogenesis (Kehrer 1993; Kranl et al. 2005). Hence, the antioxidants and antimicrobial agents have been widely used to maintain the quality, acceptability, and safety of foods (Koleva et al. 2003). Nevertheless, synthetic antioxidant and antimicrobial agents in food products are under strict regulation due to the potential health hazards (Kranl et al. 2005). Therefore, the search for natural antioxidants and antimicrobial as alternatives to synthetic counterpart is of great interest.

Seaweeds or marine macroalgae are potential renewable resources in the marine environment. About 6000 species of seaweeds have been identified and are grouped into different classes including green (Chlorophytes), brown (Pheophytes), and red (Rhodophytes) algae (Abbott 1995). Seaweeds have been widely used for the production of hydrocolloids such as agar, carrageenan. and alginates. Additionally, marine algae extracts were demonstrated to have strong antioxidant properties (Kuda et al. 2005), protective effects against liver injury caused by carbon tetrachloride, antimicrobial activity and antiviral properties (Newman et al. 2003). However, algae grown in Thailand are still underutilized. Only small portion has been used as food, animal fodder, fertilizers, and for the production of hydrocolloids (Aungtonya and Liao 2002).

In recent years, many marine algae extracts have been demonstrated to have antioxidant (Kuda et al. 2005) as well as antimicrobial properties (Ely et al. 2004). Seaweeds are considered to be a rich source of antioxidants (Cahyana et al. 1992). The potential antioxidant compounds were identified as some pigments (e.g., fucoxanthin, astaxanthin, carotenoid) and polyphenols (e.g., phenolic acid, flavonoid, tannins) (Yoshie et al. 2000). Those compounds are widely distributed in plants or seaweeds and are known to exhibit antioxidative activities via reactive oxygen species scavenging activity and the inhibition of lipid per-oxidation (Athukorala et al. 2003; Heo et al. 2005; Siriwardhana et al. 2003). Plant extracts have been reported to possess antimicrobial activity, mainly from phenolic compounds or essential oil. The extracts from seaweed have been reported to exhibit the antimicrobial activity and can be used as natural antimicrobial agent (Bennamara et al. 1999; Ely et al. 2004). Moreover, chlorophylls in plants have been also reported to have pro-oxidant activity, providing protection of vegetable edible oils by preventing autoxidation (Wanasundara and Shahidi 1998)

However, there exist few reports on the antioxidant and antimicrobial activities of seaweeds from southern coast of Thailand. To maximize the utilization of seaweeds, particularly brown seaweeds which are common in southern Thailand, the uses of their extracts as natural antioxidant and antimicrobial should be focused. The brown algae, being in the division of Pheophyta, are brownish in color due to the large amounts of the caroteniod fucoxanthin masking the remaining pigments, chlorophyll a and c, β-carotene, and other xanthophylls. Cell walls are composed of alginic acid which can be extracted as algin or alginate, used for various industrial purposes. The brown algae range in size from small filaments to the largest marine algae. Most members of this division are almost exclusively of marine occurrence. Most of the brown algae grow in the intertidal belt and the upper littoral region (Abbott 1995). Two brown algae, Sargassum polycystum and Turbinaria ornata, which are common in southern part of Thailand (Lewmanomont 1995) were focused in this present study.

Sargassum polycystum (Fig. 1a) proliferates in tropical waters. In Thailand, it is found in Gulf of Thailand and Andaman Sea. In Chumporn Province, it has been used as food, fertilizer for palm tree (Lewmanomont 1995; Yangthong et al. 2009). T. ornata (Fig. 1b) is widely distributed from the Indo-Pacific through the Caribbean (Ang, 1986). T. ornata has invaded large areas on the intertidal and shallow sub-tidal shores in French Polynesia and Hawaii (Smith et al. 2002). It is also common and abundant on the shore at Samui Island (Mayakun and Prathep 2005) and Koh Pling, Sirinart Marine National Park, Phuket Island, Southern Thailand (Prathep 2005), where it occurs over a wide range from sheltered to very exposed shores and from the intertidal to shallow sub-tidal zones.

Figure 1. Sargassum polycystum (a) and Turbinaria ornata (b) (Lewmanomont 1995).

Therefore, in this present study, chemical property as well as antioxidant activity in various systems and antimicrobial activity of brown algae including S. polycystum and T. ornata were investigated. The knowledge of their antioxidant and antimicrobial activities could potentially elevate their beneficial value as food and additives, and expand their dietary market.

MATERIALS AND METHODS

Materials, Chemicals, and Enzyme

Collection and Preparation of Seaweed

Brown seaweeds, T. ornata and S. polycystum, were collected freshly from Samui Island, Suradthanee Province, Thailand. Samples collected were washed thoroughly with tap water, packed in polyethylene bag and transported to the Department of Material Product Technology, Prince of Songkla University, Hat Yai within 5 h. Upon arrival, seaweeds were dried at 35 °C for 24 h using an air-force oven. The dried seaweeds were powdered using a blender (National, MX-T2GN, Taipei, Taiwan). Then powder was sieved using a screen with a diameter of 0.5 mm. The seaweed powder was placed in the polyethylene bag and stored at 4 °C until use.

Chemicals and Enzyme

Sodium chloride (NaCl), urea, and sodium dodecylsulfate (SDS) were purchased from Univar (Worksafe, Australia). Glycerol, α-chymotrypsin, and Coomassie Brilliant Blue G250 were purchased from Wako Pure Chemical Industry, Ltd. (Tokyo, Japan). Methanol was obtained from Merk (Darmstadt, Germany). 2,2-Azino-bis(3-ethylbenzothiazoline-6-sulfonic acid) diammonium salt (ABTS), 6-hydroxy-2,5,7,8-tetramethylchroman-2-carboxylic acid (Trolox), 2,2-diphenyl-l-picryl hydrazyl (DPPH), linoleic acid, β-mercaptoethanol (β-ME), and L-α-phosphatidylcholine (L-α-lecithin) were purchased from Sigma Chemical Co. (St. Louis, MO, USA). Acrylamide, N,N,N′N′- tetramethylethylenediamine (TEMED), 2-thiobarbituric acid, bis-acrylamide, Ferric chloride hexahydrate, 2-thiobarbituric acid, butylated hydroxytoluene (BHT), and potassium persulfate were procured from Fluka Chemical Co. (Buchs, Switzerland).

Extraction and Characterization of Seaweed Extracts

Preparation of Seaweed Extracts

Seaweed powder (5 g) was mixed with 150 ml of solvents, either methanol or ethanol following the method of Terada et al. (1987). The mixtures were homogenized at 10,000×g for 2 min using IKA LABOTECHNIK homogenizer (model T25 basic, ULTRA TURREX®, Selangor Malaysia). The homogenate was then stirred continuously at room temperature for 30 min. The mixtures were centrifuged at 5,000×g for 10 min at room temperature using a Sorvall Model RC-5B Plus refrigerated centrifuge (Newtown, CT, USA) to remove undissolved debris. A portion of extract was subjected to chlorophyll removal as per the method of Lanfer-Marquez et al. (2005). The extract was mixed with petroleum ether at a ratio of 5:2 (v/v) at room temperature. The extraction was repeated for 3 times. The layer of petroleum ether was drawn off. The solvent in the extracts without and with chlorophyll removal was removed by a rotary evaporator (Model Rotavapor-R, Brinkmann, Switzerland) at 40 °C. The volume of evaporated extract was adjusted to 10 ml using the same solvent.

Chlorophyll Content Determination

Total chlorophyll content was determined spectrophotometrically according to the method of the AOAC (2000). Methanolic and ethanolic extracts of seaweed with and without chlorophyll removal prepared as previously described were dehydrated with anhydrous sodium sulfate. Immediately, the pigments were quantified spectrophotometrically at 660 and 642.50 nm. For the blank, methanol or ethanol was used instead of extracts. Total chlorophyll content (TCC) was calculated using the following equation:

$$\text{TCC (mg/gdryextract)} = 7.12(A_{660}) + 16.8(A_{642}).$$

Total Phenolic Content Determination

Total phenolic content was determined with Folin–Ciocalteu reagent according to the method of Slinkard and Singleton (1997). One ml of seaweed extract was added with 200 μl of reagent (the mixture of Folin–Ciocalteu reagent and deionized water, 1:1 (v/v)) and mixed thoroughly. After 3 min, 3 ml of 2 % Na_2CO_3 was added. The mixture was allowed to stand at room temperature for 30 min. The absorbance was measured at 760 nm using a UV-16001 spectrophotometer (Shimadzu, Kyoto, Japan). The concentration of total phenolic compounds in seaweed extract was calculated from the standard curve of catechin with the range of 0.01–0.1 mg/ml and expressed as mg catechin/g dry seaweed.

Study on Antioxidative Activity of Seaweed Extracts

Ethanol and methanol extract from both seaweeds without and with chlorophyll removal were subjected to determination of antioxidative activity.

DPPH Radical Scavenging Activity

DPPH radical scavenging activity was determined as described by Wu et al. (2003) with a slight modification. Sample (1.5 ml) was added with 1.5 ml of 0.15 mM DPPH in 95 % methanol. The mixture was then mixed vigorously and allowed to stand at room temperature in dark for 30 min. The absorbance of resulting solution was measured at 517 nm using an UV-1601 spectrophotometer (Shimadzu, Kyoto, Japan). The control was prepared in the same manner except that the distilled water was used instead of sample. The sample blank was prepared by using ethanol or methanol instead of DPPH solution. The standard curve was prepared using Trolox in the range of 10–60 μM. The activity was expressed as μmol Trolox equivalents (TE)/ml.

ABTS Radical Scavenging Activity

ABTS radical scavenging activity was determined as per the method of Arnao et al. (2001) with a slight modification. The stock solutions included 7.4 mM ABTS solution and 2.6 mM potassium persulfate solution. The working solution was prepared by mixing the two stock solutions in equal quantities and allowing them to react for 12 h at room temperature in dark. The solution was then diluted by mixing 1 ml of ABTS solution with 30 ml of methanol to obtain an absorbance of 1.1 ± 0.02 at 734 nm using an UV-1601 spectrophotometer (Shimadzu, Kyoto, Japan). ABTS solution was freshly prepared for each assay. Sample (150 μl) was mixed with 2.85 ml of ABTS solution, and the mixture was left at room temperature for 2 h in dark. The sample blank was prepared by using methanol instead of ABTS solution. The absorbance was then measured at 734 nm using the spectrophotometer. The standard curve of Trolox ranging from 50 to 600 μM was prepared. The activity was expressed as μmol Trolox equivalent (TE)/ml.

Reducing Power

Reducing power was determined according to the method of Wu et al. (2003) with a slight modification. Diluted sample (1 ml) was mixed with 1 ml of 0.2 M phosphate buffer (pH 6.6) and 1 ml of 1 % potassium ferricyanide. The mixtures were incubated at 50 °C for 20 min, followed by addition of 1 ml of 10 % trichloroacetic acid. An aliquot (1 ml) of reaction mixture was added with 1 ml of distilled water and 200 μl of 0.1 % $FeCl_3$. The absorbance of resulting solution was measured at 700 nm. Increased absorbance of the reaction mixture indicates the increased reducing power. The sample blank was prepared by using methanol instead of $FeCl_3$solution.

Determination of Antioxidative Activity of Seaweed Extract in Different Systems

Methanolic extracts of T. ornata with and without chlorophyll removal at different concentrations were tested in different systems. BHT at levels of 50 and 200 mg/L was used for comparison purpose. For the control, methanol was added instead of the extract or BHT.

Lecithin Liposome System

Antioxidative activity of seaweed extracts in lecithin liposome system was determined according to the method of Frankel et al. (1996). Lecithin was suspended in deionized water at a concentration of 8 mg/ml. The mixture was stirred with glass rod, followed by sonicating for 30 min using a sonicating bath (Transsonic 460/H, Elma, Germany). Seaweed extracts were mixed with lecithin liposome to obtain the final concentrations of 100, 200, and 500 mg/L. The liposome suspension was then sonicated for 2 min. To initiate the assay, 20 μl of 0.15 M cupric acetate was added. The mixture was shaken at 120 rpm using a shaker (UNIMAX 1010, Heidolph, Germany) at 37 °C in dark for 0, 6, 12, 18, 24, 30, 42, and 48 h. Liposome oxidation was monitored by determining thiobarbituric acid-reactive substances (TBARS). TBARS values were calculated from the standard curve of (0–3 mg/L malonaldehyde (MDA)) and expressed as mg MDA/ml liposome.

Linoleic Oxidation System

Antioxidative activity of seaweed extracts in linoleic oxidation system was tested as described by Sakanaka et al. (2004). Seaweed extracts were mixed with 10 ml of 50 mM linoleic acid in 99.5 % ethanol to obtain the final concentrations of 100, 200, and 500 mg/l and the mixture was kept at 40 °C in dark. During incubation, aliquots of the reaction mixtures were taken for measurement of the oxidation using the ferric thiocyanate method every day for totally 10 days. To 50 μl of the reaction mixture, 2.35 ml of 75 % ethanol, 50 μl of 30 % ammonium thiocyanate, and 50 μl of 20 mmol/L ferrous chloride solution in 3.5 % HCl were added and mixed thoroughly. After 3 min, the absorbance of the colored solution was measured at 500 nm. For the control, methanol was added instead of antioxidant in the assay system. BHT at level of 50 and 200 mg/L was also used.

Study on Antimicrobial Activity of Seaweed Extracts

Microorganisms

Bacillus subtilis, Staphylococcus aureus,Salmonella enteritidis, and Aspergillus niger were obtained from the Department of Microbiology, Prince of Songkla University, Hat Yai, Thailand. The microorganisms were sub-cultured in slant plate count agar for B. subtilis, S. aureus, and S. enteritidis and potato dextrose agar (PDA) for A. niger and kept at 4 °C until use.

Determination of Antimicrobial Activity of Seaweed Extracts

Antimicrobial activity was tested according to the method of Baydar et al. (2005). The microorganism sub-cultured in slant (1 loop) was incubated in tubes of nutrient broth (10 ml) at 37 °C for 18–24 h to obtain approximately 10^7–10^8 colony forming units/ml (CFU/ml). The solutions (0.1 ml) were spread on plate count agar (PCA) for B. subtilis, S. aureus, and S. enteritidis and potato dextrose agar (PDA) for A. niger. Sterile disks (5 mm) prepared from Whatman No. 4 filter paper were absorbed with 50 μl of either methanolic or ethanolic seaweed extract at different concentrations (0, 100, 200, 300, 400, and 500 mg/L). The disks were dried at 40 °C for 30 min and placed onto surface of each microorganism plate and incubated at 37 °C for 24 h. The inhibition zones were measured as the diameter of disk.

Statistical Analysis

All experiments were run in triplicate. All data were subjected to Analysis of Variance (ANOVA), and the differences between means were evaluated by Duncan's Multiple Range Test. SPSS statistic program (SPSS 11.0 for window, SPSS Inc., Chicago, IL, USA.) was used for data analysis.

RESULTS AND DISCUSSION

Extraction and Chemical Properties of Seaweed Extracts

Total Phenolic Content and Total Chlorophyll Content of Brown Seaweed Extracts

Total phenolic contents (TPC) of methanolic and ethanolic extracts of T. ornata and S. polycystum are presented in Table 1. Without chlorophyll removal, both methanolic and ethanolic of T. ornata contained higher TPC than those of S. polycystum. For the same seaweeds, methanolic extract had the greater TPC than did ethanolic counterpart. Various phenolic compounds including quercetin, myricetin (flavonoles); genistin, daidzein (iso-flavones); hesperidin (flavanones), and lutein (flavones) were found in methanolic extract of red, brown, and green seaweeds (Yoshie et al. 2000; Yangthong et al. 2009; Santoso et al. 2004).

Table 1. Total phenolic content (TPC) of methanolic and ethanolic extract of T. ornata and S. polycystum without and after chlorophyll removal

Extracts	Total phenolic content* (mg catechin/g dry seaweed)	
	T. ornata	*S. polycystum*
Methanolic		
After chlorophyll removal	2.07 ± 0.06^{bA}**	0.54 ± 0.09^{aB}
Without chlorophyll removal	2.18 ± 0.01^{aA}	0.59 ± 0.01^{aB}
Ethanolic		
After chlorophyll removal	1.03 ± 0.04^{dA}	0.22 ± 0.01^{cB}
Without chlorophyll removal	1.25 ± 0.01^{cA}	0.32 ± 0.00^{bB}

* Mean ± SD (n = 3)

** Different letters within the same column indicate significant differences ($p < 0.05$) and different capital letters within the same row indicate significant differences ($p < 0.05$)

After chlorophyll removal, TPC of the same extract decreased, except for methanolic extract of S. polycystum, in which chlorophyll removal had no impact on TPC. During extraction of chlorophylls using petroleum ether, some phenolics, particularly non-polar, could be removed together with chlorophyll, leading to the lower TPC remained in the extract. Yan et al. (1999) reported some active compounds from brown seaweed, which were identified as phylopheophytin in Eisenia bicyclis and fucoxantine in Hizikia fusiformis. Methanol has been intensively used to extract plant phenols, and its extraction efficiency was generally higher than ethanol (Kumaron and Karunakaran 2007). Lim et al. (2002) reported that phenolic compound of methanolic extract from Sargassum siliquastrum had higher than did ethyl acetate, chloromethane, and buthanol.

Total chlorophyll content (TCC) of different seaweed extracts is shown in Table 2. After chlorophyll removal, methanolic and ethanolic extracts had lower TCC ($p < 0.05$). Since chlorophyll contains several non-polar moieties, it can be extracted out by petroleum ether upon chlorophyll removal (Lanfer-Marquez et al. 2005; Wong 1989). Efficacy in chlorophyll removal varied with seaweeds and solvents used. This was possibly due to different polarities of the solvents used to prepare the extracts which could result in different partitioning abilities of petroleum ether used to remove the chlorophyll and thus leading to different chlorophyll extraction efficiencies. In addition, chlorophyll constituted in various seaweeds might be in different forms so that the extraction efficacy could be varied (Wong 1989). Methanolic extract of S. polycystum had the decrease in TCC by 41.32 % after chlorophyll removal.

Table 2. Total chlorophyll content of methanolic and ethanolic extract of T. ornata and S. polycystum without and after chlorophyll removal

Extracts	Total chlorophyll content* (μg chlorophyll/g dry seaweed)	
	T. ornata	***S. polycystum***
Methanolic		
After chlorophyll removal	$20.79 \pm 0.29^{dA**}$	13.25 ± 0.10^{dB}
Without chlorophyll removal	35.23 ± 0.04^{bA}	32.09 ± 0.15^{bA}
Ethanolic		

After chlorophyll removal	25.14 ± 0.27[cA]	21.57 ± 0.05[cA]
Without chlorophyll removal	37.77 ± 0.44[aA]	37.18 ± 0.09[aA]

* Mean ± SD (n = 3)

** Different letters within the same column indicate significant differences ($p < 0.05$) and different capital letters within the same row indicate significant differences ($p < 0.05$)

Antioxidative Activities of Brown Seaweed Extracts

DPPH Radical Scavenging Activity

DPPH radical scavenging activity of methanolic and ethanolic extracts from T. ornata and S. polycystum without and with chlorophyll removal at different concentrations is shown in Fig. 2. For extract without chlorophyll removal, methanolic extract of T. ornata and ethanolic extract of S. polycystum exhibited the highest DPPH radical scavenging activity. At the same concentration tested, methanolic extract of S. polycystum had the lowest activity. Due to the different TPC between methanolic extract of T. ornata and ethanolic extract of S. polycystum(Table 1), it was suggested that TPC was not correlated well with DPPH radical scavenging activity. It was also presumed that different types of phenolic compounds with different antioxidant activities were presented in both extracts. Polyphenolic constituent in seaweed was capable of functioning as free radical scavengers (Chew et al. 2008). Polyphenols such as phlorotannins, which are bi-polar in nature, were found in brown seaweeds (Burtin 2003). For each extract used, DPPH radical scavenging activity increased with increasing concentration ($p < 0.05$). DPPH is one of the compounds that have a proton-free radical with a characteristic absorption, which decreases significantly on exposure to proton radical scavengers (Yamaguchi et al. 1998). It is well accepted that the DPPH radical scavenging by antioxidants is attributable to their hydrogen-donating ability (Chen and Ho 1995). The reduction capability of DPPH radicals was determined by the decrease in its absorbance at 517 nm induced by antioxidant (GülGin et al. 2004). Therefore, DPPH radical scavenging activity varied with seaweed species. Methanolic extract of three Indian brown seaweeds (Turbinaria conoides, Padina tetrastomatica, and Sargassum marginatum) had different DPPH radical scavenging activities. T. conoides extracts showed higher activity than those of P. tetrastomatica and S. marginatum (Chandini et al. 2008).

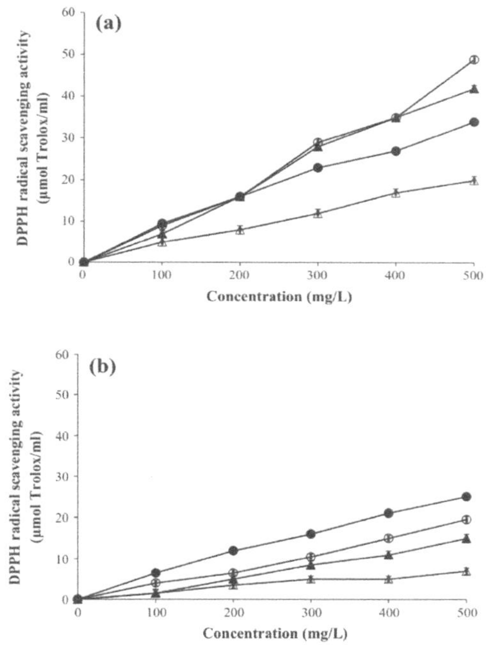

Figure 2. DPPH radical scavenging activity (µmol TE/ml) of methanolic (ME) and ethanolic (EE) extracts of T. ornata and S. polycystum without (a) and with (b) chlorophyll removal at various concentrations. Bars represent the standard deviation from triplicate determinations.

Table 3. Inhibition zone of different microorganisms in the presence of methanolic and ethanolic extracts of T. ornata and S. polycystum with and without chlorophyll removal at 500 mg/L

Seaweeds	Extracts	*B. subtilis*	*S. aureus*	*S. enteritidis*	*A. niger*
T. ornata	ME/W	–	6.0	–	–
	ME/W/O	–	6.0	–	–
	EE/W	–	6.0	–	–
	EE/W/O	–	6.0	–	–

S. polycystum	ME/W	–	7.0	–	–
	ME/W/O	–	6.0	–	–
	EE/W	–	6.0	–	–
	EE/W/O	–	6.5	–	–

The value is the distance (mm) across the zone of inhibition and the disk (diameter 5 mm)

Values are the mean values of triplicate determinations

– no inhibition; B. subtilis: Bacillus subtilis; S. aureus: Staphylococcus aureus; S. enteritidis: S. enteritidis; A. niger: Aspergillus niger; ME/W: methanolic extract with chlorophyll removal; ME/W/O: methanolic extract without chlorophyll removal; EE/W: ethanolic extract with chlorophyll removal; EE/W/O: ethanolic extract without chlorophyll removal

After chlorophyll removal, all extracts exhibited the lower DPPH radical activity. Nevertheless, the activity increased as the concentration used increased ($p < 0.05$). With chlorophyll removal, methanolic extract of T. ornata had the highest DPPH radical scavenging activity, whereas methanolic extract of S. polycystum exhibited the lowest activity at all concentrations tested ($p < 0.05$). The result suggested that chlorophylls contributed to antioxidative activity of seaweed extracts. Endo et al. (1985) reported antioxidative activity of chlorophyll-a, followed by chlorophyll-b and pheophytin. Additionally, chlorophyll removal using petroleum ether might remove some antioxidative compounds from the extracts. As a result, antioxidative compounds in the extracts could be affected by chlorophyll removal process.

ABTS Radical Scavenging Activity

Methanolic and ethanolic extracts of T. ornata and S. polycystum without and with chlorophyll removal at different concentrations showed different ABTS radical scavenging activities (Fig. 3). ABTS radical scavenging activity of all extracts increased with increasing concentrations ($p < 0.05$). For the extracts without chlorophyll removal (Fig. 3a), methanolic extract of T. ornata exhibited the highest activity ($p < 0.05$), whereas methanolic extract of S. polycystum showed the lowest activity ($p < 0.05$). For the extract of T. ornata, methanolic extract had the higher ABTS radical scavenging activity than did ethanolic counterpart. Nevertheless, ethanolic extract of S. polycystum showed the higher activity than did methanolic counterpart ($p < 0.05$). Different results between DPPH and ABTS radical scavenging activities indicated the differences in mode of action of antioxidants, particularly in term of specificity in radical scavenging. Both of seaweed extracts without chlorophyll removal had higher ABTS radical scavenging activity than did those with chlorophyll removal, indicating that some antioxidative compounds could be removed by petroleum ether used in chlorophyll removing process. Furthermore, removal of chlorophylls, which had antioxidative activity, resulted in the lower activity. Hagerman et al. (1998) reported that molecular weight, the number of aromatic rings, and nature of hydroxyls groups substitution, rather than the specific functional groups, determine antioxidative activity of phenolic compounds. ABTS· assay is an excellent tool for determining the antioxidant activity of hydrogen-donating antioxidants (scavengers of aqueous phase radicals) and of chain breaking antioxidants (scavenger of lipid peroxyl radical) (Leong and Shui 2002). For the extracts with chlorophyll removal, ethanolic extract of T. ornata had lower ABTS radical scavenging activity than did methanolic counterpart ($p < 0.05$) (Fig. 3b). No differences were noticeable between methanolic and ethanolic extracts of S. polycystum. This reconfirmed that chlorophyll removal affected the types of antioxidative compounds and composition of extracts.

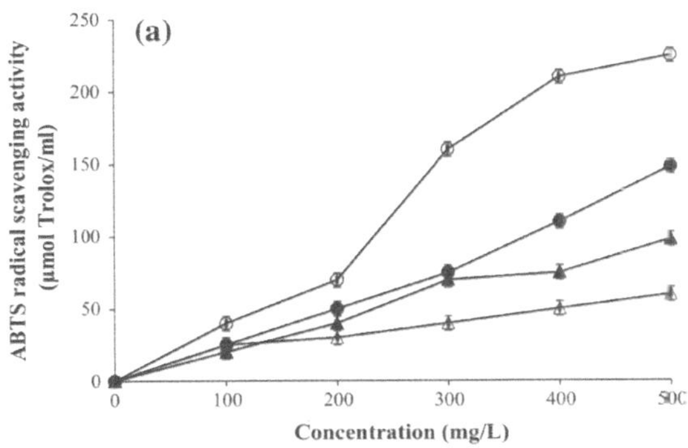

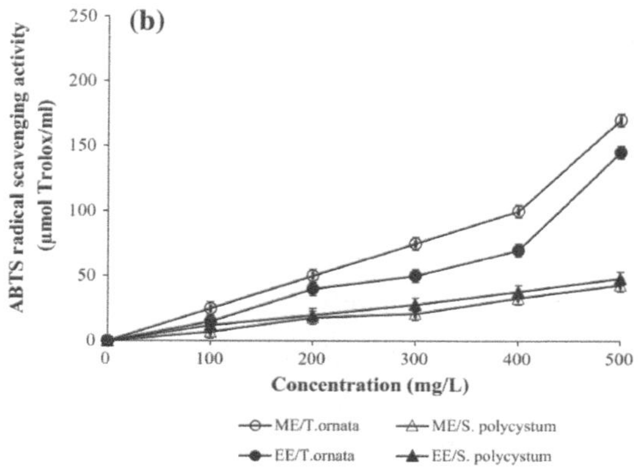

Figure 3. ABTS radical scavenging activity (µmol TE/ml) of methanolic (ME) and ethanolic (EE) extracts of T. ornata and S. polycystum without (**a**) and with (**b**) chlorophyll removal at various concentrations. Bars represent the standard deviation from triplicate determinations.

Reducing Activity Power

Among all extracts without chlorophyll removal, methanolic extract of T. ornata had the highest reducing activity power (RAP) and ethanolic extract of S. polycystum showed the lowest RAP ($p < 0.05$) (Fig. 4). After chlorophyll removal, slight decrease in RAP was generally found. In general, extracts from T. ornta possessed the greater RAP than did those of S. polycystum. For the same extract, RAP was not absolutely in accordance with DPPH and ABTS radical scavenging activities. The result suggested that different extracts contained antioxidative compounds with different functions in inhibiting lipid oxidation. Kuda et al. (2005) and Chandini et al. (2008) reported that methanolic extracts of T. conoides and P. tetrastomatica had higher reducing power than did S. marginatum. RAP of all extracts increased as the concentrations increased up to 500 mg/L ($p < 0.05$). RAP indicated that all extracts were capable of donating the electrons to the radicals, in which propagation could be terminated or retarded.

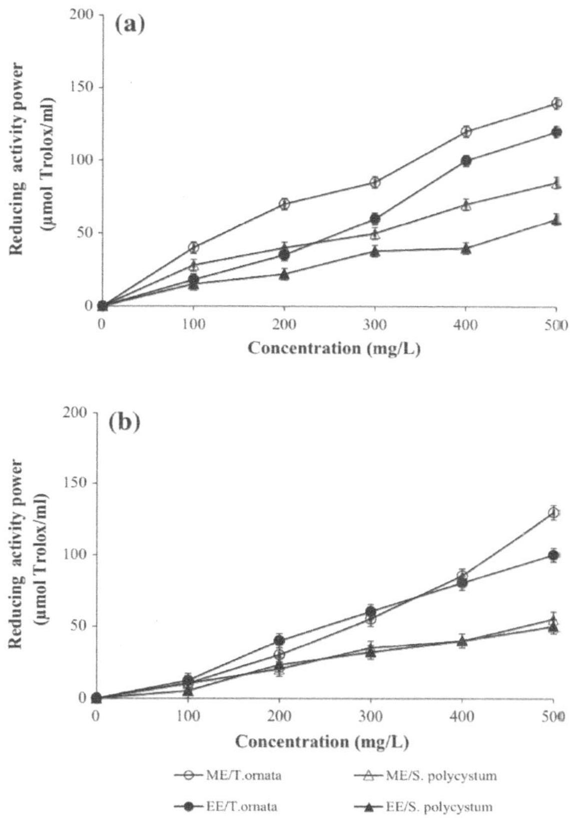

Figure 4. Reducing activity power (µmol TE/ml) of methanolic (ME) and ethanolic (EE) extracts of T. ornata and S. polycystum without (a) and with (b) chlorophyll removal at various concentrations. Bars represent the standard deviation from triplicate determinations.

The Correlation between Antioxidative Activities Tested by Different Assays

The correlations between antioxidative activities determined by different assays and reported as Trolox equivalent antioxidant capacity (TEAC) were observed (Fig. 5). Their linear correlations could be described as $TEAC_{ABTS} = 5.1562TEAC_{DPPH}$ ($R^2 = 0.8732$), $TEAC_{RAP} = 0.7186TEAC_{ABTS}$ ($R^2 = 0.7178$), and $TEAC_{RAP} = 3.8178TEAC_{DPPH}$ ($R^2 = 0.7150$). Zhao et al. (2006) also reported the correlation between antioxidative activity of barley extract tested by DPPH and ABTS radical scavenging activities and reducing power (R^2 ranging from 0.902 to 0.992). Nevertheless, Śanchez et al. (2007) reported the best correlation between total polyphenol contents and antioxidant capacity determined by ABTS and DPPH radical scavenging activity method ($R^2 = 0.8927$ and 0.8052, respectively) in virgin olive oil. The result revealed that brown seaweed extracts possessed the ability in donating electron as well as the capability of scavenging various radicals. It could be inferred that the extracts could be used as antioxidants. Since methanolic extract of T. ornata had the highest antioxidative activity determined by all assays, it was used to prevent lipid oxidation in different systems.

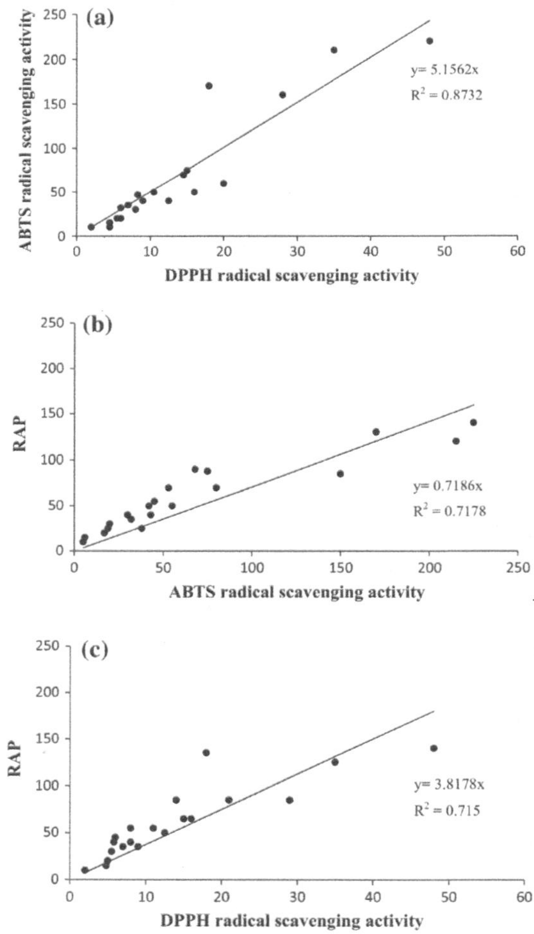

Figure 5. Correlation between ABTS and DPPH radical scavenging activity

(a), RAP and ABTS radical scavenging activity (b) and RAP and DPPH radical scavenging activity (c) of the extract of T. ornata and S. polycystum without and with chlorophyll removal.

Antioxidative Activity of Seaweed Extract in Different Systems

Lecithin Liposome System

Antioxidative activities of methanolic extract of T. ornata at different concentrations in lecithin liposome system as monitored by the formation of TBARS during incubation at 37 °C are shown in Fig. 6. Generally, the increase in liposome oxidation was observed when incubation time increased up to 30 h ($p < 0.05$). During the first 6–30 h of incubation, the sharp increases in TBARS were found in the control. Systems containing seaweed methanolic extract had lower TBARS than did the control throughout the incubation of 48 h. TBARS values in liposome system added with 50, and 200 mg/L BHT was very low throughout the incubation period of 48 h ($p < 0.05$). When the methanolic extract without chlorophyll removal was used (Fig. 6a), the liposome system added with 200 ppm extract generally had lower TBARS than did that with 100 ppm extract. However, when the extract was used at 500 ppm, the liposome system showed higher TBARS than the others. This was more likely attributable to the pro-oxidative effect of chlorophyll containing in the seaweed extract (Wanasundara and Shahidi 1998). TBARS values represent the content of secondary lipid oxidation products, mainly aldehydes (or carbonyls), that contribute to off-flavors in oxidized oils (Wanasundara and Shahidi 1998). The decrease in TBARS was found after 30 h of incubation. This might be due to the losses in secondary products with low molecular weight, leading to lower amount of such products (Stanhne 1995). The extract with chlorophyll removal exhibited the higher antioxidant activity than did that without chlorophyll removal in this system ($p < 0.05$). This was probably due to the pro-oxidative effect of chlorophylls and their derivatives in this system. Wanasundara and Shahidi (1998) reported that ethanolic green tea extract (GTE) with chlorophyll showed pro-oxidative activity in seal blubber oil (SBO) and menhaden oil (MHO). However, after dechlorophyllization, the extract exhibited antioxidant activity in both SBO and MHO oil. However, He and Shahidi (1997) found that GTE, despite the presence of chlorophyll, had an antioxidant effect when applied to white muscles of mackerel. Therefore, antioxidant/pro-oxidant activity of GTE was system-dependent (He and Shahidi 1997). The results indicated that methanolic extract of T. ornata could inhibit lipid oxidation, irrespective of chlorophyll removal.

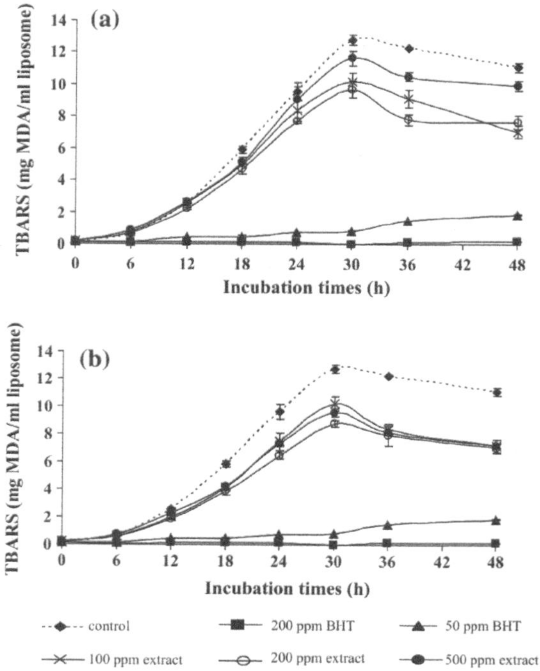

Figure 6. Changes in TBARS (mg MDA/ml liposome) of lecithin liposome system containing methanolic extract of T. ornata without (a) and with (b) chlorophyll removal at different levels. Bars represent the standard deviation from triplicate determinations.

Linoleic Oxidation System

The oxidation of linoleic acid as monitored by the increase in absorbance at 500 nm (A_{500}) in systems containing methanolic extract of T. ornata at different levels during the incubation at 40 °C for 10 days is shown in Fig. 7. All samples had a continuous increase in A_{500} throughout 10 days of incubation and the control showed the highest rate of increase. During incubation, for the systems added with the extract without chlorophyll removal (Fig. 7a), no significant difference in A_{500} was found, compared with the control, except after 7 days of incubation, when the lower A_{500} was found in sample added with the extract. For system containing the extract with chlorophyll removal at all concentrations had the lower oxidation as evidenced by lower A_{500}, compared with the control ($p < 0.05$) (Fig. 7b). The results showed that after dechlorophyllization the seaweed extract exhibited higher antioxidant activity as evidenced by the lower A_{500} ($p < 0.05$) when compared to the extract without chlorophyll removal. Reaction medium which contains unsaturated fatty acids promotes the decomposition of chlorophyll to release the metal-free, behaving as prooxidant at higher concentrations (Usuki et al. 1984). Santoso et al. (2004) reported that methanolic extract from seaweed contained not only polyphenolic compounds but also other compounds such as polyunsaturated fatty acid (e.g., eicosapentaenoic acid), minerals (e.g., copper, iron), and pigment (e.g., chlorophyll). These components might interact with polyphenolic compounds or affect directly to the emulsion system; therefore, they might act like a pro-oxidant. Linoleic acid is a C18:2 fatty acid with two double bonds at C9 and C12 and is one of the major unsaturated fatty acids in vegetable oils. The isomeric structures of hydroperoxides and decomposition volatile compounds from linoleic acid or methyl linoleate by auto-oxidation and photosensitized oxidation have been studied extensively (Frankel 1985). Lipid peroxidation is thought to proceed via radical-mediated abstraction of hydrogen atoms from methylene carbons in polyunsaturated fatty acids (Rajapakse et al. 2005). In food-related systems, anti-oxidant activity is mainly due to the chain-breaking inhibition of lipid peroxidation (Morales and Jimennez-Perez 2001). Therefore, brown seaweed extract could retard lipid oxidation in the emulsion system. However, the efficiency was lower than BHT at both levels (50 and 200 mg/L).

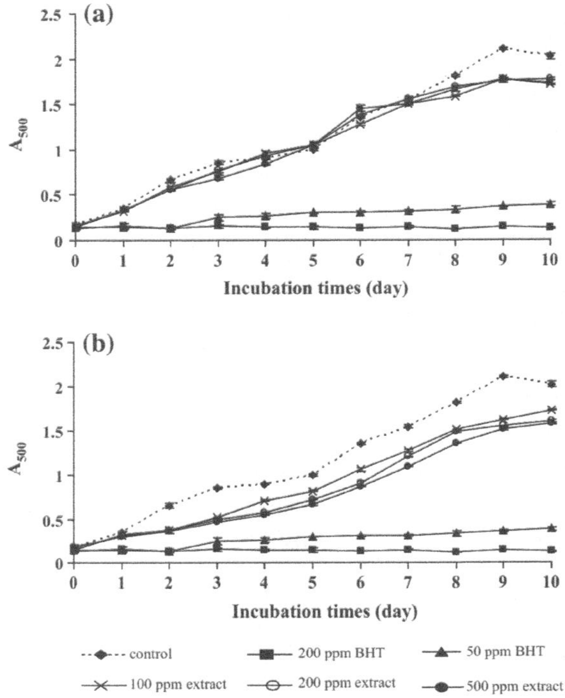

Figure 7. Changes in A_{500} of linoleic acid system containing methanolic extract of T. ornatawithout (a) and with (b) chlorophyll removal at different levels. Bars represent the standard deviation from triplicate determinations.

Antimicrobial Activity of Brown Seaweed Extracts

Methanolic and ethanolic extracts of T. ornata and S. polycystum without and with chlorophyll removal at 500 mg/L was able to inhibit only gram-positive bacteria (S. aureus) but could not inhibit gram-negative bacteria (S. enteritidis), gram-positive bacteria, spore-forming rods (B. subtilis), and fungi (A. niger) (Table 3). Similar clear zone diameter was observed among all extracts tested, but methanolic extract of T. ornata tended to exhibit the highest inhibition against S. aureus. Antimicrobial activity of extract could be because phenols reacted primarily with the phospholipids component of cell membrane (Ely et al. 2004). Different active compounds in different brown seaweeds extracts such as spatane diterpenoids, 19-accetoxy-5, 15,18-trihydrxyapata-

13,16-diene, from Stoechospermum marginatum (de Silva et al. 1982) methoxybifurcarenone, from Cystoseira tamariscifolia (Bennamara et al. 1999) and phlorotannins, from Ecklonia kurome (Nagayama et al. 2002) were reported.

CONCLUSION

Methanolic extracts of brown seaweeds, T. ornata and S. polycystum, exhibited effectively antioxidative activity. The activity increased with increasing the total phenolic content in the extracts. Antioxidants in seaweed extracts possessed the capability of scavenging the radicals together with reducing power. Therefore, the extracts of those seaweeds could be a source of natural antioxidants, which could retard lipid oxidation in different model systems, regardless of chlorophyll removal. The extracts at proper concentration could inhibit S. aureus, but were not able to inhibit the growth of B. subtilis, S. enteritidis, and A. niger.

ACKNOWLEDGMENTS

This work was supported by Thai government and Prince of Songkla University. The TRF Senior Research scholar program was also acknowledged.

AUTHOR'S CONTRIBUTIONS

SB and TP formulated the hypothesis and designed the study. SR conducted the experiments and analysis of data. SR wrote the manuscript with assistance from SB and TP. All authors read and approved the final manuscript.

REFERENCES

1. Abbott IA (1995) Taxonomy of economic seaweeds: reference to some Pacific species, vol V. California Sea Grant College, Uni. Calif. La Jolla, California
2. Ang JPO (1986) Analysis of the vegetation structure of a Sargassum community in the Philippines. Mar Ecol-Prog Ser 28:9–19
3. AOAC (2000) Official methods of analysis, 17th edn. Association of Official Chemists, Gaithersberg, Maryland

4. Arnao MB, Cano A, Acosta M (2001) The hydrophilic and lipophilic contribution to total antioxidant activity. Food Chem 73:239–244
5. Athukorala Y, Lee KW, Song CB, Ahn CB, Shin TS, Cha YJ, Shahidi F, Jeon YJ (2003) Potential antioxidant activity of marine red alga Grateloupia filicina extracts. J Food Lipids 10:251–265
6. Aungtonya C, Liao LM (2002) Marine flora (alga and sea grasses) in the reference collection of the Phuket Marine Biology Center, Thailand. Phuket Mar Biol Center Resour 64:65–68
7. Baydar NG, Ozkan G, Sagdis O (2005) Total phenolic contents and antimicrobial activity of grape (Vistis vinfera L.) extracts. Food Control 15:335–339
8. Bennamara A, Abourriche A, Berrada M, Charrouf M, Chaib N, Boudouma M, Garneau FX (1999) Methoxybifurcarenone: an antifungal and antibacterial meroditerpenoid from the brown alga Cystoseira tamariscifolia. Phytochemistry 52:37–40
9. Burtin P (2003) Nutritional value of seaweeds. Electron J Agric Food Chem 2:498–503
10. Cahyana AH, Shuto Y, Kinoshita Y (1992) Pyropheophytin a as an antioxidative substance from the marine alga, Arame (Eisennia bicyclis). Biosci Biotechnol Biochem 56:1533–1535
11. Chandini SK, Ganesan P, Bhaskar N (2008) In vitro antioxidant activities of three selected brown seaweeds of India. Food Chem 107:707–713
12. Chen CW, Ho CT (1995) Antioxidant properties of polyphenols extracted from green and black tea. J Food Lipid 2:35–46
13. Chew YL, Lim YY, Omar M, Khoo KS (2008) Antioxidant activity of three edible seaweeds from two areas in South East Asia. LWT-Food Sci Technol 41:1067–1072
14. de Silva SMS, Gamage SKT, Savitrikumar N, Balasubramanium S (1982) Antibacterial activity of extract from the brown seaweed Stoechospermum marginatum. Phytcochemistry 21:944–945
15. Ely R, Supriya T, Naik CG (2004) Antimicrobial activity of marine organisms collected off the coast of South East India. J Exper Mar Biol Ecol 309:121–127
16. Endo Y, Usuki R, Kaneda T (1985) Antioxidant effects of chlorophyll

and pheophytin on the auto-oxidation of oils in dark. I: comparison of the inhibitory effects. J Am Oil Chem Soc 62:1375–1378

17. Frankel EN (1985) Chemistry of auto-oxidation: mechanism, products and flavor significance. In: Min DB, Smouse TH (eds) Flavor chemistry of fats and oils. Champaign, Illinois, pp 1–37
18. Frankel EN, Huang SW, Aeschbach R, Prior E (1996) Antioxidant activity of a rosemary extract and its constituents, carnosic acid, carnosol, and rosmarinic acid, in bulk oil and oil-in-water emulsion. J Agric Food Chem 44:131–135
19. GülÇin İ, Sat Gİ, Beydemir ş, Elmastaş M, Küfrevioğlu Öİ (2004) Comparison of antioxidant activity of clove (Eugenia caryophylata thumb) buds and lavender (Lavandula stoechas L.). Food Chem 87:393–400
20. Hagerman AE, Riedl KM, Jones GA, Sovik KN, Ritchard NT, Hartzfeid PW (1998) High molecular weight plant polyphenolics tannins as biological antioxidants. J Agric Food Chem 46:1887–1892
21. He Y, Shahidi F (1997) Antioxidant activity of green tea and its catechins in a fish meat model system. J Agric Food Chem 45:4262–4266
22. Heo SJ, Park EJ, Lee KW, Jeon YJ (2005) Antioxidant activities of enzymatic extracts from brown seaweeds. Biores Technol 96:1613–1623
23. Kehrer JP (1993) Free radicals as mediators of tissue injury and disease. CRC Crit Rev Toxicol 23:21–48
24. Koleva II, Linssen JPH, Beek TAV, Evstatieva LN, Kortenska V, Handjieva N (2003) Antioxidant activity screening of extracts from Sideritis species (Labiatae) grown in Bulgaria. J Sci Food Agric 83:809–819
25. Kranl K, Schlesier K, Bitsch R, Hermann H, Rohe M, Böhm V (2005) Comparing antioxidative food additives and secondary plant products use of different assays. Am J Clin Nutr 93:171–175
26. Kuda T, Tsunekawa M, Goto H, Araki Y (2005) Antioxidant properties of four edible algae harvested in the Noto Peninsuka, Japan. J Food Comp Anal 96:625–633

27. Kumaron A, Karunakaran RJ (2007) In vitro antioxidant properties of methanol extract of five Phollanthus species from India. LWT-Food Sci Technol 40:344–352
28. Lanfer-Marquez UM, Barros RMC, Sinnecker P (2005) Antioxidant activity of chlorophylls and their derivatives. Food Res Int 38:885–891
29. Leong LP, Shui G (2002) An investigation of antioxidant capacity of fruits in Singapore markets. Food Chem 76:69–75
30. Lewmanomont K (1995) Algae in Thailand. Office of Environmental Policy and Planning, Ministry of Science, Technology and Environment, Bangkok
31. Lim SN, Cheung PCK, Ooi VEC, Ang PO (2002) Evaluation of antioxidative activity of extracts from a brown seaweed Sargassum siliquastrum. Food Chem 50:3862–3866
32. Mayakun J, Prathep A (2005) Seasonal variations in diversity and abundance of macroalgae at Samui Island, Surat Thani Province, Thailand. Songklanakarin J Sci Technol 27(Suppl 3):653–663
33. Morales FJ, Jimennez-Perez S (2001) Free radical scavenging capacity of Maillard reaction products as related to color and fluorescence. Food Chem 72:119–125
34. Nagayama K, Iwamura Y, Shibata T, Hirayama I, Nakamura T (2002) Bacteriocidal activity of phlorotannins from the brown alga Ecklonia kurome. J Antimicrob Chemother 50:889–893
35. Newman DJ, Cragg GM, Snader KM (2003) Natural products as source of new drugs over the period 1981–2002. J Nat Prod 66:1022–1037
36. Prathep A (2005) Spatial and temporal variations in diversity and percentage cover of macroalgae at Sirinart Marine National Park, Phuket province, Thailand. Sci Asia 31:225–233
37. Rajapakse N, Mendis E, Jung WK, Je JY, Kim SK (2005) Purification of a radical scavenging peptide from fermented mussel sauce and its antioxidant properties. Food Res Int 38:175–182
38. Sakanaka S, Tachibana Y, Ishihara N, Juneja LR (2004) Antioxidant activity of egg-yolk protein hydrolysates in a linoleic acid oxidation system. Food Chem 86:99–103

39. Śanchez CS, Gonźalez AMT, Garćıa-Parrilla MC, Granados JJQ, Garćıa de la Serrana HL, Martınez MCL (2007) Different radical scavenging tests in virgin olive oil and their relation to the total phenol content. Anal Chem Acta 593:103–107
40. Santoso J, Yoshie-Stark Y, Suzuki T (2004) Anti-oxidant activity of methanol extracts from Indonesian seaweeds in an oil emulsion model. Fish Sci 70:183–188
41. Siriwardhana N, Lee KW, Kim SH, Ha JW, Jeon YJ (2003) Antioxidant activity of Hizikia fusiformis on reactive oxygen species scavenging and lipid peroxidation inhibition. Food Sci Technol Int 9:339–346
42. Slinkard K, Singleton VL (1997) Total phenol analyses: automation and comparison with manual methods. Am J EcolVittic 28:49–55
43. Smith JE, Hunter CL, Smith CM (2002) Distribution and reproductive characteristics of nonindigenous and invasive marine algae in the Hawaiian Islands. Pac Sci 56:299–315
44. Stanhne LH (1995) Dried sausages fermented with Staphylococcus pylorus at different temperatures and different ingredient levels-part II volatile components. Meat Sci 41:193–209
45. Terada S, Maeda Y, Masui T, Suzuki Y, Ina K (1987) Comparison of caffenine and catechin components in infusion of various tea and tea drinks. Nippon Shokuhin Kogyo Gakkaishi Japan 34:20–27
46. Usuki R, Suzuki T, Endo Y, Kaneda TK (1984) Residual amounts of chlorophylls and pheophytins in refined edible oils. J Biol Chem 48:991–994
47. Wanasundara UN, Shahidi F (1998) Antioxidant and pro-oxidant activity of green tea extracts in marine oils. Food Chem 63:335–342
48. Wong DWS (1989) Mechanism and theory in food chemistry. Van Nostrand Reinhold, New York
49. Wu HC, Chen HM, Shiau CY (2003) Free amino acids and peptides as related to antioxidant properties in protein hydrolysates of mackerel (Scomber austriasicus). Food Res Int 36:949–957
50. Yamaguchi T, Takamura H, Matoba T, Teara J (1998) HPLC method for evaluation of the free radical-scavenging activity of food by using 1,1-diphenyl-2-picrylhydrazyl. Biosci Biotechnol Biochem

62:1201–1204

51. Yan X, Nagata T, Fan X (1999) Antioxidative activities in some common seaweeds. Plant Foods Hum Nutr 52:253–262
52. Yangthong M, Hutadilok-Towatan N, Phromkunthong W (2009) Antioxidant activities of four edible seaweeds from the southern coast of Thailand. Plant Foods Hum Nutr 64:218–223
53. Yoshie Y, Wang W, Petillo D, Suzuki T (2000) Distribution of catechins in Japanese seaweeds. Fish Sci 66:998–1000
54. Zhao H, Dong J, Lu J, Chen J, Li Y, Shan L, Line Y, Fan W, Gu G (2006) Effects of extraction solvent mixtures on antioxidant activity evaluation and their extraction capacity and selectivity for free phenolic compounds in barley (Hordeum vulgare L.). J Agric Food Chem 54:7277–7286

CHAPTER

16

BIOGAS FROM MACROALGAE: IS IT TIME TO REVISIT THE IDEA?

Adam D Hughes[1], Maeve S Kelly[1], Kenneth D Black[1,2], and Michele S Stanley[1,2]

[1]Scottish Association for Marine Science, Scottish Marine Institute, Department of Ecology, Oban, Argyll PA37 1QA, Scotland.

[2]Scottish Association for Marine Science, Scottish Marine Institute, Marine BioEnergy Scotland, Oban, Argyll PA37 1QA, Scotland.

ABSTRACT

The economic and environmental viability of dedicated terrestrial energy crops is in doubt. The production of large scale biomass (macroalgae) for biofuels in the marine environment was first tested in the late 1960's. The culture attempts failed due to the engineering challenges of farming offshore. However the energy conversion via anaerobic digestion was successful as the biochemical composition of macroalgae makes it an ideal feedstock. The technology for the mass production of macroalgae has developed principally in China and Asia over the last 50 years to such a degree that it is now the single largest product of aquaculture. There has also been significant technology transfer and macroalgal cultivation is now well tried and tested in Europe and America. The inherent advantage of production of biofuel feedstock in the marine environment is that it does not compete with food production for land or fresh water. Here we revisit the idea of the large scale cultivation of macroalgae at sea for subsequent anaerobic digestion to produce biogas as a source of renewable energy, using a European case study as an example.

KEYWORDS Biogas, Methane, Anaerobic digestion, Seaweed, Macroalgae, Aquaculture

INTRODUCTION

Growing terrestrial crops for biofuel may make a negligible contribution to net greenhouse gas emissions [1, 2] and may cause other environmental impacts while reducing freshwater resources and food security [3]. Given these limitations there has been renewed / increased interest in aquatic and marine production for biofuels [4, 5]. This interest can be divided into two principal components: biofuels derived from macroalgae (seaweed) and biofuels derived from microalgae (single cell plants). Microalgal derived biofuels have received much attention as a source for biodiesel [6, 7, 8], however production costs are an order of magnitude too expensive [3]. Although there is currently enormous research investment into the bulk production of microalgae for biodiesel, photo bioreactors are unlikely to be economically competitive for bioenergy production, and culture in outdoor ponds is only suited to regions with a relatively high number of sunlight hours and even then may still be uncompetitive in the biofuels market [9].

Macroalgae as a source of bioenergy first received intensive scrutiny as part of the US Ocean Food and Energy Farm project as proposed by Wilcox [10], initiated in 1973 and lasting over a decade [11]. It resulted in the construction of ocean farms for cultivation of the giant kelp *Macrocystis*[12]; reviewed by Kelly and Dworjanyn, [13]. While farming this species of seaweed in this truly offshore environment presented many technical challenges, the biogasification of macroalgal biomass gave excellent results [10, 12, 14, 15]. This and subsequent research highlights some of the major advantages of macroalgae over other sources of biofuels (see Table 1).

Table 1. Environmental and societal risk associated with terrestrial biofuels (after Koh and Ghazoul,[16]) and macroalgae biofuels

Environmental and societal advantages of macroalgae production for biofuels	
Net GHG emissions from land-use change	The culture of macroalgae for biofuel would be entirely marine based and would not have the associated GHG emissions associated with land use change
Threats to biodiversity	Macroalgae cultivation takes place in the water column above the seabed. Impacts of large scale macroalgae production on benthic biodiversity are currently unquantified. Likely impacts will include shading and competition for nutrients. However, most production will be in waters where the seabed is deeper than the photic zone, and where terrestrial nutrient run off creates hypernutrified water. It is likely that biodiversity would increase in the vicinity of macroalgae farms as a result of increased habitat structural complexity.
Impacts on food prices	Currently most macroalgae cultivation is for human consumption. Large scale production of macroalgae for biofuels is bound to distort this market. However the impacts on the supply of macroalgae to human food chain is likely to be small due to a clear market segregation and the far higher value of macroalgae as food compared to the price of energy.
Competition for water resources	Mass cultivation of macroalgae has a zero freshwater requirement and only modest amounts are required in anaerobic digestion

With microalgae much of the research interest has focused on their conversion to liquid biofuels such as ethanol [17, 18, 19, 20]. However, in this review we focus on anaerobic digestion of cultivated macroalgae for the production of biogas. Since this original gasification / culture research was conducted there have been substantial advances in macroalgal cultivation and offshore engineering. However the concept of ocean farming for biogas production has received relatively little attention in the 21st century.

In coastal temperate regions this technology may have significant potential to meet local energy demand, particularly in areas where the expansion of terrestrial biofuel production is limited due to high percentages of net primary productivity (NPP) already being appropriated for human use, such as the north Atlantic Coast of North America and Europe, and the western seaboard of South America (Figure 1). In this review the wild harvest of seaweeds for biofuel is not considered a viable option either in terms of potential yield or its environmental impact. Although macroalgal harvest for high value products takes place in some countries, very careful management is required to prevent serious ecosystem damage [21, 22, 23, 24] hence it would be impossible to justify harvest on the massive scale necessary [25, 26, 27] to make a significant energy contribution. Macroalgal forests, as with other biogenic structures in the marine environment such as corals and seagrasses, are considered to be biodiversity hot spots providing important habitat to a wide range of organisms including fish and birds [28]. It is worth noting that in Norway where approximately 170,000 tonnes pa *Laminaria hyperborea*are harvested, even a 4–5 year rotation is not always sufficient to allow recovery [29]. In addition, as wild stocks are generally dispersed around coastlines this would result in high costs of transport to processing plants.

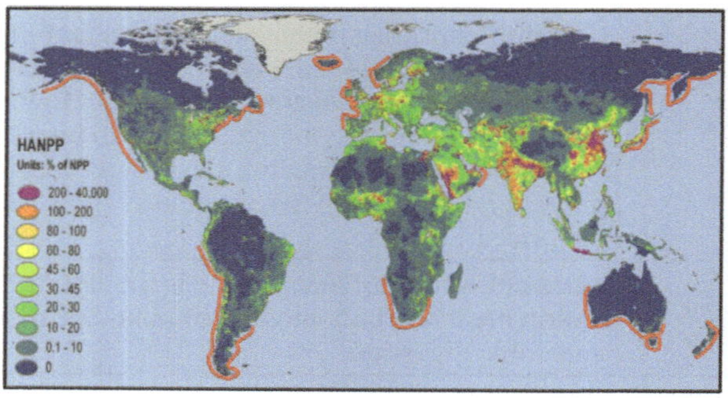

Figure 1. Natural distribution of shallow water macroalgae (red line) indicating coastal areas with the potential to culture macroalgae for biogas, and human appropriation of net primary production as a percentage of local net primary productivity (NPP) (NASA 2004 [30]). Redrawn from Santelices (2007) [31] and NASA Earth Observatory 2004.

MACROALGAL CULTURE

Currently over 100 species of macroalgae are used for food, in medicine, or as fertilizer and in the processing of phycolloids and chemicals [32]. Although used for millennia, their domestication only began in the twentieth century as a fuller understanding of their life cycle was achieved [33]. Several species are now in culture on a large-scale in east Asia. China is the world's largest producer of cultivated seaweed, mostly grown on long-line systems where hatchery produced seedlings are transplanted to sea on ropes suspended vertically from a horizontal top-line. The large brown *L. japonica* known as haidai or 'sea-strap' and originally introduced to China from Japan is the world's most cultivated species by volume and value. It was the first seaweed to be subjected to the entire process of seeding, tending and planting out and to have the status of a marine plant crop [32]. Global production of *L. japonica* alone in 2010 was 5.14 million tonnes with a value of 3.01 billion USD [34]; it is grown primarily for food but also for iodine and alginates. Its fast growth and high productivity make this and several other species of brown macroalgae particularly suited to culture for energy crops. Estimates of macroalgal primary productivity rates, in terms of carbon capture during photosynthesis, are approximately 1600 g $Cm^{-2}y^{-1}$[35], comparing favourably to a global net primary productivity of crop land of 470 g $Cm^{-2}y^{-1}$[36].

Selective breeding of macroalgae began in China in the 1960s with *Laminaria* species. This has resulted in a number of varieties that show the enhancement of desirable characteristics over wild varieties. These characteristics include:

- Increased frond growth rate at higher temperatures, resulting in a longer frond and higher production (20-58% higher)
- A higher (8-40%) iodine content as compared with the natural population
- A lower water content

Since the early seventies these selectively bred strains have been widely adopted by the *Laminaria* cultivation industry in North China. There are thus good prospects for the development of strains having traits desirable for biofuel production, such as increased sugar content or altered seasonality of production cycles [37].

MACROALGAL PRODUCTION

In Europe, hatchery raised macroalgae have been cultured successfully on long-line systems, similar to those used for mussel production. Positioned adjacent to salmon cages in Scottish sea lochs [38], a 100 m horizontal long-line bearing vertical strings carrying seaweeds every 50 cm, indicated average yields of >50 kg (native *Saccharina latissima*) per horizontal meter of long-line. If this were extrapolated to consider 40 such 100 m longlines, then yields of 200 t wet weight ha^{-1} (approximately 20 t dry weight) would be obtainable. This is comparable to yields achieved in China without fertiliser (H. Liu pers.comm. citing China Fish Annals, 2003). However if macroalgal crops are to make a significant contribution to fuel supply then very large areas would have to be farmed. MacKay (2009) [39] makes it clear that biomass energy will need to be a country-scale activity to make a meaningful contribution to UK energy needs. This will require significant changes in societal attitudes to use of the marine environment and, in many countries, regulatory changes. Inshore areas are already under significant pressure so the culture of macroalgae at the scale required for biofuel production must be largely located on continental shelves. Globally there is a very large amount of continental shelf suitable for such a massive aquaculture expansion; presently aquaculture occupies only about 0.04% of continental shelf area [40]. However culturing seaweed in an European offshore environment will require the development of more mechanised technologies for outplanting and harvest than the labour intensive methods on which the large-scale culture in Asia currently depends. This in turn may lead to the development of more specialised vessels than the mussel/salmon-farm work boats currently employed. The growth rate and productivity of seaweeds, grown on a large and dense scale, and in a different nutrient regime (offshore) to that of the inshore waters (Scottish sea lochs) has yet to be verified.

SEAWEED TO BIOGAS: ANAEROBIC DIGESTION

Macroalgae can be converted to biofuels by various processes including thermal treatment [41] and fermentation [19, 42] but the most direct route to obtaining biofuel from macroalgae is via its anaerobic digestion (AD) to biogas (~ 60% methane). Methane can be used to produce heat and electricity or compressed for use as a transport fuel. Research conducted in the 1980's [43, 44] still provides a bench mark for biogas yields for a number of macroalgal species, but since this time there have been developments in AD technology and an enormous increase in its use.

In comparison to terrestrial biomass crops, macroalgae contain little cellulose and no lignin and therefore undergo a more complete hydrolysis.

Gas yield is related both to ash content (and its inverse relationship with volatile solids content) and the level of storage sugars; and, as seaweed biochemical composition varies with season, gas yield will vary [45, 46]. The C:N ratio is also an important part of optimizing digester diet and strengthens the argument for the co-digestion of seaweeds with other more N rich substrates, for example waste food or agricultural slurries. Biogas yields are also dependent on a wide range of other variables such as inoculum, digester system configuration and feed stock composition.

Perhaps the most realistic estimate of the true industrial potential of methane production from macroalgae were obtained by Matusi (2006) [47] using a commercial scale 4 stage anaerobic digester for over 150 days, with a daily input between 0.2-1.0 tonnes of seaweed and a retention time of 15 to 25 days. This resulted in an average production of 22 m^3 of methane per tonne wet weight of brown seaweed (*Laminaria* sp). The potential energy yield from the AD of marine biomass compares favourably with that of terrestrial crops (Figure 2) on an energy per unit area basis. These figures are based on a production of 22 m^3 of methane per tonne of macroalgae wet weight, and a production of 200 tonnes of macroalgae ha $^{-1}$. However recent advances suggest there is still potential for further optimising biogas yields through co-digestion with a more nitrogenous substrate [48, 49] and manipulation of the microbial composition of the inoculum [50].

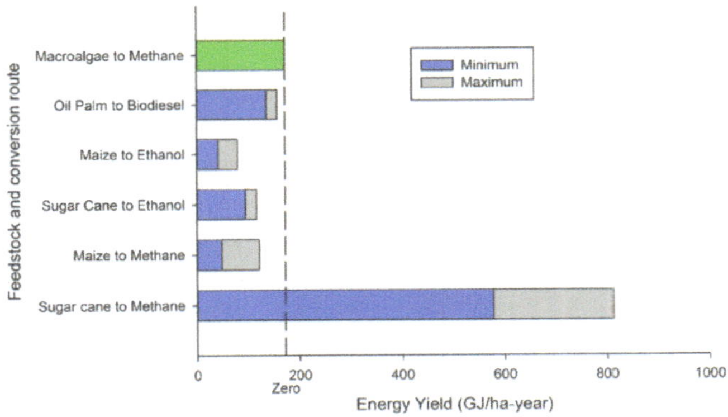

Figure 2. The energy production of biofuel crops ha^{-1}based on macroalgal production of 200 t ww ha^{-1}, terrestrial biofuel crop estimates are from Shilton and Guieysse (2010)[3].

BIOENERGY POTENTIAL – A QUESTION OF SCALE

If we use a realistic estimate of macroalgal production [26] (200 t ha^{-1}) and a conservative estimates of biogas yield after conversion (22 m^3 tonne wet weight (ww)) yielding 171 GJ ha^{-1} we can see that to make a significant contribution to bioenergy targets there will need to be macroalgal cultivation on a massive and unprecedented scale. For example if all of the brown algae currently produced in culture (6.8 million tonnes p.a. [34]) was converted to biogas using the parameters above it would yield approximately 5.7 PJ which is approx. 0.06% of the UK total energy demand for 2010 (9518 PJ [51]). To meet 1% of UK total energy demand would require an area of cultivation of approximately 5440 km^2. This is equivalent to half of the entire global area currently used for aquaculture production. However, if this is put in context of available space, this area accounts for only approximately 3% of the UK territorial waters (161200 km^2). By comparison with terrestrial biofuel production in the UK, to produce 1% of the UK's total energy demand using maize to methane would require a land area of 7700 km^2, equivalent to 18% of the UK's cropland (45000 km^2[52]). Although neither scenario seems attractive, such comparisons clearly illustrate the potential advantages of scale in moving UK biofuel production into the marine environment.

At a regional level large-scale macroalgal culture for biofuels offers real potential for rural coastal communities. A good example would be for the Isle of Mull on the west coast of Scotland which has no domestic gas supply, the main fuels being heating oil and electricity. The Isle of Mull has 1278 households [53] and the average UK annual domestic gas consumption is 57.6 GJ [54], so to provide all the households on Mull with gas would require the methane from 430 ha of macroalgae cultivation, the extent of which can be seen on Figure 3. We envisage this would both increase local employment and improve rural fuel security. The cultivation of macroalgae for biofuels could be developed through a modular approach in a European/American context where the distances between culture, processing and AD facilities are minimized, similar to that proposed for Japan [55]. Seaweed farms would supply local, coastal processing facilities where high value products can be first extracted from the crop before transfer to an on-site or a shared AD facility. From here the biogas produced could be piped directly to augment the local natural gas supply.

Figure 3. Area needed to grow enough seaweed to meet domestic gas requirements for homes on the Isle of Mull, West coast of Scotland, based on a production of 200 wet tonnes ha −1.

ENVIRONMENTAL IMPACT

Environmental impacts of large scale seaweed farms may arise from; changes to local hydrodynamics and resulting sedimentation patterns, benthic impacts from increased organic matter supply, changes to water column nutrient availability and from shading of the sea-floor (in shallow sites). Although we anticipate some types of interactions may well be positive [25] a measure of the extent and nature of interactions with fish, cetaceans and birds as well as other users of the marine environment for aquaculture, fisheries, energy generation and shipping is required.

During the growth cycle a portion of the macroalgae and the associated biota from the culture lines will be lost to the benthos either through erosion of the blade tips or shearing of cultured material creating an organically enriched zone [56]. In 'fed' aquaculture, that of fin-fish for example, where high energy feeds are supplied to the system, measurement of the extent of the zone of deposition is required [49] and has in turn led to the development of regulatory tools [57]. Although the macroalgal cultures are not 'fed' i.e. artificially supplied with additional nutrients or fertilised, the extent and effect of the zone of organic enrichment should be described. In enclosed water bodies, there may be competition for dissolved nutrients with phytoplankton but, in more open shelf systems, nutrient supply is likely to be sufficient provided that farms are spatially arranged for optimal nutrient exchange. In any event, nutrients taken up by macroalgal culture, on the scale required for biofuel production, would be far less than that produced by agricultural, urban sources and fin-fish aquaculture. If macroalgae is subjected to the AD process then a proportion of the nitrogen may be lost through denitrification depending on the conditions in the reactor. Digestates are typically higher in ammonia and lower in organic nitrogen than ingestates [58]. The digestate will most likely be used in fertilisers and so find its way back into the hydrological cycle.

There may also be a number of positive benefits; the macroalgal farms effectively acting as no-take zones for mobile gear fisheries and thus enhancing less destructive static gear fisheries within the cultivation zone and providing spill over benefits to adjacent waters [59]. In addition, providing the crop is not removed in its entirety at the end of the cycle it will provide a refuge and a substrate to enhance local biodiversity. The digestate after AD may be either a valuable by-product or an expensive waste. This will depend on a number of factors including its contaminant metal burden and whether the macroalgae has been mixed with other organic waste streams in the digestor. A study on the AD of lipid-extracted microalgal biomass [60] suggested that 80% of the nitrogen in the biomass was recoverable as ammonium/ammonia from the liquid supernatant fraction, and that the remaining nitrogen in the solid digestate fraction had a 40% bioavailability when applied to soil. A similarly detailed analysis of the fate of nitrogenous emissions following AD of macroalgal biomass is required. Overall the global effect of using macroalgal culture for biofuel is likely to be positive and an initial full life cycle analysis of biomethane production from offshore cultivation of macroalgae has shown a 69% reduction in fossil fuel utilization when compared to natural gas, a 54% reduction in greenhouse gas emissions and an improvement in the marine eutrophication index [61].

MAKING IT PAY

Costing the culture of large amounts of seaweed in a European context is currently highly uncertain as there are too many unconstrained parameters, such as scalability, location and the degree of mechanisation readily achievable. However, our analysis based on inshore production suggest that at 2011 wellhead value for natural gas (US $3.95 [62] per thousand cubic feet (equivalent to £0.09 m^3)) based on a production of 20 tonnes dry weight (dw) ha $^{-1}$ the production costs for macroalgal biogas would have to be less than £400 ha $^{-1}$ to be competitive with fossil fuels without additional subsidy. It is unlikely that in the short term such production costs could be achieved. However under the UK Renewable Heat Incentive 2011 [63] scheme injection of biomethane into the natural gas grid attracts a price of £0.068 kWh. This is equivalent to £3230 hectare which would make the cultivation of macroalgae for methane production highly competitive. In addition the identification and extraction of higher value products, prior to AD, is advisable, as is the quantification of how the prior extraction affects biogas yield. Added value could be achieved by processing part of the crop for human and animal foodstuffs, and food supplements, for its mineral content for animal feeds, as an organic slow release fertilizer, and potential bio-active compounds [64].

CONCLUSION

Our analysis of growth data from hatchery-raised macroalgal sporelings outplanted to conventional long-line systems in Scotland suggests there are no major biological obstacles to the culture process in a European context. A fuller understanding of the impacts and performance of native macroalgae grown in dense large-scale cultures can only be achieved through pilot scale trials. Technological advancement is required to mechanise the outplanting and harvest process. The biological gasification of macroalgae was well proven in the later decades of the 20th century and AD technology has sufficiently matured to offer a range of possibilities to further optimize methane yields. Compared to first generation biofuels, macroalgae have inherent advantages that make them environmentally sustainable. Given that fossil fuel prices are likely to increase and that macroalgal production costs will inevitably fall as production is expanded and intensified, it is prudent to develop the technology required to obtain significant quantities of biofuel from marine biomass in time to help meet Europe's energy needs and climate change targets.

ACKNOWLEDGEMENTS

MSK, KDB and MSS acknowledge funding for the BioMara project (http://www.biomara.ac.uk). The Biomara project is supported by the European Regional Development Fund through the INTERREG IVA Programme, Highlands and Islands Enterprise, Crown Estate, Northern Ireland Executive, Scottish Government and Irish Government. ADH received funding from the MASTS pooling initiative (The Marine Alliance for Science and Technology for Scotland), MASTS is funded by the Scottish Funding Council (grant reference HR09011) along with contributing institutions, and from the European Union's Seventh Framework Programme (FP7/2007-2013) project AT~SEA under grant agreement n° 280860. Their support is gratefully acknowledged.

AUTHORS' CONTRIBUTIONS

This review was conceived, researched and written by ADH, MSK, KDB and MSS. All authors have read and approved the final manuscript.

REFERENCES

1. ABC: Argyll and bute housing strategy 2011–2016. In *Book Argyll and Bute housing strategy 2011–2016*. Argyll and Bute Council; 2011.
2. Adams JM, Gallagher JA, Donnison IS: Fermentation study on Saccharina latissima for bioethanol production considering variable pre-treatments. *J Appl Phycol* 2009, 21: 569-574. 10.1007/s10811-008-9384-7
3. Adams JMM, Toop TA, Donnison IS, Gallagher JA: Seasonal variation in Laminaria digitata and its impact on biochemical conversion routes to biofuels. *Bioresour Technol* 2011, 102: 9976-9984. 10.1016/j.biortech.2011.08.032
4. Adani F, Scievano A, Boccasile G: In Anaerobic digestion: opportunity for agriculture and environment. In *Anaerobic digestion:*

opportunity for agriculture and environment. Milan; 2008:27-37.

5. Anon: Departmental note: support for renewable heat technologies in the domestic and non domestic sectors. In *Book departmental note: support for renewable heat technologies in the domestic and non domestic sectors*. UK Gov; 2012.
6. Bastianoni S, Coppola F, Lezzi E, Colacevich A, Borghini F, Focardi S: Biofuel potential production from the Orbetello lagoon macroalgae: a comparison with sunflower feedstock. *Biomass Bioenergy* 2008, 32: 619-628. 10.1016/j.biombioe.2007.12.010
7. Bird KT, Benson PH: *Seaweed cultivation for renewable resources*. Amsterdam; New York: Elsevier; 1987.
8. Brennan L, Owende P: Biofuels from microalgae-a review of technologies for production, processing, and extractions of biofuels and co-products. *Renew Sust Energ Rev* 2010, 14: 557-577. 10.1016/j.rser.2009.10.009
9. Brennan L, Owende P: Biofuels from microalgae—a review of technologies for production, processing, and extractions of biofuels and co-products. *Renew Sustain Energy Rev* 2010, 14: 557-577. 10.1016/j.rser.2009.10.009
10. Cecchi F, Pavan P, MataAlvarez J: Anaerobic co-digestion of sewage sludge: application to the macroalgae from the Venice lagoon. *Resour Conserv Recycl*1996, 17: 57-66. 10.1016/0921-3449(96)88182-1
11. Chisti Y: Biodiesel from microalgae. *Biotechnol Adv* 2007, 25: 294-306. 10.1016/j.biotechadv.2007.02.001
12. Christie H, Fredriksen S, Rinde E: Regrowth of kelp and colonization of epiphyte and fauna community after kelp trawling at the coast of norway. *Hydrobiologia*1998, 375–376: 49-58.
13. Chynoweth DP, Fannin KF, Srivastava VJ: Biological gasification of marine algae. In *Seaweed cultivation for renewable resources. Developments in aquaculture and fisheries science*. Edited by: Bird KT, Benson PH. Elsevier, Amsterdam; 1987:287-303.
14. Chynoweth DP, Klass DL, Ghosh S: Anaerobic digestion of kelp. In *Biomass conversion processes for energy and fuels*. Edited by: Sofer SS, Zaborsky OR. New York: Plenum Press; 1981:315-318.

15. Cromey CJ, Nickell TD, Treasurer J, Black KD, Inall M: Modelling the impact of cod (Gadus morhua L) farming in the marine environment-CODMOD. *Aquaculture*2009, 289: 42-53. 10.1016/j.aquaculture.2008.12.020

16. Cumashi A, Ushakova NA, Preobrazhenskaya ME, D'Incecco A, Piccoli A, Totani L, Tinari N, Morozevich GE, Berman AE, Bilan MI, *et al.*: A comparative study of the anti-inflammatory, anticoagulant, antiangiogenic, and antiadhesive activities of nine different fucoidans from brown seaweeds. *Glycobiology* 2007, 17: 541-552.

17. DEFRA: Crop areas, yeilds and production,livestock populations and the size of the argricultural workforce:2011 UK final results. In *Book crop areas, yeilds and production,livestock populations and the size of the argricultural workforce:2011 UK final results.* Department for Environmental Food and rural Affairs, UK; 2012.

18. Duarte CM, Holmer M, Olsen Y, Soto D, Marba N, Guiu J, Black K, Karakassis I: Will the oceans help feed humanity? *BioScience* 2009, 59: 967-976. 10.1525/bio.2009.59.11.8

19. Duarte CM, Middelburg JJ, Caraco N: Major role of marine vegetation on the oceanic carbon cycle. *Biogeosciences* 2005, 2: 1-8.

20. FAO: *The state of world fisheries and aquaculture - 2010 (SOFIA).* 2010.

21. Field CB, Campbell JE, Lobell DB: Biomass energy: the scale of the potential resource. *Trends Ecol Evol* 2008, 23: 65-72. 10.1016/j.tree.2007.12.001

22. Frank ED, Han J, Palou-Rivera I, Elgowainy A, Wang MQ: Methane and nitrous oxide emissions affect the life-cycle analysis of algal biofuels. *Environ Res Lett* 2012., 7

23. Gell FR, Roberts CM: Benefits beyond boundaries: the fishery effects of marine reserves. *Trends Ecol Evol* 2003,18(9):448-455. 10.1016/S0169-5347(03)00189-7

24. Gibbs HK, Johnston M, Foley JA, Holloway T, Monfreda C, Ramankutty N, Zaks D: Carbon payback times for crop-based biofuel expansion in the tropics: the effects of changing yield and technology. *Environ Res Lett* 2008., 3: 10.1088/1748-9326/3/3/034001

25. Goh CS, Lee KT: A visionary and conceptual macroalgae-based third-generation bioethanol (TGB) biorefinery in Sabah, Malaysia as an underlay for renewable and sustainable development. *Renew Sust Energ Rev* 2010, 14: 842-848. 10.1016/j.rser.2009.10.001

26. Haug A, Jensen A: Seasonal variations in the chemical composition of Alaria esculenta. Laminaria saccharina:Laminaria hyperborea and Laminaria digitata from Northern Norway. *Reports of the Norwegian Institute of Seaweed Research No. 4* 1954.

27. Hughes AD, Black KD, Campbell I, Heymans JJ, Orr KK, Stanley MS, Kelly MS: Comments on 'prospects for the use of macroalgae for fuel in Ireland and UK: an overview of marine management issues'. *Marine Policy*

28. Imhof M, Bounoua L: Human consumption of primary productivity. In *Book human consumption of primary productivity 2011*. Earth Observatory NASA; 2004.

29. John RP, Anisha GS, Nampoothiri KM, Pandey A: Micro and macroalgal biomass: a renewable source for bioethanol. *Bioresour Technol* 2011, 102: 186-193. 10.1016/j.biortech.2010.06.139

30. Kelly M, Dworjanyn S: *The potential of marine biomass for anaerobic biogas production*. The potential of marine biomass for anaerobic biogas production: sThe Crown Estate; 2008:103.

31. Koh LP, Ghazoul J: Biofuels, biodiversity, and people: understanding the conflicts and finding opportunities. *Biol Conserv* 2008, 141: 2450-2460. 10.1016/j.biocon.2008.08.005

32. Kraan S: Mass-cultivation of carbohydrate rich macroalgae, a possible solution for sustainable biofuel production. *Mitig Adapt Strat Glob Chang* 2010, 1-20.

33. Langlois J, Sassi J-F, Jard G, Steyer J-P, Delgenes J-P, Hélias A: Life cycle assessment of biomethane from offshore-cultivated seaweed. *Biofuels, Bioprod Biorefin* 2012. n/a-n/a

34. Lee S-M, Kim GH, Lee J-H: Bio-gas production by co-fermentation from the brown algae, Laminaria japonica. *J Ind Eng Chem* 2012,18(4):1512-1514. 10.1016/j.jiec.2012.02.014

35. Leese TM: The conversion of ocean farm kelp to methane and other products.In *Clean fuels from biomass, sewage, urban refuse, agricultural wastes; Proceedings of the Symposium, Orlando, Fla.,*

January 27-30, 1976. (A77-37652 17-44). Chicago: Institute of Gas Technology; 1976:253-266. Research sponsored by the American Gas Association and ERDA

36. Lorentsen SH, Sjotun K, Gremillet D: Multi-trophic consequences of kelp harvest.*Biol Conserv* 2010, 143: 2054-2062. 10.1016/j.biocon.2010.05.013
37. MacKay D: *Sustainable energy without the hot air*. UIT Cambridge Ltd., Cambridge; 2009.
38. MacLeay I, Harris K, Annut A: Digest of united kingdom energy statistics 2011. In *Book digest of united kingdom energy statistics 2011*. National Statistics publication; 2011.
39. Mata TM, Martins AA, Caetano NS: Microalgae for biodiesel production and other applications: a review. *Renew Sust Energ Rev* 2010, 14: 217-232. 10.1016/j.rser.2009.07.020
40. Matsui JT, Amano T, Koike Y, Saiganji A, Saito H: Methane fermentation of seaweed biomass. In *American institute of chemical engineers*. San Francisco; 2006.
41. *Natural Gas*. http://205.254.135.7/dnav/ng/ng_pri_sum_dcu_nus_a.htm
42. Nkemka VN, Murto M: Evaluation of biogas production from seaweed in batch tests and in UASB reactors combined with the removal of heavy metals. *J Environ Manage* 2010, 91: 1573-1579. 10.1016/j.jenvman.2010.03.004
43. North WJ: Oceanic farming of Macrocystis, the problems and non-problems. Seaweed cultivation for renewable resources. In *Seaweed cultivation for renewable resources. Volume 16*. Edited by: Bird KT, Benson PH. Amsterdam: Elsevier; 1987:39-68.
44. Ortiz J, Romero N, Robert P, Araya J, Lopez-Hernandez J, Bozzo C, Navarrete E, Osorio A, Rios A: Dietary fiber, amino acid, fatty acid and tocopherol contents of the edible seaweeds Ulva lactuca and Durvillaea antarctica. *Food Chem* 2006, 99:98-104. 10.1016/j.foodchem.2005.07.027
45. Ortiz M: Dynamic and spatial models of kelp forest of Macrocystis integrifolia and lessonia trabeculata (SE pacific) for assessment harvest scenarios: short-term responses. *Aquat Conserv Mar Freshwat Ecosyst* 2010, 20: 494-506. 10.1002/aqc.1126

46. Rodger ANS: *Sea-based integrated multi-trophic aquaculture: investigation of a fish, bivalve and macroalgal co-culture system.* SAMS PhD Thesis: University of the Highlands and Islands; 2010.
47. Sanderson CJ: *Reducing the environmental impact of fish cage farming through the cultivation of seaweeds.* SAMS PhD Thesis: University of the Highlands and Islands; 2006.
48. Santelices B: The discovery of kelp forests in deep-water habitats of tropical regions. *Proc Natl Acad Sci* 2007, 104: 19163-19164. 10.1073/pnas.0708963104
49. Scharlemann JPW, Laurance WF: Environmental science - How green are biofuels? *Science* 2008, 319: 43-44. 10.1126/science.1153103
50. Shilton A, Guieysse B: Sustainable sunlight to biogas is via marginal organics.*Curr Opin Biotechnol* 2010, 21: 287-291. 10.1016/j.copbio.2010.03.008
51. Singh J, Cu S: Commercialization potential of microalgae for biofuels production. *Renew Sust Energ Rev* 2010, 14: 2596-2610. 10.1016/j.rser.2010.06.014
52. Springer YP, Hays CG, Carr MH, Mackey MR: Toward ecosystem-based management of marine macroalgae-the bull kelp, Nereocystis luetkeana.*Oceanogr Mar Biol Annu Rev* 2010, 48: 1-41.
53. Thompson SA, Knoll H, Blanchette CA, Nielsen KJ: Population consequences of biomass loss due to commercial collection of the wild seaweed Postelsia palmaeformis. *Mar Ecol Prog Ser* 2010, 413: 17-31.
54. Tseng CK: Some remarks on the kelp cultivation industry of china. *Dev Aquac Fish Sci* 1987, 16: 147-155.
55. *Typical domestic energy consumption figures.*
56. Vasquez J: Production, use and fate of chilean brown seaweeds: resources for a sustainable fishery. *J Appl Phycol* 2008, 20: 457-467. 10.1007/s10811-007-9308-y
57. Wargacki AJ, Leonard E, Win MN, Regitsky DD, Santos CNS, Kim PB, Cooper SR, Raisner RM, Herman A, Sivitz AB, *et al.*: An engineered microbial platform for direct biofuel production from brown macroalgae. *Science* 2012, 335: 308-313. 10.1126/

science.1214547

58. Wilcox HA: The U.S. Navy's Ocean food and energy farm project. In *Book the U.S. Navy's Ocean food and energy farm project* Edited by: Monney NT. 83-104. City; 1977:83–104
59. Williams PJL, Laurens LML: Microalgae as biodiesel & biomass feedstocks: review & analysis of the biochemistry, energetics & economics. *Energ Environ Sci* 2010, 3: 554-590. 10.1039/b924978h
60. Wise DL, Augenstein DC, Ryther JH: Methane fermentation of aquatic biomass.*Resour Recover Conserv* 1979, 4: 217-237. 1979, 4:217–237 10.1016/0304-3967(79)90002-7
61. Wu CY, Pang SJ: World seaweed resources - an authoritative reference system.*Book world seaweed resources - An authoritative reference system* 2006. City
62. Yokoyama S, Jonouchi K, Imou K: Energy production from marine biomass: fuel cell power generation driven by methane produced from seaweed. *Proc World Acad Sci Eng Tech* 2007, 22: 320-323.
63. Zhang JH, Hansen PK, Fang JG, Wang W, Jiang ZJ: Assessment of the local environmental impact of intensive marine shellfish and seaweed farming-application of the MOM system in the Sungo Bay. *China. Aquaculture*2009,287(3-4):304-310.
64. Zhou D, Zhang LA, Zhang SC, Fu HB, Chen JM: Hydrothermal liquefaction of macroalgae Enteromorpha prolifera to Bio-oil. *Energy Fuel* 2010, 24: 4054-4061. 10.1021/ef100151h

CHAPTER

17

DIRECT BIOCONVERSION OF BROWN ALGAE INTO ETHANOL BY THERMOPHILIC BACTERIUM DEFLUVIITALEA PHAPHYPHILA

Shi-Qi Ji, Bing Wang, Ming Lu, and Fu-Li Li

Key Laboratory of Biofuels, Shandong Provincial Key Laboratory of Synthetic Biology, Qingdao Institute of Bioenergy and Bioprocess Technology, Chinese Academy of Sciences, Qingdao, People's Republic of China

ABSTRACT

Background

Brown algae are promising feedstocks for biofuel production with inherent advantages of no structural lignin, high growth rate, and no competition for land and fresh water. However, it is difficult for one microorganism to convert all components of brown algae with different oxidoreduction potentials to ethanol. *Defluviitalea phaphyphila* Alg1 is the first characterized thermophilic bacterium capable of direct utilization of brown algae.

Results

Defluviitalea phaphyphila Alg1 can simultaneously utilize mannitol, glucose, and alginate to produce ethanol, and high ethanol yields of 0.47 g/g-mannitol, 0.44 g/g-glucose, and 0.3 g/g-alginate were obtained. A rational redox balance system under obligate anaerobic condition in fermenting brown algae was revealed in *D. phaphyphila* Alg1 through genome and redox analysis. The excess reducing equivalents produced from mannitol metabolism were equilibrated by oxidizing forces from alginate assimilation. Furthermore, *D. phaphyphila* Alg1 can directly utilize unpretreated kelp powder, and 10 g/L of ethanol was accumulated within 72 h with an ethanol yield of 0.25 g/g-kelp. Microscopic observation further demonstrated the deconstruction process of brown algae cell by *D. phaphyphila* Alg1.

Conclusions

The integrated biomass deconstruction system of *D. phaphyphila* Alg1, as well as its high ethanol yield, provided us an excellent alternative for brown algae bioconversion at elevated temperature.

KEYWORDS: Brown algae, Redox balance, Bioethanol, Marine thermophile, *Defluviitalea phaphyphila*

BACKGROUND

To settle worldwide energy crisis and environmental problems, increased interests have been focused on aquatic and marine production for biofuels, which mainly includes biofuels derived from macroalgae and microalgae. Brown algae are a large group of marine seaweeds including almost 1800 species of macroalgae with a characteristic olive-green to dark brown color derived from fucoxanthin [1]. Some well-known aquaculture or wild species are from the genera of *Laminaria*, *Saccharina*, *Macrocystis*, *Sargassum*, and so on. The technology for the mass production of macroalgae has been developed significantly in China and Asia over the last 50 years [2]. Notably, China contributes 72 % of global aquaculture-based macroalgae production, including the genera of *Laminaria* (reclassified as *Saccharina* for some species, brown algae), *Undaria* (green algae), *Porphyra*, and *Gracilaria* (red algae) [3]. Brown algae have complex sugar composition, mainly including alginate, mannitol, and laminarin [3]. Alginate is the unique structural polysaccharides in brown algae, which consists of two uronic acids, namely, α-l-guluronate (G) and β-d-mannuronate (M) [4]. The content of alginate varied from 20 to 40 % of dry weight among different species [5, 6]. Mannitol and laminarin are considered as reserve carbohydrates in many brown algae species, which are mostly accumulated in summer. Mannitol is a sugar alcohol form of mannose, while laminarin is a linear polysaccharide of mannitol-containing β-1,3-linked glucose [7, 8]. The content of mannitol and laminarin in some species can reach as high as 25 and 30 %, respectively, at the beginning of autumn [9].

The inherent advantages of brown algae for biofuel production mainly include the structural advantage of containing no lignin, high growth rate, and no competition with food production for land or fresh water [1, 10, 11]. They have been used for anaerobic digestion to produce biogas and liquid biofuel production. The direct bioconversion of brown algae to produce bioethanol cannot be easily achieved because of their diverse carbohydrate components. It is difficult for one microorganism to ferment all saccharides for biofuel production. Although glucose released from the hydrolysis of glucan could be easily assimilated through glycolysis by candidate strains, mannitol catabolism needs additional enzymes before entering glycolysis which include d-mannitol phosphotransferase (PTS) permease which transports d-mannitol into cells with the formation of mannitol-1-phosphate, and one mannitol-1-phosphate dehydrogenase (MPDH) (mannitol degradation I, MetaCyc Pathway Database, http://www.metacyc.org/) [12]. One reduced nicotinamide adenine dinucleotide (NADH) or reduced nicotinamide adenine dinucleotide phosphate (NADPH) was produced in this process of oxidizing mannitol-1-phosphate to fructose-6-phosphate. Moreover, the saccharification of brown algae requires one microorganism to secrete several polysaccharide depolymerizing enzymes such as alginate lyase and laminarinase. Through the endolytical and exolytical cleavages by alginate lyase and oligoalginate lyases, alginate was degraded into unsaturated monosaccharide (spontaneously rearranged into 4-deoxy-L-erythro-5-hexoseulose uronic acid, DEH) [13]. Subsequently, a DEH reductase and one 2-keto-3-deoxy-d-gluconate (KDG) kinase converted DEH into 2-keto-3-deoxy-6-phosphogluconate (KDPG) with a consumption of one NADH or NADPH, and then KDPG was directly assimilated through the Entner–Doudoroff (ED) pathway [14, 15]. In general, most ethanologenic microorganisms did not contain these genes encoding alginate depolymerizing enzymes. Thus, acid or enzymatic pretreatments were needed to decompose their structural polysaccharides to release monomer sugars from brown algae biomass [11, 16, 17, 18, 19]. Moreover, the combined metabolism of alginate and mannitol also needs a well evolved redox system to balance the reducing equivalents, especially under anaerobic fermentation conditions [20]. Thus, in a direct bioconversion process for ethanol production, one microorganism need to secrete multiple enzymes to depolymerize polysaccharides, uptake the released sugars, metabolize the sugars, and balance the redox state of the cells. Due to these limitations, only a few natural microorganisms exhibit all the features desired for direct bioconversion of brown algae as far as we know. However, several natural strains showing partial desirable properties have been reported [17, 21, 22]. In general, they could only utilize glucan and/or mannitol released from brown algae after acid or enzyme pretreatment for bioethanol production. To generate a viable organism with better performance in brown algae bioconversion, attempts to genetically engineer natural strains were reported. *For example*, *Escherichia coli* was engineered to a microbial platform for bioethanol production directly from brown macroalgae by introducing a DNA fragment from *Vibrio splendidus* encoding alginate transport and metabolism and ethanol synthesis genes (*pdc* and *adhB*) from *Zymomonas mobilis* [23]. Recently, a synthetic yeast platform (*S. cerevisiae*) has been constructed to be able to metabolize DEH and mannitol, however, without alginate degradation ability [20]. The procedure of genetic engineering is extremely complex, and besides it is hard to meet all the requirements which favored direct fermentation processes.

In our previous work, an untapped thermophilic bacterial resource was identified from coastal sediments [24, 25, 26]. Strain *Defluviitalea phaphyphila* Alg1 is one of the species isolated from this environment [27]. Genome analysis indicated that strain Alg1 has an integrated brown algae-degrading system. In this work, the potential of *D. phaphyphila* Alg1 in direct bioconversion of brown algae to ethanol was investigated and evaluated.

METHODS

Culture Media and Microorganisms

S. japonica was purchased from Tuandao seafood market in Qingdao, China. The seaweed was dried under sunlight and then ground into powder by a knife mill. The powder was filtered through a 200-mesh sieve. The basal medium (BM) consisted of 0.1 g/L of KH_2PO4, 0.1 g/L of K_2HPO4, 1 g/L of $NaHCO_3$, 2 g/L of NH_4Cl, 30 g/L of sea salt, 0.5 g/L of l-cysteine, 0.2 % of yeast extract, and 0.0001 % (w/v) of resazurin. The BM medium containing 1 g/L of alginate (ABM) was sterilized at 115 °C for 20 min and used as inoculum medium. The previously characterized type strain *D. phaphyphila* Alg1 (= CGMCC 1.5199^T = JCM 30481^T) was used as fermentation strain [27]. A seed culture (50 ml ABM medium in anaerobic bottle) was prepared from a single colony of Alg1 from plate and incubated at 60 °C for 24 h until the optical density at 600 nm (OD_{600}) reached 0.8–0.9. For fermentation experiments, a culture was inoculated with 5 % seed culture (v/v).

Ethanol Fermentation of *D. Phaphyphila* Alg1

Fermentations were carried out in 250 ml bottle containing 100 ml BM at 60 °C and obligate anaerobic conditions by supplementing 1 % (w/v) of each substrate including glucose, mannitol, and alginate, respectively. A mixture of alginate, mannitol, and glucose (at a ratio of 5:8:1) with 3 % total sugar was used to simulate the main components of brown algae [23], and was tested in the same way as above. Fermentation of *S. japonica* powder was carried out in 1-L bioreactors. The fermentation medium consisted of 0.5 % yeast extract and 5 % milled kelp powder in BM. A constant flow of pure nitrogen was bubbled into the medium to remove dissolved oxygen and to maintain anaerobic conditions through the whole fermentation process. The fermentation was carried out under static conditions without stirring. Preliminary experiments have suggested that mechanical mixing was not favored by *D. phaphyphila* Alg1 in kelp fermentation. Samples were taken periodically, and were frozen at −20 °C until further analysis. All experiments were carried out in triplicate.

Genome Analysis of Ethanol Synthesis Pathway and Redox Balance Analysis

The Whole Genome Shotgun project of *D. phaphyphila* Alg1 has been deposited at DDBJ/EMBL/GenBank under the accession number of JWID00000000. The version described in this paper is version JWID01000000.

The redox balance of anaerobic metabolism of glucose, mannitol, alginate, and their mixture was analyzed as follows. The cells growing in different substrates were collected in late exponential phase. An Amplite™ Colorimetric NAD/NADH Ratio Assay Kit (AAT Bioquest Inc., Sunnyvale, CA, USA) was used to measure the intracellular total NAD/NADH amount and NAD/NADH ratio in culture cells according to the manual. Briefly, NAD in the cell lysate was extracted with NAD extraction solution and converted to NADH through enzyme reaction, and then recognized by the NADH probe to give a yellow-color dye after reaction, which has the absorbance at 460 nm. The amount of the dye generated is directly proportional to the concentration of NAD or NADH in the cell lysate.

Microscopic Observation of Cell Wall Morphology Change

The morphological changes of brown algae during fermentation were examined using optical microscope (Olympus BX51, Tokyo, Japan) and transmission electron microscope (TEM) (Hitachi H-7650, Tokyo, Japan). Briefly, brown algae were cut into 2 cm in width and 5 cm in length, and were added into the fermentation bottle. After 120 h of fermentation, the remaining brown algae specimens were cut into 0.1–0.2 cm pieces. For optical microscopy, brown algae pieces were stained by 0.1 % crystal violet for 10 min and rinsed two times with phosphate buffer. For transmission electron microscopy, cells were fixed by 2.5 % glutaraldehyde solution and 1 % osmic acid. Tissue was dehydrated in graded acetone starting with 30, 50, 70, 80, 90, 95 % for 30 min each and with 3 rinses in 100 % acetone. Infiltrate with acetone/Spurr resin (7:3) for 5 h, acetone/Spurr (3:7) overnight and 100 % Spurr resin for 5 h. Embed tissue in fresh Spurr resin and place in 65 °C oven and polymerize for 24 h. Leica EM UC6 (Vienna, Austria) was used to prepare ultrathin sections (70 nm) to examine the cell ultrastructure, and uranyl acetate (2 %) was used to stain the sections.

Analytical Methods for Cell Growth, Products, and Enzyme Activities

Cell growth was monitored by measuring OD_{600} using a lambda 25 UV/Vis spectrometer (PerkinElmer, Shelton, USA). In the case of alginate as substrate, solid particles in the medium prevented the determination of cell density with the spectral photometer, therefore, the cell growth was assessed by measuring the total protein concentrations using the Bradford assay kit (Biomed, Beijing, China) with bovine serum albumin as the standard [28]. The concentrations of mannitol, glucose, and fermentation products were determined using high-performance liquid chromatography (HPLC) equipped with an Aminex HPX-87H column (Bio-Rad, Hercules, USA). Filtered and degassed H_2SO_4 (5 mM) was used as eluent at a flow rate of 0.6 ml/min and the column temperature was controlled at 65 °C. Quantification of alginate was performed using the carbazole sulfuric acid method with sodium alginate as the standard [29].

The alginate lyase and laminarinase activities were determined during the fermentation of kelp powder. Alginate lyase activity was assayed by measuring the increase in absorbance at 235 nm (A_{235}) of the reaction products (unsaturated uronates) for 3 min at 70 °C in a quartz cuvette containing 2 ml of 0.2 % alginate in 100 mM acetate/sodium acetate buffer (pH 5.8) and 10 µl fermentation supernatant. One unit of activity was defined as an increase of 0.1 in A_{235} per minute. Laminarinase activity was assayed by incubating laminarin solution (1 % w/v, in 50 mM Tris/HCl, pH 7.3) with fermentation supernatant at 70 °C for 3 min. The release of reducing sugars was measured by the dinitrosalicylic (DNS) colorimetric method [30]. One unit of laminarinase activity was defined as 1 mg of reducing sugar equivalent to 1 mg of glucose per minute.

RESULTS AND DISCUSSION

Fermentation of Glucose, Mannitol, and Alginate for Ethanol Production

The growth and ethanol fermentation properties of *D. phaphyphila* Alg1 were evaluated by employing alginate, mannitol, or glucose (laminarin monomer) as substrates. When glucose (1 %, w/v) was used as sole carbon substrate, Alg1 showed vigorous growth after inoculation, and all glucose was exhausted in less than 48 h. The maximal cell concentration (OD_{600}) reached 1.2 at 48 h (Fig. 1a). After 108 h cultivation, 3.8 g/L ethanol and 0.4 g/L acetic acid were produced, with an ethanol-to-acetate ratio of 9.5:1 (Fig. 1a). The ethanol yield was 0.38 g/g-glucose which accounted for 74 % of the maximal yield of 0.51 g/g-glucose [31]. In the case of mannitol (1 %, w/v) as substrate, the cell density could reach to the highest OD_{600} of 1.4 after 108 h cultivation (Fig. 1b). The concentrations of ethanol and acetic acid could be as high as 4.3 and 0.3 g/L, respectively, with an ethanol-to-acetate ratio around 14:1 (Fig. 1b). The ethanol yield from mannitol was about 0.44 g/g-mannitol, which accounted for 86 % of the theoretical maximal yield of 0.51 g/g-mannitol [8]. Mannitol was exhausted until 108 h indicating a slower substrate assimilation rate than glucose. Indeed, from the reported literatures, mannitol normally cannot be fermented under strictly anaerobic condition by *Zymobacter palmae*, *Enterobacter* sp. JMP3 and some yeast species, while glucose can be easily converted to ethanol under the same condition. Thus, the utilization of mannitol is more complex, and has higher requirements on microorganisms. In addition, in contrast to those of *Z. palmae* (0.38 g-ethanol/g-mannitol) and *Enterobacter* sp. (0.29 g-ethanol/g-mannitol) [8, 22], *D. phaphyphila* Alg1 showed the highest ethanol conversion rate of 0.44 g-ethanol/g-mannitol, which suggests Alg1 is a potential strain for mannitol bioconversion. Alginate could not be fully dissolved in the medium for the gelling property in the presence of divalent ion. Thus, total protein concentration of cells was used to monitor the growth of *D. phaphyphila*Alg1 in the case of alginate fermentation. Cells showed vigorous growth in first 48 h, suggesting alginate is favored by Alg1, and a total of 7.6 g/L alginate was consumed after 108 h (Fig. 1c). Totally, 2.7 g/L ethanol and 3 g/L acetic acid were produced, with an ethanol-to-acetate ratio of 0.9:1. It can be concluded that *D. phaphyphila* Alg1 could successfully convert alginate into ethanol and acetate, and its products contained more acetic acid than those from glucose and mannitol. It is worth to mention that no natural strains have been reported to ferment alginate, which might attribute to the fact that alginate needs a complex system for decomposition, and the resulting uronic acid is not easily converted by strains without ED pathway. This indicates that *D. phaphyphila* Alg1 has good potential for ethanol production from seaweed compared to other reported strains [1, 11, 21, 32].

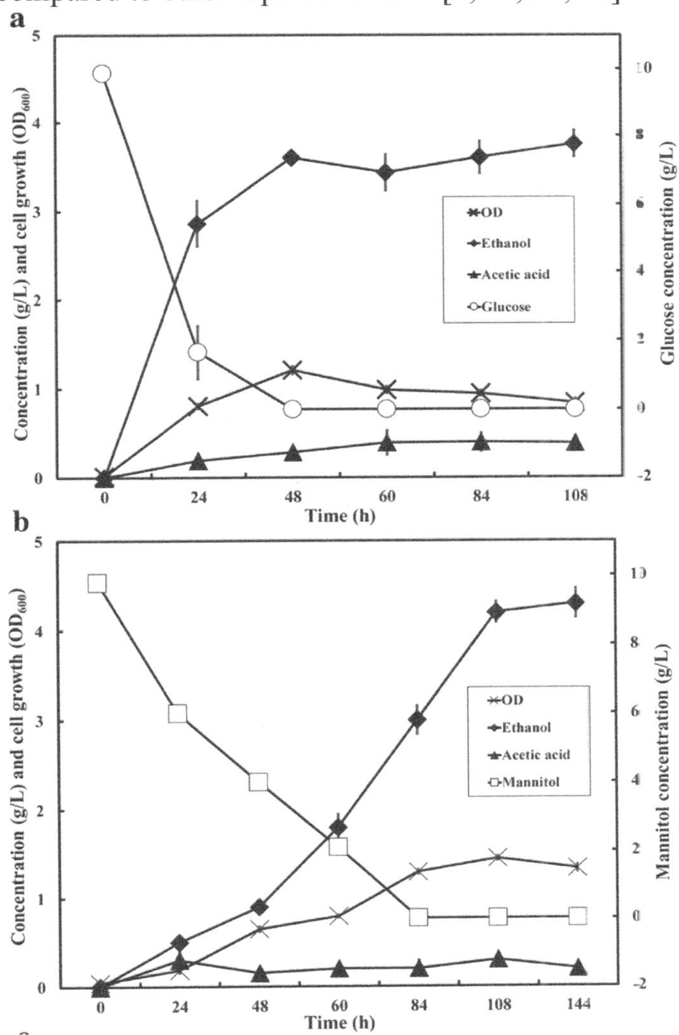

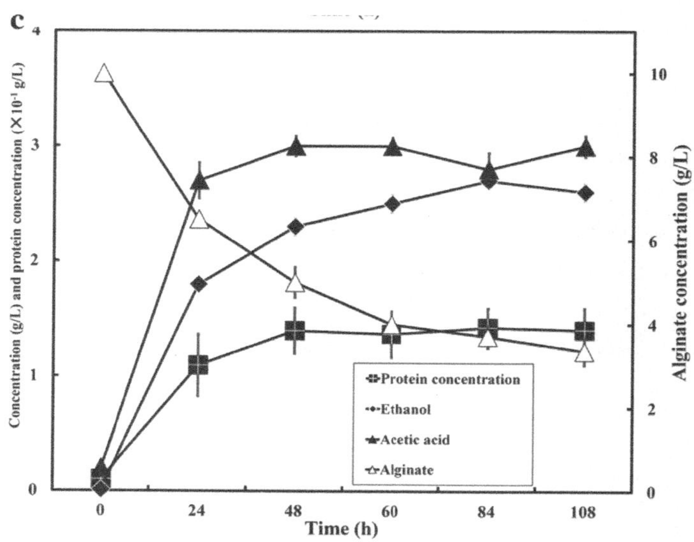

Figure 1. Microbial growth, substrate consumption, and ethanol production of *D. phaphyphila*Alg1 with the substrates of glucose (1 % w/v, a), mannitol (1 % w/v, b), and alginate (1 % w/v, c).

One of the most important characteristics shared by Alg1 is that it is a thermophilic strain producing high concentration of ethanol, which was very rare in the previous reports. The effects of cultivation temperature on ethanol yields were evaluated in this work. The growth temperatures of *D. phaphyphila* Alg1 were from 45 to 65 °C based on the previous work [27]. Accordingly, three temperatures (45, 50, and 60 °C) were chosen to estimate the effects on ethanol yields. As shown in Fig. 2, *D. phaphyphila* Alg1 showed higher ethanol productivities at both 50 and 60 °C than at 45 °C. The highest ethanol yields for glucose and alginate were at 60 °C, which were 0.44 g/g-glucose and 0.3 g/g-alginate. The ethanol yields for mannitol did not show significant difference at 50 and 60 °C, which were 0.44 g/g-mannitol and 0.42 g/g-mannitol, respectively. It can be concluded that Alg1 showed better substrate conversion rates between 50 and 60 °C. Thus, *D. phaphyphila* Alg1 turned out to be a unique thermophilic ethanologenic bacterial strain with high ethanol yields from glucose, mannitol, and alginate.

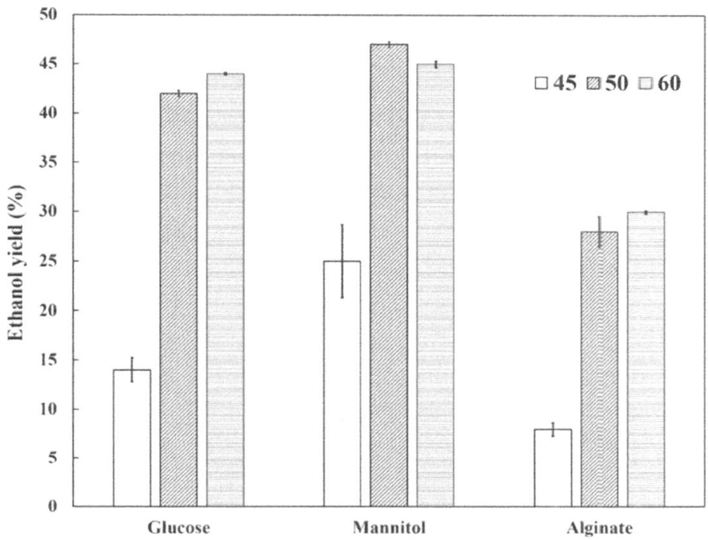

Figure 2. Ethanol yields of *D. phaphyphila* Alg1 from glucose, mannitol, and alginate under different temperatures.

Genome Analysis of Ethanol Synthesis Pathway and Redox Balance Analysis

Ethanol synthesis from the three major components of brown algae-glucose, mannitol, and alginate was analyzed from genome information. As shown in Fig. 3, the ethanologenesis starts with the formation of pyruvate through EMP (Embden–Meyerhof–Parnas) pathway and ED pathway, and pyruvate is then converted to acetyl-CoA by pyruvate:ferredoxin oxidoreductase (PFO) encoded by Dp0079 and Dp1929. Lastly, ethanol was produced from acetyl-CoA by a bifunctional alcohol and aldehyde dehydrogenase (AdhE, Dp0288) which shared 67 % of identity with the AdhE of *Thermoanaerobacterium saccharolyticum* with consumption of 2 NADH [33], and acetate was produced by a phosphate acetyltransferase (PAT, Dp0236) and an acetate kinase (Dp0235) without consumption of NADH (Fig. 3). Through the genome analysis, it can be concluded that *D. phaphyphila* Alg1 utilizes a pathway containing PFO to assimilate pyruvate. This pathway, which was also found in a famous engineered thermophilic bacterium *Thermoanaerobacterium saccharolyticum* with high ethanol titer, is different from that in previously described microbes with homoethanol fermentation [34]. According to the metabolic pathway, to produce 2 ethanol, the net consumptions of NADH for glucose and mannitol were 2 NADH and 1 NADH, respectively (Fig. 3). To produce 2 acetate, 2 NADH and 3 NADH were generated in the catabolism of glucose and mannitol, respectively. In the case of alginate assimilation, the feasible pathway of producing ethanol from this polysaccharide was rather rare in microorganisms. To decompose alginate, it was firstly depolymerized into DEH, and then NADH-dependent DEH reductases (DehR) were employed to produce KDG; lastly, KDG was fed into the ED pathway to produce pyruvate and glyceraldehyde-3-phosphate (G3P) by 2-keto-3-deoxygluconate kinase (KdgK) and KDPG aldolase (KdpgA), which has been discussed in our previous work [27]. During this process, 1 NADH was consumed by DehR, and thus would cause the deficiency of reducing equivalents. To generate 2 ethanol, 4 NADH were consumed, while NADH maintained a balance in generating 2 acetate. The vigorous growth in mannitol, glucose, and alginate under obligate anaerobic condition indicated that *D. phaphyphila* Alg1 could well balance the redox level.

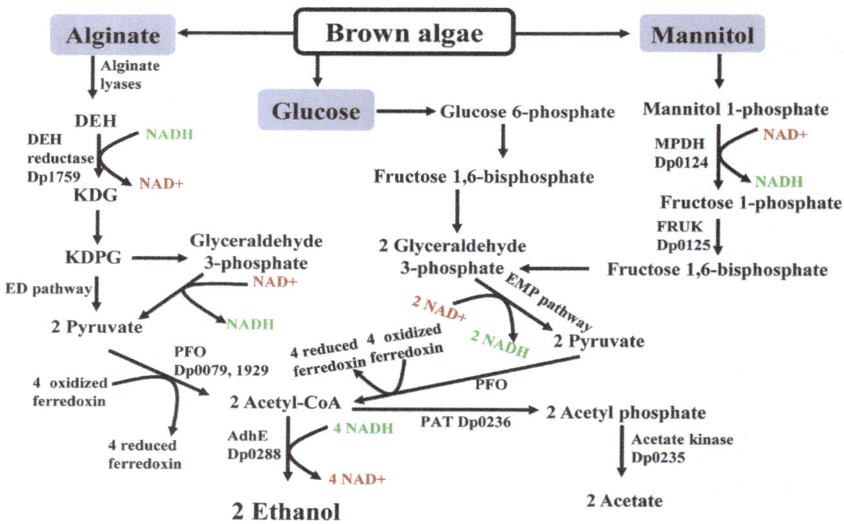

Figure 3. Genome analysis of the metabolic pathways and redox balance of ethanol synthesis in *D. phaphyphila* Alg1. *EMP* Embden–Meyerhof–Parnas; *ED* Entner–Doudoroff; *AdhE* bifunctional alcohol and aldehyde dehydrogenase; *PFO* pyruvate ferredoxin oxidoreductase; *PAT* phosphate acetyltransferase; *MPDH* mannitol-1-phosphate dehydrogenase; *FRUK* 1-phosphofructokinase.

To figure out the redox balance in metabolism of mannitol, glucose, alginate, and their mixture, the NAD/NADH ratio in the cells was analyzed (Table 1). In accordance with the results of genome analysis, the low NAD/NADH ratio (0.63 ± 0.15) in mannitol metabolism indicated that excessive reducing equivalents were produced, while the high NAD/NADH ration (7.93 ± 1.07) in alginate assimilation indicated a deficiency of reducing equivalents (Table 2). Alg1 showed a middle NAD/NADH ratio (3.48 ± 0.57) when fermenting the mixture of alginate, mannitol, and glucose (5:8:1) simulating the main components of brown algae. Through these results, it can be concluded that the different oxidoreduction potentials in substrates of glucose, mannitol, and alginate finally affected the metabolic flux into ethanol pathway. Ethanol production highly consums reducing equivalents. On the one hand, the lack of NADH could explain a low ethanol-to-acetate ratio in alginate degradation, which means production of acetic acid was a direct response to deficiency of reducing power. On the other hand, the consumption of excess reducing equivalents is also important in balancing the redox potential. The redox balance is hard to control especially in engineered strains. For example, *Sphingomonas* sp. A1 was engineered to produce ethanol from alginate, and oxygen supply should be strictly controlled to maintain the redox balance between energy supplement and ethanol production [6]. Even under precise control, *Sphingomonas* sp. strain A1 could only reach an ethanol yield of 0.25 g/g-alginate which was lower than that of strain Alg1 (0.3 g/g-alginate).

Table 1. The redox states of *D. phaphyphila* Alg1 in fermenting different sugars of brown algae

	NAD + NADH	NAD	NAD/NADH
Mannitol	1.77 ± 0.12	0.60 ± 0.16	0.63 ± 0.15
Glucose	1.75 ± 0.70	1.17 ± 0.42	2.51 ± 0.51
Alginate	2.70 ± 0.52	2.45 ± 0. 44	7.93 ± 1.07
Sugar mixture[a]	2.18 ± 0.15	1.69 ± 0.53	3.48 ± 0.57

[a]Sugar mixture contained alginate, mannitol and glucose with a ratio of 5:8:1

Table 2. Microorganisms that convert brown seaweed biomass to ethanol

Microorganism	**Characteristics**	**Major sugars utilized**	**Pretreatment**	**References**
*D. phaphyphila*Alg1	Thermophilic Nat	Mannitol, alginate, laminarin	Un-pretreatment	This study
Zymobacter palmae	Nat	Mannitol, laminarin	Thermal acid hydrolysis	[8, 39]
Pichia angophorae	Nat	Monosaccharide	Thermal acid hydrolysis	[32]
Saccharomyces cerevisiae	Nat	Glucose	Laminarinase	[17]
*Escherichia coli*BAL1611	Eng	Mannitol, alginate, glucose	Un-pretreatment	[23]
*Escherichia coli*KO11	Eng	Glucose, mannitol	Acid and hydrolytic enzymes	[11]
*Sphingomonas*sp. A1	Eng	Alginate	N/A	[6]
*Saccharomyces cerevisiae*BAL3215	Eng	Mannitol, glucose, DEH	N/A	[20]

Nat stands for natural strains, *Eng* stands for engineered strains, and *N/A* means not applicable

The different oxidoreduction potentials of glucose, mannitol, and alginate make the bioconversion of brown algae very difficult. The deficiency of reducing equivalents in alginate metabolism, and the highly reducing equivalents consumption during ethanol formation can easily cause serious redox imbalance. No natural microorganisms have been reported to be capable of converting alginate to ethanol as far as we know. As mentioned above, mannitol is also hard to be converted to ethanol under anaerobic condition in *Z. palmae*, *Enterobacter* sp. JMP3, and some yeast species. Mannitol fermentation would produce excess electrons in contrast to alginate. Under anaerobic condition, these excess reducing equivalents could not be neutralized by oxygen, especially when the microorganisms lack transhydrogenase (an enzyme converting catabolic reducing equivalent NADH to anabolic reducing equivalent NADPH) [8, 35]. However, the redox problem was well solved in *D. phaphyphila* by integrating two complementary pathways which were hard to proceed individually to reach equilibrium. These results provide insights into mechanisms of redox balancing in fermentation of brown algae, and will assist future metabolic engineering efforts on *D. phaphyphila*.

Ethanol Fermentation of Brown Algae by *D. Phaphyphila* Alg1

An artificial mixed sugar simulating the main components of brown algae with a ratio of alginate:mannitol:glucose = 5:8:1 (3 % total sugar) was used to evaluate the fermentation properties of Alg1, which would help us to have a comprehensive understanding of the fermentation process.

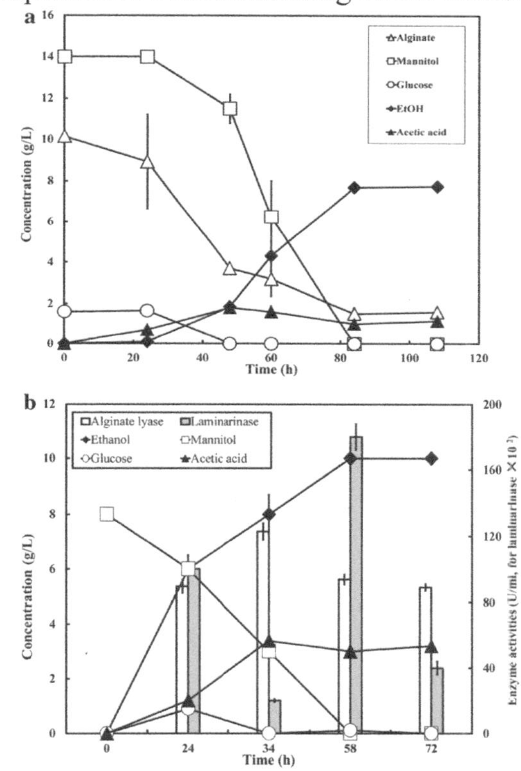

Figure 4. Fermentation properties of *D. phaphyphila* Alg1 using mix sugars (alginate:mannitol:glucose = 5:8:1) simulating main components of brown algae (a) and kelp powder (b) as substrates.

As shown in Fig. 4a, a total of 7.8 g/L ethanol and 1.2 g/L acetic acid were produced with a consumption of 24.2 g/L total sugars. The ethanol-to-acetate ratio and ethanol yield were calculated to be 6.5:1 and 0.32 g/g-total sugar, respectively. The contents of alginate, mannitol, and glucose decreased quickly after 24 h fermentation, and a simultaneous consumption of them was observed during 24–48 h. It can be concluded that *D. phaphyphila* Alg1 could simultaneously saccharify and ferment the three main components of brown algae, and although these components have different oxidoreduction potentials, reducing equivalents are well balanced and metabolic flux is directed into ethanol production in this strain.

We further used 5 % (w/v) kelp powder (dry weight 48 g) to evaluate the ethanol production ability of *D. phaphyphila* Alg1 in 1-L bioreactors. Through 72 h fermentation, 10 g/L of ethanol and 3.2 g/L of acetic acid were produced (Fig. 4b). Mannitol was released to the medium immediately after the kelp powder was added with a concentration of 8 g/L. Initially, glucose was undetectable, while an increasing amount of glucose was observed after the initial cell growth, suggesting that the secreted hydrolases caused the degradation of laminarin and the release of glucose. This hypothesis process was confirmed in the enzyme activity assay (Fig. 4b). At the end of fermentation, a total of 40 g/L kelp powder was consumed, corresponding to an ethanol yield of 0.25 g/g-kelp.

As an emerging feedstock for liquid biofuel production, brown algae bioconversion has been extensively studied. Some of the studies on converting brown seaweed to ethanol are listed in Table 2. Both natural and engineered strains were employed for bioconversion. For most studies, a multiple-step process with acid and/or enzymatic pretreatments before fermentation was used. For example, Kim et al. used *E. coli* KO11 to ferment *S. japonica* with acid pretreatment and enzymatic hydrolysis, and the study resulted in an ethanol yield of 0.4 g/g of sugars [11]. In spite of the high ethanol yield of strain KO11, it still lacks the ability of biomass saccharification. The inability of alginate and mannitol fermentation by natural ethanologenic microorganisms can be concluded from Table 2. The low ethanol yield of yeast was likely due to its inability in utilizing mannitol. The molecular basis of this inability of mannitol assimilation in *S. cerevisiae* was illustrated very recently [36]. Some specific engineered strains have displayed excellent performance in assimilating unpretreated brown algae. The ethanol yield of *E. coli* strain BAL1611 was 0.29 g/g-seaweed at 25 °C [23]. Compared with these previous studies, there are two impressive characteristics of *D. phaphyphila* Alg1 in brown algae bioconversion. Firstly, *D. phaphyphila* Alg1 is an obligate anaerobic and thermophilic bacterium, which can ferment brown algae under elevated temperatures up to 60 °C. Thermophilic bacteria could offer the advantages for ethanol production such as low contamination risk [37]. Secondly, *D. phaphyphila* Alg1 can directly assimilate unpretreated brown algae with high ethanol conversion rate, and can simultaneously ferment the mannitol, laminarin, and alginate, which are highly advantageous over most microorganisms. In a word, *D. phaphyphila* Alg1 possesses not only extracellular polymer-degrading enzymes such as alginase and laminarinase for saccharification but also good ethanol production ability, which would potentially provide it advantages in bioconversion process. Similar to the consolidated bioprocessing (CBP) process in the conversion of lignocellulose into biofuel in one step, the direct conversion of brown algae to ethanol has potential to lower the cost of biomass processing due to elimination of operating and capital costs associated with dedicated enzyme production and/or thermal acid hydrolysis, and a more effective biomass solubilization could be achieved [38]. More than that, the ethanol yields of these strains used in the two-step methods were lower than that of *D. phaphyphila* because of their inability of utilizing monomers of alginate and/or mannitol.

Biodegradation Characteristics of Brown Alga Cell Wall

To elucidate the biodegradation characteristics of *D. phaphyphila* Alg1 on brown algae, strips of *S. japonica* was used as substrate. The morphological changes induced by incubation with *D. phaphyphila* Alg1 were examined by both transmission electron microscope (TEM) and optical microscope to provide direct insight into the structure modification of the biomass. Before fermentation, a high electron density of the cell walls observed by TEM indicated the cell integrity (Fig. 5a, c). Similarly, ordered cell walls could be easily recognized by optical microscope after being stained with azaleine (Fig. 5e), and the cell showed high mechanical strength. After 120 h fermentation by *D. phaphyphila* Alg1, the morphology of residual brown algae changed dramatically. The electron density of the cell walls became very light (Fig. 5b, d). From optical microscopy, the initial ordered cell structures were destroyed, and were subsequently presented as collapsed and incomplete cell structures (Fig. 5f). Obviously, the cell wall components sustaining cell morphology which are mainly composed of alginate and glucan were largely degraded during fermentation. In addition to microscopic observation, the activities of corresponding depolymerizing enzymes during fermentation of brown seaweed were analyzed. As shown in Fig. 4b, alginate lyase and laminarinase activities were detected. The alginate lyase activity increased rapidly before 24 h and reached a maximum level after 34 h of incubation at 123 U/mL. This activity maintained at a high level during fermentation, indicating Alg1 has an evolved thermostable alginolytic system which has been proved in previous work [27]. The laminarinase activity was consistent with the appearance of glucose. High activity was detected at 24 h, and the highest concentration of glucose was achieved at the same time, which meant that laminarinase caused the release of glucose from laminarin. The high concentration of glucose might cause a negative feedback inhibition of the laminarinase gene expression. In the next 10 h, no more laminarinases were produced, and the formerly synthesized laminarinases gradually lost their activities. Therefore, the laminarinase activity became weaker at 34 h. At the same time, glucose was consumed immediately, and was exhausted at 34 h. The inhibition of laminarinase gene was relieved accordingly, and more laminarinases were synthesized to produce more glucose. The result is that laminarinase activity undergoing a short decrease achieved the maximum activities of 1.8 U/mL at 58 h (Fig. 4b). However, only small amount of glucose was detected at this time for most of laminarin had been hydrolyzed. These results further confirmed the decomposition of unpretreated brown algae cell walls and their polysaccharides by strain Alg1.

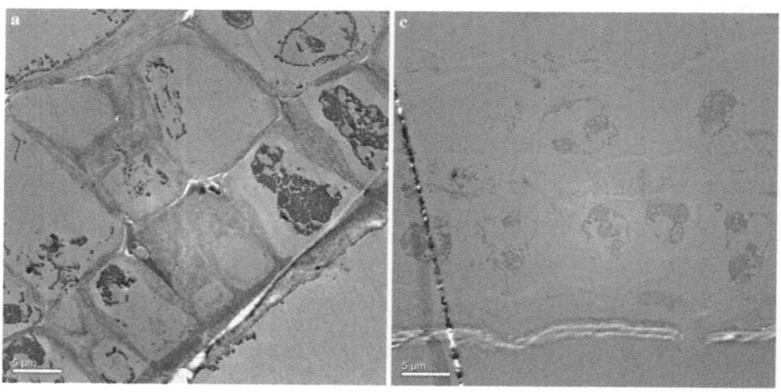

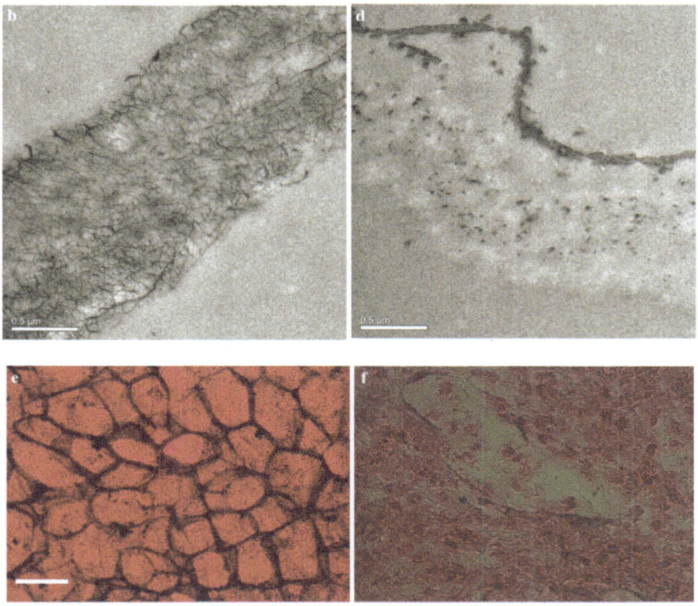

Figure 5. Transmission electron microscope (TEM) and optical microscope images of brown algae samples, unfermented brown algae (**a**, **b**, **e**), and brown algae after fermentation (**c**, **d**, **f**). **e**, **f** stained with azaleine, and *scale bars* 100 μm.

CONCLUSIONS

Defluviitalea phaphyphila Alg1 showed remarkable ethanol production ability in assimilating brown algae. High ethanol yields of 0.44, 0.47, and 0.3 g/g of sugars were obtained by fermenting glucose, mannitol, and alginate at 60 °C, respectively. An ethanol yield of 0.25 g/g-biomass was obtained in fermenting unpretreated kelp. Genome and redox analysis revealed a well evolved redox balance system in Alg1 in fermenting brown algae. The elucidation of the redox balances of substrates with different oxidoreduction potentials would help us increase ethanol productivity. This novel bacterium provides a potential gene-manipulable platform for high-value chemical production from brown algae at elevated temperature.

AUTHORS' CONTRIBUTIONS

SQJ and FLL conceived and designed the study. SQJ, BW, and ML performed experiments and analyzed the data. SQJ and FLL interpreted the data and wrote the manuscript. All authors read and approved the final manuscript.

ACKNOWLEDGEMENTS

This work was supported by grants from the National Natural Science Funds of China (No. 41506155), the Science and Technology Develop Project of Shandong province (No. 2014GHY115027), the Shandong Province Natural Science Funds for Distinguished Young Scholar (No. JQ201507), and the Qingdao Institute of Biomass Energy and Bioprocess Technology Director Innovation Foundation for Young Scientists (No. Y37207210B).

REFERENCES

1. Adams JM, Gallagher JA, Donnison IS. Fermentation study on Saccharina latissima for bioethanol production considering variable pre-treatments. J Appl Phycol. 2008;21:569–74.
2. Blumenkr N, Asboehan G. New method for quantitative-determination of uronic acids. Anal Biochem. 1973;54:484–9.
3. Bradford MM. A rapid and sensitive method for the quantitation of microgram quantities of protein utilizing the principle of protein-dye binding. Anal Biochem. 1976;72:248–54.
4. Caspi R, Altman T, Billington R, Dreher K, Foerster H, Fulcher CA, Holland TA, Keseler IM, Kothari A, Kubo A, et al. The MetaCyc database of metabolic pathways and enzymes and the BioCyc collection of pathway/genome databases. Nucleic Acids Res. 2014;42:D459–71.
5. Cho Y, Kim H, Kim SK. Bioethanol production from brown seaweed, Undaria pinnatifida, using NaCl acclimated yeast. Bioproc Biosyst Eng. 2013;36:713–9.
6. Chujo M, Yoshida S, Ota A, Murata K, Kawai S. Acquisition of the ability to assimilate mannitol by Saccharomyces cerevisiae through dysfunction of the general corepressor Tup1-Cyc8. Appl Environ

Microbiol. 2015;81:9–16.

7. Daroch M, Geng S, Wang GY. Recent advances in liquid biofuel production from algal feedstocks. Appl Energ. 2013;102:1371–81.
8. Enquist-Newman M, Faust AME, Bravo DD, Santos CNS, Raisner RM, Hanel A, Sarvabhowman P, Le C, Regitsky DD, Cooper SR, et al. Efficient ethanol production from brown macroalgae sugars by a synthetic yeast platform. Nature. 2014;505:239–43.
9. Erdei B, Barta Z, Sipos B, Reczey K, Galbe M, Zacchi G. Ethanol production from mixtures of wheat straw and wheat meal. Biotechnol Biofuels. 2010;3:16.
10. Fleming M, Manners DJ, Masson AJ. The enzymic degradation of laminarin. Biochem J. 1967;104:32–3.
11. Horn SJ, Aasen IM, Østgaard K. Ethanol production from seaweed extract. J Ind Microbiol Biot. 2000;25:249–54.
12. Horn SJ, Aasen IM, Østgaard K. Production of ethanol from mannitol by Zymobacter palmae. J Ind Microbiol Biot. 2000;24:51–7.
13. Hughes AD, Kelly MS, Black KD, Stanley MS. Biogas from Macroalgae: is it time to revisit the idea? Biotechnol Biofuels. 2012;5:86.
14. Jang JS, Cho Y, Jeong GT, Kim SK. Optimization of saccharification and ethanol production by simultaneous saccharification and fermentation (SSF) from seaweed, Saccharina japonica. Bioproc Biosyst Eng. 2012;35:11–8.
15. Jensen A, Haug A. Geographical and seasonal variation in the chemical composition of Laminaria hyperborea and Laminaria digitata from the Norwegian coast. Rep Norw Inst Seaweed Res No. 14, Department of Biotechnology, Trondheim: Akademisk trykningssentral; 1956.
16. Ji SQ, Meng DD, Zhang KD, Li FL. Diversity of thermophilic microorganisms and their roles in carbon cycle. In: Li F-L, editor. Thermophilic microorganisms. Wymondham: Caister Academic Press; 2015. p. 13–46.
17. Ji SQ, Wang B, Lu M, Li FL. Defluviitalea phaphyphila sp nov., a novel thermophilic bacterium that degrades brown algae. Appl Environ Microbiol. 2016;82:868–77.

18. Ji SQ, Wang SA, Tan Y, Chen XH, Schwarz W, Li FL. An untapped bacterial cellulolytic community enriched from coastal marine sediment under anaerobic and thermophilic conditions. FEMS Microbiol Lett. 2012;335:39–46.
19. Kim NJ, Li H, Jung K, Chang HN, Lee PC. Ethanol production from marine algal hydrolysates using Escherichia coli KO11. Bioresource Technol. 2011;102:7466–9.
20. Kloareg B. Isolation and analysis of cell-walls of the brown marine-algae Pelvetia canaliculata and Ascophyllum nodosum. Physiol Veg. 1984;22:47–56.
21. Lee J, Li P, Lee J, Ryu HJ, Oh KK. Ethanol production from Saccharina japonica using an optimized extremely low acid pretreatment followed by simultaneous saccharification and fermentation. Bioresour Technol. 2013;127:119–25.
22. Lee SM, Lee JH. Ethanol fermentation for main sugar components of brown-algae using various yeasts. J Ind Eng Chem. 2012;18:16–8.
23. Liu HY, Wang GC. Fermentative hydrogen production from macro-algae Laminaria japonica using anaerobic mixed bacteria. Int J Hydrogen Energ. 2014;39:9012–7.
24. Lo J, Zheng T, Hon S, Olson DG, Lynd LR. The bifunctional alcohol and aldehyde dehydrogenase gene, adhE, is necessary for ethanol production in Clostridium thermocellum and Thermoanaerobacterium saccharolyticum. J Bacteriol. 2015;197:1386–93.
25. Miller GL. Use of dinitrosalicylic acid reagent for determination of reducing sugar. Anal Chem. 1959;31:426–8.
26. Olson DG, McBride JE, Shaw AJ, Lynd LR. Recent progress in consolidated bioprocessing. Curr Opin Biotechnol. 2012;23:396–405.
27. Preiss J, Ashwell G. Alginic acid metabolism in bacteria. 1. Enzymatic formation of unsaturated oligosaccharides and 4-deoxy-L-erythro-5-hexoseulose uronic acid. J Biol Chem. 1962;237:309–16.
28. Preiss J, Ashwell G. Alginic acid metabolism in bacteria. 2. Enzymatic reduction of 4-deoxy-L-erythro-5-hexoseulose uronic acid to 2-keto-3-deoxy-D-gluconic acid. J Biol Chem. 1962;237:317–21.

29. Roesijadi G, Jones SB, Snowden-Swan LJ, Zhu Y. Macroalgae as a biomass feedstock: a preliminary analysis. Richland: Pacific Northwest National Laboratory (PNNL); 2010.

30. Shaw AJ, Podkaminer KK, Desai SG, Bardsley JS, Rogers SR, Thorne PG, Hogsett DA, Lynd LR. Metabolic engineering of a thermophilic bacterium to produce ethanol at high yield. Proc Natl Acad Sci USA. 2008;105:13769–74.

31. Takeda H, Yoneyama F, Kawai S, Hashimoto W, Murata K. Bioethanol production from marine biomass alginate by metabolically engineered bacteria. Energy Environ Sci. 2011;4:2575–81.

32. Taylor MP, Eley KL, Martin S, Tuffin MI, Burton SG, Cowan DA. Thermophilic ethanologenesis: future prospects for second-generation bioethanol production. Trends Biotechnol. 2009;27:398–405.

33. Vandijken JP, Scheffers WA. Redox Balances in the Metabolism of Sugars by Yeasts. FEMS Microbiol Lett. 1986;32:199–224.

34. Wang B, Ji SQ, Lu M, Li FL. Biochemical and structural characterization of alginate lyases: an update. Curr Biotech. 2016. doi:10.2174/2211550104666150723231423.

35. Wang B, Ji SQ, Tian XX, Qu LY, Li FL. Brassicibacter thermophilus sp nov a thermophilic bacterium isolated from coastal sediment. Int J Syst Evol Microbiol. 2015;65:2870–4.

36. Wang J, Kim YM, Rhee HS, Lee MW, Park JM. Bioethanol production from mannitol by a newly isolated bacterium, Enterobacter sp. JMP3. Bioresour Technol. 2013;135:199–206.

37. Wargacki AJ, Leonard E, Win MN, Regitsky DD, Santos CN, Kim PB, Cooper SR, Raisner RM, Herman A, Sivitz AB, et al. An engineered microbial platform for direct biofuel production from brown macroalgae. Science. 2012;335:308–13.

38. Wei N, Quarterman J, Jin Y. Marine macroalgae: an untapped resource for producing fuels and chemicals. Trends Biotechnol. 2013;31:70–7.

39. Yamasaki M, Ogura K, Hashimoto W, Mikami B, Murata K. A structural basis for depolymerization of alginate by polysaccharide lyase family-7. J Mol Biol. 2005;352:11–21.

CHAPTER

18

SELECTION OF YEAST STRAINS FOR BIOETHANOL PRODUCTION FROM UK SEAWEEDS

Emily T. Kostas[1], Daniel A. White[2], Chenyu Du[1], and David J. Cook[1]

[1]The University of Nottingham, Loughborough, UK

[2]Plymouth Marine Laboratory, Prospect Pl, Plymouth, UK

ABSTRACT

Macroalgae (seaweeds) are a promising feedstock for the production of third generation bioethanol, since they have high carbohydrate contents, contain little or no lignin and are available in abundance. However, seaweeds typically contain a more diverse array of monomeric sugars than are commonly present in feedstocks derived from lignocellulosic material which are currently used for bioethanol production. Hence, identification of a suitable fermentative microorganism that can utilize the principal sugars released from the hydrolysis of macroalgae remains a major objective. The present study used a phenotypic microarray technique to screen 24 different yeast strains for their ability to metabolize individual monosaccharides commonly found in seaweeds, as well as hydrolysates following an acid pre-treatment of five native UK seaweed species (*Laminaria digitata*, *Fucus serratus*, *Chondrus crispus*, *Palmaria palmata* and *Ulva lactuca*). Five strains of yeast (three *Saccharomyces* spp, one *Pichia* sp and one *Candida* sp) were selected and subsequently evaluated for bioethanol production during fermentation of the hydrolysates. Four out of the five selected strains converted these monomeric sugars into bioethanol, with the highest ethanol yield (13 g L^{-1}) resulting from a fermentation using *C. crispus* hydrolysate with *Saccharomyces cerevisiae*YPS128. This study demonstrated the novel application of a phenotypic microarray technique to screen for yeast capable of metabolizing sugars present in seaweed hydrolysates; however, metabolic activity did not always imply fermentative production of ethanol.

KEYWORDS: Macroalgae, Phenotypic microarray, Bioethanol, Fermentation, Yeast

INTRODUCTION

The European Commission has stated that European countries rely too heavily on imports of gas and fuel and energy sources need to be more diverse (www.europa.eu). The consumption of fossil fuels and effects of global warming are causing constraints on our planet, and the search to find alternative, sustainable and cleaner burning sources of energy has become an extensive area of research (Jeong et al. 2010).

Bioethanol derived from sugar-based biomass in a fermentation process is a potential way of generating energy-rich transportation fuels (Karakashev et al. 2007). First generation feedstock materials such as corn and sugarcane have already been widely exploited (Bothast and Schlicher 2005), and the current commercial production of bioethanol, mainly in the USA and Brazil, is based around the use of terrestrial plants such as maize or sugar cane. Controversy, however, has arisen over the utilisation of such substrates which are essential food sources, and attention has been diverted to lignocellulosic biomass (2nd generation biomass) which comprises mainly of agricultural waste products (Lu et al. 2009). However, the production of fuel from second generation crops has proven to be problematic with energy and chemical inputs required to breakdown the recalcitrant lignocellulosic complex highlighted as a significant technical challenge to the use of such biomass for ethanol production (Taherzadeh and Karimi 2008). The release of fermentable sugars from lignocellulosic materials usually therefore requires a physico-chemical pre-treatment to breakdown the native plant structures and enable cellulose degrading enzymes to have access to their substrates. Such pre-treatments have been found to generate a range of inhibitory compounds (e.g. furans, phenolic compounds or organic acids) which can impact on the viability of yeast in a subsequent fermentation processes (Singh et al. 2011; Mukherjee et al. 2014).

Macroalgae (seaweeds) have gained much attention in recent years as promising bioethanol feedstocks due to their fast growth rate and large biomass yields, seaweeds offer certain advantages over terrestrial crops which have previously been considered as favourable feedstocks for bioethanol production (Adams et al. 2009). Several species of seaweed are known to be rich in carbohydrates, and they contain little to no lignin (Yanagisawa et al. 2011), implying that less energy intensive processes might be established to produce bioethanol from seaweed.

Despite the many advantages seaweeds possess over terrestrial energy crops, there are potential bottlenecks which could hinder the development of bioethanol production from hydrolysates derived from seaweed. The aquaculture and adequate supply of desirable seaweed species for bioethanol production is certainly a factor. Additionally, the diverse array of carbohydrates typically found in seaweeds is also a further technical obstacle to their exploitation. This means that specialized enzyme blends will be required to liberate fermentable sugars from seaweeds and that specialized fermentative organisms are needed which can utilize the majority of the sugars present. Although seaweeds contain carbohydrates that are also found in terrestrial plants (e.g. some starch and cellulose), the majority of carbohydrate present is in the form of polysaccharides such as laminarin (glucose polymer with β1-3 linkages), mannitol (C6 sugar alcohol), agar (agarobiose polymer) and carrageenan (D-galactose and 3,6-anhydro-D-galactose monomers); hence, biomass processing which has been developed for terrestrial plants cannot be directly applied to seaweeds (Jung et al. 2013). Conversion methods to release monosaccharides from their various sources and structures and the choice of microorganisms for bioethanol production from seaweed are necessary pre-requisites for the development of a viable process utilizing seaweed as a feedstock. Attempts have been made to identify natural strains or to generate genetically engineered microorganisms with the ability to ferment sugars inherent to seaweed into ethanol. Horn et al. (2000a) revealed the possibility of simultaneously fermenting laminarin and mannitol extracts from the brown species *Laminaria hyperborea* to ethanol using *Pichia angophorea*, where 0.43 g ethanol g^{-1} of substrate (both laminarin and mannitol) was recovered. Further research by Horn et al. (2000b) demonstrated that the bacterium *Zymobacter palmae* was able to ferment synthetic mannitol under different oxygen regimes, yielding 0.38 g ethanol g^{-1} mannitol. Additionally, Wargacki et al. (2012) engineered an *Escherichia coli* strain that was capable of co-fermenting glucose, mannitol and alginate from the brown algal species *Saccharina japonica*, yielding 0.41 g ethanol g^{-1} of carbohydrate sugars (alginate, mannitol and glucan). Similarly, Enquist-Newman et al. (2014) developed the first *S. cerevisiae* synthetic biology platform with the ability to produce ethanol from mannitol and the alginate monomer 4-deoxy-l-erythro-5-hexoseulose uronate (DEHU), achieving yields up to 83 % of the maximum theoretical yield from these sugars. These findings reveal that all major polysaccharides in seaweeds could be converted to ethanol and other value-added chemicals.

As the interest in bioethanol production from seaweeds is increasing, the search to find wild-type microorganisms with the ability to ferment sugars specific to the breakdown of seaweed has proven to be somewhat challenging. Natural strains may lack the relevant metabolic pathways for the conversion of certain sugars into ethanol (Roesijadi et al. 2010), and therefore, the maximum conversion of the mixed array of monosaccharides found in seaweed hydrolysates into bioethanol cannot be fully achieved. However, phenotypic screening of yeast strains may be employed and can be a useful tool with the potential to identify strains with desirable traits for bioethanol production (Greetham et al. 2014a, b). This was recently highlighted in the work of Wimalasena et al. (2014), where candidate strains for efficient second generation bioethanol production were identified by employing a phenotypic microarray (PM) analysis to measure tolerance to stresses on *Saccharomyces* spp yeast during fermentation. There are many advantages of utilizing a PM approach, which include the high output of informative data in a short amount of time, and avoiding the need to use traditional yeast methodologies (Greetham et al. 2014a, b). The PM technique gives an insight into the yeast cell's metabolic activity against various media (e.g. individual monosaccharide carbon sources). Although this is not directly indicative of fermentation activity, the approach can identify microorganisms which are able to metabolize different carbon sources. Even when an organism does not produce ethanol as a major metabolic product, it remains useful to identify organisms which can assimilate and metabolize the sugars which are liberated from macroalgae. Whilst not the focus of the current research, this could identify novel routes to the production of other platform chemicals.

This study reports the first application of PM analysis to the screening and selection of yeast strains able to metabolize a range of monosaccharides originating from the breakdown of seaweed polysaccharides. Strains which displayed promising metabolic activity when cultured on a wide cross-section of such carbon sources were further screened for ability to metabolize hydrolysates prepared from the acid hydrolysis of five native UK seaweeds. Following these initial screening processes, selected yeast strains were examined in laboratory-scale fermentations of hydrolysates derived from seaweed in which fermentation progression and ethanol formation was determined.

MATERIALS AND METHODS

Two brown algae (*Laminaria digitata* and *Fucus serratus*), two red algae (*Chondrus crispus*and *Palmaria palmata*) and green alga (*Ulva lactuca*) were collected from the coast around Plymouth, UK, during the spring of 2013. The seaweed samples were rinsed in distilled water to remove salt and debris, before being dried in a fan oven at 80 °C for a minimum of 48 h until perceived to be dry. The seaweed samples were then milled using a ball mill (Fritsch, Germany) to obtain a fine homogeneous powder and stored away from direct sunlight and moisture until further use.

Compositional Analysis of the Seaweed Species

Moisture content was measured by drying 5 g of each seaweed species in a convection oven at 105 °C for 90 min (Santos et al. 2003). After heating, the samples were placed in a desiccator for 30 min to prevent rehydration before being weighed again.

Ash content was determined by heating in a muffle furnace at 580 °C for 24 h.

A Thermo Flash Nitrogen Analyzer (ThermoFischer Scientific, USA) was used to determine the protein content of each species. Sample (50 mg) was sealed in a tin capsule and combusted at approximately 1800 °C. Combustion gases were passed into a reduction reactor (at 680 °C and containing reduced copper) where nitrogen oxides are converted to elemental nitrogen. Carbon dioxide, sulphur dioxide and water were removed via filters of soda lime, magnesium perchlorate and a molecular sieve. The effluent stream was passed through a nitrogen separation column (50 °C) and into a thermal conductivity detector. Quantitation was achieved with Eager 300 software using an L-aspartic acid standard. Protein was determined using the N × 6.25 conversion factor.

Lipid content was determined using an adaption of the Folch method (Folch et al. 1957). Powdered sample (400 mg) was added to a 50-mL glass centrifuge tube. To each sample, 12 mL of dichloromethane/methanol (2:1, *v/v*) was added and left for 2 h with occasional agitations. The glass tubes were then centrifuged at 1000 rpm for 5 min or until a pellet had formed at the bottom of the tube. Using a glass syringe, the upper organic phase was removed and transferred into a clean 50-mL centrifuge glass tube were 2.5 mL KCl (0.88 %, *w/v*) was added before being inverted, vortexed and centrifuged at 1000 rpm for 5 min. The lower organic phase was removed using a glass syringe and transferred into a pre-weighed glass tube. The lower organic phase was dried with nitrogen gas and left uncapped in a fume cupboard overnight until all the liquid had evaporated.

Carbohydrate content was measured following a modified version of the Dubois assay (Dubois et al. 1956) where 1 mL of 12 M H_2SO_4 was added to 30 mg of seaweed in a heat resistant screw cap glass tube and incubated at 37 °C for 1 h. Water (11 mL) was added to the sample to dilute the acid strength to 1 M, following which, samples were incubated at 100 °C for 2 h. Liberated monosaccharides (mannitol, fucose, arabinose, galactose, glucose and xylose) were analyzed by HPLC as described in section 2.6. HPLC analysis was used instead of the colourimetric phenol-sulphuric acid method as for the purpose of this study as it was important to be able to identify individual monosaccharides in the different seaweed species and whether different yeast strains could utilise specific sugars.

Acid Hydrolysis and Preparation of Seaweed Hydrolysates

For the production of seaweed hydrolysates, acid hydrolysis was conducted where 200 g of seaweed powder was treated with 2 L of 5 % (*w*/*v*) H_2SO_4 in 2 L Duran bottles. The reaction took place in a benchtop autoclave at 121 °C for 15 min. After hydrolysis, the liquid fraction was separated from the residue by filtration using a Büchner funnel and flask connected to a pump. The seaweed hydrolysates were adjusted to pH 7 using NaOH and filter sterilized before being analyzed for monosaccharides and concentrations of potential yeast-inhibitory compounds (see below). The remaining solid seaweed residues were washed repeatedly with excess water until a neutral pH was achieved, dried overnight at 60 °C and stored for future work. Free amino nitrogen (FAN) was determined by following the European Brewing Convention ninhydrin spectrophotometric method, which utilises glycine as a standard (Analytica—EBC, 2000).

Yeast Strain Screen Using the Biolog-Omnilog

Yeast Strains and Growth Conditions

All 24 yeast strains which were screened in this investigation (listed in Table 1) were taken from glycerol stocks stored at −80 °C and inoculated into a liquid medium containing 10 g L^{-1} yeast extract, 20 g L^{-1} peptone, and 20 g L^{-1} glucose (YPD) and placed in an orbital shaker (180 rpm; CERTOMAT BS-1 incubator, Germany) at 30 °C under aerobic conditions for 48 h. One hundred microliters of these cells were then spread on YPD slopes and incubated at 30 °C under aerobic conditions for a further 48 h.

Table 1. List of yeast strains investigated in this study

Microorganism	Source	Equivalent strain designations/Ref	ID in this study
Kluyveromyces marxianus	TP	ATCC 22296, CBS 5671, NRRL Y8287, Phaff 71-15	TP1
Candida veronae	TP	–	TP2
Pichia Mexicana	TP	ATCC 28874, CBS 5815, NRRL Y17672, Phaff 69-32	TP3
Candida utilitis var thermophila	TP	CBS 1517, NRRL Y-1082, Phaff 74-62, DBVPG 7306	TP4
Candida mogii	TP	–	TP5
Candida sorboxylosa	TP	Phaff 91-491.3	TP6
Candida stellate	TP	Phaff 72-1034	TP7
Candida utilis	TP	ATCC 9950, CBS 5609, Phaff 74-61	TP8
Candida santjackobensis	TP	ATCC 58898, CBS 8183	TP9
Pichia stipitis	TP	ATCC 58785, CBS 6054, NRRL Y11545	TP10
Candida valida	TP	Phaff 94-150.3	TP11
Pichia anomala	TP	CBS 5759, NRRL Y-366, Phaff 76-71 T	TP12
Saccharomyces cerevisiae	NCYC	Melbourne No. 1 strain, NCTC 4919	NCYC 192
Hanseniaspora valbyensis	NCYC	CBS 479, NRRL Y-1626, ATCC 10631, IFO 0670, NCTC 478	NCYC 17
Candida tenuis	NCYC	ATCC 58782, CBS 4285, CCRC 21772, NRRL Y-17106	NCYC 2545
Candida succiphila	NCYC	ATCC 46049, CBS 8003, CCRC 21410, IAM 12489, IFO 1911, JCM 9445, KL30, NRRL Y-11998	NCYC 1403
Saccharomyces cerevisiae	NCYC	CBS 1200, ATCC 4126, CCRC 21494, MUCL 39497, CLIB 409	NCYC 2592
Candida lyxosophila	NCYC	CBS 8194, CCY 29-173-1, JCM 7532	NCYC 2379

Pichia guilliermondii	NCYC	ATCC 46036, CCRC 21697, CCRC 22093, DBVPG 6571, IFO 10106, NRRL Y-2075	NCYC 443
Candida shehatae var lignosa	NCYC	CBS 4705, ATCC 58779, CCRC 21774, IGC 3590, JCM 9837, NRRL Y-12856, NRRL Y-17027	NCYC 2389
Candida arabinofermentans	NCYC	CBS 8468, NRRL YB-2248	NCYC 2916
Candida shehatae	BSYC	–	BSYC 1
Saccharomyces cerevisiae	EL	Liti et al. (2009)	YPS128
Saccharomyces cerevisiae	EL	Liti et al. (2009)	Y12

TP courtesy of Trevor Phister, *NCYC* National Collection of Yeast Cultures, *BSYC* Brewing Science Yeast Culture, *EL* courtesy of Ed. Louis

Phenotypic Microarray Analysis (Biolog-Omnilog) Using Synthetic Minimal Media and Seaweed Hydrolysates

The Biolog-Omnilog phenotypic microarray (PM) system (Biolog, Hayward, USA) was used according to the manufacturer's instructions with minor modifications according to Greetham et al. (2014a) to screen for yeast strains that can metabolise different carbon sources, in particular monosaccharides liberated from the acid hydrolysis of seaweed species. Synthetic minimal media containing different individual carbon sources (arabinose, fucose, galactose, glucose, mannitol, rhamnose and xylose) at 6 % (*w*/*v*) were prepared and supplemented into 0.67 % (*w*/*v*) yeast nitrogen base (YNB), 2.6 μL of yeast nutrient supplement mixture (NSx48-24 mM Adenine-HCl, 4.8 mM L-histidine HCl monohydrate, 48 mM L-leucine, 24 mM L-lysine-HCl, 12 mM L-methionine, 12 mM L-tryptophan and 14.4 mM uracil) and 0.2 μL of dye D. Reverse osmosis (RO) sterile distilled water was added to make a final volume of 60 μL, which was then aliquoted into individual wells of the PM array plates. Fifty-six microlitres of seaweed hydrolysate which was generated in section 2.3 was supplemented with 0.4 μL of dye D due to the hydrolysates' darker colour and then aliquoted into the wells of the PM plates. Sixty microlitres of hydrolysate without dye D was also aliquoted into specific wells in order to act as a control. Yeast strains were prepared according to the method described by Greetham, et al. (2014a).

The conditions were made anaerobic by vacuum packing the PM plates in PM gas bags (Biolog) before being placed in the Omnilog reader at 30 °C for 95 h. The Omnilog reader was programmed to photograph the PM plates every 15 min for 95 h, converting the pixel density of each well into a signal data value that represents cell metabolic output and dye colour change. The reduction of the dye, producing a colour change from colourless to violet, results from the cells' metabolic activity when in contact with the different media in each well; this has been defined here as redox signal intensity. The redox signal intensity data was assembled in the Biolog software and exported using Microsoft Excel. All assays were conducted in triplicate and the mean values are presented.

Fermentations of Hydrolysates Derived From the Acid Pre-Treatment of Seaweed Using Selected Yeast Strains

Yeast strains which appeared to be more metabolically active by having a redox signal intensity value (RSI) of ≥ 20 on the seaweed specific sugars and seaweed hydrolysates from the yeast strain screening assay were directly used for bioethanol fermentations. Twenty-five-millilitre fermentations of the seaweed acid hydrolysate produced from section 2.3 were conducted in glass serum bottles (30 mL; Wheaton, USA) using a method adapted from Quain et al. (1985) and Powell et al. (2003). The bottles were made anaerobic by sealing the vessels with rubber septa and a one-way valve was used in order to facilitate the expulsion of CO_2 produced during the fermentation process as sugars are converted to ethanol. The fermentation vessels containing seaweed hydrolysate were inoculated with the selected yeast strain at a pitching rate of ca 10^7 cells/mL. The vessels were incubated at 30 °C (MIR-253 incubator, Sanyo Electric Co., Japan) with magnetic stirring set at 120 rpm, and the progression of the fermentation was monitored by tracking the weight loss of the vessels at frequent intervals resulting from the removal of CO_2. The end of the fermentation was indicated by the vessels reaching constant mass. Samples were taken at the end of the fermentation for sugar and ethanol analysis via HPLC. All fermentations were carried out in triplicate.

Analysis of Monosaccharides, Inhibitors and Bioethanol

The monosaccharide concentrations were quantified via HPLC using Dionex ICS-3000 Reagent-Free Ion Chromatography, electrochemical detection using ED 40 and computer controller. The CarboPacTM PA 20 column (3 × 150 mm) was used, and the mobile phase was 10 mM NaOH with a flow rate of 0.5 mL min^{-1}. The injection volume was 10 μL and the column temperature was 30 °C. Authentic standards of monosaccharides (mannitol, fucose, arabinose, galactose, glucose and xylose) with concentrations within range of 1 to 0.0625 g L^{-1} were used for monosaccharide quantification.

The analysis of inhibitors (5-hydroxy-methyl-furfural (HMF), furoic acid, furfural, vanillic acid, vanillin, ferulic acid and *p*-coumaric acid) were quantified by HPLC using UV detection at 270 nm (2695 HPLC system and 996 Photodiode Array Detector, Waters, USA) and a Techsphere ODS C18 column (5 μm, 4.6 mm × 250 mm; HPLC Technologies, UK) was used at room temperature. The sample volume was 10 μL, and the mobile phase was a gradient of methanol in 1 % acetic acid at an overall flow rate of 1.0 mL min^{-1}. The methanol concentration was increased from 20 to 50 % over 30 min with a 100 % methanol column cleaning phase and a 9-min re-equilibration period. Data were recorded using Millennium Chromatography software (Waters, USA).

Ethanol yields produced during fermentation were quantified by HPLC using an AS-2055 Intelligent Auto-sampler and a PU-1580 Intelligent HPLC Pump (Jasco, Japan). The Rezex ROA Organic Acid H+ organic acid column (5 μm, 7.8 mm × 300 mm; Phenomenex, UK) was operated at ambient temperature with a mobile phase of 0.005 N H_2SO_4 was used at a flow rate of 0.5 mL min^{-1}. A Refractive Index cell (RI-2031 Intelligent Refractive Index detector, Jasco, Japan) was used for detection, and the injection volume was 10 μL. Data were acquired using the Azur software package v. 4.6.0.0 (Datalys, France). Prior to HPLC analysis, all samples and standards were filtered using Whatman GD/X syringe filters (GF/C 25 mm filter diameter/1.2 μm pore size; Whatman, UK). All analyses were conducted in triplicate.

RESULTS

Biochemical Composition of Native UK Seaweed Species

The gross chemical compositions of the five seaweed species investigated in this study are listed in Table 2. The ash content was relatively high across all species, ranging from 18.8–25.7 % dry weight (*d/w*), which is substantially higher than the amounts present in most terrestrial plants as seaweeds have the ability to easily absorb inorganic substances from their environment (Yanik et al. 2013). Determination of protein content revealed that protein was within the range of 9.6–26.8 % (dry wt) and the lipid content was between 0.48–3.3% (dry wt) across all species. Carbohydrate content was measured as the total of the monosaccharides, glucose, galactose, mannitol, fucose, xylose and arabinose, analysed by HPLC following an acid hydrolysis. Amongst the five seaweed species screened in this study, *P. palmata* had the highest carbohydrate content, 39.4 % ± 1.00 of the dry material, which likely reflects the hydrolysis of agar, a major component located in the extracellular matrix of red seaweed (Noseda et al. 1999). *L. digitata* had the lowest measured carbohydrate content of 21.7 %.

Table 2. Proximate composition of seaweed species used in this study

		Composition % (dry weight basis)			
Seaweed species	**Ash**	**Protein**	**Lipid**	**Carbohydrate[a]**	**Moisture**
L. digitata	24.3 ± 0.38	26.8 ± 0.19	1.9 ± 0.09	21.7 ± 0.68	12.1 ± 0.39
F. serratus	18.8 ± 0.58	9.6 ± 0.72	2.8 ± 0.38	26.4 ± 0.75	10.6 ± 0.06
C. crispus	19 ± 1.02	19.9 ± 0.27	0.48 ± 0.25	21.8 ± 1.57	12.6 ± 0.21
P. palmata	25.7 ± 0.31	22.9 ± 0.16	3.3 ± 0.60	39.4 ± 1.00	5.2 ± 0.01
U. lactuca	21.5 ± 0.29	16.4 ± 0.14	1.0 ± 0.23	23.8 ± 0.80	10.0 ± 0.01

Data are the mean ± SD of three measurements

[a]Carbohydrate was estimated as the sum of monosaccharides arabinose, galactose, glucose, xylose, fucose and mannitol. It is assumed that the unaccounted for dry matter is principally polysaccharide material either nor broken down under the hydrolysis conditions employed or not quantified against authentic standards during HPLC analysis

Analysis of the Hydrolysates Generated From Acid Pre-Treatment

The composition of hydrolysates derived from the seaweed using a 5 % (*w*/*v*) H_2SO_4 pre-treatment in a benchtop autoclave at 120 °C for 15 min can be seen in Table 3. The range and concentrations of monosaccharides liberated into the hydrolysate differed according to seaweed species. The red species *C. crispus* and *P. palmata* and the green seaweed *U. lactuca* yielded the highest concentrations of glucose, 2.02, 2.99 and 5.30 g L^{-1}, respectively, which most likely resulted from the hydrolysis of cellulose. *C. crispus* and *P. palmata* produced the highest yields of galactose (4.24 and 6.84 g L^{-1}, respectively) which would most likely be a product of carrageenan hydrolysis. Brown seaweed species *L. digitata* and *F. serratus* only released 1.51 and 0.95 g L^{-1} of glucose, respectively, but yielded 2.99 and 3.77 g L^{-1} of fucose and 3.49 and 4.38 g L^{-1} of mannitol, respectively. Interestingly, *P. palmata* yielded 3.04 g L^{-1} of fucose, a sugar which has not previously been reported or identified in this particular red algal species. Hydrolysates derived from seaweed were analysed for presence of known yeast-inhibitory compounds (a range of phenolic compounds, organic acids and furan compounds), which can be formed in the pre-treatment of lignocellulosic biomass. Such compounds are generated during the acid hydrolysis of biomass at high temperatures and can negatively impact on ethanol production during fermentation (Meinita et al. 2012). HMF and furfural were the only compounds present in the hydrolysates (Table 3) but were present at low levels and were not detectable in the hydrolysate of *F. serratus*. HMF derives from the breakdown of hexose sugars and furfural from the breakdown of pentose sugars. *C. crispus* generated the highest amount of HMF (2.13 g L^{-1}) which is consistent with its high glucose and galactose contents (C-6 sugars).

Table 3. Analysis of monosaccharides, yeast-inhibitory compounds and free amino nitrogen (FAN) present in the hydrolysates of *L. digitata*, *F. serratus*, *C. crispus*, *P. palmata* and *U. lactuca* after 5 % H_2SO_4 treatment at 120 °C for 15 min in a benchtop autoclave

Seaweed species	Liberated monosaccharides (g L^{-1})							Inhibitors (g L^{-1})		Others (g L^{-1})
	Arabinose	Fucose	Galactose	Glucose	Mannitol	**Xylose**	**Total**	**HMF**	**Furfural**	**FAN**
L. digitata	0.07 ± 0.01	2.29 ± 0.11	0.57 ± 0.03	1.51 ± 0.08	3.49 ± 0.08	0.55 ± 0.03	8.48 ± 0.34	0.02 ± 0.02	–	0.24 ± 0.001
F. serratus	0.08 ± 0.00	3.77 ± 0.10	0.91 ± 0.03	0.95 ± 0.03	4.38 ± 0.09	1.00 ± 0.03	11.09 ± 0.30	–	–	0.38 ± 0.004
C. crispus	0.14 ± 0.01	3.04 ± 0.15	6.84 ± 0.17	2.02 ± 0.07	0.01 ± 0.02	1.31 ± 0.06	13.36 ± 0.47	2.13 ± 0.16	–	0.36 ± 0.01
P. palmata	0.05 ± 0.01	0.08 ± 0.00	4.24 ± 0.21	2.99 ± 0.15	0.04 ± 0.00	8.71 ± 0.42	16.13 ± 0.79	0.02 ± 0.00	0.25 ± 0.01	0.38 ± 0.01
U. lactuca	2.15 ± 0.00	0.11 ± 0.02	0.78 ± 0.00	5.30 ± 0.03	0.06 ± 0.00	3.52 ± 0.02	11.91 ± 0.08	0.09 ± 0.00	0.05 ± 0.00	0.16 ± 0.003

Data are the mean ± SD of three measurements

Levels of free amino nitrogen (FAN) are also reported for each of the seaweed hydrolysates (Table 3). This is because amino acids represent a significant nitrogen source required for yeast growth and metabolism during fermentation. FAN levels were within the range 0.16 to 0.38 g L^{-1}. The minimum FAN content normally specified for brewing fermentations is 0.15 g L^{-1}, so the seaweed hydrolysates appeared to contain sufficient quantities of amino acids to avoid the need for supplementary nitrogen sources to aid fermentation.

Phenotypic Microarray Analysis of Yeast Strains

Metabolic Output of Yeast Strains Grown on Synthetic Minimal Media with Different Carbon Sources

The metabolic outputs (defined here as redox signal intensity) of 22 yeast strains on minimal media containing individual carbon sources: glucose, galactose, mannitol, xylose, fucose and rhamnose (at 6 % *w*/*v*) were measured (Fig. 1a–f). These carbon sources were chosen because they are the most abundant in sugar composition across the three taxonomical groups of macroalgae. There was measurable metabolic output for all strains utilising glucose after 50 h of incubation at 30 °C, with considerable variation observed between the strains (2.3–98.7 RSI) (Fig. 1). Some strains, however, exhibited extremely low RSI levels when utilising glucose as a sole carbon source, namely *Candida shehatae* var *lignosa* NCYC 2389 and *Pichia stipitis* TP10, and were as a result removed from further study. The majority of the strains displayed a high metabolic output when utilising galactose (Fig. 1b), whereas the strains exhibited very poor metabolic output when utilising fucose (Fig. 1d). The metabolic outputs of all strains screened here appear to be similar in response when utilising mannitol, xylose and rhamnose (Fig. 1b, e, f).

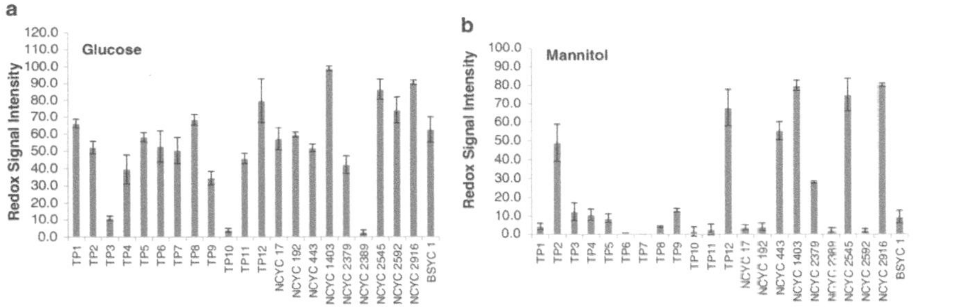

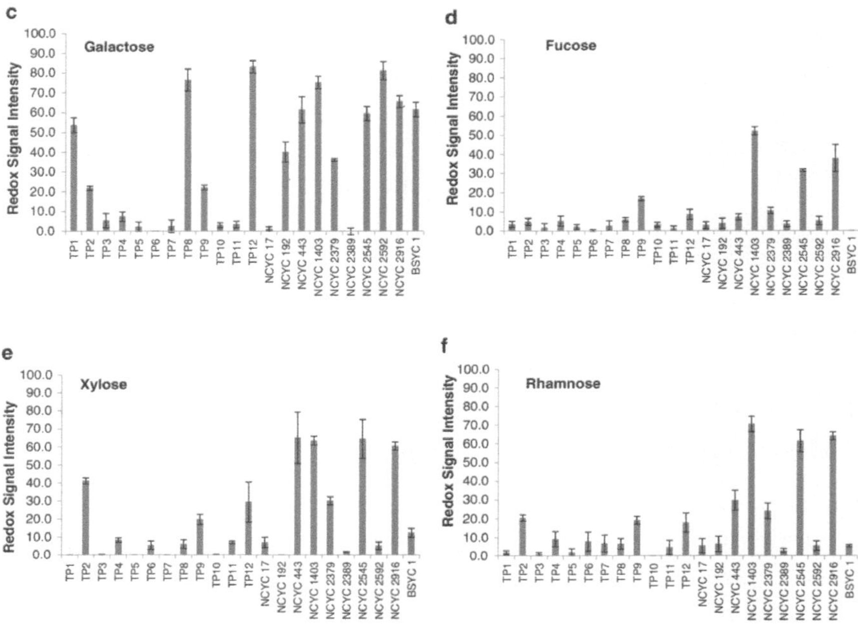

Figure 1. Phenotypic microarray analysis (redox signal intensity) at 50 h for strains on minimal media containing a glucose, b mannitol, c galactose, d fucose, e xylose and f rhamnose at 6 % (*w*/*v*) inclusion. Data are the mean ± SD of three replicates. See Table 1 for description of yeast strains.

In order to have a better understanding of each yeast strains' ability to metabolise unusual carbon sources, the PM assay results were adjusted to show each strain's metabolic output for these sugars in relation to their utilisation of glucose. Interestingly, strains which had relatively high utilisation of glucose (≥79.7) also displayed high metabolic outputs when utilising galactose, mannitol, fucose, xylose and rhamnose. These were: *C. succiphila* NCYC 1403, *P. anomala* TP12, *C. tenuis* NCYC 2545 (Table 4).

Table 4. Phenotypic microarray results (RSI at 50 h) for each yeast strain investigated in this study on minimal media containing: galactose, mannitol, fucose, xylose and arabinose (6 % *v/w*) relative to the corresponding glucose value (RSI). Data are the mean ± SD of three replicates. See Table 1 for description of yeast strains

		Monosaccharide (% RSI in comparison to glucose)				
Yeast ID	**Glucose (RSI)**	**Galactose**	**Mannitol**	**Fucose**	**Xylose**	**Rhamnose**
TP1	66.0	81.8	6.1	5.6	0.0	3.0
TP2	52.0	42.3	93.6	9.6	79.5	39.7
TP3[a]	10.7	53.1	112.5	18.8	3.1	9.4
TP4	39.3	19.5	26.3	13.6	22.0	23.7
TP5	58.0	4.0	14.4	4.0	0.0	3.4
TP6	52.7	0.0	0.6	0.6	10.8	15.2
TP7	50.3	5.3	0.0	5.3	0.0	13.9
TP8	68.3	112.2	5.4	8.8	9.3	9.8
TP9	34.3	65.0	36.9	49.5	58.3	56.3
TP10[a]	3.7	90.9	36.4	90.9	9.1	0.0
TP11	45.7	8.0	5.1	2.9	16.1	10.2
TP12	79.7	104.6	84.9	10.9	37.2	22.6
NCYC 17	57.0	2.3	5.8	5.3	12.9	9.9
NCYC 192	59.3	68.0	6.2	6.7	0.0	11.2
NCYC 443	51.7	119.4	106.5	14.2	126.5	58.1
NCYC 1403	98.7	76.4	80.7	53.0	64.5	71.6
NCYC 2379	42.0	86.5	65.9	25.4	72.2	57.9
NCYC 2389[a]	2.3	0.0	71.4	157.1	85.7	128.6
NCYC 2545	86.3	69.1	86.5	36.7	74.9	71.4
NCYC 2592	74.0	109.9	2.3	7.2	7.2	7.7
NCYC 2916	90.3	72.7	88.6	42.1	67.2	71.2
BSYC 1	62.3	98.9	14.4	0.0	20.3	9.1

[a]Removed from study due to low metabolic output on glucose

Metabolic Output of Yeast Strains Grown on Seaweed Hydrolysates

Three yeast strains, *C. tenuis* NCYC 2545, *C. succiphila* NCYC 1403 and *P. anomala* TP12 were able to metabolise the majority of the individual sugars as evaluated using PM analysis. They were thus selected for further experiments to assess their abilities to metabolise the acid hydrolysates generated from five seaweed species (Fig. 2a–e). *S. cerevisiae* NCYC 2592 was also included in the trial as a reference distiller's production strain (www.ncyc.co.uk), expected to provide efficient conversion of glucose and galactose to ethanol. Additionally, two *S. cerevisiae* Y12 and YPS128 strains (Liti et al. 2009) were also included as reference strains. *S. cerevisiae* Y12 and YPS128 are wild yeast derived from clean lineages that have had no alternations to their genomes from human influences or domestication events (Liti et al. 2009). Assays revealed that all strains cultured on *L. digitata*, *F. serratus* and *U. lactuca* hydrolysates were metabolically active (Fig. 2a, b, e). However, assays with *C. tenuis* NCYC 2545 revealed a decrease in RIS after 20 h from ca 50–60 RIS to ca 10 RIS in both brown seaweed hydrolysates (Fig. 2a, b). Although this phenomenon has been observed in other microarray assays before, it is unknown why it occurs but could be down to cell lysis and loss of the irreversibly reduced dye (Darren Greetham, personal communication). Due to this result, *C. tenuis* NCYC 2545 was not included in future work.

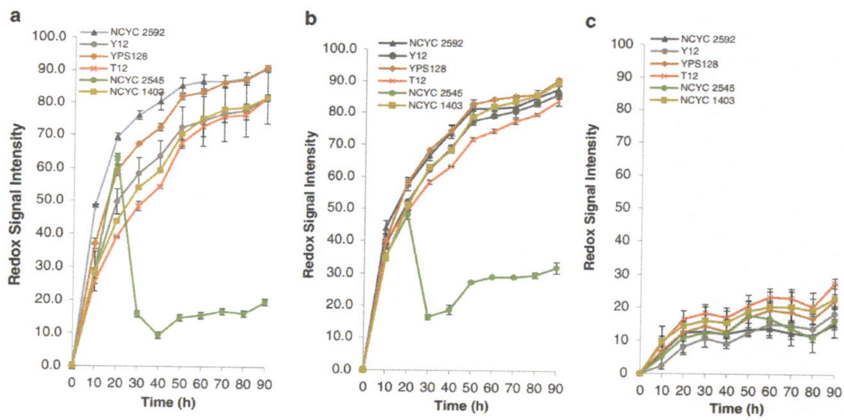

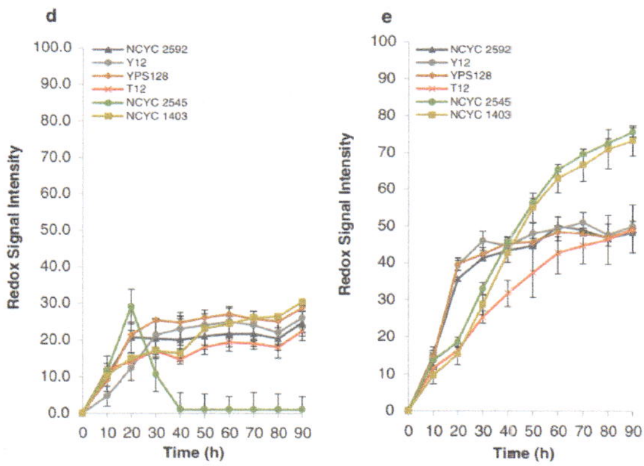

Figure 2. Phenotypic microarray analysis (redox signal intensity) for yeast on a*L. digitata*hydrolyste, b*F. serratus* hydrolysate, c*C. crispus* hydrolysate, d*P. palmata* hydrolysate, e*U. lactuca* hydrolysate. Data are the mean ± SD of three replicates. See Table 1 for description of yeast strains.

Screening for Bioethanol Production from Seaweed Hydrolysates in Laboratory-Scale Fermentations

Trial fermentations were conducted with the yeast strains *S. cerevisiae* NCYC 2592, *C. succiphila* NCYC 1403, *P. anomala* TP12 and *S. cerevisiae* strains YPS128 and Y12 to determine whether they could produce ethanol as a carbon end point. Hydrolysates derived from the five seaweed species post-acid pre-treatment were neutralised and fermented using each strain individually at 30 °C.

Fermentation progression was monitored as weight loss due to the evolution of carbon dioxide, a metabolic product of the fermentation of any monosaccharide (Fig. 3a–e). There were clear differences in the fermentation progression for each hydrolysate using the strains employed in this trial. Overall, fermentation of the *L. digitata* and *F. serratus* hydrolysates by all strains did not perform well as there were only small weight losses after 120 h (Fig. 3a, b). *P. palmata* and *C. crispus* fermentations on the contrary progressed better, with *P. anomala* TP12 performing better on the *C. crispus* hydrolysate, albeit with a longer lag-phase and longer attenuation time (Fig. 3c, d). *S. cerevisiae* YPS128 displayed a similar fermentation profile grown on the *C. crispus* hydrolysate yet *S. cerevisiae* Y12 did not readily ferment this substrate.

All three *S. cerevisiae* strains demonstrated significant weight loss on the *U. lactuca* hydrolysate, whereas it took up to 24 h for *P. anomala* TP12 to start to ferment this hydrolysate (Fig. 3e).

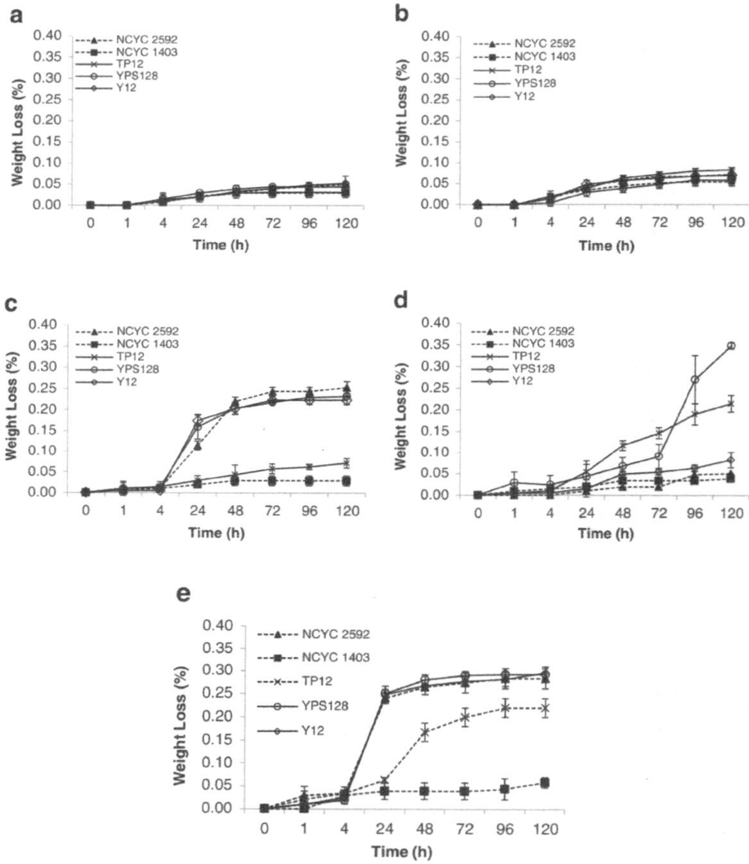

Figure 3. Fermentation progression of yeast strains on a *F. serratus* hydrolysate, b *L. digitata*hydrolysate, c *P. palmata* hydrolysate, d *C. crispus* hydrolysate, e *U. lactuca* hydrolysate by yeast strains, as indicated by vessel weight loss resulting from the release of carbon dioxide. Data are the mean ± SD of three replicates. See Table 1 for description of yeast strains.

Utilization of monosaccharides pre- and post-fermentation and the mean ethanol yields achieved from each strain on the five hydrolysates was determined. Overall, each strain apart from *C. succiphila* NCYC 1403 managed to produce ethanol; however, the yields varied between strain and hydrolysate (Table 5). The highest yield of ethanol, ca 13 g L^{-1}, was produced from the fermentation of a hydrolysate derived from *C. crispus* by *S. cerevisiae* YPS128, and it appears that this strain was able to consume the majority of the available monosaccharides that were present in the hydrolysate. Interestingly, strains *S. cerevisiae* Y12 and *P. anomala* TP12 also consumed the vast majority of monosaccharides, however, were only able to yield ca 4.5 and 3 g L^{-1} of ethanol, respectively. The highest yields of ethanol produced from the hydrolysates of *P. palmata* and *U. lactuca* were ca 7 g L^{-1}, resulting from the use of *Saccharomyces* spp. Interestingly, the fucose which was present in the *C. crispus* hydrolysate (at a concentration of ca 3 g L^{-1}) and not in the brown seaweed hydrolysates, was partially consumed by strains *S. cerevisiae* NCYC 2592, Y12 and *P. anomala* TP12 during the fermentations. Disappointingly, the five strains hardly yielded any ethanol post-fermentation of hydrolysates derived from the two brown seaweeds (0.00–1.55 g L^{-1} ethanol).

Table 5. Utilisation of monosaccharides in hydrolysates of *L. digitata*, *F. serratus*, *C. crispus*, *P. palmata* and *U. lactuca* and production of bioethanol by different yeast strains post-fermentation. Data are the mean ± SD of three replicates

			Yeast strain				
			NCYC 1403	NCYC 2592	TP12	Y12	**YPS128**
Hydrolysate	**Monosaccharide/ ethanol (g L^{-1})**	**Start**	**End**	**End**	**End**	**End**	**End**
L. digitata	Mannitol	3.49 ± 0.08	3.40 ± 0.07	3.46 ± 0.11	3.44 ± 0.48	3.42 ± 0.48	3.48 ± 0.11
	Fucose	2.29 ± 0.11	2.25 ± 0.07	2.23 ± 0.13	2.28 ± 0.24	2.30 ± 0.24	2.31 ± 0.01
	Arabinose	0.07 ± 0.01	0.06 ± 0.00	0.07 ± 0.02	0.02 ± 0.05	0.08 ± 0.05	0.07 ± 0.00
	Galactose	0.57 ± 0.03	0.55 ± 0.04	0.02 ± 0.01	0.05 ± 0.01	0.05 ± 0.01	0.00 ± 0.00
	Glucose	1.51 ± 0.08	1.49 ± 0.08	0.02 ± 0.01	0.01 ± 0.00	0.01 ± 0.00	0.01 ± 0.00
	Xylose	0.55 ± 0.03	0.56 ± 0.02	0.37 ± 0.03	0.14 ± 0.10	0.14 ± 0.10	0.41 ± 0.00
	Ethanol	–	0.09 ± 0.12	1.24 ± 0.12	1.26 ± 0.06	1.22 ± 0.05	1.55 ± 0.17
F. serratus	Mannitol	4.38 ± 0.08	4.36 ± 0.55	4.35 ± 0.56	4.32 ± 0.50	4.38 ± 0.11	4.37 ± 0.13
	Fucose	3.77 ± 0.11	3.58 ± 0.40	3.58 ± 0.28	3.75 ± 0.42	3.76 ± 0.05	3.77 ± 0.09
	Arabinose	0.08 ± 0.01	0.04 ± 0.01	0.04 ± 0.00	0.06 ± 0.02	0.04 ± 0.00	0.04 ± 0.00
	Galactose	0.91 ± 0.03	0.88 ± 0.13	0.00 ± 0.00	0.00 ± 0.01	0.00 ± 0.00	0.02 ± 0.00
	Glucose	0.95 ± 0.08	0.89 ± 0.12	0.01 ± 0.00	0.00 ± 0.00	0.02 ± 0.00	0.01 ± 0.00
	Xylose	1.00 ± 0.03	0.94 ± 0.13	0.82 ± 0.12	0.07 ± 0.04	0.80 ± 0.02	0.97 ± 0.09
	Ethanol	–	0.00 ± 0.03	0.58 ± 0.05	0.46 ± 0.04	0.78 ± 0.17	1.05 ± 0.07

P. palmata	Mannitol	0.04 ± 0.02	0.02 ± 0.03	0.02 ± 0.00	0.03 ± 0.02	0.01 ± 0.00	0.03 ± 0.08
	Fucose	0.08 ± 0.15	0.05 ± 0.01	0.04 ± 0.00	0.05 ± 0.15	0.06 ± 0.00	0.07 ± 0.04
	Arabinose	0.05 ± 0.00	0.04 ± 0.02	0.05 ± 0.00	0.03 ± 0.01	0.05 ± 0.00	0.04 ± 0.00
	Galactose	4.24 ± 0.17	4.22 ± 0.04	0.39 ± 0.01	4.23 ± 0.17	0.40 ± 0.01	0.23 ± 0.23
	Glucose	2.99 ± 0.07	2.97 ± 0.02	0.00 ± 0.00	2.96 ± 0.07	0.00 ± 0.00	0.00 ± 0.12
	Xylose	8.71 ± 0.06	8.70 ± 0.02	8.69 ± 0.14	8.70 ± 0.06	8.68 ± 0.25	6.20 ± 0.00
	Ethanol	–	0.00 ± 0.01	5.21 ± 0.27	1.69 ± 0.05	6.57 ± 1.20	6.08 ± 0.40
C. crispus	Mannitol	0.01 ± 0.02	0.00 ± 0.03	0.01 ± 0.00	0.01 ± 0.02	0.00 ± 0.04	0.01 ± 0.01
	Fucose	3.04 ± 0.15	2.35 ± 0.10	1.37 ± 0.00	0.05 ± 0.00	0.25 ± 0.00	0.39 ± 0.00
	Arabinose	0.14 ± 0.01	0.11 ± 0.00	0.13 ± 0.01	0.04 ± 0.00	0.11 ± 0.00	0.15 ± 0.01
	Galactose	6.84 ± 0.17	6.84 ± 0.20	6.63 ± 0.11	0.36 ± 0.01	0.50 ± 0.03	1.32 ± 0.01
	Glucose	2.02 ± 0.07	1.97 ± 0.09	0.00 ± 0.10	0.00 ± 0.00	0.01 ± 0.00	0.03 ± 0.04
	Xylose	1.31 ± 0.06	0.79 ± 0.04	0.84 ± 0.02	0.43 ± 0.13	0.51 ± 0.01	0.43 ± 0.02
	Ethanol	–	0.00 ± 0.02	2.90 ± 0.02	3.26 ± 0.01	4.43 ± 0.18	12.84 ± 0.70
U. lactuca	Mannitol	0.06 ± 0.00	0.05 ± 0.05	0.06 ± 0.00	0.04 ± 0.07	0.03 ± 0.01	0.06 ± 0.04
	Fucose	0.11 ± 0.02	0.09 ± 0.00	0.10 ± 0.00	0.11 ± 0.00	0.01 ± 0.01	0.11 ± 0.00
	Arabinose	2.15 ± 0.00	2.15 ± 0.04	1.98 ± 0.08	2.03 ± 0.14	2.15 ± 0.03	1.98 ± 0.02
	Galactose	0.78 ± 0.00	0.71 ± 0.01	0.70 ± 0.03	0.04 ± 0.00	0.01 ± 0.00	0.04 ± 0.00
	Glucose	5.30 ± 0.07	5.28 ± 0.04	0.05 ± 0.00	0.00 ± 0.00	0.03 ± 0.00	0.00 ± 0.00
	Xylose	3.51 ± 0.02	3.49 ± 0.06	2.29 ± 0.04	2.43 ± 0.04	2.71 ± 0.02	2.06 ± 0.01
	Ethanol	–	0.26 ± 0.02	6.14 ± 0.22	3.49 ± 0.06	6.80 ± 0.34	7.31 ± 0.50

DISCUSSION

The strains used in this study (mainly *Candida* and *Pichia* strains) for the initial screen on various carbon sources were selected based upon the fact that these species of yeast have the ability to ferment C5 sugars such as xylose (present in seaweeds) (Hahn-Hägerdal et al. 2007). Therefore, it was of interest to examine whether these yeasts could utilise the sugars which are only found in seaweeds and whether they would produce ethanol as a result.

In the PM method, changes in colour of a redox sensitive dye are monitored as yeast cells respire on selected media arranged in a 96-well plate format (Bochner et al. 2001). The colour development is a measure of the yeast's metabolic activity and essentially indicates its ability to metabolise the sugar present in the assay. It is not possible to determine the metabolic products (e.g. ethanol) directly, just metabolic activity. The majority of the strains screened in this study displayed an increase in the rate of metabolic output, defined as redox signal intensity (RSI), in the presence of glucose and galactose (Fig. 1a, c) compared to the other tested monosaccharides (fucose, mannitol, xylose, arabinose and rhamnose). This finding was as expected since these monosaccharides are favourable carbon sources for yeast (Flores et al. 2000). Interestingly, the *Candida* and *Pichia* spp examined in this study displayed an increase in metabolic activity, albeit at varying intensities, when utilising mannitol, xylose or rhamnose as sole carbon sources (Fig. 1b, e, f) compared to *Saccharomyces*, *Kluyveromyces* and *Hanseniaspora* spp in this study. *Candida* spp and *Pichia* spp have been previously reported to be able to metabolise and ferment pentose sugars (Dien et al. 1996; Jarosz et al. 2008); however, only 1 % of all known xylose utilising yeasts are known to display the potential of fermenting xylose to ethanol (McMillan and Boynton 1994). *Pichia angiophorae* has already been shown to be able to ferment both mannitol and laminarin in seaweed extracts to ethanol (Horn et al. 2000a). *C. tenuis* NCYC 2545, *C. succiphila* NCYC 1403 and *P. anomala* TP12 exhibited high RSI levels on all monosaccharides analysed in relation to their corresponding glucose values, which implies that they are able to metabolise these sugars (Table 4). What is interesting is that they are the only species which appear to be metabolically active on synthetic media containing fucose, with an $RSI \geq 20$ (Fig. 1d). Fucose is a hexose sugar and is the fundamental subunit of the fucoidan polysaccharide found in brown seaweeds (Ale et al. 2011). To date, there has been no evidence to suggest that fucose can be metabolised by yeast and converted into ethanol during a fermentation. Results here revealed that three *Candida* spp displayed metabolic activity in the presence of fucose as a sole carbon source. This suggests that these strains may possibly contain a fucose utilisation pathway or surface transporter, enabling the uptake of fucose into the cell. A gene has been identified in the *C. albicans* genome, C2_09280C_B, of which is predicted to be either a membrane transporter, member of the fucose:proton symporter (FHS) family or part of a major facilitator superfamily (MFS), and orthologous genes have been identified in other *Candida*, *Aspergillus* and *Saccharomyces* species (www.candidagenome.org). It would be of great interest to analyse whether the same or orthologous genes associated with fucose uptake or metabolism can also be identified in the *Candida* spp examined in the present study.

The metabolic activities of the shortlisted strains *C. succiphila* NCYC 1403, *P. anomala* TP12, *C. tenuis* NCYC 2545 and *S. cerevisiae* NCYC 2592 cultured on red seaweed hydrolysates could not be reliably interpreted using the present methodology. This is because the colour of the hydrolysates interfered with detection of generation of the tetrazolium dye (Fig. 2c, d). This is undoubtedly a limitation on the current application of the PM. If this approach is to be employed in future work, it would be necessary to experiment with the use of other reporter dyes such as DAPI, dihydroethidium or GFP.

This study used hydrolysates generated from the thermo-chemical pre-treatment of different seaweed species. There had been no previous reports of the utilisation of hydrolysates produced from the acid pre-treatment of seaweeds for bioethanol production, until Mutripah et al. (2014) fermented the hydrolysate of the rhodophyte *Palmaria palmata* with *Saccharomyces cerevisiae*, where 17.3 mg of ethanol per gramme of seaweed was produced. Typically, the objective of applying an acid (or in some cases alkaline) pre-treatment step is to modify the chemical structure of the biomaterial by removing the more recalcitrant complexes. Hydrolysis of polysaccharides into their monosaccharide constituents is typically a further step that is sequential to the acid pre-treatment, by the addition of selective enzymes (Meinita et al. 2012); however, addition of enzyme is expensive and it is necessary to apply the correct enzyme blend to suit the particular species of seaweed that is being processed. The liquid fraction generated from the acid pre-treatment step is usually discarded. Analysis of the hydrolysates derived from a pre-treatment of seaweed indicates that they contained a cocktail of available monosaccharides which could be fermented to ethanol, without the need to apply an enzyme hydrolysis step, and contained low levels of yeast-inhibitory compounds which should not unduly influence yeast performance during fermentation (Table 5). According to the compositional analysis undertaken in the study, *P. palmata* had the highest carbohydrate content (39.4 %) and actually yielded the highest concentration of total sugars in its hydrolysate post-pre-treatment (ca 16.1 g L^{-1}). Although it appears that this particular species could be an ideal candidate for bioethanol production, its abundance in UK waters is not as high as that of certain brown kelp species, such as *L. hyperborea* and *A. nodosum* (Wilkinson 1995). *L. digitata* had the lowest measured carbohydrate content of 21.7 % (dry wt), which could in part be due to the low levels of phaeophycean polysaccharides, in particular laminarin and fucoidan, as seaweeds were harvested in May for this study. A study by Adams et al. (2011) revealed that the most suitable month to harvest *L. digitata* for bioethanol production was July, as this is when carbohydrate yields are highest. Further, some specialised monosaccharides such as mannuronic and guluronic acids from the polysaccharide alginate might have been released but were not quantified using the HPLC method applied. Alginate is extremely difficult to hydrolyse and under the conditions utilised in this study is unlikely to have liberated significant quantities of these particular monosaccharides. The main cell wall constituents that make up brown seaweeds are typically cellulose microfibrils which are embedded in an amorphous matrix of alginate and fucoidan polysaccharides (Kloareg et al. 1986). Mannitol is often found as a terminating side chain on laminarin, a polymer of β-(1-3)-D-glucose that can be easily extracted from brown seaweed (Horn et al. 2000a) and the fucose yields have resulted from the breakdown of fucoidan in the brown seaweed hydrolysates.

Saccharomyces cerevisiae spp are traditionally used in fermentations for the production of commercial bioethanol, as strains can adequately consume hexose sugars and resist any osmotic and ethanol toxicity stresses which may arise during the process (Zhao and Bai 2009). However, their inability to utilise pentose sugars, despite having the relevant pathways in their genomes (Hahn-Hägerdal et al. 2007), limits the full potential of their use in the production of bioethanol.

The yeast strains selected for fermentation trials performed significantly better on the red seaweed hydrolysates than the other hydrolysates produced from the brown and green seaweeds, with the exception of *C. succiphila* NCYC 1403, which failed to produce any ethanol. *Candida* spp are obligate aerobes and are Crabtree-negative microorganisms, implying that they respire and produce biomass independent of the sugar concentration as long as oxygen is available (Fredlund et al. 2004). This explains the extremely low levels of ethanol produced in the anaerobic environment encountered during a fermentation. Despite the fact that strain *P. anomala* TP12 is also a Crabtree-negative yeast, it demonstrated the ability to produce bioethanol in an anoxic environment from all five seaweed hydrolysates, with the highest ethanol yield of ca 3.5 g L^{-1} being produced from the green *U. lactuca* hydrolysate. There are reports that have revealed this strain to be ethanol-producing and potentially resistant to high ethanol concentrations (Passoth et al. 2013; Tao et al. 2011). Interestingly, a study by Ruyters et al. (2015) revealed that *P. anomala* could be a candidate strain for ethanol production from lignocellulosic hydrolysates, as ethanol yields were similar to those with *S. cerevisiae* and *P. anomala* displayed tolerance to high levels of ethanol. The group also found that *P. anomala*needed a longer fermentation time, which is in agreement with the results obtained from this study when *P. anomala* TP12 was inoculated into the seaweed hydrolysates (Fig. 3). Although the oxygen-sensitive nature of this strain will undoubtedly limit its fermentation performance, it appears that *P. anomala* TP12 and possibly other *P. anomala* strains are compatible not only with lignocellulosic hydrolysates but also with seaweed.

The relatively poor fermentation performance of the five strains on the brown seaweed hydrolysates resulted in the lowest ethanol yields (<2 g L^{-1} ethanol for all strains). This is due to the low levels of fermentable sugars in the hydrolysate, namely glucose and galactose, which together amounted to 2.08 and 1.86 g L^{-1} in the *L. digitata* and *F. serratus* hydrolysates, respectively, principally as a result of harvesting in May. In order to make bioethanol production commercially viable from brown seaweed species, harvesting at the appropriate time period of the year for maximum sugar quantities and hydrolysis with enzymes will most likely be required, as acid hydrolysis alone was not sufficient to generate an adequate yield of fermentable sugars. Polysaccharides such as cellulose would therefore need to be targeted by selective cellulase enzymes in order to release glucose in high yields.

Wild yeast strains *S. cerevisiae* Y12 and YPS128, which were included in this study as reference strains due to the fact that their genomes have remained unaltered from human influences and domestication events, were both able to consume the vast majority of monosaccharides (including pentose sugars) that were present in the hydrolysate of *C. crispus*. However, *S. cerevisiae* Y12 only generated 4.43 g L^{-1} of ethanol, whereas *S. cerevisiae* YPS128 produced the highest yield of ethanol in this study of 12.84 g L^{-1} on the *C. crispus* feedstock. *S. cerevisiae* Y12 has been reported to be extremely sensitive to the presence of inhibitory compounds which arise as a result of the thermo-chemical pre-treatment of biomass (Wimalasena et al. 2014). A haploid segregant population derived from these yeasts is available, and it would be interesting to screen for variance in utilisation of these sugars in that population, as research has revealed that populations of haploid segregants have a varied response to any given environment (Greetham et al. 2014b). Hydrolysate from the pre-treatment of *C. crispus* was shown to have the highest yield of HMF at ca 2 g L^{-1}, which could explain *S. cerevisiae* Y12's poor fermentation progress in this instance. Conversely, *S. cerevisiae* YPS128 displays tolerance to inhibitory compounds (Wimalasena et al. 2014), and such phenotypes are highly desirable for bioethanol production. *P. anomala* TP12 was also able to consume the majority of the monosaccharides present in the hydrolysate of *C. crispus*; however, only 3.26 g L^{-1} of ethanol was generated. It is thus likely possible that the sugars were converted to other carbon end point products in addition to ethanol. In order to enhance the performance of strains *S. cerevisiae* Y12 and *P. anomala* TP12 during the fermentation of *C. crispus* hydrolysate, it would be worthwhile attempting to detoxify the hydrolysate by selective removal of inhibitory compounds. Many methods have been developed in order to remove toxic compounds from hydrolysates, as reviewed by Mussatto and Roberto (2004).

Interestingly, only the fucose which was present in the *C. crispus* hydrolysate (at a concentration of ca 3 g L^{-1}) was partially consumed by strains *S. cerevisiae* NCYC 2592, Y12 and *P. anomala* TP12 during fermentation (Table 4). This calls into question the correct assignment of the HPLC peak to fucose in this instance. This monosaccharide is typically only found in phaeophytes, not rhodophytes. It may be concluded that the *C. crispus* hydrolysate contained a monosaccharide which demonstrated chromatographic similarity to the fucose standard under experimental conditions. Likewise, it eluted within the same retention window as the peak assigned to fucose in brown seaweed hydrolysates. However, LC-MS studies would be required to further confirm this identification.

In conclusion, in this study, we have demonstrated a novel application of the PM assay to evaluate the ability of yeast strains to metabolise seaweed-derived monosaccharides (individually) as well as entire hydrolysates of five UK seaweed species. The most suitable yeast strains to utilise sugars derived from macroalgae and convert them into bioethanol were *S. cerevisiae* spp with the highest concentration of ethanol (ca 13 g L^{-1}) resulting from the fermentation of the acid hydrolysate of *C. crispus* by *S. cerevisiae* YPS128. Shortlisted stain *P. anomala* TP12 was able to utilise the majority of the monosaccharides present in the hydrolysate of *C. crispus* including fucose, generating ca 3.5 g L^{-1} ethanol, which was similar to yields of ethanol produced by *S. cerevisiae* strains NCYC 2592 and Y12 from the same hydrolysate. This reveals that *P. anomala* TP12 could be a possible candidate strain for further research. The present work was focused on strain selection and did not seek to optimise ethanol yields from the fermentation of the seaweed hydrolysates. This would be required in order to make the industrial production of ethanol from seaweed economically viable. Industrial production of ethanol from seaweeds would require the generation of more efficient strains that can ferment the majority of sugars contained in seaweed polysaccharides. Although ethanol was the only product screened for in this study, other high value, carbon end point products such as xylitol, succinic and acetic acids could also be screened for and sourced from the growth of similar or other microorganisms on seaweed hydrolysates.

ACKNOWLEDGMENTS

The authors gratefully acknowledge the financial support of the Biotechnology and Biological Science Research Council (BBSRC) and CRODA International plc for funding this research and Miss E. Kostas's studentship. We also thank Drs D. Greetham and T. Wimalasena for their help and input with the phenotypic microarray (Biolog) screen and Dr T. Phister and Prof. E. Louis for providing yeast strains.

REFERENCES

1. Adams JM, Gallagher JA, Donnison IS (2009) Fermentation study on Saccharina latissima for bioethanol production considering variable pre-treatments. J Appl Phycol 21:569–574
2. Adams JM, Toop TA, Donnison IS, Gallagher JA (2011) Seasonal variation in Laminaria digitata and its impact on biochemical conversion routes to biofuels. Bioresour Technol 102:9976–9984
3. Ale MT, Mikkelsen JD, Meyer AS (2011) Important determinants for fucoidan bioactivity: a critical review of structure-function relations and extraction methods for fucose-containing sulfated polysaccharides from brown seaweeds. Mar Drugs 9:2106–2130
4. Bochner BR, Gadzinski P, Panomitros E (2001) Phenotype microarrays for high-throughput phenotypic testing and assay of gene function. Genome Res 11:1246–1255
5. Bothast RJ, Schlicher MA (2005) Biotechnological processes for conversion of corn into ethanol. Appl Biochem Biotechnol 67:19–25
6. Dien BS, Kurtzman CG, Saha BC, Bothast RJ (1996) Screening for L-arabinose fermenting yeasts. Appl Biochem Biotechnol 57–58:233–242
7. Dubois M, Gilles KA, Hamilton JK, Rebers PA, Smith F (1956) Colorimetric method for determination of sugars and related substances. Anal Chem 28:350–356
8. Enquist-Newman M, Faust AME, Bravo DD, Santos CNS, Raisner RM, Hanel A, Sarvabhowman P, Le C, Regitsky DD, Cooper SR, Peereboom L, Clark A, Martinez Y, Goldsmith J, Cho MY, Donohoue PD, Luo L, Lamberson B, Tamrakar B, Kim EJ, Villari JL, Gill A, Tripathi SA, Karamchedu P, Paredes CJ, Rajgarhia V, Kotlar HK, Bailey RB, Miller DJ, Ohler NL, Swimmer C, Yoshikuni Y (2014) Efficient ethanol production from brown macroalgae sugars by a synthetic yeast platform. Nature 505:239–243
9. Flores CL, Rodríguez C, Petit T, Gancedo C (2000) Carbohydrate

and energy-yielding metabolism in non-conventional yeasts. FEMS Microbiol Rev 24:507–529

10. Folch J, Lees M, Stanley GHS (1957) A simple method for the isolation and purification of total lipids from animal tissues. J Biol Chem 226:497–509
11. Fredlund E, Druvefors UÄ, Olstorpe MN, Passoth V, Schnürer J (2004) Influence of ethyl acetate production and ploidy on the anti-mould activity of Pichia anomala. FEMS Microbiol Lett 238:133–137
12. Greetham D, Wimalasena T, Kerruish DWM, Brindley S, Ibbett RN, Linforth RL, Tucker G, Phister TG, Smart KA (2014a) Development of a phenotypic assay for characterisation of ethanologenic yeast strain sensitivity to inhibitors released from lignocellulosic feedstocks. J Ind Microbiol Biotechnol 41:931–945
13. Greetham D, Wimalasena TT, Leung K, Marvin ME, Chandelia Y, Hart AJ, Phister TG, Tucker GA, Louis EJ, Smart KA (2014b) The genetic basis of variation in clean lineages of Saccharomyces cerevisiae in response to stresses encountered during bioethanol production. PLoS ONE 9, e103233
14. Hahn-Hägerdal B, Karhumaa K, Fonseca C, Spencer-Martins I, Gorwa-Grauslund MF (2007) Towards industrial pentose-fermenting yeast strains. Appl Biochem Biotechnol 74:937–953
15. Horn SJ, Aasen IM, Ostgaard K (2000a) Ethanol production from seaweed extract. J Ind Microbiol Biotechnol 25:249–254
16. Horn SJ, Aasen IM, Ostgaard K (2000b) Production of ethanol from mannitol by Zymobacter palmae. J Ind Microbiol Biotechnol 24:51–57
17. Jarosz A, Kordowska-Wiater M, Targonski Z (2008) Biotransformation of L-arabinose to arabitol by yeasts from genera Pichia and Candida. Bio Technologia 1:177–188
18. Jeong TS, Um BH, Kim JS, Oh KK (2010) Optimizing dilute-acid pretreatment of rapeseed straw for extraction of hemicellulose. Appl Biochem Biotechnol 161:22–33
19. Jung KA, Lim SR, Kim Y, Park JM (2013) Potentials of macroalgae as feedstocks for biorefinery. Bioresour Technol 135:182–190

20. Karakashev D, Thomsen AB, Angelidaki I (2007) Anaerobic biotechnological approaches for production of liquid energy carriers from biomass. Biotechnol Lett 29:1005–1012
21. Kloareg B, Demarty M, Mabeau S (1986) Polyanionic characteristics of purified sulphated homofucans from brown algae. Int J Bio Macromol 8:380–386
22. Liti G, Carter DM, Moses AM, Warringer J, Parts L, James SA, Davey RP, Robert IN, Burt A, Koufopanos V, Tsai IJ, Bergman CM, Bensasson D, O'Kelly MJT, van Oudenaarden A, Barton DBH, Bailes E, Nguyen AN, Jones M, Quail MA, Goodhead I, Sims S, Smith F, Blomberg A, Durbin R, Louis EJ (2009) Population genomics of domestic and wild yeasts. Nature 458:337–341
23. Lu XB, Zhang YM, Angelidaki I (2009) Optimization of H2SO4-catalyzed hydrothermal pretreatment of rapeseed straw for bioconversion to ethanol: focusing on pretreatment at high solids content. Bioresour Technol 100:3048–3053
24. McMillan JD, Boynton BL (1994) Arabinose utilization by xylose-fermenting yeasts and fungi. Appl Biochem Biotechnol 45:569–584
25. Meinita MDN, Kang JY, Jeong GT, Koo HM, Park SM, Hong YK (2012) Bioethanol production from the acid hydrolysate of the carrageenophyte Kappaphycus alvarezii (cottonii). J Appl Phycol 24:857–862
26. Mukherjee V, Steensels J, Lievens B, Van de Voorde I, Verplaetse A, Aerts G, Willems KA, Thevelein JM, Verstrepen KJ, Ruyters S (2014) Phenotypic evaluation of natural and industrial Saccharomyces yeasts for different traits desirable in industrial bioethanol production. Appl Microbiol Biotechnol 98:9483–9498
27. Mussatto, Roberto (2004) Alternatives for detoxification of diluted-acid lignocellulosic hydrolysates for use in fermentative processes: a review. Bioresour Technol 93:1–10
28. Mutripah S, Meinita MDN, Kang JY, Jeong GT, Susanto AB, Prabowo RE, Hong YK (2014) Bioethanol production from the hydrolysate of Palmaria palmata using sulfuric acid and fermentation with brewer's yeast. J Appl Phycol 26:687–693
29. Noseda MD, Tulio S, Duarte ME (1999) Polysaccharides from the red seaweed Bostrychia montagnei: chemical characterization. J

Appl Phycol 11:35–40

30. Passoth V, Tabassum MR, Nair HA, Olstorpe M, Tuikova I, Stahlberg J (2013) Enhanced ethanol production from wheat straw by integrated storage and pre-treatment (ISP). Enzyme Microb Technol 52:105–110

31. Powell CD, Quain DE, Smart KA (2003) The impact of brewing yeast cell age on fermentation performance, attenuation and flocculation. FEMS Yeast Res 3:149–157

32. Quain DE, Box WG, Walton EF (1985) An inexpensive and simple small-scale laboratory fermenter. Lab Pract 34:84, CAS

33. Roesijadi G, Jones S, Snowden-Swan L, Zhu Y (2010) Macroalgae as a biomass feedstock: a preliminary analysis, PNNL 19944. Pacific Northwest National Laboratory, Richland

34. Ruyters S, Mukherjee V, Verstrepen KJ, Thielevein JM, Willems KA, Lievens B (2015) Assessing the potential of wild yeasts for bioethanol production. J Ind Microbiol Biotechnol 42:39–48

35. Santos M, Jiménez J, Bartolomé B, Gómez-Cordovés C, del Nozal MJ (2003) Variability of brewer's spent grain within a brewery. Food Chem 80:17–21

36. Singh A, Nigam PS, Murphy JD (2011) Renewable fuels from algae: an answer to debatable land based fuels. Bioresour Technol 102:10–16

37. Taherzadeh MJ, Karimi K (2008) Pretreatment of lignocellulosic wastes to improve ethanol and biogas production: a review. Int J Mol Sci 9:1621–1651

38. Tao N, Gao Y, Liu Y (2011) Isolation and characterization of a Pichia anomala strain: a promising candidate for bioethanol production. Braz J Microbiol 42:668–675

39. Wargacki AJ, Leonard E, Win MN, Regitsky DD, Santos CNS, Kim PB, Cooper SR, Raisner RM, Herman A, Sivitz AB, Lakshamanaswamy A, Kashiyama Y, Baker D, Yoshikuni Y (2012) An engineered microbial platform for direct biofuel production from brown macroalgae. Science 335:308–313

40. Wilkinson M (1995) Information review on the impact of kelp harvesting. Scottish Natural Heritage

41. Wimalasena TT, Greetham D, Marvin ME, Liti G, Chandelia Y, Hart AJ, Louis EJ, Phister TG, Tucker GA, Smart KA (2014) Phenotypic characterisation of Saccharomyces spp. yeast for tolerance to stresses encountered during fermentation of lignocellulosic residues to produce bioethanol. Microb Cell Fact 13:47
42. Yanagisawa M, Nakamura K, Ariga O, Nakasaki K (2011) Production of high concentrations of bioethanol from seaweeds that contain easily hydrolyzable polysaccharides. Process Biochem 46:2111–2116
43. Yanik J, Stahl R, Troeger N, Sinag A (2013) Pyrolysis of algal biomass. J Anal Appl Pyrolysis 103:134–141
44. Zhao X, Bai F (2009) Mechanisms of yeast stress tolerance and its manipulation for efficient fuel ethanol production. J Biotechnol 144:23–30

CHAPTER

19

BIOREFINERY OF THE GREEN SEAWEED ULVA LACTUCA TO PRODUCE ANIMAL FEED, CHEMICALS AND BIOFUELS

Paul Bikker[1], Marinus M. van Krimpen[1], Piet van Wikselaar[1], Bwee Houweling-Tan[2], Nazareno Scaccia[2], Jaap W. van Hal[3], Wouter J. J. Huijgen[3], John W. Cone[4], and Ana M. López-Contreras[2]

[1]Livestock Research, Wageningen University and Research Centre, De Elst 1, 6708 WD Wageningen, The Netherlands

[2]Food and Biobased Research, Wageningen University and Research Centre, Bornse Weilanden 9, 6708 WG Wageningen, The Netherlands

[3]Biomass & Energy Efficiency, Energy research Centre of the Netherlands (ECN), Westerduinweg 3, 1755 LE Petten, The Netherlands

[4]Animal Nutrition Group, Wageningen University, De Elst 1, 6708 WD Wageningen, The Netherlands

ABSTRACT

The growing world population demands an increase in animal protein production. Seaweed may be a valuable source of protein for animal feed. However, a biorefinery approach aimed at cascading valorization of both protein and non-protein seaweed constituents is required to realise an economically feasible value chain. In this study, such a biorefinery approach is presented for the green seaweed *Ulva lactuca* containing 225 g protein ($N \times 4.6$) kg^{-1} dry matter (DM). The sugars in the biomass were solubilized by hot water treatment followed by enzymatic hydrolysis and centrifugation resulting in a sugar-rich hydrolysate (38.8 g L^{-1}sugars) containing glucose, rhamnose and xylose, and a protein-enriched (343 g kg^{-1} in DM) extracted fraction. This extracted fraction was characterized for use in animal feed, as compared to *U. lactuca* biomass. Based on the content of essential amino acids and the in vitro N (85 %) and organic matter (90 %) digestibility, the extracted fraction seems a promising protein source in diets for monogastric animals with improved characteristics as compared to the intact *U. lactuca*. The gas production test indicated a moderate rumen fermentation of *U. lactuca* and the extracted fraction, about similar to that of alfalfa. Reduction of the high content of minerals and trace elements may be required to allow a high inclusion level of *U. lactuca*products in animal diets. The hydrolysate was used successfully for the production of acetone, butanol, ethanol and 1,2-propanediol by clostridial fermentation, and the rhamnose fermentation pattern was studied.

KEYWORDS: Seaweed, *Ulva lactuca*, Animal feed, In vitro digestibility, Biobutanol, Cascading biorefinery, ABE fermentation

INTRODUCTION

The expected increase in the world population and in standards of living in developing countries is expected to create an increasing demand for animal-derived protein (FAO 2006). New initiatives, including the use of novel protein sources for both human and animal nutrition, are required to produce a sufficient amount of high-quality human edible protein (Boland et al. 2013). Intact seaweed as well as seaweed components are considered potential novel protein sources for animal nutrition (Holdt and Kraan 2011). Seaweeds offer advantages compared to traditional terrestrial feed materials, including higher productivity (biomass produced per unit of surface), no competition for arable land and lower fresh water consumption (Van den Burg et al. 2013). Among seaweed species available in European temperate Atlantic waters, *Ulva* spp. have been extensively characterized (Fleurence et al. 1995), showing a high crude protein content, up to 44 % of dry matter (DM) (Holdt and Kraan 2011). Biomass from *Ulva* spp. is extensively available since it represents the main seaweed in mass of algal growth (green tides), causing negative effects on the environment and tourism at coastal areas, that necessitate harvesting (Briand and Morand 1997). In addition, *Ulva* spp. are cultivated successfully in integrated multi-trophic aquaculture (IMTA) systems enabling scalable controlled cultivation conditions (Marinho et al. 2013; Robertson-Andersson et al. 2008) and removal of excess nutrients from N- and P-rich wastewater from land-based aquaculture (Lawton et al. 2013). In terms of protein supply, therefore, *Ulva* spp. are promising for further assessment of their potential application in animal nutrition. In this assessment, amino acid composition and protein digestibility are important parameters. On the other hand, seaweeds may contain factors limiting their use, e.g. high levels of minerals as sodium, potassium and chloride, and heavy metals. These may affect animal performance and health, and food safety and need to be addressed (Ventura et al. 1994; Gardiner et al. 2008; Moroney et al. 2012; Makkar et al. 2016).

Based on current information on the costs and benefits, offshore seaweed production in the North Sea, primarily for use as animal feed ingredient, is not economically feasible (Van den Burg et al. 2016). In order to develop an economically feasible seaweed value chain, a cascading biorefinery approach aimed at valorization of both protein and non-protein seaweed constituents has been proposed, including use as nutrient source in animal feed and for production of biofuels (van Hal et al. 2014; Van den Burg et al. 2016). Until now, only a few studies of such biorefinery processes have been reported, including recent studies using *Gracilaria* species (Francavilla et al. 2013, 2014). However, consequences for the nutritive value of residues for inclusion in animal diets have not been addressed and require further attention.

Ulva spp. contain a significant amount of polysaccharides, varying from 15 to 65 % of the total DM (Kraan 2013). These polysaccharides include ulvans, sulphated polysaccharides with rhamnose, uronic acids and xylose as major components, as well as glucans including starch. In previous studies, the use of polysaccharides from *Ulva* spp. as feedstock for the production of acetone, butanol and ethanol (ABE) and 1,2-propanediol (1,2-PD) by fermentation has been described (Potts et al. 2012; van der Wal et al. 2013). However, the efficiency of conversion of different sugars in the hydrolysate, especially rhamnose, into ABE and 1,2-PD was not addressed. Moreover, no attention was paid to the valorisation of the residue fractions. Therefore, the aims of the present study were to fractionate the green seaweed *Ulva lactuca*using aqueous pre-treatment followed by enzymatic hydrolysis, evaluate the potential of the liquid fraction for fermentative production of chemicals and fuels by *Clostridium beijerinckii* in comparison with several control media, and to evaluate the solid extracted fraction for animal feed in comparison to intact *U. lactuca*. The fermentability of the hydrolysate to ABE and 1,2-PD was directly related to the sugar composition of the hydrolysate and its nutrient content. Moreover, we hypothesised that the fractionation would improve the nutritional value of the extracted fraction, making it more suitable for feed application than intact *U. lactuca*.

MATERIALS AND METHODS

To valorize the protein and sugars in *U. lactuca*, a cascading biorefinery scheme was developed in which the sugars are solubilized and fermented to biofuels and platform chemicals, and the protein-rich extracted fraction is evaluated as animal feed ingredient in comparison to intact *U. lactuca* (Fig. 1).

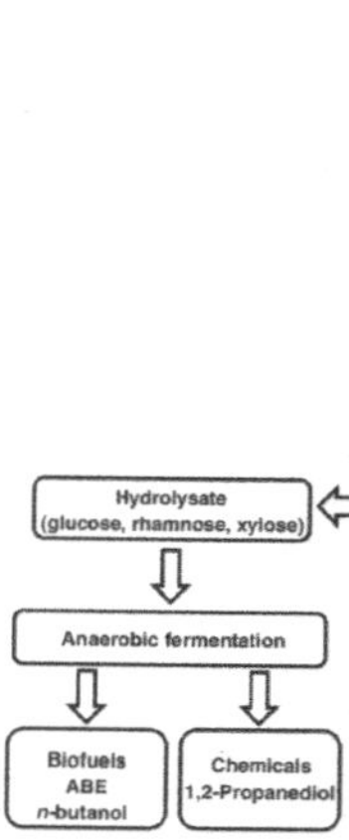

Figure 1. Cascading biorefinery for the green seaweed *U. lactuca*.

Source and Storage of *U. Lactuca*

The biomass was harvested in May at the Irish coast near Galway. Directly after harvesting, the biomass was washed with tap water, freeze-dried and stored at room temperature (RT).

Pre-treatment and Hydrolysis Conditions

The pre-treatment and hydrolysis of the seaweed biomass was performed according to van der Wal et al. (2013) with some modifications, as follows: *U. lactuca* was milled at 2 mm particle size and 139 g of material was suspended at 20 % (*w/v*) in demineralized water. The suspension was subjected to thermal treatment at 150 °C for 10 min, followed by enzymatic hydrolysis with a commercial cellulase cocktail (GC220; Dupont Industrial Biosciences) for 24 h at 50 °C with continuous stirring. The cellulase cocktail was added at 0.3 mL g^{-1} DM of *U. lactuca*. The supernatant obtained (560 mL) was recovered after centrifugation at 10,000 × *g* for 15 min and stored at −20 °C until further use. The insoluble pellet (further referred to as extracted fraction) was freeze-dried and stored at RT.

Analysis of the Biomass

Freeze-dried *Ulva* was milled over a 2-mm screen. Solvent and water extractives were determined by successive extraction with toluene/ethanol (2:1 *v/v*), 95 % ethanol and boiling water. Sugar content was determined in the *Ulva* samples by hydrolysis of the milled *Ulva* with 12 M H_2SO_4 at 30 °C for 1 h, followed by dilution to 1 M H_2SO_4 and hydrolysis for 3 h at 100 °C. The hydrolysate was neutralised and analysed for neutral sugars using high-performance anion-exchange chromatography (HPAEC, ICS3000; Dionex, USA) equipped with a CarboPac PA1 (250 × 4 mm) column with a CarboPac PA1 (50 × 4 mm) guard column (Dionex), and pulsed amperometric detection, using fucose as an internal standard (no fucose was found in *Ulva* samples), as described previously (van den Oever et al. 2003).

Nutrient Determination in *U. Lactuca* and Extracted Fraction

Prior to analyses, *U. lactuca* and extracted fraction were ground using a laboratory mill (Peppink 200 AN, Netherlands) equipped with a 1-mm sieve. The samples were analysed using official methods described to determine moisture (DM), nitrogen (crude protein), ether extract, ash, crude fibre, starch, total sugar, free sugars, fatty acids, neutral detergent fiber (NDF), acid detergent fiber (ADF), acid detergent lignin (ADL), minerals and amino acids. Briefly, DM was determined by drying to a constant weight at 103 °C (ISO 6496 1999). The N content was measured using the Kjeldahl method with $CuSO_4$ as catalyst (ISO 5983 2005). Ether extract (crude fat) was determined gravimetrically after hydrolysis with HCl and extraction with light petroleum (boiling point 40–60 °C) (ISO 6496 1999). Crude ash was determined by combustion to a constant weight at 550 °C (ISO 5984 2002). The part of the crude ash that was not soluble in HCl was determined as Ash-HCl. Neutral detergent fiber, ADF and ADL were analyzed after a treatment with acid detergent reagent followed by gravimetric ash procedure (ISO 13906 2008). Starch was enzymatically determined (ISO 15914 2004). Samples were first extracted with 40 % ethanol to remove non-structural sugars, followed by a two-step hydrolysis with DMSO at 100 °C and concentrated HCl at 60 °C. Starch was quantitatively converted into glucose by amyloglucosidase and spectrometrically measured at 340 nm using the hexokinase method. Total sugars were extracted in dilute ethanol and determined after inversion with the Luff Schoorl method (EC 159/2009 2009). Amino acids were analysed after hydrolysis with 6 M HCl for 23 h (ISO 13903 2005). Tryptophan was analysed after hydrolysis with BaOH by HPLC fluorimetric detection (ISO 13904 2005). Macro and micro minerals (P, Ca, Cu, Fe, Mg, Zn, K, Mn, Na, S) were analysed by ICP-OES (Optima 7300 DV; PerkinElmer, USA) after acid digestion (NEN-EN 15510 2007) and micro minerals (Cd, Pb, Ni, As, Hg, Co, Se) were analysed by ICP-MS (NexION 300D; PerkinElmer, USA) after acid digestion (NEN-EN 15763 2010 and NEN-EN 14627 2010). Chloride was determined by potentiometric analysis (ISO 6496 1999). The fatty acid composition was analysed by gas chromatography (Agilent 6890N, USA; NPR-CEN-ISO/TS 17764-1&2: 2006) and sugars were quantified by HPAEC after complete hydrolysis, as described above.

In Vitro Digestion

In vitro incubations were performed according to a modified Boisen two- and three-step method (Boisen and Fernandez 1997). The three-step in vitro incubation simulated the digestive process in the stomach, small intestine and large intestine of a pig and estimated the total tract digestibility. For the three-step in vitro incubation, substrates (1 g) were incubated in beakers with 75 mL of a 0.1 M phosphate buffer solution ($Na_2HPO_4{\cdot}2H_2O$ 0.99 g L^{-1} and $NaH_2PO_4{\cdot}2H_2O$ 14.72 g L^{-1}; pH 6.0) and a HCl solution (30 mL, 0.2 M). The pH was adjusted to 2.0 with 1 M HCl or 10 M NaOH. Fresh pepsin solution (1 mL, 25 g L^{-1}, porcine pepsin 2000 FIP U g^{-1}; Sigma P7000) was added and each beaker was covered with a glaze and placed in an incubator (Marius Instrumenten, the Netherlands, type 90A) at 39 °C for 2 h under constan stirring. Subsequently, 30 mL of a 0.2 M phosphate buffer ($Na_2HPO_4{\cdot}2H_2O$ 4.83 g L^{-1} and $NaH_2PO_4{\cdot}2H_2O$ 11.37 g L^{-1}; pH 6.8) and 12 mL of a 0.6 M NaOH solution were added. The pH was adjusted to 6.8 with 1 M HCl or 10 M NaOH. Fresh pancreatin solution (1 mL, 100 g L^{-1}pancreatin, Porcine pancreas grade VI; SigmaP-1750) was added and hydrolysis was continued for 4 h under the same conditions. Then 30 mL of a 0.2 M EDTA solution was added and the pH adjusted to 4.8 with 30 % acetic acid. After that, 0.5 mL of a mix of cell wall degrading enzymes (Viscozyme; Sigma V2010) was added. Hydrolysis was continued for another 18 h under the same conditions. A two-step in vitro incubation was performed without the 18-h incubation with Viscozyme to simulate the digestive process in the stomach and small intestine and estimate the ileal digestibility of the substrates. After hydrolysis, the residues were collected by filtration of the slurries on a nylon gauze (37 μm) folded in a Būchner porcelain funnel. The sample was washed twice by acetone (99.5 %) followed by ethanol (96 %). Then the cloth with the residue was temporarily placed on a clean paper to evaporate the remaining ethanol/acetone overnight. The residue was scraped off the nylon cloth and collected in a pre-weighed jar. The two-step in vitro incubations were conducted 4-fold, of which two replicates were used to determine N-digestibility and two replicates to determine DM and OM digestibility. The three-step in vitro incubations were conducted in duplicate to determine DM and OM digestibility.

To investigate potential fermentation in the rumen, gas production on the *U. lactuca* samples was measured after incubation in rumen fluid as described by Cone et al. (1996). The rumen fluid was obtained from dairy cows kept on a ration of maize and grass silage. Rumen fluid was taken 2 h after the morning feeding and collected in a warm insulated flask filled with CO_2. Rumen fluid was filtered through cheese cloth and mixed (1:2 *v/v*) with an anaerobic buffer/mineral solution containing per liter 8.75 g $NaHCO_3$, 1.00 g NH_4HCO_3, 1.43 g Na_2HPO_4, 1.55 g KH_2PO_4, 0.15 g $MgSO_4 \cdot 7H_2O$, 0.52 g Na_2S, 0.017 g $CaCl_2 \cdot 2H_2O$, 0.015 g $MnCl_2 \cdot 4H_2O$, 0.002 g $CoCl_2 \cdot 6H_2O$, 0.012 g $FeCl_3 \cdot 6H_2O$ and 1.25 mg resazurin. The residue on the muslin was discarded. All manipulations were done under continuous flushing with CO_2.

Fermentations were conducted in 250-mL serum bottles in which 400 mg organic matter was incubated in 60 mL of buffered rumen fluid saturated with CO_2. The bottles were placed in a shaking water bath with 50 rpm at 39 °C. Each sample was incubated in triplicate with a blank run (rumen fluid without sample) in duplicate in each of the three series. To compare the potential fermentation of *Ulva* with known feedstuffs, the gas production was measured for palm kernel expeller, sugar beet pulp, alfalfa meal and grass silage.

Fermentation by *Clostridium Beijerinckii* and Analysis of Metabolites

The laboratory strain *Clostridium beijerinckii* NCIMB 8052 was stored as spore suspension and cultivated as previously described (López-Contreras et al. 2000). For the preparation of pre-cultures, spores were heat-shocked and placed into CM2 medium, composed per litre of 2.5 g yeast extract, 1 g KH_2PO_4, 0.85 g $K_2HPO_4 \cdot 3H_2O$, 2.9 g NH_4Ac, 0.1 g *p*-aminobenzoic acid, 1 g $MgSO_4 \cdot 7H_2O$ and 6.6 mg $FeSO_4 \cdot 7H_2O$. Cultures were prepared under anaerobic conditions in serum flasks, with culture volumes of 30 mL, and incubated at 37 °C without shaking. As carbon sources, stock solutions of glucose, xylose, rhamnose or mixes of these were prepared and sterilised separately and added to the medium at the following concentrations: 42.2 g glucose L^{-1} for culture CM2-G, 39.7 g rhamnose L^{-1} for culture CM2-R and 23.3 g glucose L^{-1}, 13.8 rhamnose L^{-1} and 5.2 xylose L^{-1} for the CM2-G/R/X culture. The hydrolysate-based cultures (H) contained 15.4 g glucose L^{-1}, 11.5 g rhamnose L^{-1} and 1.8 g xylose L^{-1}. The sugar concentrations in the control cultures are standard concentrations used at our laboratory, at which fermentation by *C. beijerinckii* is optimal, approximately 40 g L^{-1}. In the CM2-G/R/X cultures, the ratio of sugars in the hydrolysate (culture H) was mimicked.

Sugars and fermentation products were determined in clear culture supernatants from samples taken during the growth experiments and stored at −20 °C. Organic acids, solvents and sugars were analyzed by high-performance liquid chromatography (HPLC) as previously described (van der Wal et al. 2013). Separation of propionic acid and 1,2-propanediol was performed using Dionex RSLC equipment (Dionex Corporation, USA) consisting of an Ultimate 3000 RS (Rapid Separation) pump and an Ultimate 3000 autosampler, a refractive index detector (Waters model 2414) and an UV absorbance detector (Waters model 2487). The separation was carried out using a Bio-Rad Aminex HPX-87H column at 30 °C using an isocratic run of 60 min with an eluent flow rate of 0.6 mL min^{-1}. As internal standard, valeric acid at 100 mM was used.

RESULTS

Ulva Lactuca Composition and Fractionation

The main components in the *U. lactuca* biomass were sugars, ash and protein (Tables 1, 2 and 3). The total sugar content of the *U. lactuca* sample was approximately 24 % of DM and consisted mainly of glucose, rhamnose and xylose (Table 1). The latter two are the main constituents of ulvan (Ray and Lahaye 1995). The protein (amino acid) content of the *U. lactuca* sample (Table 3) was 26.3 %.

Table 1. Starch, total monomeric sugar and fibre content (% of DM) of *U. lactuca* and extracted fraction

	Starch	Rha	Gal	Glc	Xyl	**NDF**	**ADF**	**ADL**
Ulva lactuca	4.2	9.0	0.7	11.3	2.9	25.9	13.5	6.9
Extracted fraction	0.3	1.7	0.2	3.4	0.5	20.3	17.9	10.6

For *U. lactuca* and the extracted fraction, fructose, saccharose, lactose, raffinose, stachyose, maltose, verbascose and maltotriose were below the detection limit of 0.1 % DM

Rha rhamnose, *Gal* galactose, *Glc* glucose, *Xyl* xylose, *NDF* neutral detergent fibre, *ADF* acid detergent fibre, *ADL* acid detergent lignin

Table 2. Content of crude ash, minerals and trace elements (kg^{-1} DM) of *U. lactuca* and extracted fraction

Element	***U. lactuca***	**Extracted fraction**	**Soybean meal**[a]
Ash (g)	173	160	65
Macro minerals			
P (g)	2.56	1.79	7.3
Ca (g)	20.3	26.5	3.2
K (g)	11.5	6.8	25.2
Mg (g)	24.2	14.5	3.4
Na (g)	10.7	10.6	0.2
Cl (g)	9.6	11.8	0.3
S (g)	50.5	27.0	4.1
Trace elements			
Cu (mg)	22	41	17
Fe (mg)	353	658	270
Mn (mg)	86	74	46
Zn (mg)	17	39	55
Ni (mg)	8.5	12.0	n.a.
As (mg)	5.8	7.3	n.a.
Co (μg)	271	503	344
Se (μg)	<100	109	n.a.
Cd (μg)	257	411	n.a.
Pb (μg)	956	1825	n.a.
Hg (μg)	<10	19	n.a.

n.a. not available

[a]CVB (2007)

Table 3. Nitrogen and total amino acid (AA) content (g kg^{-1} DM), individual AA content (g (100 g^{-1}) of total AA) and N to protein conversion factor of *Ulva Lactuca* and extract fraction, compared to literature and soybean meal

Amino acids	*U. lactuca*	Extracted fraction	*U. lactuca*, literature[a]	Soybean meal[a]
Lysine	4.6	3.3	4.7–7.4	6.3
Methionine	2.2	2.2	1.1–5.5	1.4
Cysteine	1.1	1.0	0.5–2.2	1.5
Threonine	4.7	5.9	4.6–6.5	3.9
Tryptophan	0.7	1.0	–	1.3
Leucine	7.4	8.6	7.5–9.2	7.8
Isoleucine	4.0	4.6	4.0–6.1	4.7
Histidine	1.0	1.1	0.5–2.8	2.7
Phenylalanine	5.4	6.4	2.5–10.2	5.3
Tyrosine	3.6	4.6	3.6–5.5	3.7
Arginine	7.6	5.4	4.0–9.5	7.6
Asparagine + aspartic acid	13.8	11.8	9.2–12.2	11.7
Serine	4.9	5.9	4.0–6.8	5.2
Glutamic + glutamic acid	13.2	11.6	9.8–13.0	18.0
Glycine	5.9	6.3	6.1–7.5	4.4
Alanine	7.8	8.2	8.0–9.0	4.5
Valine	5.8	6.6	2.8–8.5	4.9
Hydroxyproline	0.7	0.4	–	–
Proline	5.5	4.9	3.8–7.0	5.2
Essential AA[b]	40.5	45.5	42.7–50.9	43.5
Total AA (g kg^{-1} DM)	262.7	401.3	–	459.9
AA-N (g kg^{-1} DM)[c]	39.5	57.6	–	–
Total N (g kg^{-1} DM)	48.7	72.7	11.3–43.5	–
Non-protein N, % of total N	18.8	20.7	–	–
N-Protein factor, K_P[d]	4.62	4.72	–	-
N-Protein factor, K_A[d]	5.69	5.95	–	–

[a]Mai et al. (Mai et al. 1994) (Ireland, no time of harvest indicated), Wong et al. (Wong and Cheung 2000) (Hong Kong, December harvest), Ortiz et al. (Ortiz et al. 2006) (Chile, November harvest), Yaich et al. (Yaich et al. 2011) (Tunisia, July harvest) and Tabarsa et al. (Tabarsa et al. 2012) (Persian Gulf, April harvest). Soybean meal based on CVB (2007)

[b]Essential amino acids for monogastric species: lysine,

methionine + cysteine, threonine, tryptophan, leucine, isoleucine, histidine, phenylalanine + tyrosine and valine

[c]Based on N content of each individual amino acid (Sosulski and Imafidon 1990)

[d]N to protein conversion factor, K_P as ratio between sum of anhydrous AA residues and total N, K_A as ratio between sum of anhydrous AA residues and N recovered from AA residues (AA-N) as described by Mariotti et al. (2008)

The ash content of *U. lactuca* was 17.3 % of the DM (Table 2). Sulphur is an important component in the ash and it is mostly derived from the sulphated polysaccharide ulvan. The S content in the original seaweed biomass was higher than in the extracted fraction, indicating that a large part of the ulvan polymer has been preferentially solubilised during the pre-treatment and enzymatic hydrolysis of the biomass.

Sugar extraction with aqueous treatment at 150 °C for 10 min released 59.6 % of total sugars in *U. lactuca*. Glucose and rhamnose were the main components of the carbohydrate fraction present in this extract, with a small amount of xylose. Enzymatic hydrolysis was carried out to improve release of monomeric sugars. The percentage of individual sugars solubilized after enzymatic treatment was 78.4, 53.7 and 66.7 % of glucose, rhamnose and xylose, respectively. These percentages correspond to 22, 12 and 4.8 g L^{-1} of glucose, rhamnose and xylose, respectively, present in the hydrolysate.

Results in Table 3 show that the solubilization of the polysaccharides in the biomass resulted in a solid fraction (extracted fraction) enriched in total protein, with a content of 40.1 % of DM.

Evaluation of *U. Lactuca* and Extracted Fraction as Animal Feed Ingredient

Composition

The contents of starch, NDF and sugars after hydrolysis were higher in *U. lactuca* compared to the extracted fraction, whereas the contents of ADF and ADL were higher in the extracted fraction (Table 1). The contents of minerals and trace elements, including heavy metals, in *U. lactuca* and extracted fraction are shown in Table 2. Contents in soybean meal, being the most important protein source in diets of monogastric farm animals, have been included for comparison. The most abundant macro minerals in *U. lactuca* were S, Mg and Ca, with lower contents of K, Na and Cl. In the extracted fraction, the content of K, Mg and S was up to 50 % less than in *U. lactuca*. The content of trace elements was up to two times higher in the extracted fraction compared to the intact *Ulva*.

The amino acid pattern of *U. lactuca* and extracted fraction, the N content and the calculated N to protein conversion factor are included in Table 3, in comparison with data published in the literature and the amino acid contents of soybean meal as a major protein source in feed for pigs and poultry.

The amino acid profile of the extracted fraction differed somewhat from that of the dried *U. lactuca*, with a higher content of essential amino acids, apart from lysine, and a lower content of arginine, glutamine plus glutamic acid and asparagine plus aspartic acid.

The total N content in the extracted fraction was higher than in the intact *U. lactuca* due to partial removal of the carbohydrate fraction. The N to protein conversion factor (K_A) in the samples used in this study was 5.7–6.0 for amino acid nitrogen to protein, reflecting the mean N content of the amino acids. The conversion factor for total N to protein (K_P) was 4.6–4.7, largely due to the presence of approximately 20 % of N as non-amino acid N.

The fatty acid profile of *U. lactuca* in the current study mainly consisted of saturated fatty acids (SFA, 46.9 %), with lower proportions of mono (MUFA) and poly (PUFA) unsaturated fatty acids (19.4 and 25.1 %, respectively) of the total fatty acid profile (Table 4). The most abundant fatty acids were C16:0 and C18:1. The proportion of PUFA was relatively high in the *U. lactuca*but lower in the extracted fraction.

Table 4. Fatty acid pattern (g (100 g^{-1}) FA[a]) of *U. lactuca* and extracted fraction

Fatty acids		*U. lactuca*	Extracted fraction	*U. lactuca*,literature[b]	Soybean meal[a]
C14:0	Myristic acid	0.5	0.9	1.1–5.5	0.2
C16:0	Palmitic acid	39.8	51.9	14.0–59.4	11.0
Iso-C16:0	Iso-Palmitic acid	3.8	4.4	na	na
C16:1n7	Palmitoleic acid	0.9	1.2	0.7–6.9	0.2
C18:0	Stearic acid	0.5	0.6	1.9–8.4	4.0
C18:1n9	Oleic acid	1.4	1.5	2.6–27.8	22.0
C18:1n	Other isomers	17.1	20.4	na	na
C18:2n6	Linoleic acid	2.4	1.5	2.4–8.3	54.0
C18:2	Trans isomers	5.7	2.9	1.7	na
C18:3n3	α-Linolenic acid	10.9	5.5	2.8-4.4	8.0
C18:4n3	Stearidonic acid	10.4	4.1	0.4	na
C20:1	Eicosenoic acid	<0.3	<0.3	1.5–4.2	na
C20:5n3	Eicosapentaenoic acid	1.4	0.3	1.0–5.0	na
C22:0	Behenic acid	2.4	2.6	0.3–4.2	na
C24:0	Lignoceric acid	<0.3	<0.3	9.5	na
Not identified FA		8.5	5.5	11.3	0.4
SFA[a]		46.9	60.3	33.8–69.0	15.2
MUFA[a]		19.4	23.0	5.1–36.7	22.2
PUFA[a,c]		25.1	11.1	6.7–24.8	62.0
Total FA (g kg^{-1} DM)		21.1	34.3		12.5

[a]*FA* fatty acid, *SFA* saturated fatty acids, *MUFA* monounsaturated fatty acids, *PUFA*polyunsaturated fatty acids

[b]Data from Ortiz et al. (2006), Yaich et al. (2011) and Tabarsa et al. (2012); *na* not available, if one value is reported, only one reference was available for this fatty acid

[c]C18:2 trans isomers not included

In Vitro Degradation

The in vitro simulated ileal OM and N digestibility and total tract OM digestibility are presented in Table 5 in comparison to that of soybean meal. The simulated ileal OM digestibility of *Ulva* extracted fraction of 86.9 % was 20 percentage points higher than intact *U. lactuca* and slightly higher than that of soybean meal (Table 5). The N digestibility of *Ulva* extracted fraction was 5 % points higher than for *U. lactuca* and much lower than soybean meal. The simulated total tract OM digestibility was 7 percentage points different (not significant) between the two *Ulva* products and 8–15 percentage points lower than soybean meal.

Table 5. In vitro simulated ileal and total tract dry matter (DM), organic matter (OM) and nitrogen (N) digestibility (%) in *U. lactuca* and extracted fraction, as compared to soybean meal as reference feed stuff

	Ileal		**Total tract**
	OM	*N*	**OM**
Ulva lactuca	67.2a	79.9a	82.8a
Ulva extracted fraction	86.9c	84.7b	90.1a
Soybean meal	84.2b	98.0c	98.5b
SEM	0.61	0.79	1.67
P value	<0.001	0.001	0.016

Results in columns without the same letter are significantly different ($P<0.05$). SEM, Pooled standard error of mean

Gas Production Test

The results of the in vitro rumen fermentation experiment using the gas production technique (Cone et al. 1996) are shown in Fig. 2 and Table

6, showing that the maximum cumulative gas production of the *Ulva* products was below that of the reference feed materials. In addition, the rate of gas production was relatively slow, as indicated by the higher time to reach half of the maximum gas production (T_{half}) and the less steep slope (shape) compared to other products. No significant differences between *U. lactuca* and the extracted fraction were observed.

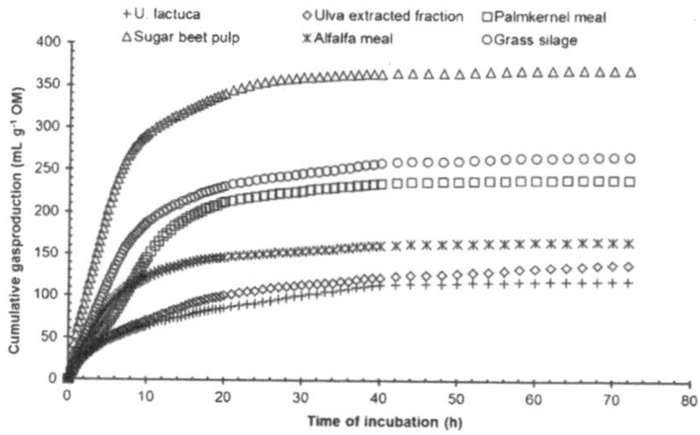

Figure 2. Cumulative gas production of *U. lactuca*, *U. lactuca* extracted fraction and other feed ingredients for ruminants.

Table 6. Calculated asymptotic maximum gas production (Max.), time at which half of this gas production (T_{half}) was reached and parameter determining the shape of the curve (Shape) for different substrates in the gas production test

	Max. (mL g^{-1} OM)	T_{half} (h)	Shape	CP, % of OM[a]	NDF, % of OM	Corr. max. (mL g^{-1} OM)[a]
U. lactuca	150a	10.6c	0.98a	27	31	218
Extracted fraction	172a	11.0c	1.02a	41	24	274
Palm kernel meal	248b	8.3bc	1.85c	16	67	287
Sugar beet pulp	380c	4.5a	1.49b	9	39	404
Alfalfa	174a	5.3ab	1.32ab	21	48	228
Grass silage	290b	7.0ab	1.38b	22	51	345
SEM	21.6	1.22	0.12	–		–
P value	<0.001	0.008	0.002	–		–

Results in columns without the same letter are significantly different ($P<0.05$)

[a]Maximum gas production corrected with an increase of 2.5 mL per percent crude protein (CP) in organic matter (OM) of the feed materials used as substrate, according to Cone and van Gelder (1999). For *U. lactuca* and extracted fraction, the N to protein conversion factor of 4.62 and 4.72, respectively, was used (Table 4)

Fermentation of *U. Lactuca* Hydrolysate to Acetone, Butanol, Ethanol and 1,2-Propanediol

Clostridium beijerinckii produced butanol as the major end-product on the media containing glucose or a mixture of glucose and rhamnose as carbon sources (Table 7). The end concentration of ABE produced in these cultures was similar, 10.8 g L^{-1} ABE vs. 9.8 g L^{-1} ABE on glucose and on the glucose/rhamnose/xylose mixture, respectively.

In the rhamnose-only grown cultures (culture CM2-R in Table 7), the major end-product was 1,2-propanediol. A small amount of acetone, butanol and propionic acid (<1 g L^{-1}, not shown) was produced in these cultures, indicating that not all rhamnose was converted into 1,2-propanediol.

In the hydrolysate-based cultures (Table 7, culture H), almost all glucose and xylose were consumed. However, the rhamnose utilisation was poor, and only 2.8 g L^{-1} (approximately 25 % of the initial amount) was consumed. The yields of 1,2-PD produced from rhamnose in the hydrolysate are similar to those observed in the control cultures. The hydrolysate without further addition of nutrients supported ABE production up to 7.5 g L^{-1} ABE, similar levels as would be expected in control cultures with the same sugar content.

Table 7. Fermentation of control media and *U. lactuca* hydrolysate by *C. beijerinckii*

Culture	Sugars consumed (g L^{-1})			Products (g L^{-1})						Yields	
	Glucose	**Rhamnose**	**Xylose**	**Acetone**	**Butanol**	**Ethanol**	**1,2-PD**	**Acetic acid**[a]	**Butyric acid**	**g ABE g^{-1}total sugars consumed**	**g 1,2-PD g^{-1}rhamnose consumed**
CM2-G	32.3	–	–	2	8.5	0.3	–	0.8	0.4	0.3	–
CM2-R	–	7.5	–	0.5	0.2	0.1	2	2.1	1.1	0.1	0.3
CM2-G/R/X	20	7.3	4.5	2.1	7.5	0.2	2.8	0.6	0.3	0.3	0.4
Hydrolysate	14.7	2.8	1.6	2.1	5	0.4	1	<0.1	0.3	0.4	0.3

The sugar concentration at the start of the fermentation was 42.2 g glucose L^{-1} for CM2-G, 39.7 g L^{-1} rhamnose for CM2-R and 23.3 g glucose L^{-1}, 13.8 g rhamnose L^{-1} and 5.2 g xylose L^{-1} for CM2-G/R/X cultures. The hydrolysate-based cultures contained 15.4 g glucose L^{-1}, 11.5 g rhamnose L^{-1} and 1.8 g xylose L^{-1}. The data correspond to $t = 72$ h of fermentation for the CM2 cultures and to $t = 148$ h of fermentation for the hydrolysate cultures

[a]Acetic acid is consumed in the CM2 cultures

DISCUSSION

Biomass Composition and Fractionation

The composition of the *U. lactuca* biomass corresponds well with other literature data on this species. In general, the sugar content of green seaweeds is lower than that in other seaweeds and in other biomass types, such as lignocellulosic feedstocks, in which sugars may account for up to 70 % of the DM of the biomass (Sorek et al. 2014). It is well known that the chemical composition of seaweeds shows a strong variation influenced by the season and the growth location. Several studies on *Ulva* sp. harvested at the coast of Brittany (France) showed that the content of sugars and total carbohydrates decreased while the protein content increased from spring to autumn (Briand and Morand 1997; Robic et al. 2009). Abdel-Fattah and Edrees (1973) showed that for *U. lactuca* harvested at the Mediterranean coast, rhamnose varied from 1.5 % (November) to 28 % (April), while total proteins ranged from 8.7 % (April) to 33.8 % (August). These studies suggest that *Ulva* biomass harvested in spring, as used in this study, is relatively rich in carbohydrates compared to protein content. However, it is difficult to draw general conclusions on the influence of the time of harvest on the seaweed composition. Therefore, future macroalgae biorefineries need to have the flexibility to deal with variations caused by species, cultivation and harvesting conditions.

After pre-treatment and hydrolysis, the total concentration of monosaccharides in the *U. lactuca* hydrolysate reached 38.8 g L^{-1}, making it suitable for fermentation purposes. This concentration of sugars in the hydrolysate was higher than that reported by van der Wal et al. (2013), most probably due to the highedry matter content of biomass used for the pre-treatment (20 vs. 10 %).

Use of *Ulva* and Extracted Fraction as Animal Feed Ingredient

Minerals and Trace Elements

Ulva lactuca and the extracted fraction contained high amounts of minerals compared to soybean meal. In the extracted fraction, the content of K, Mg and S was lower, whereas the content of trace elements was up to two times higher compared to the intact *U. lactuca*. This differential effect of extraction on minerals and trace elements presumably reflects differences in their solubility and form in which they are present in the dried *U. lactuca*. Trace elements complexed with proteins and carbohydrates presumably have a low solubility and were concentrated in the pellet after centrifugation whereas inorganic salts may have a higher solubility and may be partly removed with the liquid fraction.

In a review, MacArtain et al. (2007) reported levels of macro minerals in *U. lactuca* that were two to three times higher and Fe and Zn contents approximately five times higher than the values in our study. Because of their accumulation capacity, algae are considered a valuable indicator for the assessment of heavy metals in coastal areas (Haritonidis and Malea 1999; Boubonari et al. 2008). The contents of some heavy metals in *Ulva* sp. are largely determined by the concentrations in the sediment and seawater, as reflected by the large variation in the contents of Fe, Pb, Zn, Cu and Cd in samples of *U. lactuca* from different geographic areas, while season of harvesting affected the heavy metal contents in these samples as well (Haritonidis and Malea 1999).

The high content of minerals and (heavy) metals may limit the use of *U. lactuca* in animal diets, depending on the animal species, its requirements and availability of minerals in regional feed materials. According to Directive 2002/32 EC (2002), legal limits for heavy metals in seaweed as feed ingredient are 40 ppm for As, of which the maximum is 2 ppm of inorganic As, 1 ppm for Cd, 10 ppm for Pb and 0.1 ppm for Hg. Provided that As in *U. lactuca* is largely (>90 %) converted to a non-toxic organic form (Holdt and Kraan 2011), these legal limits to heavy metals would not directly restrict the inclusion of *U. lactuca* used in this study, but levels should be carefully monitored because of the large variation as discussed above. No legal limits are applicable for the use of macro minerals. However, because of the high content of inorganic matter, inclusion of a substantial amount of *U. lactuca* (e.g. 5–10 %) as a protein source would considerably increase the contents of macro minerals in the animal feed, especially S, Ca, Mg, Na and Cl, as compared to the commonly used soybean meal in monogastric diets. Hence, use of *U. lactuca* or extracted fraction would reduce the need for inclusion of supplementing the diet with Ca, Na and Cl. Supplementation of Mg and S is not commonly required for monogastric diets, but may be valuable in diets for grazing ruminants in tropical regions (Machado et al. 2015). Considering the high osmotic capacity of specific minerals, the use of this *U. lactuca* sample would increase the risk of low faecal consistency, diarrhoea and wet droppings in pigs and poultry as shown for high inclusion of an *Ascophyllum nodosum* residue in pigs (Whittemore and Percival 1975). In addition, interaction between minerals and trace elements may reduce the absorption of specific nutrients from the digestive tract. For example, the high content of Ca and Mg may contribute to complexation with (phytate)-P, thus reducing the P availability of the animals (NRC 2005).

Furthermore, the high S content is of particular concern in ruminants since ruminally produced hydrogen sulphide from dietary sulphur is toxic for the central nervous system (polioencephalomalacia) (NRC 2005). High S may contribute to osmotic diarrhoea in non-ruminants, although it has not been described to what extent sulphated carbohydrates contribute to these phenomena. A total dietary content of 3.5 and 4 g kg^{-1} is regarded safe for cattle, and pigs and poultry, respectively (NRC 2005), suggesting that inclusion of these *Ulva*products should not exceed 5–10 %. In conclusion, the high content of minerals and trace elements may limit a high inclusion level of *U. lactuca* and extracted fraction in animal diets. Close monitoring of heavy metals to assure food safety is required, whereas reduction of the mineral fraction would reduce the risk of osmotic diarrhoea and toxicosis in farm animals. In cultivated *Ulva* species, and cultivated seaweeds in general, seaweed composition could be better controlled and the levels of metals expected in the biomass are lower than those found in wild grown seaweeds (Dr. M.H. Abreu, personal communication). This would increase the potential of using cultivated seaweeds for feed and food applications, e.g. from integrated aquaculture systems as presently being developed (Abreu et al. 2011; Cole et al. 2014).

Protein and Amino Acids

The amino acid profile of *U. lactuca* (Table 3) fitted well in the range of earlier published values for this species. The variation in published amino acid profiles is quite substantial. The relative sum of essential amino acids in soybean meal was between that of *U. lactuca* and extracted fraction.

Furthermore, *U. lactuca* and extracted fraction were relatively low in lysine, tryptophan and histidine and relatively rich in methionine and threonine. These results suggest that *U. lactuca* and extracted fraction can be a good protein source in monogastric diets, provided that lysine and tryptophan, often among the first limiting amino acids (Edmonds et al. 1985), are adequately supplemented. The higher (essential) amino acid content in the extracted fraction, due to partial removal of the carbohydrate fraction after hydrolysis and centrifugation, makes this fraction more valuable as protein rich feed ingredient.

Many studies, including those cited in Table 3, determined the crude protein content as 6.25× N content. However, the use of this factor overestimates the real protein content in many seaweed species (Lourenço et al. 2002; Angell et al. 2016). We distinguished between the N to protein factors K_P, and K_A, in line with literature (Angell et al. 2016; Mariotti et al. 2008). The factor K_A reflects the ratio between amino acids and N from amino acids. The value of K_A (5.69 and 5.95 for *U. lactuca* and extracted fraction) is determined by the N content of the individual amino acids (Sosulski and Imafidon 1990) and hence by the amino acid pattern of the protein. It indicates that the mean N content in amino acids is slightly higher in *U. lactuca* (17.6 %) than in extracted fraction (16.8 %). The value of K_P reflects the ratio between amino acids and total N and hence is reduced in the presence of substantial amounts of non-protein N. The total N to protein conversion factor (K_P) of 4.6–4.7 was similar for *U. lactuca* and extracted fraction, and in good agreement with a median value of 4.68 as determined for green seaweeds in the recent review of Angell et al. (2016). Our data confirm that this lower value should be used to avoid overestimation of the contribution of *Ulva* to the amino acid supply of monogastric animals. The low conversion factor is largely due to the presence of approximately 20 % of N as non-amino acid N, which can be used as N source by microbiota in ruminating animals, but not by pigs and poultry. The variation in *Ulva* N content (Table 3) and in N to protein conversion factor between studies (Angell et al. 2016) indicate that determination of the amino acid content in *U. lactuca* samples is the preferred method to obtain insight in the value of specific batches to be used in monogastric animal diets.

Fatty Acids

The total FA content of *U. lactuca* and extracted fraction was 21 and 34 g kg^{-1} DM (Table 4). The low fat content was in agreement with other studies into *U. lactuca* (Khotimchenko et al. 2002; Ortiz et al. 2006; Tabarsa et al. 2012) and green seaweed species in general (Holdt and Kraan 2011), although some others reported a lipid content up to 8 % in DM (Yaich et al. 2011). The latter may be related to geographic location, climate, nutrient availability and stage and season of harvest since Mercer et al. (1993) reported a lipid content of 4.0 and 8.4 % in *U. lactuca* harvested in Ireland in May and February, respectively. The most abundant fatty acids were C16:0 and C18:1, with substantial amounts of C18:2, 3 and 4. These results were largely within the range of studies in the literature as summarised in Table 4. As such, *U. lactuca* is a valuable source of n-3 fatty acids and may contribute to an optimal balance between n-3 and n-6 fatty acids. It is not a major source of the specific marine fatty acids C20:5 and C22:6, as some brown and red seaweeds (Holdt and Kraan 2011). The higher fat content contributes to a higher energy content of the extracted fraction compared to *U. lactuca*. However, the proportion of PUFA was lower in the extracted fraction, suggesting that the pre-treatment and hydrolysis applied to the biomass caused a relatively higher solubilisation and removal of PUFA compared to SFA. Alternatively, part of the PUFA might have been lost due to oxidation under the conditions of increased moisture and temperature during the extraction process. This would make the extracted fraction a less valuable source of n-3 PUFA. The overall results obtained on PUFAs composition indicate that *U. lactuca* and extracted fraction are not a major source of fatty acids in animal diets. More insight in fatty acid digestibility is required.

In Vitro Digestibility

The in vitro simulated ileal OM digestibility of *Ulva* extracted fraction of 86.9 % was 20 percentage units higher than for intact *U. lactuca* and even slightly higher than for soybean meal (Table 5). Presumably, the enzymatic pre-treatment of *U. lactuca* with the cellulase cocktail hydrolysed poorly digestible carbohydrates and thus increased the in vitro digestibility of the extracted solid fraction in comparison to the intact *U. lactuca*. In addition, the N digestibility was enhanced by 5 % points in the extracted fraction, presumably because of the release of cell wall bound or encapsulated protein during the pre-treatment hydrolysis, thus improving the protein digestibility in monogastric species. The simulated total tract digestibility was only 7 percentage units different (not significant) between the two *Ulva*products. The large difference between simulated ileal and total tract OM digestibility of *U. lactuca* indicates that a large portion of the ileally indigestible carbohydrates were broken down by the cell wall degrading enzymes (Viscozyme; Sigma V2010) in the in vitro system. This suggests that hindgut fermentation of structural carbohydrates in *U. lactuca* may contribute to the energy supply of pigs. Despite the improved digestibility of the *Ulva* extracted fraction, the simulated ileal N and total tact OM digestibility were 8–13 percentage units lower than for soybean meal. These differences may be caused by the presence of glycoproteins or phenolic compounds. Wong and Cheung (2001) reported a negative correlation between total phenolic compounds and in vitro protein digestibility in different seaweed species, including *U. lactuca*, presumably because of formation of insoluble complexes. Fleurence (1999) observed a negative correlation between the glycoprotein content in *Ulva armoricana* and its in vitro protein digestibility. Furthermore, binding or encapsulation by poorly digestible cell wall polysaccharides, including ulvan, glucuronan and xyloglucan, may hamper the accessibility of proteins and OM digestibility (Lahaye and Robic 2007). *Ulva lactuca* cell wall polysaccharides are poorly digestible by purified cellulase and poorly fermentable by human colonic microbiota, whereas the use of endoxylanase may substantially contribute to further degradation (Bobin-Dubigeon et al. 1997). The simulated ileal N and total tract in vitro OM digestibility of soybean meal were approximately 10 and 6 % higher than the in vivo values in CVB (2007). Consequently, the ileal N and total tract OM digestibility of the *Ulva* products may also be overestimated and between 70 and 80 %. Based on these results, it is expected that intact *U. lactuca* and *Ulva* extracted fraction will have a moderate to good contribution to the nutrient supply of monogastric species, respectively. We are not aware of in vivo digestibility studies with *U. lactuca* in monogastric animals. An in vivo digestibility study with chickens (Ventura et al. 1994) reported a true metabolisable energy (TMEn) value of *Ulva rigida* of 5.7 MJ kg^{-1} DM, which was only 40 % of the gross energy value. The low TME value presumably can be attributed to the high content of indigestible polysaccharides and the lack of fermentative capacity in birds. The polysaccharides significantly increase digesta viscosity and layer thickness near the intestinal wall, thereby reducing ileal nutrient digestibility, resulting in a low digestible energy value (Choct et al. 1996; Flourie et al. 1984). The limitations of the use of *Ulva* sp. in poultry is confirmed in several studies, showing that low (up to 3 %) dietary *Ulva*inclusion levels did not affect performance levels of broilers, whereas high (10 % or more) *Ulva*inclusion levels negatively affected feed intake, body weight gain and feed conversion ratio (Abudabos et al. 2013; Ventura et al. 1994). Thus, based on digestibility, *U. lactuca* may be a better feed ingredient for pigs than for poultry, whereas the extracted fraction seems a promising ingredient for further evaluation in both species. Results of an in vitro digestibility study showed promising results for proteins that were isolated from *U. lactuca* (Wong and Cheung 2001) by extraction under alkaline conditions. Subsequently, phenolic compounds were removed from the extracted proteins. This procedure resulted in an ingredient low in indigestible polysaccharides and high in crude protein (76 %, as N × 6.25) with an in vitro protein digestibility of 85.7 % (Wong and Cheung 2001), indicating that the protein digestibility could be substantially improved by separating the proteins from the intact *Ulva* seaweed and interacting compounds.

Gas Production Test

The lower in vitro rumen degradation expressed in a lower rate and maximum cumulative gas production of the *Ulva* products compared to the reference feed materials are in agreement with results of Dubois et al. (2013) and can be explained by the low NDF content and the relatively high content of ADL and protein. The fermentation of protein as substrate causes a lower gas production by the microbial fermentation compared to that of carbohydrates (Cone and van Gelder 1999). These authors estimated that for each percent of protein, cumulative gas production after 72 h is reduced by 2.5 mL g^{-1} OM and proposed a correction of this magnitude for comparison of feed materials with widely differing protein content. The results, including the corrected maximum gas production, indicate that the rumen fermentation of *U. lactuca* was in the range of alfalfa and below that of grass silage. These results are in line with studies of Arieli et al. (1993) in sheep and Ventura and Castañón (1998) in goat who concluded that the nutritive value of *U. lactuca* is similar to a medium-quality alfalfa hay, with a higher protein content. The corrected gas production of the extracted fraction was somewhat higher than that of *U. lactuca*, suggesting a beneficial effect of the enzyme treatment on the fermentation of the residue. Moreover, digestibility may be higher in ruminants adapted to seaweeds in their ration, as demonstrated in vitro for Orkney sheep by Greenwood et al. (1983). It is recommended to further validate these in vitro results with studies in farm animals before practical application.

Fermentation of *U. Lactuca* Hydrolysate to ABE and 1,2-Propanediol

The *U. lactuca* hydrolysate as such supported growth and production of ABE and 1,2-PD by *C. beijerinckii*. This is consistent with our previous study on hydrolysate of *U. lactuca* from a different source (van der Wal et al. 2013). *Ulva* species are rich in proteins and salts, which are partially solubilised during the pre-treatment and enzymatic hydrolysis and can serve as nutrients for microbial fermentation.

The metabolism of glucose by Clostridia for ABE production has been studied before; however, the metabolism of rhamnose and of rhamnose-containing mixes is not well known. The production of 1,2-PD from rhamnose has been reported for a number of microorganisms (Saxena et al. 2010), and in Clostridial species, it is considered to occur in a pathway analogous to that in *Escherichia coli*, *Salmonella typhimurium* and *Caldicellulosiruptor saccharolyticus*(Bennett and San 2001; Forsberg et al. 1987; van de Werken et al. 2008). The consumption of rhamnose in the control cultures was low, corresponding to 19 % of the initial amount in the medium (culture CM2-R, Table 7). This is most likely due to the lower energy yield from the rhamnose compared to that in the glucose to ABE metabolism (Forsberg et al. 1987) and not to product toxicity, as in these cultures the ABE level was very low.

When grown on a mixture of glucose, rhamnose and xylose, both 1,2-PD and ABE were produced. Glucose, xylose and part of the rhamnose were converted into ABE (Table 7, culture CM2-G/R/X). The rhamnose consumption in these cultures was again incomplete (53 % of the initial amount), resulting in production of 2.8 g L^{-1} 1,2-PD. In this case, most probably, growth inhibition due to high ABE concentration took place, but again the lower metabolic efficiency of the rhamnose conversion route may have played a role in the limited substrate utilisation. To our knowledge, the toxicity of 1,2-PD to bacterial cultures has not been characterised. However, end concentrations of this metabolite up to approximately 10 g L^{-1} have been observed in our laboratory (van der Wal et al. 2013), and therefore we expect that the end concentrations produced in the cultures CM2-R and CM2-G/R/X (Table 7) are not inhibitory.

Since only on rhamnose-grown cultures production of small concentrations of propionic acid has been observed (results not shown), it is possible that the clostridial strain used has some catabolic pathway for conversion of 1,2-PD into this organic acid. In *Clostridium phytofermentas*, a route for the catabolism of 1,2-PD into propionic acid and *n*-propanol has been proposed (Petit et al. 2013). Currently, we are studying this subject in our laboratory since propionic acid and *n*-propanol represent interesting products with applications in the food industry and as solvent, respectively.

Biorefinery Concept

In this study, we have developed a valorisation strategy of *U. lactuca* biomass as feedstock for several products using a cascading biorefinery approach. The polysaccharides from *U. lactuca*biomass were hydrolysed and used as substrate for the fermentative production of industrially relevant components with application as fuels (ABE) and chemicals (ABE, 1,2-PD). The hydrolysate from *U. lactuca* contained a high concentration of fermentable sugars, making it interesting as a substrate for fermentation processes in general. The extracted fraction showed improved value as animal feed ingredient because of the increased amino acid content, ileal digestibility and rumen fermentation compared to intact *U. lactuca*, although the high mineral content requires further attention. As a next step, the feasibility of the proposed cascading *U. lactuca*-based biorefinery will have to be assessed by an economic and environmental system analysis.

ACKNOWLEDGMENTS

The authors wish to thank Declan Haniffy and Stefan Kraan from Ocean Harvest Ltd for the supply of the *U. lactuca* and useful discussions, Hector Capella Monsonís for assistance in the pre-treatment tests and Saskia van

Laar for assistance with the in vitro gas production measurements. This research was financially supported by the Dutch Ministry of Economic Affairs through AgentschapNL (currently RvO) (project nr. EOS LT 08027), the Triple P @ Sea Program of the Wageningen UR (http://www.wageningenur.nl/en/About-Wageningen-UR/Strategic-plan/TriplePSea-Coastal-and-Marine-resources.htm) and the European Commission through the EU ITN project RENESENG (grant 607415).

REFERENCES

1. Abdel-Fattah AF, Edrees M (1973) Seasonal changes in the constituents of Ulva lactuca. Phytochemistry 12:481–485
2. Abreu MH, Pereira R, Yarish C, Buschmann AH, Sousa-Pinto I (2011) IMTA with Gracilaria vermiculophylla: productivity and nutrient removal performance of the seaweed in a land-based pilot scale system. Aquaculture 312:77–87
3. Abudabos AM, Okab AB, Aljumaah RS, Samara EM, Abdoun KA, Al-Haidary AA (2013) Nutritional value of green seaweed (Ulva lactuca) for broiler chickens. Ital J Anim Sci 12(2), e28
4. Angell AR, Mata L, de Nys R, Paul NA (2016) The protein content of seaweeds: a universal nitrogen-to-protein conversion factor of five. J Appl Phycol 28:511–524
5. Arieli A, Sklan D, Kissil G (1993) A note on the nutritive-value of Ulva lactuca for ruminants. Anim Prod 57:329–331
6. Bennett GN, San KY (2001) Microbial formation, biotechnological production and applications of 1,2-propanediol. Appl Microbiol Biotechnol 55:1–9
7. Bobin-Dubigeon C, Hoebler C, Lognone V, Dagorn-Scaviner C, Mabeau S, Barry JL, Lahaye M (1997) Chemical composition, physico-chemical properties, enzymatic inhibition and fermentative characteristics of dietary fibres from edible seaweeds. Sci Aliments 17:619–639
8. Boisen S, Fernandez JA (1997) Prediction of the total tract digestibility of energy in feedstuffs and pig diets by in vitro analyses. Anim Feed Sci Technol 68:277–286
9. Boland MJ, Rae AN, Vereijken JM, Meuwissen MPM, Fischer ARH, van Boekel M, Rutherfurd SM, Gruppen H, Moughan PJ,

Hendriks WH (2013) The future supply of animal-derived protein for human consumption. Trends Food Sci Technol 29:62–73

10. Boubonari T, Malea P, Kevrekidis T (2008) The green seaweed Ulva rigida as a bioindicator of metals (Zn, Cu, Pb and Cd) in a low-salinity coastal environment. Bot Mar 51:472–484
11. Briand X, Morand P (1997) Anaerobic digestion of Ulva sp. 1. Relationship between Ulva composition and methanisation. J Appl Phycol 9:511–524
12. Choct M, Hughes RJ, Wang J, Bedford MR, Morgan AJ, Annison G (1996) Increased small intestinal fermentation is partly responsible for the anti-nutritive activity of non-starch polysaccharides in chickens. Bri Poultry Sci 37:609–621
13. Cole AJ, de Nys R, Paul NA (2014) Removing constraints on the biomass production of freshwater macroalgae by manipulating water exchange to manage nutrient flux. PLoS One 9(7), e101284
14. Cone JW, van Gelder AH (1999) Influence of protein fermentation on gas production profiles. Anim Feed Sci Technol 76:251–264
15. Cone JW, vanGelder AH, Visscher GJW, Oudshoorn L (1996) Influence of rumen fluid and substrate concentration on fermentation kinetics measured with a fully automated time related gas production apparatus. Anim Feed Sci Technol 61:113–128
16. CVB (2007) Veevoedertabel 2007, Centraal Veevoederbureau. Lelystad, The Netherlands
17. Dubois B, Tomkins NW, Kinley RD, Bai M, Seymour S, Paul NA (2013) Effect of tropical algae as additives on rumen in vitro gas production and fermentation characteristics. Am J Plant Sci 4:34–43
18. EC 159/2009 (2009) Commission Regulation of 27 January 2009 laying down the methods of sampling and analysis for the official control of feed
19. Edmonds MS, Parsons CM, Baker DH (1985) Limiting amino acids in low protein corn soybean meal diets fed to growing chicks. Poultry Sci 64:1519–1526
20. European Commission (2002) Directive 2002/32/EC of the European Parliament and of the Council of 7 May 2002 on undesirable substances in animal feed—Council statement. OJ L

140, 30.5.2002, 10–22

21. FAO (2006) World agriculture: towards 2030/2050. Interim Report. Food and Agriculture Organisation, Rome
22. Fleurence J (1999) Seaweed proteins: biochemical, nutritional aspects and potential uses. Trends Food Sc. Technol 10(1):25-28 doi:10.1016/S0924-2244(99)00015-1
23. Fleurence J, LeCoeur C, Mabeau S, Maurice M, Landrein A (1995) Comparison of different extractive procedures for proteins from the edible seaweeds Ulva rigida and Ulva rotundata. J Appl Phycol 7:577–582
24. Flourie B, Vidon N, Florent C, Bernier JJ (1984) Effect of pectin on jejunal glucose-absorption and unstirred layer thickness in normal man. Gut 25:936–941
25. Forsberg CW, Donaldson L, Gibbins LN (1987) Metabolism of rhamnose and other sugars by strains of Clostridium acetobutylicum and other Clostridium species. Can J Microbiol 33:21–26
26. Francavilla M, Franchi M, Monteleone M, Caroppo C (2013) The red seaweed Gracilaria gracilis as a multi products source. Mar Drugs 11:3754–3776
27. Francavilla M, Pineda A, Romero AA, Colmenares JC, Vargas C, Monteleone M, Luque R (2014) Efficient and simple reactive milling preparation of photocatalytically active porous ZnO nanostructures using biomass derived polysaccharides. Green Chem 16:2876–2885
28. Gardiner GE, Campbell AJ, O'Doherty JV, Pierce E, Lynch PB, Leonard FC, Stanton C, Ross RP, Lawlor PG (2008) Effect of Ascophyllum nodosum extract on growth performance, digestibility, carcass characteristics and selected intestinal microflora populations of grower-finisher pigs. Anim Feed Sci Technol 141:259–273
29. Greenwood Y, Orpin CG, Paterson IW (1983) Digestibility of seaweeds in Orkney sheep. Proc Physiol Soc 120
30. Haritonidis S, Malea P (1999) Bioaccumulation of metals by the green alga Ulva rigida from Thermaikos Gulf, Greece. Environ Pollut 104:365–372
31. Holdt SL, Kraan S (2011) Bioactive compounds in seaweed: functional food applications and legislation. J Appl Phycol 23:543–

597

32. ISO 13903 (2005) Animal feeding stuffs—determination of amino acids content. International Organization for Standardization, Geneva
33. ISO 13904 (2005) Animal feeding stuffs—determination of tryptophan content. International Organization for Standardization, Geneva
34. ISO 13906 (2008) Animal feeding stuffs—determination of acid detergent fibre (ADF) and acid detergent lignin (ADL) contents. International Organization for Standardization, Geneva
35. ISO 15914 (2004) Animal feeding stuffs—enzymatic determination of total starch content. International Organization for Standardization, Geneva
36. ISO 5983 (2005) Animal feeding stuffs—determination of nitrogen content and calculation of crude protein content—Part 1: Kjeldahl method. International Organization for Standardization, Geneva
37. ISO 5984 (2002) Animal feeding stuffs—determination of crude ash. International Organization for Standardization, Geneva
38. ISO 6496 (1999) Animal feeding stuffs—determination of moisture and other volatile matter content. International Organization for Standardization, Geneva
39. Khotimchenko SV, Vaskovsky VE, Titlyanova TV (2002) Fatty acids of marine algae from the Pacific coast of north California. Bot Mar 45:17–22
40. Kraan S (2013) Mass-cultivation of carbohydrate rich macroalgae, a possible solution for sustainable biofuel production. Mitig Adapt Strateg Glob Chang 18:27–46
41. Lahaye M, Robic A (2007) Structure and functional properties of ulvan, a polysaccharide from green seaweeds. Biomacromolecules 8:1765–1774
42. Lawton RJ, Mata L, de Nys R, Paul NA (2013) Algal bioremediation of waste waters from land-based aquaculture using Ulva: selecting target species and strains. PloS One 8(10), e77344
43. López-Contreras AM, Claassen PAM, Mooibroek H, De Vos WM (2000) Utilisation of saccharides in extruded domestic organic

waste by Clostridium acetobutylicum ATCC 824 for production of acetone, butanol and ethanol. Appl Microbiol Biotechnol 54:162–167

44. Lourenço SO, Barbarino E, De-Paula JC, Pereira LODS, Lanfer Marquez UM (2002) Amino acid composition, protein content and calculation of nitrogen-to-protein conversion factors for 19 tropical seaweeds. Phycol Res 50:233–241
45. MacArtain P, Gill CIR, Brooks M, Campbell R, Rowland IR (2007) Nutritional value of edible seaweeds. Nutr Rev 65:535–543
46. Machado L, Kinley RD, Magnusson M, de Nys R, Tomkins NW (2015) The potential of macroalgae for beef production systems in Northern Australia. J Appl Phycol 27:2001–2005
47. Mai K, Mercer JP, Donlon J (1994) Comparative-studies on the nutrition of 2 species of abalone, Haliotis tuberculata L and Haliotis discus hannai Ino. 2. Amino-acid-composition of abalone and 6 species of macroalgae with an assessment of their nutritional-value. Aquaculture 128:115–130
48. Makkar HPS, Tran G, Heuze V, Giger-Reverdin S, Lessire M, Lebas F, Ankers P (2016) Seaweeds for livestock diets: a review. Anim Feed Sci Technol 212:1–17
49. Marinho G, Nunes C, Sousa-Pinto I, Pereira R, Rema P, Valente LP (2013) The IMTA-cultivated Chlorophyta Ulva spp. as a sustainable ingredient in Nile tilapia (Oreochromis niloticus) diets. J Appl Phycol 25:1359–1367
50. Mariotti F, Tome D, Mirand PP (2008) Converting nitrogen into protein—beyond 6.25 and Jones' factors. Crit Rev Food Sci Nutr 48:177–184
51. Mercer JP, Mai KS, Donlon J (1993) Comparative studies on the nutrition of two species of abalone, Haliotis tuberculata Linnaeus and Haliotis discus hannai Ino I. Effects of algal diets on growth and biochemical composition. Invertebr Reprod Dev 23(2–3):75–88 doi:10.1080/07924259.1993.9672298
52. Moroney NC, O'Grady MN, O'Doherty JV, Kerry JP (2012) Addition of seaweed (Laminaria digitata) extracts containing laminarin and fucoidan to porcine diets: influence on the quality and shelf-life of fresh pork. Meat Sci 92:423–429

53. NEN-EN 14627 (2010) Foodstuffs—determination of trace elements—determination of total arsenic and selenium by hydride generation atomic absorption spectrometry (HGAAS) after pressure digestion. Stichting Nederlands Normalisatie-Instituut [NL] 1–14
54. NEN-EN 15510 (2007) Animal feeding stuffs—determination of calcium, sodium, phosphorus, magnesium, potassium, iron, zinc, copper, manganese, cobalt, molybdenum, arsenic, lead and cadmium by ICP-AES. Stichting Nederlands Normalisatie-Instituut [NL] 1–29
55. NEN-EN 15763 (2010) Foodstuffs—determination of trace elements—determination of arsenic, cadmium, mercury and lead in foodstuffs by inductively coupled plasma mass spectrometry (ICP-MS) after pressure digestion. Stichting Nederlands Normalisatie-Instituut [NL] 1–18
56. NPR-CEN-ISO/TS 17764–1 and 2 (2006) Animal feeding stuffs—determination of the content of fatty acids—Part 1: preparation of methyl esters and Part 2: gas chromatographic method. International Organization for Standardization, Geneva
57. NRC (2005) Mineral tolerance of animals. National Research Council of the National Academies, second revised edition. The National Academic Press (Eds), Washington DC
58. Ortiz J, Romero N, Robert P, Araya J, Lopez-Hernandez J, Bozzo C, Navarrete E, Osorio A, Rios A (2006) Dietary fiber, amino acid, fatty acid and tocopherol contents of the edible seaweeds Ulva lactuca and Durvillaea antarctica. Food Chem 99:98–104
59. Petit E, LaTouf WG, Coppi MV, Warnick TA, Currie D, Romashko I, Deshpande S, Haas K, Alvelo-Maurosa JG, Wardman C, Schnell DJ, Leschine SB, Blanchard JL (2013) Involvement of a bacterial microcompartment in the metabolism of fucose and rhamnose by Clostridium phytofermentans. PLoS One 8(1), e54337
60. Potts T, Du JJ, Paul M, May P, Beitle R, Hestekin J (2012) The production of butanol from Jamaica bay macro algae. Environ Prog Sustain Energy 31:29–36
61. Ray B, Lahaye M (1995) Cell-wall polysaccharides from the marine green-alga Ulva rigida (Ulvales, Chlorophyta)—chemical structure of ulvan. Carbohydr Res 274:313–318

62. Robertson-Andersson D, Potgieter M, Hansen J, Bolton J, Troell M, Anderson R, Halling C, Probyn T (2008) Integrated seaweed cultivation on an abalone farm in South Africa. J Appl Phycol 20:579–595
63. Robic A, Sassi J-F, Dion P, Lerat Y, Lahaye M (2009) Seasonal variability of physicochemical and rheological properties of ulvan in two Ulva species (Chlorophyta) from the Brittany coast. J Phycol 45:962–973
64. Saxena RK, Anand P, Saran S, Isar J, Agarwal L (2010) Microbial production and applications of 1,2-propanediol. Indian J Microbiol 50:2–11
65. Sorek N, Yeats T, Szemenyei H, Youngs H, Somerville C (2014) The implications of lignocellulosic biomass chemical composition for the production of advanced biofuels. Bio Sci. doi:10.1093/biosci/bit037
66. Sosulski FW, Imafidon GI (1990) Amino acid composition and nitrogen-to-protein conversion factors for animal and plant foods. J Agr Food Chem 38:1351–1356
67. Tabarsa M, Rezaei M, Ramezanpour Z, Waaland JR (2012) Chemical compositions of the marine algae Gracilaria salicornia (Rhodophyta) and Ulva lactuca (Chlorophyta) as a potential food source. J Sci Food Agric 92:2500–2506
68. van de Werken HJG, Verhaart MRA, VanFossen AL, Willquist K, Lewis DL, Nichols JD, Goorissen HP, Mongodin EF, Nelson KE, van Niel EWJ, Stams AJM, Ward DE, de Vos WM, van der Oost J, Kelly RM, Kengen SWM (2008) Hydrogenomics of the extremely thermophilic bacterium Caldicellulosiruptor saccharolyticus. Appl Environ Microbiol 74:6720–6729
69. Van den Burg S, Stuiver M, Veenstra F, Bikker P, López Contreras A, Palstra A, Broeze J, Jansen H, Jak R, Gerritsen A, Harmsen P, Kals J, Blanco A, Brandenburg W, Van Krimpen M, Van Duijn AP, Mulder W, Van Raamsdonk L (2013) A Triple P review of the feasibility of sustainable offshore seaweed production in the North Sea, Wageningen UR LEI Report 13–077, Wageningen, September 2013, pp 1–105
70. Van den Burg S, Van Duijn AP, Bartelings H, Van Krimpen MM,

Poelman M (2016) The economic feasibility of seaweed production in the North Sea. Aquac Econ Manag. doi:10.1080/13657305.2016.1177859

71. van den Oever MJA, Bas N, van Soest LJM, Melis C, van Dam JEG (2003) Improved method for fibre content and quality analysis and their application to flax genetic diversity investigations. Ind Crops Prod 18:231–243
72. van der Wal H, Sperber BLHM, Houweling-Tan B, Bakker RRC, Brandenburg W, López-Contreras AM (2013) Production of acetone, butanol, and ethanol from biomass of the green seaweed Ulva lactuca. Bioresour Technol 128:431–437
73. van Hal JW, Huijgen WJJ, Lopez-Contreras AM (2014) Opportunities and challenges for seaweed in the biobased economy. Trends Biotechnol 32:231–233
74. Ventura MR, Castañón JIR (1998) The nutritive value of seaweed (Ulva lactuca) for goats. Small Ruminant Res 29:325–327
75. Ventura MR, Castanon JIR, McNab JM (1994) Nutritional-value of seaweed (Ulva rigida) for poultry. Anim Feed Sci Technol 49:87–92
76. Whittemore CT, Percival JK (1975) Seaweed residue unsuitable as a major source of energy or nitrogen for growing pigs. J Sci Food Agric 26:215–217
77. Wong KH, Cheung PCK (2000) Nutritional evaluation of some subtropical red and green seaweeds Part I: proximate composition, amino acid profiles and some physico-chemical properties. Food Chem 71:475–482
78. Wong KH, Cheung PCK (2001) Nutritional evaluation of some subtropical red and green seaweeds Part II. In vitro protein digestibility and amino acid profiles of protein concentrates. Food Chem 72:11–17
79. Yaich H, Garna H, Besbes S, Paquot M, Blecker C, Attia H (2011) Chemical composition and functional properties of Ulva lactuca seaweed collected in Tunisia. Food Chem 128:895–901

SECTION V

SEAWEED CULTIVATION AND ENVIRONMENT

CHAPTER

20

A GLOBAL ANALYSIS OF THE RELATIONSHIP BETWEEN FARMED SEAWEED PRODUCTION AND HERBIVOROUS FISH CATCH

E. James Hehre[1] and Jessica J. Meeuwig[2]

[1]Sea Around Us Project /Institute for oceans and Fisheries, University of British Columbia, Vancouver, BC, Canada

[2]School of Animal Biology and Centre for Marine Futures, University of Western Australia, Crawley, WA, Australia

ABSTRACT

Globally, farmed seaweed production is expanding rapidly in shallow marine habitats. While seaweed farming provides vital income to millions of artisanal farmers, it can negatively impact shallow coral reef and seagrass habitats. However, seaweed farming may also potentially provide food subsidies for herbivorous reef fish such as the Siganidae, a valuable target family, resulting in increased catch. Comparisons of reef fish landings across the central Philippines revealed that the catch of siganids was positively correlated to farmed seaweed production whilst negatively correlated to total reef fish catch over the same period of time. We tested the generality of this pattern by analysing seaweed production, siganid catch, and reef fish catch for six major seaweed-producing countries in the tropics. We hypothesized that increased seaweed production would correspond with increased catch of siganids but not other reef fish species. Analysis of the global data showed a positive correlation between farmed seaweeds and siganids in Southeast Asia (Indonesia, Malaysia, and the Philippines) but not Africa (Tanzania and Zanzibar), or the Western Pacific (Fiji). In Southeast Asia, siganid catch increased disproportionately faster with seaweed production than did reef fish catch. Low continuity, sporadic production and smaller volumes of seaweed farming may explain the differences.

INTRODUCTION

The commercial cultivation of seaweeds occurs in approximately 35 countries around the world and provides a variety of products that, in 2011, produced 21 million tons with a total annual value of US$7.35 billion [1]. Of that total, food products contributed almost US$ 5 billion [1]. Seaweed cultivation continues to expand rapidly as demand for seaweed products such as carrageenan has outstripped supply from wild resources. Although seaweed farming occurs globally, the vast majority of seaweed farming (98.9%, 18.9 million tons) is concentrated in China (60%) and Southeast Asia, including Indonesia (21%), the Philippines (9%), and Malaysia (1%). Throughout Southeast Asia, small subsistence farms (<1 ha) predominate, and their proliferation is in large part governed by both accessibility to useable habitats and proximity to markets [2]. Rising demand for seaweed products and the need for impoverished communities to develop alternative livelihoods are driving seaweed farms to expand into new locations, including onto coral reefs [2,3].

The majority of the world's coral reefs are found in developing countries with high rates of population growth [4]. Combined with social inequality, population growth has significantly increased pressures on tropical marine fisheries [2,5–8], which are an important source of revenue and protein for millions of people globally [5,9,10]. The ecological impacts of coastal population growth primarily derive from both the system loading effects of pollution and siltation and the extractive and degrading effects of resource overexploitation [11–14]. In many places, these impacts are exacerbated by the use of destructive practices like blast and cyanide fishing. Once damaged, the capacity of a reef to recover depends on several factors, including its fundamental starting condition and the degree to which the causes of reef decline have been removed. While a growing number of studies indicate that reef recovery is possible with effective implementation of coastal management and alternative livelihood programs [15–17], progress will be constrained unless human population growth rates slow and poverty is alleviated.

The benefits of seaweed farming are unclear despite the practice being advocated as a way to improve reef health through poverty alleviation and reduced fisheries exploitation [2]. Recent studies have shown that the introduction of seaweed farming does little to mitigate the effects of fisheries overexploitation, and that rather than replacing fishing, it is utilized as an additional source of income [18]. However, seaweed farms also tend to be located in easily accessible, shallow and sheltered habitats situated in close proximity to markets. As such, many of the areas in which farms are located have already been degraded by overfishing and habitat loss [19,20] and thus may not cause additional habitat degradation. However, the direct ecological impacts of seaweed farming are still debated as empirical studies have typically produced different and conflicting results. For instance, in Indonesia, Blankenhorn (2007) found that where seagrass was not cleared as part of farm establishment, seaweed farming itself had no negative effect on seagrass beds [21]. By contrast, Ekloff (2006) recommended that seaweed farming in shallow seagrass areas should be avoided and that damage to seagrass beds was mitigated only by the small scale of farms and the recovery periods dictated by generally low market prices [22].

Hehre and Meeuwig (2015) also showed that seaweed farming on degraded shallow coral reefs corresponded with lower species richness, abundance, and biomass of associated fish assemblages, despite initial speculation that farms may benefit fish assemblages by adding physical complexity and shelter, and a potential food source for herbivores [2,22–25]. These results are consistent with other studies that have shown both lower fish abundances and species richness in macroalgal-dominated versus coral-dominated habitats [26,27]. Furthermore it has been demonstrated experimentally that herbivorous fishes will avoid areas of high macroalgal biomass [28]. However, despite these findings, it is still possible seaweed farms increase rabbitfish productivity rather than standing biomass [29–31]. While the Underwater Visual Census methods used in Hehre & Meeuwig 2015 give us a measure of abundance, abundance may not necessarily reflect productivity [32,33]. Increases in siganid productivity relative to farming could be masked by the concentration of fishing effort within the farms.

Though they may provide benefits in terms of both recruitment and food, specific fisheries benefits derived from seaweed farms also remain unclear. Herbivorous fish such as siganids forage on a broad range of algae [34,35] [1,36–39]. Field studies have demonstrated that siganids play an important role as consumers of naturally-occurring macroalgae on coral reefs, and *Sargassum* in particular [36,40,41]. Additionally, siganid foraging on macroalgae has been blamed for wide-scale damage to seaweed crops throughout Southeast Asia [20].

Moreover, Hehre and Meeuwig (in review) found that although siganids feed on farmed seaweeds, the consumption of farmed seaweeds likely functions as a replacement for wild seaweeds rather than an actual subsidy to catches, given animals were smaller and less abundant within farmed areas than in areas not associated with farms. There is evidence that some species of siganids such as *Siganus canaliculatus*, *Siganus fuscescens* and *Siganus spinus* settle directly to macroalgal beds despite most siganids settling to coral-dominated habitats[26,34]. Many of the species that settle to macroalgal beds are also those that are frequently targeted by fishers, as is the case in the Philippines for *S. canaliculatus*. However, the potential of seaweed farms to enhance fisheries requires further investigation. In the context of declining fish returns and the potential for critical income for poor people, seaweed farms may be beneficial to reef fish by adding both structure to habitats homogenized by human presence and a potential food source to the environment. Additionally, in those areas already subjected to a high degree of disturbance, where the majority of substrate has already been negatively affected, the presence of any additional human disturbance may not be detectable. Indeed, it is possible that the addition of further human activities may in fact serve to benefit the underlying reef by reducing some of the most destructive activities like blast and cyanide fishing, and replacing them with less destructive ones. Limiting the structural degradation of reefs caused by destructive practices is particularly important in light of the link between coral reef decline and losses in fisheries productivity [42].

Here, we test the hypothesis that increased seaweed production is correlated with higher catches of siganids and whether siganid catches increase proportionately more quickly with seaweed production than associated reef fish catches. We use regional data from a major centre of seaweed farming in the Philippines on seaweed production and catches of the siganid, *S. canaliculatus*, to determine if a localised relationship exists and how this corresponds to reef fish catches more generally. We then collated global data on farmed seaweed and fisheries catches as reported to the FAO [1,2]. We focused on six tropical countries from three regions that both produce seaweeds and report siganid catches: Southeast Asia including Indonesia, Malaysia, and the Philippines; Africa including Tanzania, and Zanzibar; and the Western Pacific which was represented by Fiji. Combined, these regions account for 35% of global farmed seaweed production in 2011. We also extracted FAO catch data for non-herbivorous reef fish to determine whether siganid catches increased more quickly with increasing seaweed production, than general reef fish catches. These data allow us to understand how the large-scale implementation of seaweed farming throughout the tropics may influence catches of herbivores like siganids.

METHODS

Local Data

Bohol, a province located in the central Philippines, is a populous region with a substantial degree of poverty [2,3,18]. Residents of Bohol, particularly in the northern region of Danajon Bank, are highly dependent on both fishing and seaweed farming since few alternative income-generating opportunities exist [4,18,43–45]. Indeed, Danajon Bank is a major producer of farmed seaweed in the Philippines [2,5–8,45]. The majority of farming here is conducted on an artisanal scale: households are able to claim an area of up to 1 ha for farming on the reef which they are required to register with municipal agricultural officer [5,9,10,18]. Local subsistence fisheries are multispecies with a wide range of targets exploited for either direct consumption or sale, with siganids a key target for both purposes. Continued unsustainable levels of fishing effort combined with the use of illegal and destructive fishing methods, such as dynamite and cyanide, have lead to declining catches further compounding poverty (Green et al. 2000, 2004; Christie et al. 2006; Armada et al. 2009). Additionally the continued degradation of the reef structure from anthropogenic activities has lead Danajon bank to be classified as one of the most degraded reef systems in the world [11–14,46].

Seaweed production, and siganid and reef fish catch data for Bohol were extracted from the database of the Philippine Bureau of Agricultural Statistics (PBAS; [5,47]). The PBAS generates basic data on fisheries production and socioeconomic data related to agriculture and fisheries. Fisheries landings of reef fishes (in tons per year) were monitored for the years 2002–2012 as part of a government project designed to evaluate artisanal catch returns, and centered mainly on Danajon Bank [2,47].

Regional Data

Regional data on seaweed production and catches of herbivorous siganids and common reef fish species (in tonnes per year) have been compiled by the FAO since 1950 in varying levels of detail. We extracted these data for six countries in three regions: Southeast Asia (Indonesia, Malaysia, and the Philippines); Africa (Tanzania, and Zanzibar); and Western Pacific (Fiji) (Fig 1). A range of reef-associated taxa was selected (excluding siganids) as a control for trends in siganid catches, as effort data is unavailable. The inclusion of as many reef-associated species as possible was important in order to integrate the effects of changes in effort across a variety of reef fisheries independent of the gear used for extraction, which can vary within and among regions.

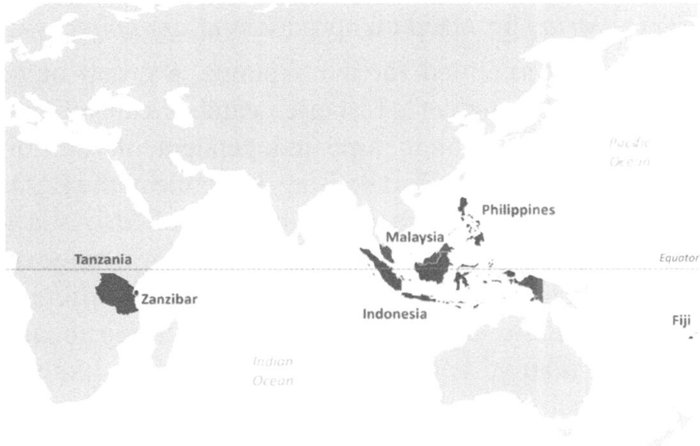

Figure 1. Countries from the three regions (Southeast Asia, Africa, and the Western Pacific) included in the global analysis of seaweed production.

Combined, these countries account for 35% of the world's seaweed production, and 58.4% when China is excluded. These countries were included in the analysis because concurrent records were available for seaweed production, siganid catches and reef fish catches for at least 15 years (FAO world fisheries and aquaculture 2012). Data were extracted from the Food and Agriculture Organization of the United Nations (FAO) database using the Fishstat J software (http://www.fao.org/fishery/statistics/software/fishstatj/en). These statistics mainly represent commercial operations as artisanal, subsistence, and recreational fisheries are not typically reported [18,48,49]. As such production levels may underestimate total landings, depending on the scale of non-commercial activities.

Our analysis of commercially farmed seaweeds included all carrageenophytes, the marine plants commonly known as red seaweeds (Rhodophyceae). The carrageenan produced by these algae is a polysaccharide used as a hydrocolloid for the manufacture of many food, pharmaceutical and industrial products. Carrageenophytes comprise nearly 50% of global landings and receive the highest prices. The carrageenophytes in the FAO database are classified as "*Eucheuma* seaweeds nei" (where nei is "not elsewhere identified") and "Spiny eucheuma" (interpreted as *Eucheuma spinosum*). *Gracilaria* red seaweeds (*Gracilaria* spp.), and elkhorn moss (*Kappaphycus alvarezii*).

Catch data were compiled for the siganids, a group of herbivores reported to FAO as "spinefeet". Fisheries catches can increase simply due to increased effort through time independent of total abundance [19,20,50–52]. However, no effort data are available in the FAO database against which changes in catch could be controlled. As a surrogate control for effort, we also extracted the catch data for a range of reef fish species (excluding siganids) representing a total of 35 different families to allow us to determine whether there was a disproportionate increase in siganid catches relative to catches of other reef fish, likely subject to similar levels of fishing effort and gears.

Modelling

We used regression analysis to examine the relationship between siganid catches and seaweed production, and between siganid catches and reef fish catches. For each variable, we calculated the percentage of the maximum value (PMV) for each year as the fraction of the highest value observed over the time series. This was done for both fish catches and seaweed production in order to generate a general trend independent of volume [21,53]. Specifically, this then allows a comparison of a standardized change in siganid catch or reef fish catch as a function of a unit change in seaweed production. Data were checked to ensure that the assumptions of linear regression in terms of distribution and homogeneity [54]. Regressions included only years where commercial seaweed farming began consistently so that initial low years prior to wide scale commercial production did not confound the results. Outliers, defined as data points diverging more than three standard deviations from the mean (there was 1 in the analysis), were also removed from the analysis. Slopes of the regression line were interpreted as a rate of increase relative to seaweed production in order to compare the relationship between siganid catch and seaweed production against other reef fish (excluding siganids) and seaweed production, which acted as a *de facto* control for increased fishing pressure across all reef fish. Differences between slopes were tested using a t test [22,54].

RESULTS

Local Analysis

Reports from the province of Bohol to the PBAS revealed that widescale farmed seaweed production for both elkhorn (*K. alvareezi*) and eucheuma (*E. spinosum*) began in 1995. Over a twelve-year time period, production increased steadily from 74,755 tonnes per annum in 2002 to 126,551 in 2011, with the exception of 2008 where mean annual production declined to 84,924 tonnes (Fig 2). A local survey of reef fish including siganids conducted by the Bohol office of the PBAS recorded landings for the same time period peaked at 825.8 tonnes in 2011, which coincided with the maximum production in seaweeds (Fig 2).

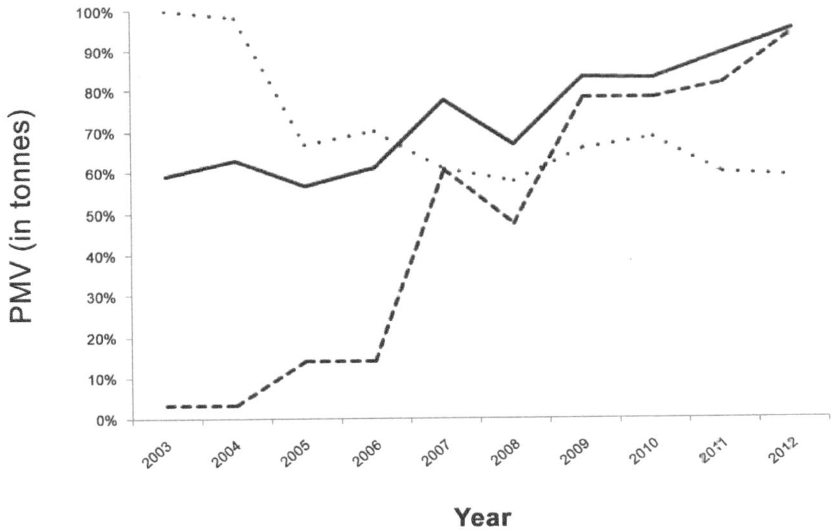

Figure 2. Temporal trends in seaweed production (solid line), siganid catch (dashed line) and reef fish catch (dotted line) as a percentage of maximum value (PMV) for the Bohol Province, Philippines.

Reef fish landings for the same time frame were highest in 2004 at 80170 tonnes (Fig 2). The relationship between siganid catch and seaweed production was significant and positive ($p = 2.77$ E-06, $R^2 = 0.89$, $n = 12$), however there was no significant relationship between reef fish catch and seaweed production ($p = 0.10$, $R^2 = 0.50$, $n = 12$): reef fish catches initially decreased relative to seaweed production and in general were steady or declining with rising seaweed production (Table 1; Fig 3). We did not compare the slopes given the non-significant relationship for reef fish catch and seaweed production (Table 1; Fig 3).

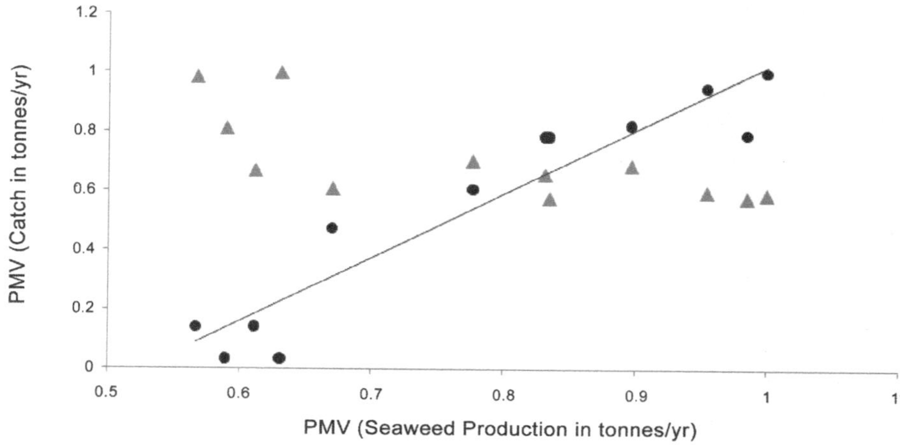

Figure 3. Comparison of the relationships between siganid catch (circles) and reef fish catch (triangles) vs. seaweed production, with all values calculated as a percentage of the maximum value (PMV) in tonnes for Bohol Province, Philippines over the period 2002–2012.

Table 1. Regression statistics for siganid and reef fish catches as a function of seaweed production respectively, including the estimated slope, intercept, coefficient of determination (R2) and p values. Where both relationships are positive and significant ($p<0.05$), slopes were compared with a t test with corresponding t-values (t), degrees of freedom (df) and p values presented. NT indicates no test. Results are presented for the regional analysis (Bohol) and for the three regions: Southeast Asia (Indonesia, Malaysia, and the Philippines), Africa (Tanzania and Zanzibar), and the Western Pacific (Fiji).

Country/Province	Siganids				Reef fish				Comparison of Slopes		
	slope	intercept	R2	p	slope	intercept	R2	p	t	df	p
Bohol	2.06	-1.05	0.89	**2.77 E-06**	-0.66	1.22	0.50	0.10			NT
Indonesia	0.80	0.26	0.78	**0.004**	0.26	0.59	0.53	**0.006**	3.97	12	<0.001
Malaysia	0.33	0.68	0.55	**0.01**	0.24	0.45	0.72	**0.002**	6.36	16	<0.001
Philippines	0.44	0.43	0.65	**5.5 E-06**	0.20	0.42	0.82	**1.0 E-06**	3.20	40	<0.002
Tanzania	-0.41	0.82	0.40	**0.001**	0.13	0.38	0.18	**0.04**			NT
Zanzibar	-0.01	0.70	0.00	0.58	-0.29	0.74	0.41	**0.03**			NT
Fiji	0.37	0.25	0.37	0.09	0.12	0.19	0.11	**0.008**			NT

doi:10.1371/journal.pone.0148250.t001

Regional Analyses

Fishbase lists 23 species of Siganid for Southeast Asia, Africa, and the Western Pacific (Table 2). Nine species are common to Southeast Asia and the Western Pacific, while only two: *Siganus argenteus* and *Siganus stellatus* are confirmed between Southeast Asia and Africa, with another three species, *Siganus guttatus*, *Siganus rivulatus* and *Siganus sutor* listed as possibly shared but unconfirmed.

Table 2. The 23 Siganid species for six countries used in the analysis and their diets as listed on Fishbase, where P is a possible but unconfirmed sighting.

Siganid species	Zanzibar	Tanzania	Fiji	Philippines	Indonesia	Malaysia	Diet
Siganus argenteus	✓	✓	✓	✓	✓	✓	Plants/Zoobenthos
Siganus canaliculatus			✓	✓	✓	✓	Plants/Zoobenthos
Siganus corallinus			✓	✓	✓	✓	Plants/Zoobenthos
Siganus doliatus			✓	✓	✓	✓	Plants
Siganus fuscescens				✓	✓	✓	Plants/Detritus/Zoobenthos
Siganus guttatus	P	P		✓	✓	✓	Plants/Zooplankton
Siganus javus				✓	✓	✓	Plants/Detritus/Zoobenthos
Siganus labyrinthodes					✓		Plants
Siganus lineatus				✓	✓	✓	Plants
Siganus luridus	✓	✓					Plants
Siganus magnificus					✓		Plants/Zooplankton
Siganus puellus				✓	✓	✓	Zoobenthos
Siganus punctatissimus			✓	✓	✓	✓	Plants
Siganus punctatus			✓	✓	✓	✓	Plants
Siganus rivulatus	✓	✓		P			Plants
Siganus spinus			✓	✓	✓		Plants
Siganus stellatus	✓	✓			✓	✓	Plants/Zooplankton
Siganus sutor	✓	✓			P	P	Plants
Siganus unimaculatus				✓			Plants/Zooplankton
Siganus uspi			✓				Plants/Zooplankton
Siganus vermiculatus			✓	✓	✓	✓	Plants
Siganus virgatus				✓	✓	✓	Plants/Zooplankton
Siganus vulpinus				✓	✓	✓	Plants/Zooplankton

doi:10.1371/journal.pone.0148250.t002

Siganid landings in Southeast Asia (Indonesia, Malaysia and the Philippines) increased at a faster rate relative to seaweed production than did reef fish landings as a proportion of the maximum value (PMV) (Fig 4). Further, Siganid catch was significantly correlated with seaweed production (Table 1; Fig 5). The other regions showed no consistent patterns.

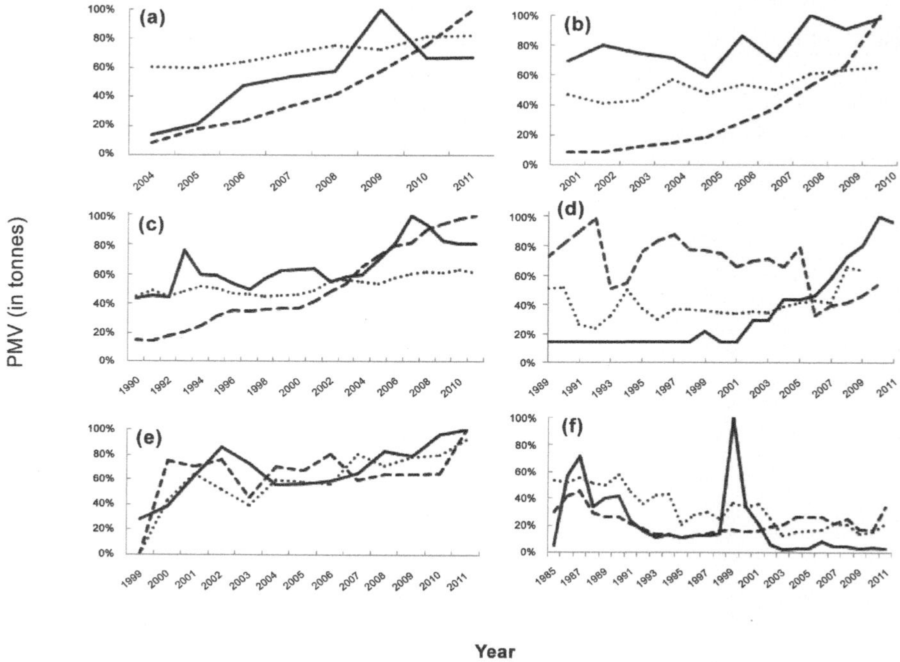

Figure 4. Temporal trends in seaweed production (solid line), siganid catch (dashed line) and reef fish catch (dotted line) as a percentage of maximum value (PMV) for the focal countries in each of the three regions: Southeast Asia (a) Indonesia, (b) Malaysia, (c) the Philippines; Africa (d) Tanzania and (e) Zanzibar; and the Western Pacific (f) Fiji.

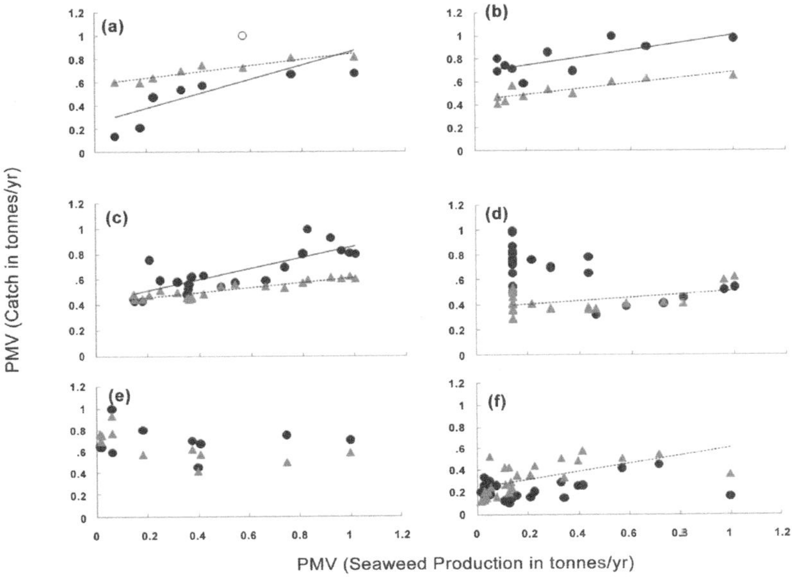

Figure 5. Comparison of the relationships between siganid catch and seaweed (circles) and reef fish catch (triangles) vs. seaweed production, with all values calculated as a percentage of maximum value (PMV) for three regions: (Southeast Asia, Africa, and the Western Pacific) (a) Indonesia, (b) Malaysia, (c) the Philippines (d) Tanzania (e) Zanzibar; and (f) Fiji. *Values with an open circle are outliers were removed from analysis but presented on the figure.

Southeast Asia

In Indonesia, reports to the FAO were initiated in 1950 under the category of "red algae" (Fig 4). In 2000, this category was made redundant and production was instead allocated to "Eucheuma nei" and "Gracilaria spp". Eucheuma nei was dominant and accounted for 92% of total red algae production on average (range: 86%-97%), with Gracilaria spp. tending to become more important through time. Given the long time series available for the combined production of Eucheuma nei and Gracilaria spp, we summed the two between 2000 and 2011 for a combined value comparable to that reported from 1950 to 1999. Seaweed production varied from approximately 10 tonnes in 1950 to 5,170,201 tonnes in 2011, exhibiting a sharp increase onward of the 1990s (Fig 4). Reporting of reef fish generally commenced in 1950, but siganid landings were only reported from 2004 to 2011, showing an annual rate of catch increase of 8% per annum, with some suggestion of a decline in the last two years (Fig 4). Mean landings of reef fish in terms of PMV increased steadily at 2% per annum from 1975 (Fig 4). There were significant positive relationships between siganid catch and farmed seaweed ($p = 0.004$; $R^2 = 0.78$, $n = 8$) (Table 1; Fig 5), as well as between reef fish catch and farmed seaweed ($p = 0.0006$; $R^2 = 0.88$, $n = 8$), (Table 1; Fig 5). Comparison of the slopes showed siganid catch increased more quickly relative to seaweed production than did reef fish catch ($t = 3.97$; $p<0.001$) (Table 1; Fig 5).

In Malaysia, reports to the FAO began in 2001 for both Spiny Eucheuma and Elkhorn with the latter accounting for about 96% of the production over this period. Annual production varied from approximately 863 and 18,000 tonnes in 2001 to a peak of 7,892 (2010) and 239,450 (2011) for Spiny Eucheuma and Elkhorn respectively (Fig 4). Both taxa showed rapid increases in production averaging approximately 9–10% per annum over this period. Reporting of reef fish generally commenced in 1950, and siganid landings were reported from 1982 to 2011 during which period there was an approximate 1% increase per annum in siganid landings. Mean catch of other reef fish increased steadily at 1.7% per annum from 1975 (Fig 4). There was a significant positive relationship between siganid catch and seaweed production ($p = 0.01$; $R^2 = 0.55$; $n = 10$) (Table 1; Fig 5) and between reef fish catch and seaweed production, ($p = 0.002$; $R^2 = 0.72$; $n = 10$) (Table 1; Fig 5). Comparison of the slopes showed siganid catch increased more quickly relative to seaweed production than did reef fish catch ($t = 6.61$; $p<0.001$)(Table 1; Fig 5).

In the Philippines, reports to the FAO began in 1965 for Elkhorn, 1974 for Spiny Eucheuma, and in 2002 for Gracilaria. Gracilaria is reported at very low levels, accounting for typically less than 0.1% of the combined production of the three seaweeds. Production of Elkhorn varied from 1,000 tonnes to 1,697,682 tonnes while Spiny Eucheuma ranged from 3,000 tonnes to 136,183 tonnes per annum, and Gracilaria from 389 to 2479 (Fig 4). Elkhorn showed a steady rise from 1980 with two distinct peaks in 1980 and 1996. Spiny Eucheuma showed a sharp increase in production from 2000 at a rate of approximately 5% per annum. Reporting of reef fish generally commenced in 1950, with siganid catches reported from 1963 to 2011, during which period there was an approximate 2% increase per annum (Fig 4). Mean landings of other reef fish increased steadily at less than 1% per annum over the same period (Fig 4). There were strong significant positive relationships between siganid catch and farmed seaweed production, and between reef fish catch and seaweed production ($p = 5.50$ E-6; $R^2 = 0.65$; $n = 22$, and $p = 1.0$ E-06; $R^2 = 0.81$; $n = 22$, respectively) (Table 1; Fig 5). Comparison of the slopes showed siganid catch increased more quickly relative to seaweed production than did reef fish catch ($t = 3.20$; $p<0.002$) (Table 1; Fig 5).

Africa

In Tanzania, reports to the FAO began in 1989 under the heading of "Eucheuma species nei" (Fig 4). Production varied from approximately 1,000 tonnes per annum in 1989 to 6,885 tonnes per annum in 2010, exhibiting a sharp increase in production of approximately 8% per annum from 2001 (Fig 4). Reporting of reef fish generally commenced in 1973 with a large increase in siganid landings between 1989 and 2005, after which landings decreased substantially (Fig 4). There was a significant but negative correlation between siganid catch and farmed seaweed production ($p = 0.0001$, $R^2 = 0.40$; $n = 23$) while reef fish catch was significantly and positively correlated to seaweed production ($p = 0.04$; $R^2 = 0.18$; $n = 23$) (Table 1; Fig 5).

In Zanzibar, reports to the FAO began in 1990 under the heading of "Spiny Eucheuma" (Fig 4). Production varied from 8,080 tonnes per annum in 1990 to 129,779 tonnes per annum in 2011, exhibiting a general steady increase of approximately 4% per annum (Fig 4). Reporting of reef fish commenced in 2000 with no clear trends in siganid landings: catches ranged from 710 tonnes per annum in 2003 to 1573 tonnes per annum in 2011 with a mean of 1096 tonnes per annum (±207 SD) and no trends through time. Mean landings of other reef fish generally increased between 2000 and 2011 at a rate of 4% per annum (Fig 4). There was no significant relationship between siganid catch and farmed seaweed production ($p = 0.58$; $R^2 = 0.005$; $n = 12$), and reef fish catch and seaweed production though significant, were negatively correlated ($p = 0.03$; $R^2 = 0.41$; $n = 12$) (Table 1; Fig 5).

Western Pacific

In Fiji, reports to the FAO on farmed seaweed production began in 1985 under the heading of "Eucheuma nei" (Fig 4). Production varied from 250 tonnes per annum in 2003 to 15,090 tonnes per annum in 1999. There was a strong boom-and-bust cycle with two major peaks in 1987 and 1999. Annual production was highly variable until 1989, when it declined to, and stabilised at a mean value of approximately 590 tonnes per year (±260 SD). Reported landings of siganids ranged between 62 and 595 tonnes per annum, reaching a general plateau from 1991 to 2011 of 112 tonnes per annum (±34.9 SD). Siganid landings peaked in 1980, six years prior to the first peak in farmed eucheuma. Mean landings of other reef fish generally declined from the 1980s onwards. There were no significant relationships between siganid catch and farmed eucheuma production ($p = 0.09$, $R^2 = 0.11$, $n = 27$) although there was a significant positive relationship between reef fish catch and seaweed production ($p = 0.008$, $R^2 = 0.37$, $n = 27$) over the period that eucheuma production was reported (1985–2011) (Table 1; Fig 5).

DISCUSSION

At the regional level in Bohol, our study documented a positive relationship between siganid catch and the production of farmed seaweed with siganid catches increasing more rapidly. This lends empirical support to the idea that more abundant food supplies may increase production of some herbivorous reef fish [2,22–25,55,56]. Such a derived benefit appears to have occurred despite the elevated levels of habitat fragmentation driven by the rapid expansion of the seaweed farming industry [1,2,22,25,37–39] and within the context of a complex mosaic of anthropogenic use in the generally degraded seascapes of Danajon Bank region [20,44]. Reefs in the area were already highly degraded from a host of anthropogenic activities before the addition of seaweed farming. However, clearing associated with the establishment of farms would serve to remove the living coral and rubble alike, along with constituent seaweeds, in order to reduce entanglement of the monolines used in farms [44,57]. However, in spite of the additional homogenisation of the substrate associated with farms, they would also serve to introduce a food source for the herbivorous rabbitfish and it is therefore possible that in this context, the establishment of seaweed farms fosters increased rabbitfish catch.

Fishing effort is a spatially and temporally heterogeneous process that generally increases as a function of human population size [50–52]. Therefore it was necessary to establish a control for inherent differences in fishing effort. The interpretation of the positive correlation between siganid catch and seaweed production as evidence of a seaweed-derived benefit to siganids, is based on the use of the reef fish /seaweed production relationship as a proxy for fishing effort. In Bohol, the relationship between reef fish and seaweed production is flat, relative to the doubling of siganid catch per unit increase in seaweed production. In context, within the same region, fisheries catch per unit effort is declining [47] due to a declining resource base. The reef fish comparison may be inappropriate if the set of species used in the comparison are unrepresentative of general fishing effort, but care was taken to incorporate species subject to similar fishing techniques and found on similar habitats. It may also be inappropriate if there has been a shift in effort towards siganids over this period. This is unlikely as there is a long history of siganid extraction in the region [47], particularly in light of the region's depleted state since the 1970s. The comparison does suggest that siganid catches are increasing disproportionately quickly relative to reef fish catches that are in decline, which provides confidence in the use of reef fish catch as a control.

We observed similar patterns regionally in Southeast Asia, where strong correlations were found between siganid catch and seaweed production, and where these relationships showed more rapid rates of increase than those based on reef fish and seaweed production. The strongest correlations between siganids and farmed red seaweeds were observed in Southeast Asia. Given the Philippines and Indonesia are respectively ranked 2nd and 3rd for global production of carrageenophytes, and Malaysia 7th, [1], this suggests a pattern of global significance. Such patterns were not, however, apparent in either Africa or the Western Pacific. Differences between seaweed production in Africa and the Western Pacific, as opposed to Southeast Asia, lie both in the duration and continuity of farming, which could have significant implications for the establishment of farmed seaweed as a food source for reef fish. In Fiji, for instance, the recurrent destruction of farming operations by typhoons has discouraged farmers from investing in infrastructure [45,58], and as a result, seaweed farming has been re-introduced on at least three occasions since the 1970s, typically on a small scale. Furthermore, fluctuating world market prices, high transportation costs to remote farming sites and an absence of local processing infrastructure all make seaweed farming less attractive than traditional fishing to many Fijians [45,58].

Similarly, in Africa, commercial seaweed farming has been both slow to establish and inconsistent in its application due to several factors, including (a) the failure of an economically valuable species of carrageenophyte, *K. alvarezii* (Elkhorn moss), and (b) societal and cultural changes associated with increased farming activities [59]. For example, *K. alvarezii*, the most profitable seaweed species, is now failing to grow in areas where it was previously cultivated due to changes in environmental conditions. These changes include warmer seas, epiphytism, and fouling [60,61]. Additionally, while initially promoted as a tool for coastal development particularly for women in Africa, further studies have shown that many women were abandoning seaweed farm because of health concerns [59,62,63] as well as in response to negative perceptions of the benefits of farming compared to the additional workload [62], and pressure over the associated cultural and societal changes brought by increased farming activities [59,64].

The scale of commercial seaweed farming may be another contributing factor to differences between Southeast Asia and Africa and the Western Pacific. Africa and the Western Pacific make relatively small contributions to worldwide production (< 1%) [1]. In Fiji, seaweed farming has occurred on a fairly small scale, with the maximum export occurring close to the inception of commercial production in 1987 and with a peak export of only 10,850 tonnes [1]. Further, in the 26 years since the commencement of commercial seaweed farming in Fiji, there has only been one increase in total production in six years (in 2000), and the overall trend has been one of general decline [1]. In both Tanzania and Zanzibar, maximum seaweed production was 129,000 and just over 100,000 tonnes respectively. These levels are orders of magnitude lower seaweed production levels in Southeast Asia. It is important to recognise that the scale of seaweed farming is to some degree a function of available habitat for farms. There is much less reef in Tanzania, Zanzibar and Fiji than in the Southeast Asian countries, 28% vs 4.8% of world's reefs [1]. Where seaweed farming occurs in Southeast Asia, it tends to be concentrated: seaweed production for Indonesia, Malaysia and the Philippines exceeds 101 t•km^{-2}, 72 t•km^{-2}, and 70 t•km^{-2}, based on total reef area by nation[65], and regions such as Bohol support intense production [47]. In Africa and the Western Pacific seaweed production for Tanzania and Fiji was only 2 t•km-2 [65], with farming occurring in low volumes compared to the scale possible based on available reefs. It may be that under these conditions, seaweed production does not increase to a threshold sufficient to support increased siganid catches. The exception is Zanzibar, which has relatively high seaweed production relative to reef area [65]. However, the comparison is inappropriate for Zanzibar as most seaweed production here occurs on sand flats and not shallow coral reefs. The implication is that in locations with low areal coverage, reef fish would be much less likely to encounter and subsequently benefit from farmed seaweed.

Differences in the relationships between siganids, reef fish, and seaweed production in Southeast Asia and Africa may also reflect differences in the ecology and feeding strategies of the siganids found in these regions [34,35,66–69]. For example, in Bohol, *S. canaliculatus* is the major siganid targeted by fisheries. It tends to settle directly on algal beds rather than coral reefs and consumes macroalgae [34], and therefore may be particularly well adapted to taking advantage of the implementation of seaweed farming. In contrast, *Siganus sutor*, common in the Indian Ocean Region and east Africa, while known to settle in algal beds like *S. canaliculatus*, exhibits a dietary preference for turf algae [70–73], suggesting that seaweed farms would not provide the same dietary benefits for *S. sutor* as they would for *S. canaliculatus*. In the absence of taxonomic resolution within the FAO global database, regional studies will help elucidate the relative importance of seaweed farming in terms of the provision of shelter vs. the provision of additional food sources.

The differences in feeding ecology of the main targeted species between the two regions may also underpin the lack of relationship between siganid catches and seaweed production in the Eastern Indian Ocean. Siganid fisheries for *S. sutor* in the Indian Ocean appear to be enhanced by the presence of algal dominated degraded reefs [70–73], reflecting *S.* sutor's dietary preference for turf algae within these systems. As seaweed farms result in the clearing of turf algae, seaweed farming may present a dietary penalty for *S. sutor* rather than a benefit as it does for *S. canaliculatus*. Such a scenario would imply that the effects of seaweed farming may vary depending on location and species, and highlights the need for further investigation into both the ecological and dietary strategies of affected fish assemblages.

Another possibility is that siganid catch may not be a direct result of seaweed farming itself, but an effect of algal domination as a result of coral reef degradation present where seaweed farming tends to occur, but initiated before commercial seaweed farming. Coral reefs in Southeast Asia show significantly higher levels of human impact from a variety of activities than either Fiji or Africa [19], and degraded reefs tend to be dominated by macroalgae [74]. As was the case in the central Philippines where the consumption of farmed red seaweeds by siganids initially appeared as a direct food subsidy (Hehre & Meeuwig in review), farms in Southeast Asia tend to be situated across wide expanses of algal dominated reefs that have subsequently been cleared for farms, leaving farmed seaweed as a replacement for areas that would have otherwise contain wild food items. Following such a system shift from algal dominated coral reefs to seaweed farms, farmed seaweed may affect fish populations in two ways. Farms may provide a replacement food source where farms have been established by clearing reefs. In this case seaweed farms, while increasing siganid catches do so only because other seaweeds have been cleared as a result of their implementation thereby resulting in either increases or maintenance of herbivorous fish. Second, farms may serve to physically concentrate populations of dependent fish, which in turn may facilitate their capture. FAO data alone do not encapsulate this information nor does it allow us to disentangle these two processes without additional surveys. It is therefore difficult to confirm whether seaweed farming provides a true subsidy to fish or whether the benefits derived from the implementation of commercial seaweed production come as a result of a decrease in other available food items.

In the face of declining returns from fisheries depleted by over-extraction, destructive fishing practices, and habitat degradation over an extended period of time [45], the potential for an expanding seaweed farming industry to enhance the productivity of a valuable food fish, the siganid, is potentially important to both artisanal fishers and seaweed farmers alike. Over the last two decades, seaweed farming has grown worldwide and become an important commodity on the world market that generates significant socio-economic benefits for marginalized coastal communities in developing countries. Higher continuity, less sporadic production and higher volumes of seaweed production may explain why siganid catch increased disproportionately faster than reef fish catches in Southeast Asia when compared to Africa and the South Pacific, and therefore, this correlative study demonstrates the potential for seaweed farming to increase siganid catch.

ACKNOWLEDGMENTS

The authors wish to thank Daniel Pauly, and the Project Seahorse Foundation for Marine Conservation for supporting this project.

AUTHOR CONTRIBUTIONS

Conceived and designed the experiments: EH JM. Performed the experiments: EH JM. Analyzed the data: EH JM. Contributed reagents/materials/analysis tools: EH JM. Wrote the paper: EH JM.

REFERENCES

1. FAO (Food and Agriculture Organization of the United Nations) Fisheries and Aquaculture Department. The state of the world's fisheries and aquaculture 2012. FAO. Rome; 2012. Available: http://www. fao. org/docrep/016/i2727e/i2727e01. pdf
2. Sievanen L, Crawford B, Pollnac R, Lowe C. Weeding through assumptions of livelihood approaches in ICM: Seaweed farming in the Philippines and Indonesia. Ocean & Coastal Management. 2005;48: 297–313.
3. Graham NAJ, Wilson SK, Jennings S, Polunin NVC, Bijoux JP, Robinson J. Dynamic fragility of oceanic coral reef ecosystems. Proceedings of the National Academy of Sciences. 2006;103: 8425–8429.
4. Munro J. The scope of tropical reef fisheries and their management. In: Polunin NC, Roberts C, editors. Chapman & Hall Fish and Fisheries Series. Dordrecht: Springer Netherlands; 1996. pp. 1–14–14.
5. Jennings S, Polunin N. Impacts of Fishing on Tropical Reef Ecosystems. Ambio. 1996;25: 44–49.
6. McManus JW. Tropical marine fisheries and the future of coral

reefs: a brief review with emphasis on Southeast Asia. Coral Reefs. 1997;16: S121–S1276.

7. Polunin NVC, Roberts CM, editors. Trophodynamics of reef fisheries productivity. London: Chapman and Hall; 1996.
8. Pauly D. From growth to Malthusian overfishing: stages of fisheries resources misuse. Traditional Marine Resource Management and Knowledge Information Bulletin. 1994;3: 7–14.
9. Allison EH, Ellis F. The livelihoods approach and management of small-scale fisheries. Marine Policy. 2001;25: 377–388.
10. Badjeck MC, Allison EH, Halls AS, Dulvy NK. Impacts of climate variability and change on fishery-based livelihoods. Marine Policy. 2010;34: 375–383.
11. Hughes TP. Catastrophes, phase shifts, and large-scale degradation of a Caribbean coral reef. Science-AAAS-Weekly Paper Edition. 1994;265: 1547–1551.
12. Knowlton N. The future of coral reefs. Proceedings of the National Academy of Sciences. 2001;98: 5419–5425.
13. Jackson JBC. Historical Overfishing and the Recent Collapse of Coastal Ecosystems. Science. 2001;293: 629–637. pmid:11474098
14. Hughes TP. Climate Change, Human Impacts, and the Resilience of Coral Reefs. Science. 2003;301: 929–933. pmid:12920289
15. Rasher DB, Hoey AS, Hay ME. Consumer diversity interacts with prey defenses to drive ecosystem function. Ecology. Ecological Society of America; 2013;94: 1347–1358.
16. MacNeil MA, Graham NAJ, Cinner JE, Wilson SK, Williams ID, Maina J, et al. Recovery potential of the world's coral reef fishes. Nature. 2015;520: 341–344. pmid:25855298
17. Graham NAJ, Jennings S, MacNeil MA, Mouillot D, Wilson SK. Predicting climate-driven regime shifts versus rebound potential in coral reefs. Nature. Nature Publishing Group; 2015;518: 94–97. pmid:25607371
18. Hill N, Rowcliffe JM, Koldewey HJ, Milner-Gulland EJ. The Interaction between Seaweed Farming as an Alternative Occupation and Fisher Numbers in the Central Philippines. Conservation Biology. 2011;26: 324–334. pmid:22098395

19. Burke L, Spalding M. Reefs at risk. World Resources Institute, Washington, DC. 2011.
20. Juanich GL. Manual on seaweed farming. ASEAN/UNDP/FAO, Manila (Philippines). ASEAN/UNDP/FAO; 1988.
21. Blankenhorn S. Seaweed farming and artisanal fisheries in an Indonesian seagrass bed-Complementary or competitive usages? PhD Thesis. 2007.
22. Eklöf JS, Henriksson R, Kautsky N. Effects of tropical open-water seaweed farming on seagrass ecosystem structure and function. Mar Ecol Prog Ser. 2006;325: 73–84.
23. Hehre EJ, Meeuwig JJ. Differential Response of Fish Assemblages to Coral Reef-Based Seaweed Farming. PLoS ONE. 2015;: 1–16.
24. Orth R, Heck K, Montfrans J. Faunal communities in seagrass beds: A review of the influence of plant structure and prey characteristics on predator-prey relationships. Estuaries. Springer-Verlag; 1984;7: 339–350.
25. Bergman KC, Svensson S, Öhman MC. Influence of algal farming on fish assemblages. Marine Pollution Bulletin. 2001;42: 1379–1389. pmid:11827126
26. Wilson SK, Fisher R, Pratchett MS, Graham NAJ, Dulvy NK, Turner RA, et al. Habitat degradation and fishing effects on the size structure of coral reef fish communities. Ecological Applications. 2010;20: 442–451. pmid:20405798
27. Williamson DH, Ceccarelli DM, Evans RD, Jones GP, Russ GR. Habitat dynamics, marine reserve status, and the decline and recovery of coral reef fish communities. Ecol Evol. 2014;4: 337–354. pmid:24634720
28. Hoey AS, Bellwood DR. Suppression of herbivory by macroalgal density: a critical feedback on coral reefs? Ecology Letters. Blackwell Publishing Ltd; 2011;14: 267–273. pmid:21265975
29. Allen KR. Relation Between Production and Biomass. Available: http://dxdoiorg/101139/f71-236. NRC Research Press Ottawa, Canada; 1971.
30. Beddington JR, Cooke JG. The potential yield of fish stocks. FAO Fisheries Technical Paper (FAO) no 242. 1983.

31. Christensen V, Pauly D. Fish production, catches and the carrying capacity of the world oceans. Naga (ICLARM QUARTERLY). 1995: 34–40.
32. Myers RA, Worm B. Rapid worldwide depletion of predatory fish communities. Nature. 2003;423: 280–283. pmid:12748640
33. Maunder MN, Sibert JR, Fonteneau A, Hampton J, Kleiber P, Harley SJ. Interpreting catch per unit effort data to assess the status of individual stocks and communities. ICES Journal of Marine Science. 2006;63: 1373–1385.
34. Hoey AS, Brandl SJ, Bellwood DR. Diet and cross-shelf distribution of rabbitfishes (f. Siganidae) on the northern Great Barrier Reef: implications for ecosystem function. Coral Reefs. 2013;32: 973–984.
35. Fox RJ, Sunderland TL, Hoey AS. Estimating ecosystem function: contrasting roles of closely related herbivorous rabbitfishes (Siganidae) on coral reefs. Mar Ecol Prog Ser. 2009;385: 261–269.
36. Fox RJ, Bellwood DR. Remote video bioassays reveal the potential feeding impact of the rabbitfish Siganus canaliculatus (f: Siganidae) on an inner-shelf reef of the Great Barrier Reef. Coral Reefs. 2008;27: 605–615.
37. Lam TJ. Siganids: their biology and mariculture potential. Aquaculture. Elsevier; 1974;3: 325–354.
38. Ogden JC, Lobel PS. The role of herbivorous fishes and urchins in coral reef communities. Environmental Biology of Fishes. 1978;3: 49–63.
39. Soliman VS, Mendoza AB, Yamaoka K. Seaweed-associated fishes of Lagonoy Gulf in Bicol, the Philippines -with emphasis on Siganids (Teleoptei: Siganidae). Kuroshio Science. 2008;2: 67–72.
40. McCook LJ. Effects of herbivory on zonation of Sargassum spp. within fringing reefs of the central Great Barrier Reef. Marine Biology. 1997;129: 713–722.
41. Mantyka CS, Bellwood DR. Direct evaluation of macroalgal removal by herbivorous coral reef fishes. Coral Reefs. 2007;26: 435–442.
42. Rogers A, Blanchard JL, Mumby PJ. Vulnerability of Coral Reef

Fisheries to a Loss of Structural Complexity. Current Biology. 2014;24: 1000–1005. pmid:24746794

43. Green S, Flores JO, Dizo-corrale JQ, Martinez RT, Nunal DRM, Armada NB, et al. The Fisheries of Central Visayas, Philippines: Status and Trends. Department of Environment and Natural Resources Regional Development Council of Region 7, Provincial, Regional and Municipal Government Units of Bohol, Cebu, Negros, Oriental, and Siquijor. 2004: 1–178.
44. Christie P, Armada NB, White AT. Coastal environmental and fisheries profile of Danajon Bank, Bohol, Philippines. Fisheries improved for Sustainable Harvest (FISH) Project. 2006;: 1–76.
45. Armada N, White AT, Christie P. Managing Fisheries Resources in Danajon Bank, Bohol, Philippines: An Ecosystem-Based Approach. Coastal Management. 2009;37: 308–330.
46. Marcus JE, Samoilys MA, Meeuwig JJ, Villongco ZAD, Vincent ACJ. Benthic status of near-shore fishing grounds in the central Philippines and associated seahorse densities. Marine Pollution Bulletin. 2007;54: 1483–1494. pmid:17645896
47. Philippine Bureau of Agricultural Statistics. In: www. bas. gov. ph [Internet]. [cited 10 Aug 2014]. Available: http://www. bas. gov. ph/
48. Norberg RA. An Ecological Theory on Foraging Time and Energetics and Choice of Optimal Food-Searching Method. The Journal of Animal Ecology. 1977;46: 511–529.
49. Harper S, Ota Y, Zeller D. Fisheries catch reconstructions: islands, part II. Harper S, Zeller D, editors. Fisheries Centre University of British Columbia Canada. 2011;19.
50. Swartz W, Sala E, Tracey S, Watson R, Pauly D. The spatial expansion and ecological footprint of fisheries (1950 to present). PLoS ONE. 2010;5.
51. Worm B, Branch TA. The future of fish. Trends in Ecology & Evolution. 2012;27: 594–599.
52. Anticamara JA, Watson R, Gelchu A, Pauly D. Global fishing effort (1950–2010): Trends, gaps, and implications. Fisheries Research. 2011;107: 131–136.
53. Worm B, Barbier EB, Beaumont N, Duffy JE, Folke C, Halpern BS,

et al. Impacts of Biodiversity Loss on Ocean Ecosystem Services. Science. 2006;314: 787–790. pmid:17082450

54. Zar JH. Biostatistical Analysis. 5 ed. River Saddle, New Jersey: New Jersey: Prentice Hall; 2010.
55. Allen DC, Vaughn CC, Kelly JF, Cooper JT, Engel MH. Bottom-up biodiversity effects increase resource subsidy flux between ecosystems. Ecology. 2012;93: 2165–2174. pmid:23185878
56. Chopin T, Buschmann AH, Halling C. Intergrating Seaweeds into Marine Aquaculture Systems: A Key Toward Sustainability. Journal of Phycology. 2001;37: 975–986.
57. Graham NAJ. Habitat complexity: coral structural loss leads to fisheries declines. Curr Biol. 2014;24: R359–61. pmid:24801184
58. Lal A, Vuki V. The historical development of seaweed farming, including roles of men and women, and prospects for its future development in Fiji. SPA Women in fisheries Information Bulletin. 2010;21: 11–16.
59. Fröcklin S, la Torre Castro de M, Lindström L. Seaweed mariculture as a development project in Zanzibar, East Africa: A price too high to pay? World Aquaculture. 2011;356: 30–39.
60. Kite-Powell H, Jiddawi N, Mmochi A. Integrated Seaweed and Shellfish Farming on Zanzibar: A Pilot Project. Final Report to the McKnight Foundation Zanzibar, Institute of Marine Science. 2005;12.
61. Msuya FE, Shalli MS, Sullivan K, Crawford B. A comparative economic analysis of two seaweed farming methods in Tanzania. The Sustainable Coastal Communities and Ecosystems Program Coastal Resources Center, University of Rhode Island and the Western Indian Ocean Marine Science Association. 2007: 27.
62. Bryceson I. Coastal Aquaculture Developments in Tanzania: Sustainable and Non-sustainable Experiences. Western Indaian Ocean Journal of Marine Science. Western Indian Ocean Marine Science Association (WIOMSA); 2002;1: 1–10.
63. la Torre-Castro de M, Lindström L. Fishing institutions Addressing regulative, normative and cultural-cognitive elements to enhance fisheries management. Marine Policy. Elsevier; 2009;34: 77–84.

64. Pettersson-Löfquist P. Social & [and] Economic Change in Zanzibar: Seaweed Farming, Tourism and Livelihood in a Coastal Community. University of Stolkhom. University, Department of social anthropology, Development studies unit; 1995.

65. Spalding M, Ravilious C, Green EP. World atlas of coral reefs. University of California Press; 2001.

66. Woodland DJ. Revision of the fish family Siganidae with descriptions of two new species and comments on distribution and biology. Indo-paciific Fish. 1990 ed.

67. Borsa P, Lemer S, Aurelle D. Patterns of lineage diversification in rabbitfishes. Molecular Phylogenetics and Evolution. 2007;44: 427–435. pmid:17383906

68. Brandl SJ, Hoey AS, Bellwood DR. Micro-topography mediates interactions between corals, algae, and herbivorous fishes on coral reefs. Coral Reefs. 2014;33: 421–430.

69. Brandl SJ, Bellwood DR. Morphology, sociality, and ecology: can morphology predict pairing behavior in coral reef fishes? Coral Reefs. 2013;32: 835–846.

70. Mcclanahan TR, Graham NAJ, Calnan JM, MacNeil MA. Toward pristine biomass: reef fish recovery in coral reef marine protected areas in Kenya. Ecological Applications. 2007;17: 1055–1067. pmid:17555218

71. Samoilys M, Kanyange N, Macharia D, Maina GW. Dynamics of rabbitfish (Siganus sutor) spawning aggregations in southern Kenya. Reef Fish Spawning …. 2013;: 33–58.

72. Robinson J, Marguerite M, Payet R, Isidore M. Investigation of the importance of reef fish spawning aggregations for the sustainable management of artisanal fisheries resources in Seychelles. Western Indian Ocean Marine Science Association (WIOMSA); 2007.

73. Vincent IV, Hincksman CM, Tibbetts IR, Harris A. Biomass and Abundance of Herbivorous Fishes on Coral Reefs off Andavadoaka, Western Madagascar. Western Indian Ocean Journal of Marine Science. Western Indian Ocean Marine Science Association; 2011;10: 83–99.

74. Bellwood DR, Hughes TP, Folke C, Nyström M. Confronting the coral reef crisis. Nature. Nature Publishing Group; 2004;429: 827–833. pmid:15215854

CHAPTER

21

BIOCONVERSION OF BALTIC SEAWEEDS INTO ORGANIC COMPOST

Izabela Michalak, Radosław Wilk, and Katarzyna Chojnacka

Department of Advanced Material Technologies, Faculty of Chemistry, Wrocław University of Science and Technology, Wrocław, Poland

ABSTRACT

Purpose

Large quantities of green seaweeds, resulting from the eutrophication, are cast ashore on the beaches of Baltic Sea (Poland). There is a necessity to collect this biomass and dispose in a cost-effective and environmental-friendly method. In the present paper, preliminary study on the possibility of composting of Baltic algae is described. After 8 weeks, algal compost and compost extract were produced. The obtained bio-products are expected to enhance plant growth and crop yield since seaweeds are known to be rich in macro- and microelement nutrients, amino acids, vitamins and plant growth promoting substances.

Methods

The obtained compost and compost extract were characterized in terms of the content of micro- and macroelements and toxic metals. The phytotoxicity of products were then biologically evaluated in germination tests on garden cress (*Lepidium sativum*) and radish (*Raphanus sativus*). In the control group, both plant species were watered with distilled water. After the experiment, dry weight of the cultivated plants was measured. Moreover, yield and the content of and micro- and macronutrients in plants was determined.

Results

Results showed that there were no phytotoxic effects in both cases and the addition of compost and extract increased the plant growth. The difference in weight for both plants between the examined groups was insignificant. The biomass of plants fertilized with the algal compost and extract was biofortified in micro- and macroelements (in particular B, Fe, Cu, Zn, Ca, K, S).

Conclusions

According to these results, composting is a useful technology both to solve waste disposal problems and to produce a valuable organic fertilizer. This could help to reduce the use of chemical fertilizers and improve physicochemical properties of the soil and to increase the plant growth.

KEYWORDS: Macroalgae, Baltic Sea, Compost, Compost extract, Natural fertilizer, Plant growth biostimulant

INTRODUCTION

Intense growth of seaweeds and their drifts, being a result of eutrophication in coastal ecosystems is observed [1, 2, 3, 4]. In some countries, algal biomass is collected in order to remove nutrients and organic matter from the coastal waters [1, 5]. Part of this biomass is used as a fertilizer in the nearby land [6], part is designated for the landfill disposal (many disadvantages, e.g., high costs, lack of the area suitable for waste settlement, leachate and biogas production during decomposition of algae) [7], part is allocated for the composting purposes [1, 2, 8, 9]. The last option opens interesting prospects for the treatment of seaweeds with the possibility of the recovery of fertilizing value of algal biomass [7]. Recycling of nutrients that minimizes their quantity imported to the farm (especially through the fertilizers) is promoted by the organic agriculture, which seeks for high quality food while respecting the environment and maintaining soil fertility by the optimal use of resources [10]. Seaweeds (fresh and dry) or its products (extracts, composts, soil conditioners) have been used in agriculture for a long time to enhance plant growth and productivity [11]. One of the main benefits of organic fertilizers is the release of available nutrients into soil, which may last for several years [12].

Different composting technologies are applied for algal biomass stabilization. Generally they are divided into three groups: (1) passive piles or windrows—material is left undisturbed and relies on natural convective air flows for aeration; (2) turned or aerated piles or windrows—air is incorporated when the material is mechanically turned and mixed and (3) in-vessel systems—optimum environmental conditions (aeration, moisture, temperature) for the quick decomposition of algal biomass [2, 9]. The obtained compost quality varies with the feedstocks that were used to produce it, the applied methods of pre-treatment of materials before composting, the composting time, the method in which the product is processed after composting etc. [13].

Co-composting of drifting and beach-cast seaweeds is a useful method of utilization of the waste biomass. Marine algae, as a component of the composting mixture, have been tested for the use as a source of elements (micro- and macroelements) and chemical substances (e.g., amino acids, vitamins, plant growth hormones, polysaccharides), which are known to have fertilizer or protective role [14]. There are four main pathways in which macroalgae can contribute to the plant health: (1) nutritional benefits (nitrogen, phosphorous, potassium, trace elements), (2) disease resistance (sulphated polysaccharides), (3) endocrine effects (cytokinins, auxins, gibberellins) and (4) soil conditioning (water holding capacity, beneficial soil biotia) [15].

The obtained algal product in the composting process can be suitable for the agricultural use as a fertilizer, soil amendment or plant growth substrate [1, 16]. By avoiding the use of a treated straw or urban and industrial sludge in the compost as co-composting materials, the produced seaweed compost shows very low content of pesticides, such as organochlorine compounds and cereal growth regulators [5]. Seaweed compost treated as organic fertilizer has several advantages over regular plant compost, especially in the case of the content of micro- (Mn, Zn, I) and macroelements (P, K, Ca, Mg), as well as the content of plant hormones such as total auxins and cytokinins (adenine) (~5 times more) and total amino acids (~7 times more) [17].

The second interesting group of algal products are extracts. They are considered as a new generation of products for sustainable agriculture—plant growth biostimulants [18, 19, 20]. These liquid products contain highly effective compounds that promote faster germination of seeds, increase the yield and the resistance to the biotic and abiotic stress. Their application may avoid excessive fertilization since they can improve mineral absorption by plants. Unlike, chemical fertilizers, extracts derived from seaweeds are biodegradable, nontoxic, non-polluting and non-hazardous to plants [18]. Seaweed extracts are known to contain nutrients that are essential for plant growth [18, 20].

The aim of the present study was to develop a simple method for the conversion of seaweed biomass into natural compost and extract. The obtained products were examined in terms of their phycotoxicity [germination tests on garden cress (*Lepidium sativum*) and radish (*Raphanus sativus*)] and multielemental composition using ICP-OES method. Additionally, the effect of algal compost as a natural organic fertilizer and compost extracts as biostimulant on the plant growth was evaluated. The yield and the mineral composition of the cultivated plants were determined.

MATERIALS AND METHODS

Raw Material for Composting

Drifting marine macroalgae were collected from the Baltic Sea near Sopot beach (Poland) in August 2013 when algal biomass was at its annual maximum. Multi-species seaweed biomass consisted of the green algae *Cladophora glomerata*, *Ulva flexuosa*, *Ulva clathrata* and the red alga *Polysiphonia fucoides*. Then the biomass has been subjected to the following processes: (1) rinsing with tap water immediately after collection to remove salt and sand, (2) removal of bigger impurities such as stones, shells and pieces of wood in a perforated rotary drum, (3) dewatering using hydraulic press (AWELD, Czech Republic), (4) drying in the rotary drum at <55 °C (Tessa, Poland) to reduce water content in algae below 15 % [20]. Afterwards, algal biomass was pelleted using a Granulator TL 700 (Gama Pardubice, Czech Republic).

Composting of Algae

About 1 ton of dry, pelleted Baltic algae underwent composting process. The process was conducted in the open air in a pile with dimensions of approximately 1.5 m × 1 m × 1.2 m, which allowed for the effect of so called "chimney aeration". Moisture was maintained at the level 60–65 % for 6 weeks. The mixture was overturned once a week. The maximum temperature achieved was 38 °C. The temperature was measured (in triplicate) during the process, at various points around the piles, at a depth of 60 cm. The maturity of the compost was evaluated by measuring the parameters such as change of the odour, colour and the structure of compost and performing the phytotoxicity test. The resulting compost was also characterized with respect to chemical composition.

Extract from the Compost

From the obtained compost, extract was produced according to the procedure described by Alvarado et al. [21]. Subsequently, 130 g of the compost was filtered using a funnel with filter paper (Whatman Filter Paper No. 2; GE Healthcare Life Sciences, Little Chalfont, England) and 300 mL of distilled water. The process was repeated 5 times. An intense brown color liquid was obtained. The filtrate was considered as 100 % [21]. The obtained extract from the compost was characterized in terms of chemical composition and phytotoxicity.

The Phytotoxicity of Compost—Germination Tests

The phytotoxicity of compost and compost extract was evaluated in seed germination of cress (*Lepidium sativum*) and radish (*Raphanus sativus*). These two plants were chosen because they can reach maturity in short time period (to 2 weeks). Tests were carried out on plates (90 mm diameter) with wool in a germinator—type Jacobsen (Laborset, Lodz, Poland) in the room temperature, according to the international norm (International Rules for Seed Testing, 2011—International Seed Testing Association).

Seeds of each plant (50) were placed on a plate (N = 4, four replications for each tested group). Then, prepared plates were put for three days into a fridge (stratification process). Tests were performed in three variants: (1) a control group—the growth of plants watered with water (2 mL); (2) compost—the growth of plants with the addition of compost (1.0 g) (3) extract obtained from the compost (100 %)—the growth of the plants watered with extract (2 mL) [22]. All the plates were watered with 2 mL of water every two days. Germination tests lasted for 9 days. At the end of the experiment, the plants were collected from each plate, dried and the dry mass was weighed. Additionally, the effect of the tested algal product (compost or extract) on the multielemental composition of cultivated plants was evaluated using ICP-OES technique (Inductively Coupled Plasma—Optical Emission Spectrometry).

Analytical Methods

The analysis of the content of C and N in algae and in the compost was made with the Vario MACRO Cube CN analyzer (Elementar Analysensysteme GmbH, Germany). The analysis of the elemental composition of algae, compost, extract from the compost as well as the cultivated plants was performed using ICP–OES iCAP 6500 Axial and Radial Vista (Thermo Scientific, Waltham, Massachusetts, USA) in accordance with PN-EN ISO/IEC 17025:2005. Before multielemental analysis, the samples [about 0.5 g of dry mass (d. m.)] underwent the mineralization (with 5.0 mL of 69 % HNO_3) in Teflon bombs using the microwave oven—Milestone Start D (Milestone S.r.l., Sorisole, Italy).

Statistical Analysis

The results were analyzed with the use of *Statistica* ver. 12, where statistical significance was established for p values < 0.05.

RESULTS AND DISCUSSION

The maturity of the obtained algal compost was evaluated by the observation of the following properties: (1) the odor of the compost—change from the intense odor of marine algae into the earthy/humus odor (the change was observed after 5 weeks of the process), (2) the color of the compost—from dark green to dark brown (after 5 weeks), (3) loosening of the structure of the pellets to the form of loose soil (after 5 weeks), (4) no forms of fibrous material (after 4 weeks).

ELEMENTAL COMPOSITION OF ALGAL BIOMASS, ALGAL COMPOST AND EXTRACT FROM THE COMPOST

Scientists working on co-composting of algal biomass suggest that the obtained products could be used as a fertilizer and soil conditioner in agriculture, as they are of a natural origin [10, 23] and constitute a rich source of micro- and macronutrients, amino acids, vitamins, plant growth hormones (auxins, cytokines) and polysaccharides such as alginates, ulvans etc. [19]. The quality of the obtained product in terms of the degree of stabilization and the final available nutrients depends mainly on the carbon to nitrogen ratio (C:N ratio) of the organic material being composted. It is an important factor, since affects the microbial compost communities [2]. An initial C:N ratio of 20:1–40:1 is recommended for rapid composting. However, C:N ratios as low as 14:1 also compost well and are practical for composting animal mortalities [24].

The content of C and N in the raw biomass of algae used in the present study, before composting process, was 20.2 ± 0.2 % d. m. and N 1.86 ± 0.02 % d. m (C:N ratio 10.8). Algae are characterized by a relatively low carbon to nitrogen ratio [2], for example C:N ratio in green macroalga *Ulva* sp. was 8.5 [1], 8.8 [25]. The C:N ratio in the obtained algal compost was equal 10.8 (the content of C was 14.3 ± 0.2 % d. m. and N 1.31 ± 0.01 % d. m.). The content of C and N decreased by 29 and 30 %, respectively in the compost when compared with the natural biomass of algae. During the composting process, carbon and nitrogen are mineralized and lost in gaseous forms mainly as CO_2, NH_3, N_2O and N_2 [26]. In Table 1, several examples of the composting of algae in piles and windrows are presented. As can be seen, C:N ratio in the final compost varies from 7 to 45.

Table 1. Co-composing of algae in piles and windrows

Composting mixture	Location of seaweeds harvesting	Composting technique	Properties of compost	References
Seaweed and cow dung (2:1)	Cape Monze and Manora, Karachi coast, Pakistan	750 kg; windrow pile; 40 days	C:N ratio 7.0 2.3 % N, 0.86 % P, 1.8 % K	[28]
Drift seaweed (a mixture of *Ulva* sp., *Codium* spp. and *Dictyota* sp.) and sawdust (a mixture of poplar, willow and pine) (4:1)	Puerto Madryn in northeastern Patagonia, Argentina	Pile: length 5 m, large 1,2 m and high 0,50 m; 10 months	C:N ratio 11.2 0.16 % N, 0.0096 % P, 0.16 % K	[34]
Algae (*Ulva* spp., *Codium vermilara*, *Dictyota dichotoma*, *Ceramium rubrum*) and lingocellulosic material (wood chips, yard trimmings and sawdust) (3:1)	Puerto Madryn in northeastern Patagonia, Argentina	Pile (C20—approx. 6 m^3; 20 months); pile (C9—approx. 14 m^3; 9 months)	C:N ratio: C20 11.9; C9 11.0 C20: 0.19 % N, 0.008 % P, K—not available (n.a.) C9: 0.30 % N, 0.008 % P, K—n.a.	[11]
Partially composted seaweed and partially composted yard trimmings (1:1)	Florida, USA	Windrow; 3 months	C:N ratio—n.a. 0.0028 % N, 0.003 % P, 0.0171 % K	[13]
Seaweed from the beach (primarily *Laminaria* sp. and *Cystoseira* sp.), fish waste and pine bark (1:1:3)	Lugo, Spain	Pile: 2 m wide at the base, 1 m high and 6 m long (approx. 10 m^3); 4 months	C:N ratio 22.6 2.1 % N, 0.63 % P, 0.69 % K	[10]

Mediterranean seagrass and seaweed residues with yard waste (1:1)	Coast of Denia, Alicante, Spain	Open windrow; 8 months	C:N ratio 33.8 0.91 % N, 0.079 % P, 0.83 % K	[29]
Seaweed from the beach (primarily *Laminaria* sp. and *Cystoseira* sp.), fish waste and pine bark (1:1:3)	Lugo, Spain	Pile (6 × 2×1.5 m); 2.5 month	C:N ratio 21.9 2.11 % N, 0.61 % P, 0.72 % K	[16]
Drift seaweed (mainly *Laminaria* spp. and *Cystoseira* spp.) with mackerel waste and pine bark (1:1:3)	Northern coast of Galicia (Spain)	Windrow; 1.5 month	C:N ratio ~22 2.0 % N, 0.60 % P, 0.70 % K	[35]
Ulva sp. and lignocellulosic substrates: poplar bark or poplar sawdust or sawdust or municipal lignocellulosic refuse (*Thuja*)	Hillion Bay area, Gulf of St Malo, Brittany, France	Pile (from 12 to 75 m^3, e.g. *Ulva*:poplar bark ~9:1—36 m^3); 3 months	C:N ratio—from 11 (*Ulva*: poplar bark ~9:1) to 45 (*Ulva*:sawdust ~4:1); *Ulva:* poplar bark ~9:1 0.57 % N, 0.09 % P, 0.76 % K	[1]
Posidonia oceanica leaf residues, with vegetable residues (from broccoli, fennel, and celery scraps) and grapevine pruning residues (1:2:2 or 3:1:1)	Alicante, Spain	Piles (approx. 3 m^3); 3 months	C:N ratio of 35–45 NPK content—n.a.	[36]

Ulva and straw (9:1) *Ulva* and wood (9:1)	Venice Lagoon (Italy)	Windrow; 20 t; ~3 months	(1) C:N ratio 12.3; 1.7 % N, 0.183 % P, 0.967 % K (2) C:N ratio 25.1; 0.9 % N, 0.107 % P, 0.771 % K	[3]

In Table 2, the multielemental composition of algal biomass, compost and extract from the compost is shown.

Table 2. Multielemental composition of algal biomass, algal compost and compost extract

Element	Wavelength (nm)	Algal biomass [mg/kg d. m., (mean ± SD; N = 3)]	Algal compost [mg/kg d. m., (mean ± SD; N = 3)]	Compost extract [mg/L, (mean ± SD; N = 3)]
Al	308.215	3324 ± 665	3846 ± 769	0.55 ± 0.08
B	249.773	175 ± 26	146 ± 22	5.16 ± 0.77
Ca	315.887	13,289 ± 2658	18,097 ± 3619	375 ± 56
Cd	228.802	0.93 ± 0.12	0.88 ± 0.11	0.01 ± 0.00
Cr	267.716	10.30 ± 1.54	26.02 ± 3.90	0.02 ± 0.00
Cu	324.754	23.85 ± 3.58	26.58 ± 3.99	0.21 ± 0.03
Fe	259.940	7727 ± 1545	8460 ± 1692	3.01 ± 0.45
K	766.491	5443 ± 1 089	4917 ± 983	149 ± 22
Mg	285.213	3822 ± 764	3949 ± 790	487 ± 73
Mn	257.610	223 ± 33	189 ± 28	6.52 ± 0.98
Mo	202.030	0.67 ± 0.10	0.66 ± 0.10	0.04 ± 0.01
Na	588.995	5209 ± 1042	4451 ± 890	981 ± 147
P	213.618	1449 ± 290	1587 ± 317	499 ± 75
Pb	220.353	10.59 ± 1.38	12.54 ± 1.63	0.02 ± 0.00
S	181.972	8963 ± 1793	7665 ± 1533	18.75 ± 2.81
Si	251.611	48.55 ± 7.28	177 ± 27	17.60 ± 2.64
Zn	213.856	124 ± 19	118 ± 18	0.50 ± 0.08

Algae are known to have a suitable content of nitrogen and potassium, but are much lower in phosphorus than traditional animal manures and the typical N:P:K ratios in chemical fertilizers [6]. In the present study, the content of N, P and K in the raw biomass of algae was as follows: N:

1.86 % d. m., P: 0.145 % (0.332 % as P_2O_5) and K: 0.544 % (0.656 % as K_2O). By comparing the macroelemental composition of algal biomass with other natural manures, such as farmyard manure (N: 0.57 %, P as P_2O_5: 0.45 %, K as K_2O: 0.52 %) and liquid manure (N: 0.42 %, P as P_2O_5: 0.23 %, K as K_2O: 0.37 %) [27], it was found that algae were a richer source of nitrogen and potassium than the compared manures. Additionally, algae contained more phosphorus than liquid manure but less than farmyard manure. The same tendency was observed in the case of the obtained algal compost that contained: N 1.31 % d.m., P: 0.159 % (0.364 % as P_2O_5) and K: 0.492 % (0.592 % as K_2O). These results are in agreement with the data presented by McHugh et al. [6]. Morand et al. [5] showed that the compost made from green and brown seaweeds, poplar ground bark and laying-hen droppings had higher content of N, P and K when compared to manure of cattle and horses (P as P_2O_5: 1.05 % w. m. (wet mass) and 0.1–0.5 % w. m. respectively, K as K_2O: 1.0 and 0.1–0.7 % w. m., total N: 0.85 and 0.3–0.7 % w. m.). The content of N, P, K in compost produced from algae ranges from 0.0028 % [13] to 2.3 % [28] for N; from 0.003 % [13] to 0.183 % [3] for P; from 0.0171 % [13] to 1.8 % [28] for K (Table 1).

The results obtained in the present study show the increase of micro- and macronutrients in the produced algal compost when compared with the raw biomass (Ca by 27 %, Cu by 11 %, Fe and P by 9.0 %). These findings are in agreement with the results presented by other authors, for example Orquin et al. [29] noted the increase of P, K, Ca and Mg in the compost produced from seagrass and seaweed mixed with municipal yard and landscape green waste, Michalak et al. [30] observed the increase of Ca, Fe, Zn, P, Mg in the compost produced from seaweed *Fucus*sp., ecological hay, grass, sawdust and spent mushroom substrate, Cole et al. [8] noticed the increase of S, Ca, Fe, Mn, Zn in the compost made from green seaweed—*Ulva ohnoi* and sugarcane bagasse biomass. Calcium and iron were the dominant elements.

One of the main disadvantage of the compost produced from seaweed might be the content of toxic metals. Macroalgae are known to accumulate toxic metals [23]. In the present work it was found that the content of toxic elements in the obtained compost was below the maximum acceptable levels: Cr (26.02 ± 3.90 < 100 mg/kg d. m. limits in the obligatory law), Cd (0.88 ± 0.11 < 5 mg/kg d. m.), Pb (12.54 ± 1.63

< 140 mg/kg d. m.) taking into account Polish regulations [31]. In the present study, from the algal compost, extract was also obtained (mineral composition is presented in Table 2). For the comparison, the examples of the multielemental composition of the extracts obtained from green seaweeds by different extraction techniques are presented in Table 3.

Table 3. The examples of the multielemental composition of extracts obtained from green seaweeds

Element	***Polysiphonia, Ulva, Cladophora***	**Baltic green seaweeds**	***Polysiphonia, Ulva, Cladophora***	**Baltic green seaweeds**	***Ulva reticulata***
	SFE [20] (mg/L)	**Extract obtained by alkaline extraction with KOH [44] (mg/L)**	**Extract obtained by MAE 60 °C [45] (mg/L)**	**Extract obtained by acidic extraction with water of pH 2 [46] (mg/L)**	**Extract obtained by heating for 45 min. in 60 °C [18] (mg/L)**
Ca	1060 ± 210	127 ± 19	365 ± 54	230 ± 34	158
Cu	6.27 ± 0.94	0.79 ± 0.12	0.11 ± 0.02	0.04 ± 0.01	1.15
Fe	9.24 ± 1.39	30.17 ± 4.52	2.87 ± 0.43	3.76 ± 0.56	5.2
K	51.76 ± 7.76	28,260 ± 5650[a]	950 ± 140	143 ± 21	175
Mg	406 ± 61	34.11 ± 5.12	322 ± 48	28.50 ± 4.40	108
Mn	6.56 ± 0.98	1.96 ± 0.29	3.07 ± 0.46	1.20 ± 0.18	n.a.
Na	965 ± 145	620 ± 90	1250 ± 250	46.20 ± 6.90	295
P	43.19 ± 6.48	6 050 ± 1210[b]	32.90 ± 4.90	10.10 ± 1.50	45
S	9300 ± 1 900	1350 ± 270	700 ± 105	532 ± 80	55
Zn	5.25 ± 0.79	6.86 ± 1.03	0.17 ± 0.03	0.65 ± 0.10	1.3

[a]KOH was used as a solvent; [b]H_3PO_4 was used to neutralize the extract to pH 7

Minerals and trace elements present in seaweed extracts enhance plant nutrition and have a critical role in plant development [32]. In the present study, the extract obtained from the algal compost was a very good source of Mg, Mn, Na, P, good of Ca and K and poor of Cu, Fe, S and Zn when compared with extracts obtained by other extraction techniques (e.g., supercritical fluid extraction (SFE), microwave assisted extraction (MAE), chemical extraction etc.). The multielemental composition of the tested seaweed extract was poorer than the compost, however, as it was indicated in the literature, are bioactive at low concentrations (diluted as 1:1000 or more) [19].

THE PHYTOTOXICITY OF THE PRODUCTS—GERMINATION TESTS

In many studies, phytotoxicity analysis is carried out first in order to evaluate the utilitarian properties of the obtained compost. In the literature, it is suggested that composting of seaweeds reduced their phytotoxicity [3, 33]. Cuomo et al. [3] examined the phycotoxicity of compost produced from *Ulva* sp. on *Lepidium sativum*. It was indicated that a complete lack of phytotoxicity and the absence of any inhibitory effects on the germination rate of *L. sativum* seeds showed that the tested compost reached maturation state and had a good final quality [3]. In the present study, the obtained results showed that there were no phytotoxic effects of algal compost and compost extract in the cultivation of garden cress (*Lepidium sa*tivum) and radish (*Raphanus sativus*). The results of the germination tests after 7 days are presented in Fig. 1: (a) for garden cress and (b) for radish.

Figure 1. Germination tests on a garden cress (*Lepidium sativum*) and b radish (*Raphanus sativus*) after 7 days.

EFFECT OF SEAWEED COMPOST AND COMPOST EXTRACT FROM THE COMPOST ON GARDEN CRESS AND RADISH YIELD

In the present work, the preliminary germination tests were also used the evaluate the effect of algal compost as a natural organic fertilizer and compost extract as biostimulant of plant on the yield of the garden cress and radish. Dry weight of the cultivated plants in the examined groups is presented in Table 4. For both plants, the differences between the groups (control, algal compost, compost extract) were not statistically significant ($p < 0.05$). The tested products slightly increased the mass of the plants, however in both cases better results were obtained for compost extract, rather than for compost. In the case of garden cress, the mass of plants in the group with algal compost was 1.0 % lower than in the control group, whereas in the group with compost extract higher by 4.1 %. For the radish, the mass of plants in both experimental groups was higher than in the control group—by 6.2 % for algal compost and by 8.8 % for compost extract.

Table 4. Dry mass of the cultivated plants in the examined groups

Group		Dry mass [g (mean ± SD); N = 4]
Garden cress (*Lepidium sativum*)	Control—distilled water	0.061 ± 0.001
	Algal compost	0.060 ± 0.001
	Compost extract	0.063 ± 0.003
Radish (*Raphanus sativus*)	Control—distilled water	0.306 ± 0.026
	Algal compost	0.325 ± 0.008
	Compost extract	0.333 ± 0.023

Literature indicates that the application of algal compost improves the germination of seeds [28], plant growth and water stress resistance [34], water holding capacity [13, 34], plant height and weight [11, 14, 35, 36], resistance to diseases on plants (e.g., tomato) [11], as well as the content of micro- and macroelements in the cultivated plants [14, 36]. Some of the examples of the action of seaweed extracts in the plant cultivation are presented below. In the work of Haq et al. [28], germination tests were performed on maize (*Zea mays*). It was shown that in the group with compost (algae and cow dung, 2:1), the germination was 78 %, whereas in the group with cow dung—60 %. Eyras et al. [34] evaluated the effect of compost produced from seaweed and sawdust (4:1) on the growth of tomato plants (*Licopersicum esculentum* var. *platense*). It was found that the increase of tomato was proportional to the compost doses (0—control, 10, 25, 50 and 100 % compost in a substratum of sand). Results showed that the addition of the compost increased plant growth and water holding capacity. In the group with 100 % compost, all plants flowered and 83 % of them bore fruit and survived the imposed water stress, whereas plants in the group with NPK fertilizer were severely affected by water stress and only one plant per treatment survived. Vendrame and Moore [13] also confirmed that substrates, designated for plant cultivation, containing seaweed compost (1:1; seaweed compost and yard trimmings compost) had greater water holding capacities than substrates containing the mixture of biosolids compost and yard trimmings compost (1:1). It was suggested that seaweed compost could be potentially substituted for peat in the growing substrate. Illera-Vives et al. [35] noted that the highest dose of the compost (66 t/ha), produced from drift seaweed, mackerel waste and pine bark, increased the tomato yields (also fruit weight and fruit diameter) when compared with other groups: organic fertilizers—broiler litter (higher by 12.2 %), slow-release NPK treatment (higher by 17.6 %) and a control treatment without fertilizers (higher by 18.3 %). These beneficial effects of the compost could result not only from the supplementation of nutrients but also from the improvement of the soil physical conditions due to high alginate content in algae, as well as the content of growth regulators, such as auxins and cytokinins. Eyras et al. [11] found that the compost produced from algae and lingocellulosic material (3:1) aged 9 and 20 months influenced significantly the total weight of tomatoes, which was higher by 35 and 65 %, respectively when compared to the control group (soil without amendment). The doses (5 and 10 kg/m^2) did not show any significant differences within compost.

EFFECT OF SEAWEED COMPOST AND COMPOST EXTRACT ON THE MULTIELEMENTAL COMPOSITION OF GARDEN CRESS AND RADISH

Seaweed compost, applied just before cultivation, provides a "stimulus" effect that makes nutrients rapidly available to the plants on a relatively long period of time. It was found that the compost is suitable for intensive cultivation both in the field and the greenhouse [5]. Among the main categories of plant biostimulants—seaweed extracts and botanicals, humic and fulvic acids, protein hydrolysates and other N-containing compounds, plant growth promoting *Rhizobacteria*, the function of the improvement of mineral composition of plant tissues (biofortification) is assigned to seaweed extracts [37].

In the literature, the effect of algal products (compost, extracts) on the mineral composition of the cultivated plants is rarely studied. Usually it is limited to the main macronutrient—N, P and K. Table 5 presents the multielemental composition of the cultivated plants, garden cress and radish, treated with algal compost and compost extract.

For both plants, the addition of algal compost increased the content of microelements such as: B: by 108 % for garden cress and by 200 % for radish, Cu: by 6.8 % and 54 %, Mn: by 25 % and 73 %, Si: by 202 % and 567 %, Zn: by 17 % and 62 % and macroelements such as: Ca: by 16 % and 53 %, respectively, K: by 290 % and by 189 %, Mg: by 26 % and 26.5 %, respectively when compared with the control group.

The soil application of the compost extract increased the content of microelements such as: B: by 134 % in garden cress and by 122 % in radish, Mn: by 20 % and 30 %, Zn: by 32 % and 54 % and macroelement K: by 276 % and 153 %, respectively, when compared with the control group.

In the literature it is suggested that algal extracts can stimulate the expression of genes that encode transporters of micronutrients (e.g. Cu, Fe, Zn) [38]. Billard et al. evaluated the effect of extract from brown seaweed—*Ascophyllum nodosum* on the multielemental composition (N, S, K, Ca, P, Mg, Fe, Na, Mn, B, Si, Cu, and Zn) of oilseed rape (*Brassica napus*). The application of *Ascophyllum* extract stimulated root growth and macronutrient uptake of N, S, K, and P. Moreover, the extract also increased the content of Mg, Mn, Na, and Cu in plants and root-to-shoot translocation of Fe and Zn. These observations were associated with an increased expression of a Cu transporter (*COPT2*) and *NRAMP3*, a gene involved in Fe and Zn translocation [38].

Table 5. Multielemental composition of the cultivated plants

Element	Garden cress (*Lepidium sativum*)			Radish (*Raphanus sativus*)		
	Control [mean ± SD (mg/kg d. m.); N = 4]	Algal compost [mean ± SD (mg/kg d. m.); N = 4]	Compost extract [mean ± SD (mg/kg d. m.); N = 4]	Control [mean ± SD (mg/kg d. m.); N = 4]	Algal compost [mean ± SD (mg/kg d. m.); N = 4]	Compost extract [mean ± SD (mg/kg d. m.); N = 4]
Al	3.16 ± 0.47	78.95 ± 11.84	17.69 ± 2.65	5.13 ± 0.77	36.96 ± 5.54	4.29 ± 0.64
B	12.82 ± 1.92	26.63 ± 3.99	29.98 ± 4.50	6.75 ± 1.01	20.23 ± 3.03	15.02 ± 2.25
Ca	4346 ± 869	5028 ± 1 006	4626 ± 925	7497 ± 1 499	11,478 ± 2 296	7448 ± 1 490
Cd	0.19 ± 0.03	0.22 ± 0.03	0.20 ± 0.03	0.24 ± 0.04	0.35 ± 0.05	0.25 ± 0.04
Cr	2.29 ± 0.34	1.06 ± 0.16	0.99 ± 0.15	0.35 ± 0.05	0.66 ± 0.10	0.41 ± 0.06
Cu	5.03 ± 0.75	5.37 ± 0.81	4.53 ± 0.68	4.22 ± 0.63	6.49 ± 0.97	5.43 ± 0.81
Fe	173 ± 26	167 ± 25	123 ± 18	88.60 ± 13.29	165 ± 25	84.69 ± 12.70
K	13,246 ± 2 649	51,719 ± 10 344	49,779 ± 9 956	9800 ± 1 960	28,285 ± 5 657	24,753 ± 4 951
Mg	5883 ± 1 177	7415 ± 1483	7560 ± 1 512	6341 ± 1 268	8024 ± 1 605	5285 ± 1 057
Mn	43.32 ± 6.50	54.18 ± 8.13	52.12 ± 7.82	32.25 ± 4.84	55.59 ± 8.34	41.78 ± 6.27
Mo	2.56 ± 0.38	2.41 ± 0.36	2.05 ± 0.31	0.77 ± 0.12	0.77 ± 0.12	0.73 ± 0.11
Na	8661 ± 1732	6917 ± 1 383	7527 ± 1 505	2867 ± 573	11,415 ± 2 283	6895 ± 1 379
P	16,827 ± 3 365	16,126 ± 3 225	14,975 ± 2 995	19,626 ± 3 925	21,243 ± 4 249	12,691 ± 2 538
Pb	1.14 ± 0.17	0.16 ± 0.02	<LLD	0.01 ± 0.00	0.13 ± 0.02	<LLD
S	11,660 ± 2332	19,220 ± 3 844	18,014 ± 3 603	33,817 ± 6 763	33,515 ± 6 703	21,749 ± 4 350
Si	51.59 ± 7.74	156 ± 23	52.55 ± 7.88	15.90 ± 2.39	106 ± 16	15.12 ± 2.27
Zn	145 ± 22	169 ± 25	191 ± 29	75.24 ± 11.29	122 ± 18	116 ± 17

<LLD—Below detection limit

In the present work we checked, which algae-based product had beneficial effect on the multielemental composition of tested plants. The content of micro- and macroelements in garden cress was mainly influenced by the application of compost, especially the content of Cu was by 16 % higher than for the group with compost extract, 8.0 % higher for Ca, 6.0 % higher for S and 4.0 % higher for K and Mn. Better results in terms of multielemental composition were obtained for the second tested plant—radish. The content of Na was 66 % higher in the group with compost, than with the extract, Ca by 35 %, Mg by 34 %, B by 26 %, Mn by 25 %, Cu by 16 %, K by 12 % and Zn by 5.0 %. Boron was a trace element that appeared to be provided by seaweed biomass in both treatments. Similar observation was noted by Winberg et al. [15], who evaluated the *Microdictyon umbilicatum* bloom biomass as an agricultural compost conditioner for native and commercial plants.

In our previous work, compost and extract obtained from the composting mixture (seaweed—*Fucus* sp., ecological hay, grass, sawdust and spent mushroom substrate) were examined on garden cress. It was found that the extract better affected the content of macroelements in plants than compost, however these differences were slight (2.0 % more of Ca, 1.0 % more of K, 11 % more of Mg and 15 % more of S). On the other side, examined compost better influenced the content of microelements (B—35 % more, Cu—15 % more, Mn—28 % more, Zn—5.0 % more and Fe ~2 times more) than the extract [30]. Mininni et al. [36] tested the feasibility of the use of compost produced from the residues of alga *Posidonia oceanica*, vegetables (broccoli, fennel and celery scraps) and grapevine pruning as partial substitutes for peat in the growing media used in the cultivation of melon and tomato. For both plant, the best results in terms of the content of micro- and macroelements were obtained for growing media consisted of the composted mixture (1:2:2)—25 % and peat—75 %. For melon the increase was observed for Ca, Cu, Mn and Zn, for tomato for K, Ca, Mg, Na, Fe, Mn and Zn when compared with the control group—peat. Morand et al. [5] indicated that as growth media, algal composts should be mixed with dirt, sand, turf or artificial supports to reduce their strength for agricultural use.

Crouch et al. [39] found that the application of seaweed concentrate (Kelpak) significantly increased the yield and the amounts of Ca, K and Mg in the leaves of lettuce receiving an adequate supply of nutrients.

These results were also confirmed by Sosnowski et al. [40], who examined the effect of the extract from *Ecklonia maxima* (Kelpak) on selected micro- and macroelements in aerial biomass of hybrid alfalfa. The application of this seaweed extract led to an increase of the content of P, K, Zn and Mn in alfalfa aerial biomass. The content of Mg, Ca, Cu and Mo did not change significantly in response to the tested extract. In the work of Abd El-Motty et al. [41] it was confirmed that the application of algal extract in different doses (0.5, 1.0 and 2.0 %) enhanced N, P, K and B content in the leaves of Keitte mango trees. Also in the case of the application of seaweed extract, obtained by the homogenization of red alga *Kappaphycus alvarezii*, an significant increase of the N, P and K uptake by grains at higher concentrations (12.5 and 15 %) was observed when compared with control (water) [42]. Mancuso et al. [43] suggested that the presence of marine bioactive substances in seaweed extracts improved stomata uptake efficiency in treated plants compared to non-treated plants. From the presented examples it can be seen that algal extracts influenced mainly the content of macroelements in the cultivated plants. The main ingredients in extracts that affect plants are phytohormones (cytokinins, auxins), which are found in the composition of most seaweed concentrates [19, 40].

CONCLUSIONS

In the present work, a method of bioconversion of Baltic seaweeds, being the result of eutrophication in the coastal areas, was proposed. From algal biomass, compost and compost extract were produced. It was proposed to apply those preparations as natural organic fertilizer and plant growth biostimulant, respectively. The obtained products were characterized in terms of the content of micro- and macroelements and toxic metals using ICP-OES technique. Based on the obtained results it may be concluded that the resulting compost and extract were of good quality in terms of nutrient content and low toxic metal content. The content of C and N in the obtained compost decreased by 29 and 30 %, respectively when compared with the natural biomass of algae. The increase of micro- and macronutrients (Ca by 36 %, Cu by 11 %, Fe and P by 9.5 %) in the compost was observed when compared with the raw biomass. The algal extract obtained from the compost was a source of Mg, Mn, Na, P, Ca and K.

The phytotoxicity of algae-based products was evaluated in the germination tests on the garden cress (*Lepidium sativum*) and radish (*Raphanus sativus*). No phytotoxic effect was recorded. Additionally it was found that the examined products slightly increased the mass of the plants. For both plants, the addition of algal compost increased the content of microelements such as: B, Cu, Mn, Si, Zn and macroelements such as: Ca, K and Mg when compared with the control group. The soil application of the algal extract increased the content of microelements such as: B, Mn, Zn and macroelement K when compared with the control group. Better elemental composition had plants treated with algal compost, rather than with extract. The obtained results show that composting technology can be used as a method to turn seaweed biomass into compost that can be used as organic amendments or fertilizers for agricultural systems.

REFERENCES

1. Abd El-Motty, E.Z., Shahin, M.F.M., El-Shiekh, M.H., Abd-El-Migeed, M.M.M.: Effect of algae extract and yeast application on growth, nutritional status, yield and fruit quality of Keitte mango trees. Agric. Biol. J. North Am. 1, 421–429 (2010)
2. Abou El-Yazied, A., El-Gizawy, A.M., Ragab, M.I., Hamed, E.S.: Effect of seaweed extract and compost treatments on growth, yield and quality of snap bean. J. Am. Sci. 8, 1–20 (2012). http://www.sciencepub.net/american
3. Alvarado, D., Buitrago, E., Sole, M., Frontado, K.: Experimental evaluation of a composted seaweed extract as microalgal culture media. Aquacult. Int. 16, 85–90 (2008)
4. Arioli, T., Mattner, S.W., Winberg, P.C.: Applications of seaweed extracts in Australian agriculture: past, present and future. J. Appl. Phycol. 27, 2007–2015 (2015)
5. Billard, V., Etienne, P., Jannin, L., Garnica, M., Cruz, F., Garcia-Mina, J.M., Yvin, J.C., Ourry, A.: Two biostimulants derived from algae or humic acid induce similar responses in the mineral content and gene expression of winter oilseed rape (Brassica napus L.). J. Plant Growth Regul. 33(2), 305–316 (2014)
6. Cole, A.J., Roberts, D.A., Garside, A.L., de Nys, R., Paul, N.A.:

Seaweed compost for agricultural crop production. J. Appl. Phycol. 28, 629–642 (2016)

7. Crouch, I.J., Beckett, R.P., Van Staden, J.: Effect of seaweed concentrates on the growth and mineral nutrition of nutrient stress lettuce. J. Appl. Phycol. 2, 269–272 (1990)
8. Cuomo, V., Perretti, A., Palomba, I., Verde, A., Cuomo, A.: Utilisation of Ulva rigida biomass in the Venice Lagoon (Italy): biotransformation in compost. J. Appl. Phycol. 7, 479–485 (1995)
9. Directive of Polish Minister of Agriculture and Rural Development: Act on fertilizers and fertilization. J. Laws. No.119, Item 765 (2008)
10. du Jardin, P.: Plant biostimulants: Definition, concept, main categories and regulation. Sci. Hortic. 196, 3–14 (2015)
11. Eghball, B.: Soil properties as influenced by phosphorus-and nitrogen-basedmanure and compost applications. Agron. J. 94, 128–135 (2002)
12. Eyras, M.C., Defossé, G.E., Dellatorre, F.: Seaweed compost as an amendment for horticultural soils in Patagonia, Argentina. Compost Sci. Util. 16, 119–124 (2008)
13. Eyras, M.C., Rostagno, C.M., Defossé, G.E.: Biological evaluation of seaweed composting. Compost Sci. Util. 6, 74–81 (1998)
14. Gopinathan, M., Thirumurthy, M.: Evaluation of phytotoxicity for compost from organic fraction of municipal solid waste and paper & pulp mill sludge. Environ. Res. Eng. Manage. 1(59), 47–51 (2012)
15. Grabowski, J.: The chemical composition of natural fertilizers. Białystok. OSCh–R. P. 1. (2009)
16. Graves, R.E., Hattemer, G.M.: Chapter 2—Composting, Part 637—Environmental Engineering, In: National Engineering Handbook (2010). http://directives.sc.egov.usda.gov/OpenNonWebContent.aspx?content=28910.wba. Accessed on 26 April 2016
17. Greger, M., Malm, T., Kautsky, L.: Heavy metal transfer from composted macroalgae to crops. Eur. J. Agron. 26, 257–265 (2007)
18. Han, W., Clarke, W., Pratt, S.: Composting of waste algae: a review. Waste Manage. 34, 1148–1155 (2014)
19. Haq, T., Khan, F.A., Begum, R., Munshi, A.B.: Bioconversion of drifted seaweed biomass into organic compost collected from the

Karachi coast. Pak. J. Bot. 43, 3049–3051 (2011)

20. HELCOM.: Development of tools for assessment of eutrophication in the Baltic Sea Balt. Sea Environ. Proc. No. 104 (2006). http://www.helcom.fi/Lists/Publications/BSEP104.pdf. Accessed on 19 April 2016
21. Illera-Vives, M., Seoane Labandeira, S., Brito, L.M., López-Fabal, A., López-Mosquera, M.E.: Evaluation of compost from seaweed and fish waste as a fertilizer for horticultural use. Sci. Hortic. 186, 101–107 (2015)
22. Illera-Vives, M., Seoane Labandeira, S., López-Mosquera, M.E.: Production of compost from marine waste: evaluation of the product for use in ecological agriculture. J. Appl. Phycol. 25, 1395–1403 (2013)
23. Irshad, M., Eneji, A.E., Hussain, Z., Ashraf, M.: Chemical characterization of fresh and composted livestock manures. J. Soil Sci. Plant Nutr. 13, 115–121 (2013)
24. Khan, W., Rayirath, U.P., Subramanian, S., Jithesh, M.N., Rayorath, P., Hodges, D.M., Critchley, A.T., Craigie, J.S., Norrie, J., Prithiviraj, B.: Seaweed extracts as biostimulants of plant growth and development. J. Plant Growth Regul. 28, 386–399 (2009)
25. Lacatusu, R., Capatana, R., Lacatusu, A.R., Meghea, A.: A new compost for organic farming. Testing by plants. SGEM2015 Conference Proceedings. 15th International Multidisciplinary Scientific GeoConference SGEM 2015, www.sgem.org, June 18–24, 2015, Book 3, Vol. 2, pp. 11–18 (2015)
26. López-Mosquera, M.E., Fernández-Lema, E., Villares, R., Corral, R., Alonso, B., Blanco, C.: Composting fish waste and seaweed to produce a fertilizer for use in organic agriculture. Proc. Environ. Sci. 9, 113–117 (2011)
27. Mancuso, S., Azzarello, E., Mugnai, S., Briand, X.: Marine bioactive ubstances (IPA extract) improve foliar ion uptake and water tolerance in potted Vitis vinifera plants. Adv. Hortic. Sci. 20, 156–161 (2006)
28. Maze, J., Morand, P., Potoky, P.: Stabilisation of 'Green tides' Ulva by a method of composting with a view to pollution limitation. J. Appl. Phycol. 5, 183–190 (1993)
29. McHugh, D.J.: Other uses of seaweeds. In: A guide to the seaweed

industry, vol. 441. Chapter 9, pp. 91–99. FAO. Fisheries technical paper, Rome (2003). http://www.fao.org/3/a-y4765e.pdf. Accessed on 20 April 2016

30. Michalak, I., Chojnacka, K.: Algal compost—toward sustainable fertilization. Rev. Inorg. Chem. 33, 161–172 (2013)
31. Michalak, I., Chojnacka, K.: Use of extract from Baltic seaweeds produced by chemical hydrolysis in plant cultivation. Przem. Chem. 92, 1046–1049 (2013). in Polish
32. Michalak, I., Górka, B., Wieczorek, P.P., Rój, E., Lipok, J., Łęska, B., Messyasz, B., Wilk, R., Schroeder, G., Dobrzyńska-Inger, A., Chojnacka, K.: Supercritical fluid extraction of algae enhances levels of biologically active compounds promoting plant growth. Eur. J. Phycol. 51, 243–252 (2016)
33. Michalak, I., Tuhy, Ł., Chojnacka, K.: Co-composting of algae and effect of the compost on germination and growth of Lepidium sativum. Pol. J. Environ. Stud. 25, 1–9 (2016)
34. Michalak, I., Tuhy, Ł., Chojnacka, K.: Extraction of seaweed with potassium lye. Przem. Chem. 93, 771–774 (2014). in Polish
35. Michalak, I., Tuhy, Ł., Chojnacka, K.: Seaweed extract by microwave assisted extraction as plant growth biostimulant. Open Chem. 13, 1183–1195 (2015)
36. Mininni, C., Bustamante, M.A., Medina, E., Montesano, F., Paredes, C., Pérez-Espinosa, A., Moral, R., Santamaria, P.: Evaluation of Posidonia seaweed-based compost as a substrate for melon and tomato seedling production. J. Hortic. Sci. Biotechnol. 88, 345–351 (2013)
37. Morand, P., Charlier, R.H., Maze, J.: European bioconversion projects and realizations for macroalgal biomass: Saint-Cast-Le-Guildo (France) experiment. Hydrobiologia 204(205), 301–308 (1990)
38. Orquin, R., Abad, M., Noguera, P., Puchades, R., Maquieira, A.: Composting of Mediterranean seagrass and seaweed residues with yard waste for horticultural purposes. Acta Hortic. 549, 29–35 (2001)
39. Rathore, S.S., Chaudhary, D.R., Boricha, G.N., Ghosh, A., Bhatt, B.P., Zodape, S.T., Patolia, J.S.: Effect of seaweed extract on the

growth, yield and nutrient uptake of soybean (Glycine max) under rainfed conditions. South Afr. J. Bot. 75, 351–355 (2009)

40. Selvam, G.G., Sivakumar, K.: Effect of foliar spray from seaweed liquid fertilizer of Ulva reticulata (Forsk.) on Vigna mungo L. and their elemental composition using SEM—energy dispersive spectroscopic analysis. Asian Pac. J. Reprod. 2, 119–125 (2013)
41. Sosnowski, J., Jankowski, K., Wiśniewska-Kadżajan, B., Jankowska, J., Kolczarek, R.: Effect of the extract from Ecklonia maxima on selected micro- and macroelements in aerial biomass of hybrid alfalfa. J. Elementol. (2014). doi:10.5601/jelem.2014.19.1.608
42. Tang, J., Wang, M., Zhou, Q., Nagata, S.: Improved composting of Undaria pinnatifida seaweed by inoculation with Halomonas and Gracilibacillus sp. isolated from marine environments. Bioresour. Technol. 102, 2925–2930 (2011)
43. Vallini, G., Pera, A., Cecchi, F., Valdrighi, M.M., Sicurani, M.A.: Compost stabilization of algal biomass drawn in eutrophic lagoon ecosystems. Compost Sci. Util. 1, 49–53 (1993)
44. Vendrame, W., Moore, K.K.: Comparison of herbaceous perennial plant growth in seaweed compost and biosolids compost. Compost Sci. Util. 13, 122–126 (2005)
45. Winberg, P.C., DeMestre, C., Willis, S.: Evaluating Microdictyon umbilicatum bloom biomass as an agricultural compost conditioner for native and commercial plants. Report to Shoalhaven City Council, 30p. (2011). http://ro.uow.edu.au/smfc/1. Accessed on 22 April 2016
46. Wosnitza, T.M.A., Barrantes, J.G.: Utilization of seaweed Ulva sp. in Paracas Bay (Peru): experimenting with compost. J. Appl. Phycol. 18, 27–31 (2005)

CITATION

CHAPTER 2

Mark L. Wells, Philippe Potin, James S. Craigie, John A. Raven, Sabeeha S. Merchant, Katherine E. Helliwell, Alison G. Smith, Mary Ellen Camire, and Susan H. Brawley (2017). Algae as nutritional and functional food sources: revisiting our understanding, *Journal of Applied Phycology*, April 2017, Volume 29, Issue 2, pp 949–982. DOI: 10.1007/s10811-016-0974-5.

CHAPTER 3

Pierrick Stévant, Céline Rebours, and Annelise Chapman. Seaweed aquaculture in Norway: recent industrial developments and future perspectives (2017). *Aquaculture International*, August 2017, Volume 25, Issue 4, pp 1373–1390. DOI: 10.1007/s10499-017-0120-7.

CHAPTER 4

Céline Rebours, Eliane Marinho-Soriano, José A. Zertuche-González, Leila Hayashi, Julio A. Vásquez, Paul Kradolfer, Gonzalo Soriano, Raul Ugarte, Maria Helena Abreu, Ingrid Bay-Larsen, Grete Hovelsrud, Rolf Rødven, and Daniel Robledo (2014). Seaweeds: an opportunity for wealth and sustainable livelihood for coastal communities. *Journal of Applied Phycology*, October 2014, Volume 26, Issue 5, pp 1939–1951. DOI: 10.1007/s10811-014-0304-8.

CHAPTER 5

Adibi M. Nor, Tim S. Gray, Gary S. Caldwell, and Selina M. Stead (2016). Is a cooperative approach to seaweed farming effectual? An analysis of the seaweed cluster project (SCP), Malaysia. *Journal of Applied Phycology*, 10 December 2016, pp 1–15. DOI: 10.1007/s10811-016-1025-y.

CHAPTER 6

Siti Fadilah, Alimuddin, Petrus Rani Pong-Masak, Joko Santoso, and Andi Parenrengi (2016). Growth, Morphology and Growth Related Hormone Level in Produced by Mass Selection in Gorontalo Waters, Indonesia. *HAYATI Journal of Biosciences*, Volume 23, Issue 1, 2016, Pages 29-34. ISSN 1978-3019. http://dx.doi.org/10.1016/j.hjb.2015.09.004.

CHAPTER 7

Carl C, de Nys R, Lawton RJ, and Paul NA (2014). Methods for the Induction of Reproduction in a Tropical Species of Filamentous *Ulva*. *PLoS ONE* 9(5): e97396. https://doi.org/10.1371/journal.pone.0097396.

CHAPTER 8

Charrier B, Rolland E, Gupta V, and Reddy CRK (2015). Production of genetically and developmentally modified seaweeds: exploiting the potential of artificial selection techniques. *Front. Plant Sci*. 6:127. DOI: 10.3389/fpls.2015.00127.

CHAPTER 9

Louisi Souza de Oliveira, Gustavo Bueno Gregoracci, Genivaldo Gueiros Zacarias Silva, Leonardo Tavares Salgado, Gilberto Amado Filho, Marcio Alves-Ferreira, Renato Crespo Pereira, and Fabiano L Thompson (2012). Transcriptomic analysis of the red seaweed *Laurencia dendroidea* (Florideophyceae, Rhodophyta) and its microbiome. *BMC Genomics*, December 2012, 13:487. DOI: 10.1186/1471-2164-13-487.

CHAPTER 10

Muhammad Ikbal Illijas, Arifuddin, Luqman Saleh, and Yutaka Itabashi (2012). Fatty acid composition and prostaglandin content of the red seaweed *gracilaria* sp. from Indonesia. *Indonesian Aquaculture Journal*, Vol.7, No.1, (2012). DOI: http://dx.doi.org/10.15578/iaj.7.1.2012.49-54.

CHAPTER 11

Yi Zhang, Jin Yuan Han, Jun Mu, Yan Feng, Xiao Jie Gu, and Yuan Xi Ji (2013). Bioactivity and constituents of several common seaweeds. *Chinese Science Bulletin*, July 2013, Volume 58, Issue 19, pp 2282–2289. DOI: 10.1007/s11434-013-5745-y.

CHAPTER 12

Snehal Subhash Shitole, Amjad Khansaheb Balange, and Shardul Sham Gangan (2014). Use of Seaweed (*Sargassum tenerrimum*) extract as gel enhancer for lesser sardine (*Sardinella brachiosoma*) surimi. *International Aquatic Research*, April 2014, 6:55. DOI: 10.1007/s40071-014-0055-9.

CHAPTER 13

Seema Patel (2012). Therapeutic importance of sulfated polysaccharides from seaweeds: updating the recent findings. 3 Biotech, September 2012, Volume 2, Issue 3, pp 171–185. DOI: 10.1007/s13205-012-0061-9.

CHAPTER 14

Andreas Grassauer, Regina Weinmuellner, Christiane Meier, Alexander Pretsch, Eva Prieschl-Grassauer, and Hermann Unger (2008). Iota-Carrageenan is a potent inhibitor of rhinovirus infection. Virology Journal, December 2008, 5:107. Doi: 10.1186/1743-422X-5-107.

CHAPTER 15

Saowapa Rattaya, Soottawat Benjakul, and Thummanoon Prodpran (2015). Extraction, antioxidative, and antimicrobial activities of brown seaweed extracts, Turbinaria ornata and Sargassum polycystum, grown in Thailand. International Aquatic Research, March 2015, Volume 7, Issue 1, pp 1–16. DOI: 10.1007/s40071-014-0085-3.

CHAPTER 16

Adam D Hughes, Maeve S Kelly, Kenneth D Black, and Michele S Stanley (2012). Biogas from Macroalgae: is it time to revisit the idea? Biotechnology for Biofuels, December 2012, 5:86. DOI: 10.1186/1754-6834-5-86.

CHAPTER 17

Shi-Qi Ji, Bing Wang, Ming Lu, and Fu-Li Li. Direct bioconversion of brown algae into ethanol by thermophilic bacterium *Defluviitalea phaphyphila*. *Biotechnology for Biofuels*, December 2016, 9:81. DOI: 10.1186/s13068-016-0494-1.

CHAPTER 18

Emily T. Kostas, Daniel A. White, Chenyu Du, and David J. Cook (2016). Selection of yeast strains for bioethanol production from UK seaweeds. Journal of Applied Phycology, April 2016, Volume 28, Issue 2, pp 1427–1441. DOI: 10.1007/s10811-015-0633-2.

CHAPTER 19

Paul Bikker, Marinus M. van Krimpen, Piet van Wikselaar, Bwee Houweling-Tan, Nazareno Scaccia, Jaap W. van Hal, Wouter J. J. Huijgen, John W. Cone, and Ana M. López-Contreras (2016). Biorefinery of the green seaweed *Ulva lactuca* to produce animal feed, chemicals and biofuels. *Journal of Applied Phycology*, December 2016, Volume 28, Issue 6, pp 3511–3525. DOI: 10.1007/s10811-016-0842-3.

CHAPTER 20

Hehre EJ, Meeuwig JJ (2016). A Global Analysis of the Relationship between Farmed Seaweed Production and Herbivorous Fish Catch. PLoS ONE. 11(2): e0148250. https://doi.org/10.1371/journal.pone.0148250.

CHAPTER 21

Izabela Michalak, Radosław Wilk, and Katarzyna Chojnacka (2016). Bioconversion of Baltic Seaweeds into Organic Compost. Waste and Biomass Valorization, 2016, pp 1–11. DOI: 10.1007/s12649-016-9738-3.

Index

A

B

C

D

E

F

G